Optimization Theory for Large Systems

Optimization Theory for Large Systems

LEON S. LASDON · CASE WESTERN RESERVE UNIVERSITY

THE MACMILLAN COMPANY

COLLIER-MACMILLAN LIMITED, LONDON

Second Printing, 1972

Library of Congress catalog card number: 78–95301

THE MACMILLAN COMPANY
866 Third Avenue, New York, New York 10022
COLLIER-MACMILLAN CANADA, LTD., TORONTO, ONTARIO

Printed in the United States of America

Preface

Many decision problems are now formulated as mathematical programs, requiring the maximization or minimization of an objective function subject to constraints. Such programs often have special structure. In linear programming, the nonzero elements of the constraint matrix may appear in diagonal blocks, except for relatively few rows or columns. Nonlinear programs may become linear if certain variables are assigned fixed values, or the functions involved may be additively separable. Some definite structure is almost always found in truly large problems, since these commonly arise from a linking of independent subunits in either time or space. By developing specialized solution algorithms to take advantage of this structure, significant gains in computational efficiency and reductions in computer memory requirements may be achieved. Such methods are mandatory for truly large problems, which cannot otherwise be solved because of time and/or storage limitations.

The past decade has seen the identification of many classes of structured problems and the development of a great many algorithms for their solution. This period was initiated by the publication of the Dantzig–Wolfe decomposition principle in 1960. However, there was significant activity even before that. Growth has been explosive, with techniques developed first for linear, then for certain classes of nonlinear programs. The literature is now very large. This forces a beginner exploring the field to search through a maze of articles in many different journals. Even after doing so, the relationships between the various methods are likely to be obscured. Clearly a need to collect together the best of this literature and unify it has existed for some time.

This book is an attempt to fill this need. It discusses some of the most important algorithms for optimizing large systems. The emphasis throughout is on developing the various methods in a straightforward and logical manner from a small set of basic ideas and principles. In so doing, relationships between various procedures are made clear. Application of many methods is illustrated by numerical examples, and there are problems at the end of each chapter. In addition, much of the available computational experience has been included, as well as comments on various computational options. This information should prove useful to those interested in applications, as should those sections dealing with formulation of problems. Chapter 2 deals entirely with how large-scale programs can arise from real-world problems, and other similar examples are found throughout.

Notes for the book have been used in a one-semester graduate course in the Operations Research Department at Case Western Reserve University and in the Faculty of Industrial and Management Engineering at the Technion, Haifa, Israel. The first chapter contains most of the background material on linear and nonlinear programming needed in the rest of the text. The two appendixes deal with convex functions and their conjugates, and with subgradients and directional derivatives of convex functions. These are based on the work of R. T. Rockafellar, are self-contained, and are of much interest in themselves. The ideas in them are used extensively in Chapters 8 and 9 but very little elsewhere.

The greatest single contributor to large-scale mathematical programming is George Dantzig. This is reflected in the contents of this volume. The contributions of Arthur Geoffrion, Ben Rosen, and Philip Wolfe also play a major role. James Schoeffler introduced me to this area. His interest and that of Mihajlo Mesarovic have stimulated mine. My thanks to Allan Waren for aid in preparing Chapter 1. I wish to express appreciation to the School of Management, Case Western Reserve University, and to the Faculty of Industrial and Management Engineering at the Technion for providing secretarial assistance. Thanks also to Miss Christine Yamamoto and Miss Ellon Waters for their excellent work in typing the first draft, and to Mrs. Orah Naor for her fine efforts in typing the manuscript. Finally, special thanks to my wife Louanne, who suffered with me and encouraged me throughout.

L. S. L.

Contents

Optimization Theory for Large Systems

1

Linear and Nonlinear Programming

The problem of mathematical programming is that of maximizing or minimizing an objective function $f(x_1 \cdots x_n)$ by choice of the vector $x = (x_1 \cdots x_n)'$. The variables x_i may be allowed to take on any values whereupon the problem is one of unconstrained minimization or they may be restricted to take on only certain allowable values, whereupon the problem is constrained. Only problems in which (1) the variables x_i can vary continuously within the region of interest and (2) the objective and constraint functions are continuous and differentiable are considered here.

If the problem is constrained, its difficulty depends critically on the nature of the constraints, i.e., linear, nonlinear, etc. We consider first the unconstrained case, then the more difficult, constrained one. The constrained case will be divided into two parts: linear constraints and linear objective function (linear programming) and at least one nonlinear constraint and/or nonlinear objective (nonlinear programming).

1.1 Unconstrained Minimization

Necessary and Sufficient Conditions for an Unconstrained Minimum. The problem here is to maximize or minimize a function of n variables, $f(x)$, with no restrictions on the variables x. Many real-life problems are of this form, where whatever constraints are present do not restrict the optimum. Also, many problems in which the constraints are binding can be converted to unconstrained problems or sequences of such problems. Since the problem of maximizing $f(x)$ is equivalent to that of minimizing $-f(x)$, only the minimization problem is considered.

A point x^* is said to be a global minimum of $f(x)$ if

$$f(x^*) \leq f(x) \tag{1}$$

for all x. If the strict inequality holds for $x \neq x^*$ the minimum is said to be *unique*. If (1) holds only for all x in some neighborhood of x^*, then x^* is said to be a local or relative minimum of $f(x)$, since x^* is only the best point in the immediate vicinity, not in the whole space.

If $f(x)$ is continuous and has continuous first and second partial derivatives for all x, the necessary conditions for a local minimum are [3]

$$\frac{\partial f(x^*)}{\partial x_i} = 0, \qquad i = 1, 2, \ldots, n \tag{2}$$

and that the matrix of second partial derivatives evaluated at x^* be positive semidefinite. Any point x^* satisfying (2) is called a stationary point of $f(x)$. Sufficient conditions for a relative minimum are that the matrix of second partial derivatives of $f(x)$ evaluated at x^* be positive definite and (2) hold.

Numerical Methods for Finding Unconstrained Minima. The most obvious approach to finding the minimum of $f(x)$ is to solve (2). These are a set of n equations, usually nonlinear, in the n unknowns x_i. Unfortunately the task of solving large sets of nonlinear equations is very difficult. The function $f(x)$ may be so complex that it is difficult even to write out (2) in closed form. Further, even if (2) could be solved, there would be no guarantee that a given solution was not a maximum, saddle point, etc., rather than a minimum. Thus other approaches must be considered.

Gradient. If $f(x)$ is continuous and differentiable, a number of minimization techniques using the gradient of $f(x)$, written $\nabla f(x)$, are available. The gradient is the vector whose ith component is $\partial f/\partial x_i$. It points in the direction of maximum rate of increase of $f(x)$ ($-\nabla f$ points in the direction of greatest decrease). The vector ∇f is, at any point x_0, normal to the contour of constant function value passing through x_0.

Steepest Descent. The method of steepest descent for finding a local minimum of $f(x)$ proceeds as follows. Start at some initial point x_0 and compute $\nabla f(x_0)$. Take a step in the direction of steepest descent, $-\nabla f(x_0)$, using a step length α_0, to obtain a new point x_1. Repeat the procedure until some stop criterion is satisfied. This process is described by the relations

$$x_0 \text{ given}$$

$$x_{i+1} = x_i - \alpha_i \nabla f(x_i) \qquad i = 0, 1, 2, \ldots \tag{3}$$

where $\alpha_i > 0$. The process will, under very mild restrictions [4] on $f(x)$, converge to at least a local minimum of $f(x)$, if the α_i are chosen so that

$$f(x_{i+1}) < f(x_i) \tag{4}$$

for all i, i.e., if the function is made to decrease at each step. Since the function is initially decreasing in the directions given by $-\nabla f(x_i)$, there always exist $\alpha_i > 0$ such that (4) is satisfied.

Step Length and Optimum Gradients. One way to find α_i satisfying (4) is to choose α_i to minimize the function

$$g(\alpha) = f[x_i - \alpha \nabla f(x_i)] \tag{5}$$

Note that x_i and $\nabla f(x_i)$ are known vectors so that the only variable in (5) is α. The adaptation of the method of steepest descent which uses (5), called the method of optimum gradients [4], is described by

$$x_0 \text{ given}$$

$$s_i = -\nabla f(x_i)$$

Choose $\alpha = \alpha_i$ by minimizing $g(\alpha)$ in (5),

$$x_{i+1} = x_i + \alpha_i s_i \tag{6}$$

Set $i = i + 1$ and repeat.

Geometrically, α_i is chosen by minimizing $f(x)$ along the direction s_i starting from x_i. At a minimum,

$$\left.\frac{dg}{d\alpha}\right|_{\alpha=\alpha_i} = s_i' \nabla f(x_i + \alpha_i s_i) = 0 \tag{7}$$

so the vector $x_i + \alpha s_i$ must be tangent to a contour at $\alpha = \alpha_i$, for dg is then zero for small changes $d\alpha$. Since $\nabla f(x_{i+1})$ is normal to the same contour, successive steps are at right angles to one another. Practical methods for carrying out the one-dimensional minimization are discussed later in this section.

Stop Criteria. Some possible stop criteria are as follows:

1. Since, at a minimum $\partial f/\partial x_i = 0$, stop when

(a) $$\left|\frac{\partial f}{\partial x_i}\right| < \epsilon \qquad i = 1, 2, \ldots, n$$

or

(b) $$\sum_{i=1}^{n} \left(\frac{\partial f}{\partial x_i}\right)^2 < \epsilon$$

2. Stop when the change in function is less than some limit η, i.e.,

$$|f(x_{i+1}) - f(x_i)| < \eta$$

Others are possible. Criterion 2 appears to be the more dependable of the two, provided it is satisfied for several successive values of i.

Local versus Global Minima. The most that can be guaranteed of this or any other iterative minimization technique is that it will find a local minimum, in general the one "nearest" the starting point x_i. To attempt to find all local minima (and thus the global minimum), the method most used is to repeat the minimization from many different initial points.

Numerical Difficulties. The fact that successive steps of the optimum gradient method are orthogonal leads to very slow convergence for some functions. If the function contours are hyperspheres (circles in two dimensions), the method finds the minimum in one step. However, if the contours are in any way eccentric, an inefficient zigzag behavior results, as shown in Figure 1-1. This occurs because, for eccentric contours, the gradient direction is generally quite different from the direction to the minimum. Many, if not most, of the functions occurring in practical applications are ill-behaved in

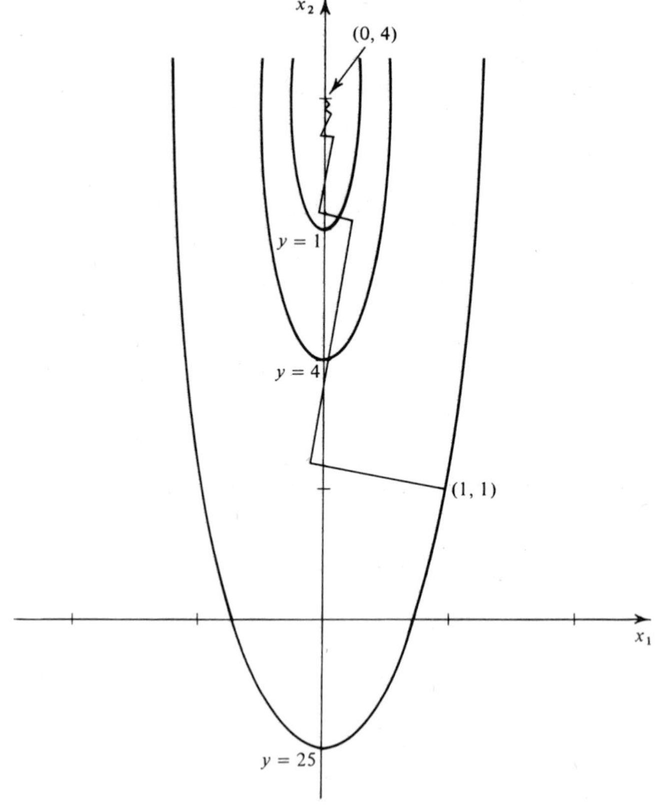

FIGURE 1-1 Equivalue lines of $y = 16x_1^2 + (x_2 - 4)^2$, normal steepest-descent method.

that their contours are eccentric or nonspherical. Thus more efficient schemes are desirable.

"Second-Order" Gradient Methods. Recently, a number of minimization techniques have been developed which substantially overcome the above difficulties. What appear to be the best of these will be described in detail. First, however, the logic behind these methods will be explained.

Since the first partial derivatives of a function vanish at the minimum, a Taylor-series expansion about the minimum x^* yields

$$f(x) \doteq f(x^*) + \tfrac{1}{2}(x - x^*)' H_f(x^*)(x - x^*) \tag{8}$$

where $H_f(x^*)$, the matrix of second partials of f evaluated at x^*, is positive definite. Thus the function behaves like a pure quadratic in the vicinity of x^*. It follows that the only methods which will minimize a general function quickly and efficiently are those which (1) work well on a quadratic and (2) are guaranteed to converge eventually for a general function. All others will be slow, at least in the vicinity of the minimum (see Figure 1-1), and often elsewhere.

Conjugate Directions. General minimization procedures can be designed which will minimize a quadratic function of n variables in n steps [5–7]. Most, if not all, are based on the ideas of conjugate directions.

The general quadratic function can be written

$$q(x) = a + b'x + \tfrac{1}{2}x'Ax \tag{9}$$

where A is positive definite and symmetric. Let x^* minimize $q(x)$. Then

$$\nabla q(x^*) = b + Ax^* = 0 \tag{10}$$

Given a point x_0 and a set of linearly independent directions $\{s_0, s_1, \ldots, s_{n-1}\}$, constants β_i can be found such that

$$x^* = x_0 + \sum_{i=0}^{n-1} \beta_i s_i \tag{11}$$

If the directions s_i are A-conjugate, i.e., satisfy

$$s_i' A s_j = 0, \qquad i \neq j, \quad i, j = 0, 1, \ldots, n-1 \tag{12}$$

and none are zero, then the s_i are easily shown to be linearly independent and the β_i can be determined from (11) as follows:

$$s_j' A x^* = s_j' A x_0 + \sum_{i=0}^{n-1} \beta_i s_j' A s_i \tag{13}$$

Using (12),

$$s_j' A x^* = s_j' A x_0 + \beta_j s_j' A s_j \tag{14}$$

and, using (10),

$$\beta_j = -(b + Ax_0)' \frac{s_j}{s_j' A s_j} \tag{15}$$

Now consider an iterative minimization procedure, starting at x_0 and successively minimizing $q(x)$ down the directions $s_0, s_1, s_2, \ldots, s_{n-1}$, where these directions satisfy (12). Successive points are then determined by the relations

$$x_{i+1} = x_i + \alpha_i s_i, \qquad i = 0, 1, \ldots, n-1 \tag{16}$$

where α_i is determined by minimizing $f(x_i + \alpha s_i)$, as in the optimum gradient method, so that

$$s_i' \nabla q(x_{i+1}) = 0 \tag{17}$$

Using (10) in (17) gives

$$s_i'(b + A(x_i + \alpha_i s_i)) = 0 \tag{18}$$

or

$$\alpha_i = -(b + Ax_i)' \frac{s_i}{s_i' A s_i} \tag{19}$$

From (16),

$$x_i = x_0 + \sum_{j=0}^{i-1} \alpha_j s_j \tag{20}$$

so that

$$x_i' A s_i = x_0' A s_i + \sum_{j=0}^{i-1} \alpha_j s_j' A s_i = x_0' A s_i \tag{21}$$

Thus (19) becomes

$$\alpha_i = -(b + Ax_0)' \frac{s_i}{s_i' A s_i} \tag{22}$$

which is identical to (15). Hence this sequential process leads, in n steps, to the minimum x^*.

Method of Fletcher and Powell. A method recently presented by Fletcher and Powell [5] is probably the most powerful general procedure now known [8] for finding a local minimum of a general function, $f(x)$. It is designed so that, when applied to a quadratic, it minimizes in n iterations. It does this by generating conjugate directions.

Central to the method is a symmetric positive definite matrix H_i which is updated at each iteration, and which supplies the current direction of motion, s_i, by multiplying the current gradient vector. An iteration is described by the following:

$$\begin{aligned} H_0 &= \text{any positive definite matrix} \\ s_i &= -H_i \, \nabla f(x_i) \end{aligned}$$

Choose $\alpha = \alpha_i$ by minimizing $f(x_i + \alpha s_i)$,

$$\begin{aligned} \sigma_i &= \alpha_i s_i \\ x_{i+1} &= x_i + \sigma_i \\ H_{i+1} &= H_i + A_i + B_i \end{aligned} \tag{23}$$

where the matrices A_i and B_i are defined by

$$\begin{aligned} A_i &= \frac{\sigma_i \sigma_i'}{\sigma_i' y_i}, \qquad y_i = \nabla f(x_{i+1}) - \nabla f(x_i) \\ B_i &= \frac{-H_i y_i y_i' H_i}{y_i' H_i y_i} \end{aligned} \tag{24}$$

Note that the numerators of A_i and B_i are both matrices, while the denominators are scalars. Fletcher and Powell prove the following:

1. The matrix H_i is positive definite for all i. As a consequence of this, the method will usually converge, since

$$\left. \frac{d}{d\alpha} f(x_i + \alpha s_i) \right|_{\alpha=0} = -\nabla f'(x_i) H_i \, \nabla f(x_i) < 0 \tag{25}$$

   i.e., the function f is initially decreasing along the direction s_i, so that the function can be decreased at each iteration by minimizing down s_i.
2. When the method is applied to the quadratic (9), then
   (a) The directions s_i (or equivalently σ_i) are A-conjugate, thus leading to a minimum in n steps.

(b) The matrix H_i converges to the inverse of the matrix of second partials of the quadratic, i.e.,

$$H_n = A^{-1}$$

When applied to a general function, H_i tends to the inverse of the matrix of second partials of the function evaluated at the minimum since, as the

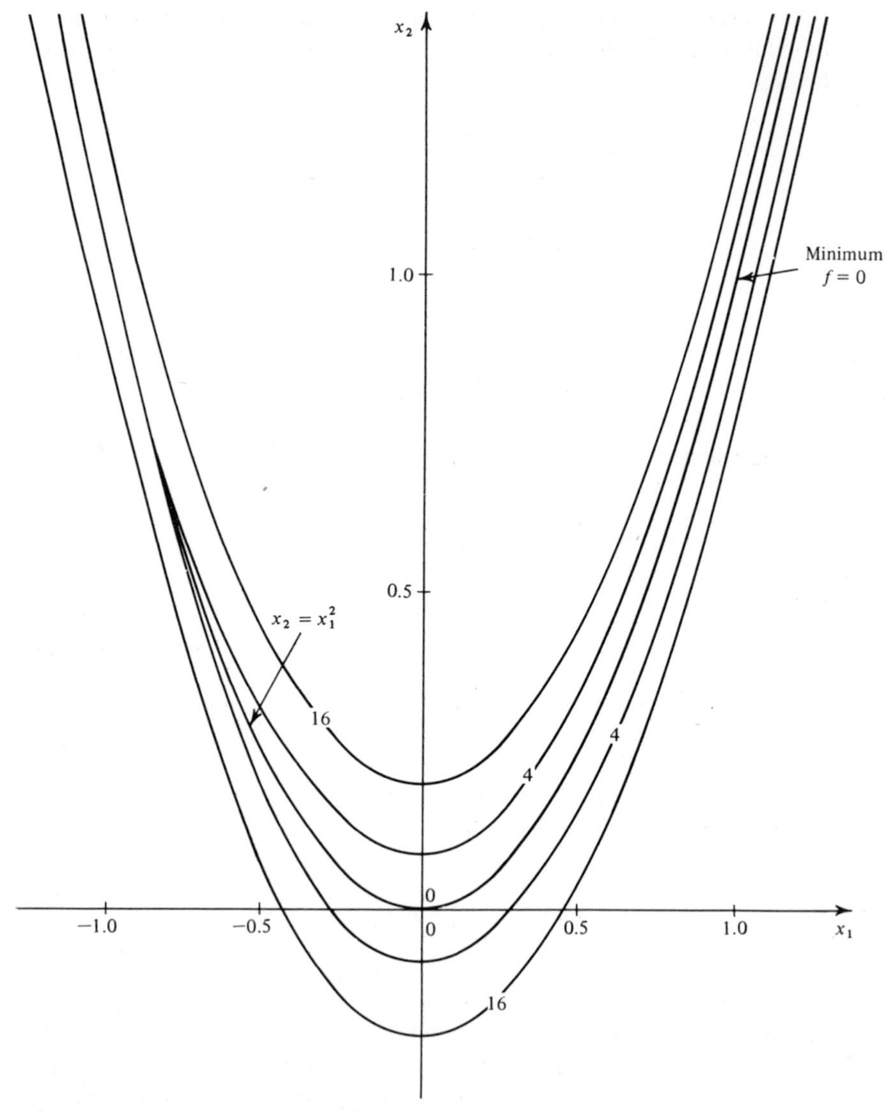

FIGURE 1-2 Contours of the Rosenbrock function.

minimum is approached, the second-order terms in the Taylor-series expansion predominate.

Numerical tests bear out the rapid convergence of this method. Consider, for example, the function

$$f(x_1, x_2) = 100(x_2 - x_1^2)^2 + (1 - x_1)^2 \tag{26}$$

called the Rosenbrock function [9], whose contours are given in Figure 1-2. The minimum is at (1, 1) and the steep curving valley makes minimization difficult. The paths taken by the optimum gradient technique and by the method of Fletcher and Powell are shown in Figures 1-3 and 1-4. The Fletcher–Powell technique follows the curved valley and minimizes very efficiently.

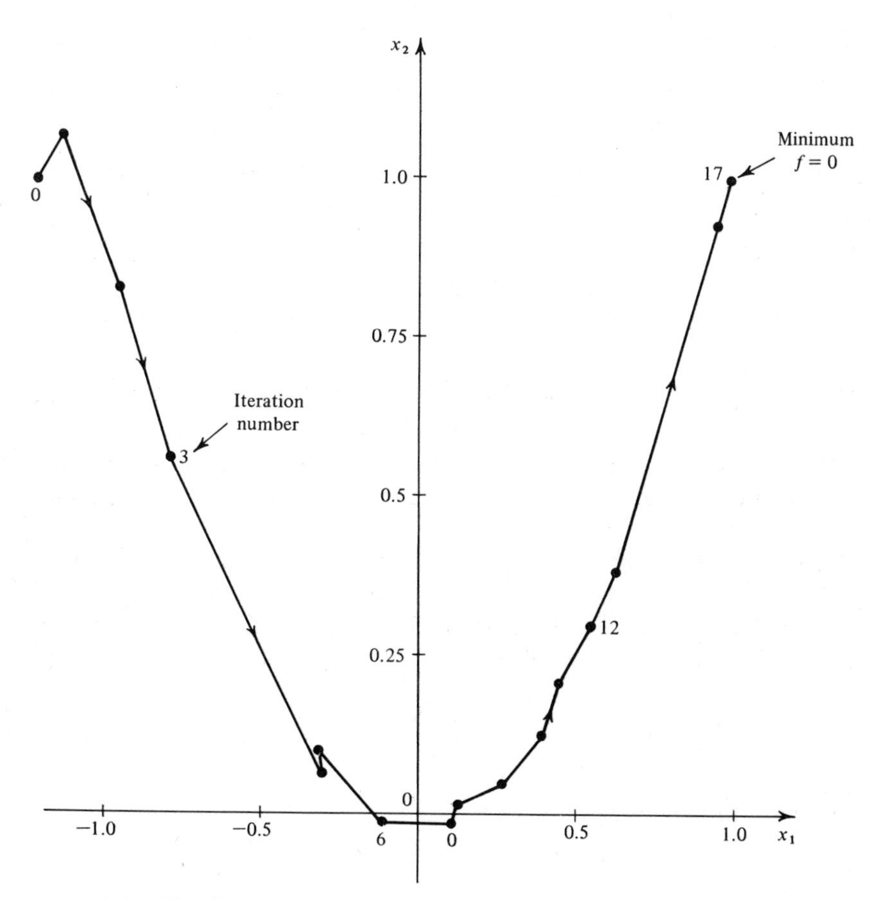

FIGURE 1-3 Fletcher–Powell method minimizing the Rosenbrock function.

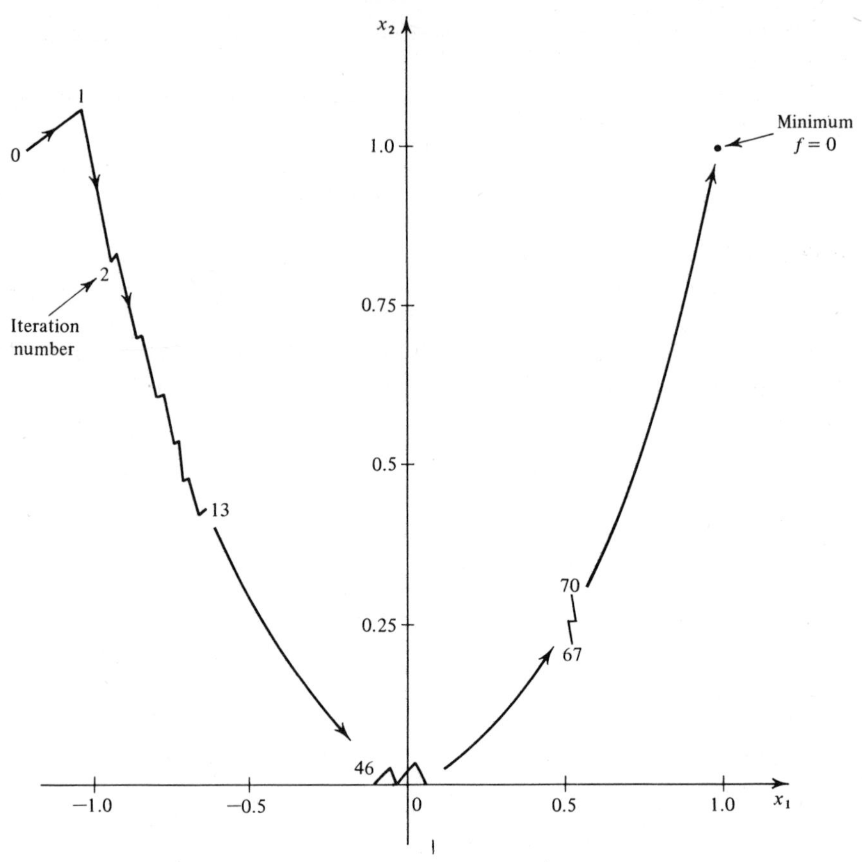

FIGURE 1-4 Optimum gradient method minimizing the Rosenbrock function.

An excellent reference, which derives this algorithm as a member of a class of methods and gives numerical comparisons with other techniques, has been prepared by Pearson [21].

Conjugate Gradient Method. Other conjugate direction minimization techniques also exist. One of these, due to Fletcher and Reeves [6], requires computation of the gradient of $f(x)$ and the storage of only one additional vector, the actual direction of search. The algorithm is

$$x_0 \text{ arbitrary}$$
$$s_0 = -\nabla f(x_0)$$

Choose α_i to minimize $f(x_i + \alpha s_i)$,

$$x_{i+1} = x_i + \alpha_i s_i$$
$$s_{i+1} = -\nabla f(x_{i+1}) + \beta_i s_i$$

where

$$\beta_i = \frac{\nabla f'(x_{i+1}) \nabla f(x_{i+1})}{\nabla f'(x_i) \nabla f(x_i)}$$

This method is not quite as efficient [8] as the Fletcher–Powell technique but requires much less storage, a significant advantage when the number of variables, n, is large.

Minimization without Derivatives. There are also a number of minimization techniques which do not require derivatives. Of these, tests performed thus far indicate that Powell's method [7] is the most efficient [8]. Each iteration requires n one-dimensional minimizations down n linearly independent directions, $s_1, s_2, \ldots, s_n$. As a result of these minimizations a new direction, s, is defined and, if a test is passed, s replaces one of the original directions, s_r. The process is usually started from the best estimate of the minimum, x_0, with the initial s_i's being the coordinate directions.

The procedure is as follows:

1. For $r = 1, 2, \ldots, n$ calculate α_r so that $f(x_{r-1} + \alpha s_r)$ is a minimum and define $x_r = x_{r-1} + \alpha_r s_r$.
2. Find the integer m, $1 \leq m \leq n$, so that $[f(x_{m-1}) - f(x_m)]$ is a maximum, and define $\Delta = f(x_{m-1}) - f(x_m)$.
3. Calculate $f_3 = f(2x_n - x_0)$ and define $f_1 = f(x_0)$ and $f_2 = f(x_n)$.
4. If either $f_3 \geq f_1$ and/or
$$(f_1 - 2f_2 + f_3)\cdot(f_1 - f_2 - \Delta)^2 \geq \tfrac{1}{2}\Delta\cdot(f_1 - f_3)^2$$
use the old directions $s_1, s_2, \ldots, s_n$ for the next iteration and use x_n for the next x_0.
5. Otherwise, defining $s = x_n - x_0$, calculate α so that $f(x_n + \alpha s)$ is a minimum. Use $s_1, s_2, \ldots, s_{m-1}, s_{m+1}, s_{m+2}, \ldots, s_n, s$ as the directions for the next iteration and $x_n + \alpha s$ for the next x_0.

One-Dimensional Minimization. All the methods discussed thus far have searched for a minimum in n-dimensional space by performing one-dimensional minimizations down a set of directions $\{s_i\}$. Thus the efficiency of any such procedure depends critically on the efficiency of the method used to solve the single-dimensional search. Three techniques are presented. The first two use polynomial interpolation, one requiring derivatives, the second only function values. The third, the Fibonacci method, also requires only function values. Unlike the interpolation methods, it does not depend on smoothness of the function being minimized, and may be applied even to discontinuous functions.

For both interpolative procedures, the variables $x_1, \ldots, x_n$ are scaled so that a unit change in any variable is a significant but not too large percentage change in that variable. For example, if a variable is expected to have a value around 100 units, then a 1-unit change would be considered significant, whereas a 10-unit change would be too large.

Cubic Interpolation. This technique, described in [6], solves the problem of finding the smallest nonnegative α, α^*, for which the function

$$g(\alpha) = f(x + \alpha s) \tag{27}$$

attains a local minimum in three stages. It uses the derivative

$$g'(\alpha) \equiv \frac{dg}{d\alpha} = s' \nabla f(x + \alpha s) \tag{28}$$

The first stage normalizes the s vector so that a step size $\alpha = 1$ is acceptable. The second stage establishes bounds on α^*, and the third stage interpolates its value.

STAGE 1. Calculate

$$\Delta = \max_j |(s)_j|$$

with $(s)_j$ component j of s and divide each component of s by Δ. This ensures that s is a reasonable change in x.

STAGE 2. Evaluate $g(\alpha)$ and $g'(\alpha)$ at the points $\alpha = 0, 1, 2, 4, \ldots, a, b$, where b is the first of these values at which either g' is nonnegative or g has not decreased. It then follows that α^* is bounded in the interval $a < \alpha^* \leq b$. If $g(1)$ is much greater than $g(0)$, divide the components of s by a factor, e.g., 2 or 3, and repeat this stage.

STAGE 3. A cubic polynomial is now fitted to the four values $g(a)$, $g'(a)$, $g(b)$, $g'(b)$, and its minimum, α_e, is taken to be the value for α^*. It is shown in [6] that the cubic has a unique minimum in the interval (a, b) which is given by

$$\alpha_e = b - \frac{g'(b) + w - z}{g'(b) - g'(a) + 2w}(b - a) \tag{29}$$

where

$$z = 3\left[\frac{g(a) - g(b)}{b - a}\right] + g'(a) + g'(b) \tag{30}$$

$$w = (z^2 - g'(a)g'(b))^{1/2} \tag{31}$$

If neither $g(a)$ nor $g(b)$ is less than $g(\alpha_e)$, then α_e is accepted as the estimate of α^*. Otherwise, according as $g'(\alpha_e)$ is positive or negative, the interpolation is repeated over the subinterval (a, α_e) or (α_e, b), respectively.

It is interesting that for small values of $g'(a)$ and $g'(b)$, the cubic has the shape that a flat metal spring would assume if fitted to the points, a, b, with slopes $g'(a)$, $g'(b)$.

Quadratic Interpolation. If derivatives are not available or are difficult to compute, then quadratic interpolation should be used in the one-dimensional minimization. The procedure can again be described in three steps.

STAGE 1. This is the same as stage 1 above.

STAGE 2. Evaluate $g(\alpha)$ at the points $\alpha = 0, 1, 2, 4, \ldots, a, b, c$, where c is the first of these values at which g has increased. Then

$$a < \alpha^* < c$$

Again, if $g(1) \gg g(0)$, then divide the components of s by a factor, e.g., 2 or 3, and repeat.

STAGE 3. A quadratic polynomial is now fitted to the three values $g(a)$, $g(b)$, $g(c)$, and its minimum, α_e, is

$$\alpha_e = \frac{1}{2}\,\frac{g(a)(c^2 - b^2) + g(b)(a^2 - c^2) + g(c)(b^2 - a^2)}{g(a)(c - b) + g(b)(a - c) + g(c)(b - a)} \tag{32}$$

If

$$g(\alpha_e) < g(b)$$

then α_e is accepted as the estimate of α^*. Otherwise, b is taken as α^*.

Fibonacci Technique. Unlike the previous technique, the Fibonacci method [20] does not use derivatives. It can thus deal with functions which are not differentiable or even continuous, functions to which the previous techniques could not be applied. It minimizes by assuming that the optimal value of the variable is within some initial interval, called the initial interval of uncertainty. It then systematically reduces this interval by evaluating the function within the interval, thus "closing in" on the optimal point. To do this without gradients, one must assume something about the function to be minimized. The Fibonacci technique proceeds on the assumption that this function is unimodal within the initial interval of uncertainty.

Unimodality. Roughly speaking, a unimodal function is one that has only one peak (max or min). More precisely, a function of one variable is said to be unimodal if, given that two values of the variable are on the same side of the optimum, the one nearer the optimum gives the better functional value, i.e., the smaller value in the case of a minimization problem. Mathematically, this is phrased as follows. Let the minimum be at x^*, and define $x_1 < x_2$. Then the function $f(x)$ is unimodal if

$$x_2 < x^* \text{ implies } f(x_2) < f(x_1)$$

and

$$x_1 > x^* \text{ implies that } f(x_1) < f(x_2)$$

This property enables us to reduce any initial interval of uncertainty by function evaluations alone. Consider the normalized interval [0, 1] and two function evaluations (henceforth called "experiments") within it, as shown in Figure 1-5. If the function is unimodal, i.e., has a single minimum, then the

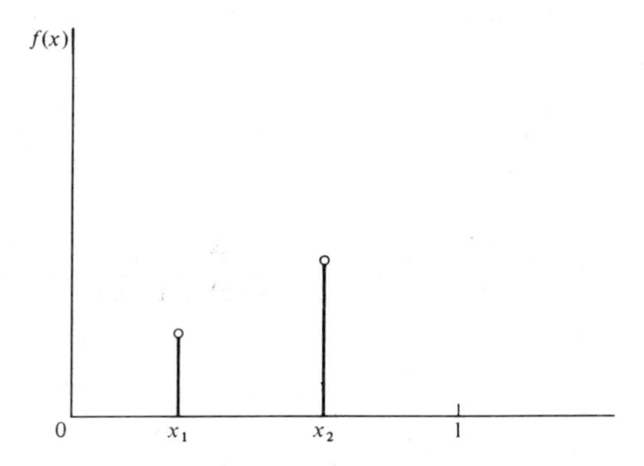

FIGURE 1-5 Initial interval of uncertainty.

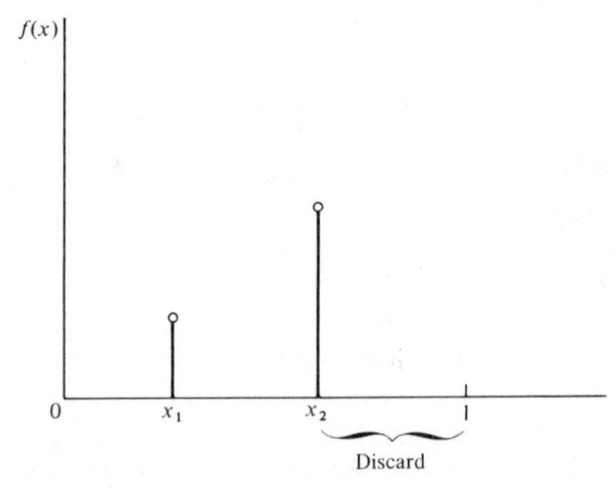

FIGURE 1-6 Reduced interval of uncertainty.

minimizing x cannot lie to the right of x_2. Thus that part of the interval can be discarded, and a new smaller interval of uncertainty results, as shown in Figure 1-6. If $f(x_1) > f(x_2)$, then the interval $[0, x_1]$ would have been discarded while if $f(x_1) = f(x_2)$, then both $[0, x_1]$ and $[0, x_2]$ can be dismissed. Moreover, if one of the original experiments remains within this new interval, only one other experiment need be placed within it in order that the process be repeated. The Fibonacci method places the experiments so that one of the original experiments always remains. The method makes use of the sequence of Fibonacci numbers, $\{F_n\}$, defined by

$$F_0 = F_1 = 1$$
$$F_n = F_{n-1} + F_{n-2}, \qquad n = 2, 3, \ldots$$

yielding the sequence 1, 1, 2, 3, 5, 8, 13, ... It assumes that n experiments are to be made, and proceeds as follows.

Method. Let the initial interval of uncertainty be L_0 and define

$$L_2^* = \frac{F_{n-1}}{F_n} L_0$$

where n is the number of experiments to be performed. Place the initial experiment L_2^* units from one end of the interval and place a second experiment L_2^* units from the other end. Discard some part of the interval using the unimodality assumption. There then remains a new smaller interval of uncertainty with one experiment left in it, that experiment being some distance in from (any) one of the ends. Place a new experiment the same distance from the other end and repeat. Stop when n experiments have been performed.

Example. The function to be minimized is shown in Figure 1-7. Let n, the number of experiments to be performed, be 5. Then

$$L_2^* = \frac{F_4}{F_5} L_0 = \tfrac{5}{8} L_0$$

Place one experiment $\frac{5}{8}L_0$ from one end, a second $\frac{5}{8}L_0$ from the other, as shown. Discard $[x_1, L_0]$. The experiment x_2 is found to be $\frac{3}{8}L_0$ in from $x = 0$. Place x_3 $\frac{3}{8}L_0$ in from $x = x_1$ and discard $[0, x_3]$. Place x_4 $\frac{1}{8}L_0$ in from x_1 and discard $[x_4, x_1]$. The experiment x_5 is now located in the middle of the remaining interval, $[x_3, x_4]$, and by past procedure we should place the last experiment right on top of it. Since this would yield no new information, we displace the last experiment by a small amount, obtaining the final interval of uncertainty $[x_2, x_4]$.

Note that, after discarding the first interval, $[x_1, L_0]$, the remaining experiment is in a distance $\frac{3}{5}L_3^* = F_3/F_4 L_3^*$ from 0, where L_3^* is the length of the remaining interval of uncertainty. This is precisely the position of the first

being minimized, minimizing it would yield a plan good only for a particular function. We can remedy this by defining

$$L_n(x_1 \cdots x_n) = \max_{1 \le K \le n} l_n(K, x_1, \ldots, x_n)$$

where L_n is the maximum final interval of uncertainty obtained over all best outcomes, K. This is independent of K, hence independent of the function being minimized. It is proved [20] that the Fibonacci search *minimizes the maximum final interval* of uncertainty; i.e., the final interval of uncertainty for the Fibonacci method is given by

$$L_n^* = \min_{x_1 \cdots x_n} L_n(x_1, \cdots, x_n) = \min_{x_1 \cdots x_n} \max_{1 \le K \le n} l_n(K, x_1, \ldots, x_n)$$

This criterion assures us that the final interval of uncertainty cannot be greater than L_n^* (assuming a unimodal function). It is a rather conservative criterion, yet leads to very effective search results. Table 1 gives the ratio of L_0/L_n^* versus n. It is seen from this that

$$\frac{L_0}{L_n^*} = F_n$$

and that the interval L_0 is thus reduced quite rapidly.

TABLE 1

Number of experiments	L_0/L_n^* for Fibonacci search	Number of experiments	L_0/L_n^* for Fibonacci search
0	1	11	144
1	1	12	233
2	2	13	377
3	3	14	610
4	5	15	987
5	8	16	1,597
6	13	17	2,584
7	21	18	4,181
8	34	19	6,765
9	55	20	10,946
10	89		

Constrained Optimization Problems

Attention is now focused on problems of constrained minimization, i.e., problems in which the variables $x_1, \ldots, x_n$ may take on only certain allowable values. Such a situation in two dimensions is shown in Figure 1-9. The

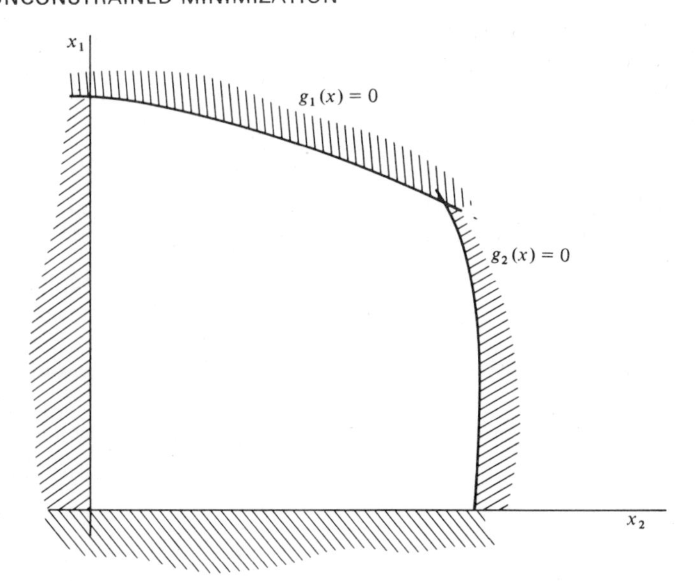

FIGURE 1-9 Constraint set.

unshaded area is the set of allowable values of x_1 and x_2, henceforth called the constraint set. Its boundaries are the curves $x_1 = 0$, $x_2 = 0$, $g_1(x) = 0$, $g_2(x) = 0$. The constraint set in Figure 1-9 is the set of all points satisfying the inequalities

$$x_1 \geq 0$$
$$x_2 \geq 0$$
$$g_1(x) \leq 0$$
$$g_2(x) \leq 0$$

A general programming problem may have equality as well as inequality constraints. The equalities often describe the operation of the system under consideration, while the inequalities define limits within which certain physical variables must lie. Thus the general problem may be written

$$\text{minimize } f(x)$$

subject to

$$g_i(x) \leq 0, \qquad i = 1, 2, \ldots, s$$
$$h_j(x) = 0, \qquad j = 1, 2, \ldots, r < n$$

When all functions f, g_i, h_j are linear, the problem is one of linear programming; if not, then nonlinear programming. The field of linear programming is by far the most fully developed and is considered first. Nonlinear programming problems are considerably more difficult and are considered later.

1.2 Linear Programming

1.2.1 Simplex Method

Geometry of Linear Programs. Consider the problem

$$\text{maximize } z = x_1 + 3x_2$$

subject to

$$\begin{aligned} -x_1 + x_2 &\leq 1 \\ x_1 + x_2 &\leq 2 \\ x_1 \geq 0, \quad x_2 &\geq 0 \end{aligned} \tag{1}$$

The constraint set is the unshaded region of Figure 1-10 defined by the

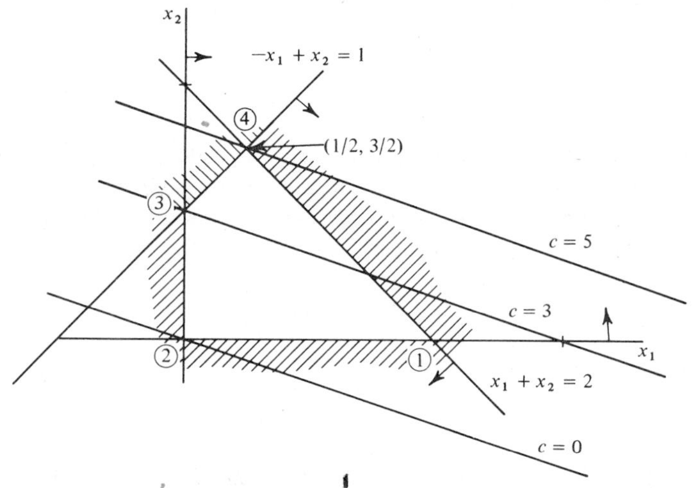

FIGURE 1-10 Geometry of a linear program.

intersections of the half-spaces satisfying the linear inequalities. The numbered points are called extreme points of the set. If the constraints are linear there are only a finite number of such points.

Contours of constant value of the objective function, z, are defined by the linear equation

$$x_1 + 3x_2 = \text{constant} = c \tag{2}$$

As c is varied, the line is moved parallel to itself. The maximum value of z is the largest c whose line has at least one point in common with the constraint set. For the figure shown, this occurs for $c = 5$ and the optimal values of x are $x_1 = \frac{1}{2}$, $x_2 = \frac{3}{2}$. Note that the maximum value occurs at an extreme point of the constraint set. If the problem had been to minimize z, the minimum is at the origin, which is again an extreme point. If the objective function had been $z = 2x_1 + 2x_2$, the line $z =$ constant would be parallel to one of the constraint boundaries, $x_1 + x_2 = 2$. In this case the maximum occurs at two extreme points, $(x_1 = \frac{1}{2}, x_2 = \frac{3}{2})$ and $(x_1 = 2, x_2 = 0)$ and, in fact, also occurs at all points on the line segment joining these extreme points.

Two additional possibilities exist. If the constraint $x_1 + x_2 \leq 2$ had been removed, the constraint set would have appeared as in Figure 1-11; i.e., the

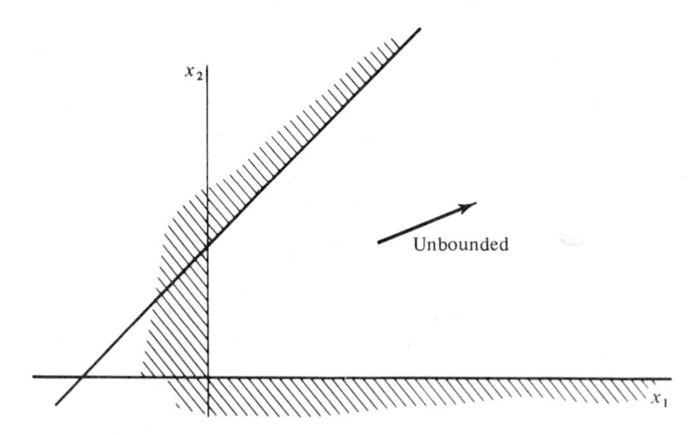

FIGURE 1-11 Unbounded minimum.

set would have been unbounded. Then max z is also unbounded, since z can be made as large as desired subject to the constraints. Of course, on the opposite extreme, the constraint set could have been empty, as in the case where $x_1 + x_2 \leq 2$ is replaced by $x_1 + x_2 \leq -1$. Thus a linear programming problem may have (1) no solution, (2) an unbounded solution, (3) a finite optimal solution, or (4) an infinite number of optimal solutions. The methods to be developed deal with all these possibilities.

The fact that the minimum of a linear programming problem always occurs at an extreme point of the constraint set is the single most important property of linear programs. It is true for any number of variables (i.e., more than two dimensions) and forms the basis for the simplex method for solving linear programs.

Of course, in many dimensions the geometrical ideas used here cannot be visualized and one must characterize extreme points algebraically. This is done in the next two sections, where the problem is placed in standard form and the basic theorems of linear programming are stated.

Standard Form for Linear Programs. A linear programming problem can always be written in the following form. Choose $x = (x_1, x_2, \ldots, x_n)$ to minimize

$$z = \sum_{j=1}^{n} c_j x_j \tag{3}$$

subject to

$$\sum_{j=1}^{n} a_{ij} x_j = b_i, \qquad i = 1, 2, \ldots, m \tag{4}$$

$$x_j \geq 0, \qquad j = 1, \ldots, n \tag{5}$$

or, in matrix form,

$$\text{minimize } cx$$

subject to

$$Ax = b$$
$$x \geq 0$$

where A is an $m \times n$ matrix of constants. If any of the equations (4) were redundant, i.e., linear combinations of the others, they could be deleted without changing any solutions of the system. If there are no solutions, or if there is only one, there can be no optimization. Thus the case of greatest interest is where the system of equations (4) is nonredundant and has at least two, hence an infinite number, of solutions. This occurs if and only if

$$n > m$$
$$\text{rank}(A) = m$$

We assume the above is true in what follows. The problem of linear programming is to first detect whether solutions exist, and, if so, to find one yielding minimum z.

Note that all the constraints in (4) are equalities and that all variables x_j are assumed to be nonnegative. It is necessary to place the problem in this form to solve it most easily (equations are easier to work with here than inequalities). If the original system is not of this form, it may easily be transformed by use of the following devices.

Slack Variables. If a given constraint is an inequality,

$$\sum_{j=1}^{n} a_{ij} x_j \leq b_i$$

then define a slack variable $x_{n+i} \geq 0$ such that

$$\sum_{j=1}^{n} a_{ij}x_j + x_{n+i} = b_i$$

and the inequality becomes an equality. Similarly, if the inequality is

$$\sum_{j=1}^{n} a_{ij}x_j \geq b_i$$

we write

$$\sum_{j=1}^{n} a_{ij}x_j - x_{n+i} = b_i$$

Note that the slacks must be nonnegative in order that the inequalities be satisfied for all x_j.

Nonnegative Variables. If, in the original formulation of the problem, a given variable, x_k, is not constrained to be nonnegative, we write it as the difference of two nonnegative variables, i.e.,

$$x_k = x'_k - x''_k, \qquad x'_k \geq 0, \quad x''_k \geq 0$$

This adds more variables to the problem, but, since nonnegativity restrictions actually simplify the solution of linear programs, it is well worth it. After the solution, we transform back to obtain x_k.

Example. Transform the following linear program into standard form:

$$\text{minimize } z = x_1 + x_2$$

subject to

$$\begin{aligned} 2x_1 + 3x_2 &\leq 6 \\ x_1 + 7x_2 &\geq 4 \\ x_1 + x_2 &= 3 \end{aligned}$$

$$x_1 \geq 0, \qquad x_2 \text{ unconstrained in sign}$$

SOLUTION. Define

$$x_2 = x'_2 - x''_2, \qquad x'_2 \geq 0, \quad x''_2 \geq 0$$

Also define slack variables $x_3 \geq 0$, $x_4 \geq 0$. Then the problem becomes

$$\text{minimize } z = x_1 + x_2' - x_2''$$

subject to

$$\begin{aligned} 2x_1 + 3x_2' - 3x_2'' + x_3 \quad &= 6 \\ x_1 + 7x_2' - 7x_2'' \quad - x_4 &= 4 \\ x_1 + x_2' - x_2'' \quad &= 3 \end{aligned}$$

$$x_1 \geq 0, \quad x_2' \geq 0, \quad x_2'' \geq 0, \quad x_3 \geq 0, \quad x_4 \geq 0$$

Basic Theorems of Linear Programming. We now proceed to generalize the ideas illustrated earlier from two to n dimensions. Proofs of the following theorems may be found in Gass [1]. First a number of standard definitions are made.

DEFINITION 1. A *feasible solution* to the linear programming problem is a vector $x = (x_1, x_2, \ldots, x_n)$ which satisfies the equations (4) and the non-negativities (5).

DEFINITION 2. A *basis matrix* is an $m \times m$ nonsingular matrix formed from some m columns of the constraint matrix A (note; since rank $(A) = m$, A contains at least one basis matrix)

DEFINITION 3. A *basic solution* to a linear program is the unique vector determined by choosing a basis matrix, setting the $n - m$ variables associated with columns of A not in the basis matrix equal to zero, and solving the resulting square, nonsingular system of equations for the remaining m variables.

DEFINITION 4. A *basic feasible solution* is a basic solution in which all variables have nonnegative values. [Note: By definition 3, at most M variables can be positive.]

DEFINITION 5. A *nondegenerate basic feasible solution* is a basic feasible solution with exactly m positive x_i.

DEFINITION 6. An *optimal solution* is a feasible solution which also minimizes z in (3).

For example, in the system

$$\begin{aligned} -x_1 + x_2 + x_3 \quad &= 1 \\ x_1 + x_2 \quad + x_4 &= 2 \\ x_i \geq 0, \quad i = 1, \ldots, 4 & \end{aligned} \tag{6}$$

obtained from (1) by adding slack variables x_3, x_4, the matrix

$$B = \begin{bmatrix} 1 & 0 \\ 0 & 1 \end{bmatrix}$$

formed from columns 3 and 4 of (6) is nonsingular, and hence is a *basis matrix*. The corresponding basic solution

$$x_1 = 0, \quad x_2 = 0, \quad x_3 = 1, \quad x_4 = 1$$

is a *nondegenerate basic feasible solution*. The matrix

$$B_1 = \begin{bmatrix} -1 & 0 \\ 1 & 1 \end{bmatrix}$$

formed from columns 1 and 4 of (6) is also a *basis matrix*. The corresponding *basic solution* is obtained by setting $x_2 = x_3 = 0$ and solving

$$\begin{aligned} -x_1 &= 1 \\ x_1 + x_4 &= 2 \end{aligned}$$

yielding $x_1 = -1$, $x_4 = 3$. This basic solution is not feasible.

The importance of these definitions is brought out by the following theorems:

THEOREM 1. *The objective function, z, assumes its minimum at an extreme point of the constraint set. If it assumes its minimum at more than one extreme point, then it takes on the same value at every point of the line segment joining any two optimal extreme points.*

This theorem is a multidimensional generalization of the geometric arguments given previously. By Theorem 1, in searching for a solution, we need only look at extreme points. It is thus of interest to know how to characterize extreme points in many dimensions algebraically. This information is given by the next theorem.

THEOREM 2. *A vector $x = (x_1, \ldots, x_n)$ is an extreme point of the constraint set of a linear programming problem if and only if x is a basic feasible solution of the constraints* (4)–(5).

Theorem 2 is true in two dimensions, as can be seen from the example of relations (1), whose constraints have been rewritten in equation form in (6).

The (x_1, x_2) coordinates of the extreme point at $x_1 = 0, x_2 = 1$ are given by the (x_1, x_2) coordinates of the basic feasible solution

$$x_1 = 0, \quad x_2 = 1, \quad x_3 = 0, \quad x_4 = 1$$

The optimal extreme point corresponds to the basic feasible solution

$$x_1 = \tfrac{1}{2}, \qquad x_2 = \tfrac{3}{2}, \qquad x_3 = x_4 = 0$$

Theorems 1 and 2 imply that, in searching for an optimal solution, we need only consider extreme points, hence only basic feasible solutions. Since a basic feasible solution has at most m of n variables positive, an upper bound to the number of basic feasible solutions is the number of ways m variables can be selected from a group of n variables, which is

$$\binom{n}{m} = \frac{n!}{(n-m)!\,m!}$$

For large n and m this is still a large number. Thus, for large problems, it would be impossible to evaluate z at all extreme points to find the minimum. What is needed is a computational scheme which selects, in an orderly fashion, a sequence of extreme points, each one yielding a lower value of z, until finally the minimum is attained. In this way we consider only a small subset of the set of all possible extreme points. The simplex method, devised by G. B. Dantzig, is such a scheme. This procedure finds an extreme point and determines whether or not it is optimal. If not, it finds a neighboring extreme point at which the value of z is less than or equal to the previous value. The process is iterated. In a finite number of steps (usually between m and $2m$) the minimum is found. The simplex method also discovers whether the problem has no finite minimal solution (i.e., min $z = -\infty$) or if it has no feasible solutions (i.e., an empty constraint set). It is a powerful scheme for solving any linear programming problem.

To explain the method, it is necessary to know how to go from one basic feasible solution (b.f.s.) to another, how to identify an optimal b.f.s., and how to find a better b.f.s. from a b.f.s. that is not optimal. We consider these questions in the following two sections. The notation and approach used is that of Dantzig [2].

Systems of Linear Equations and Equivalent Systems. Consider the system of m linear equations in n unknowns

$$\begin{aligned} a_{11}x_1 + a_{12}x_2 + \cdots + a_{1n}x_n &= b_1 \\ a_{21}x_1 + a_{22}x_2 + \cdots + a_{2n}x_n &= b_2 \\ &\vdots \\ a_{m1}x_1 + a_{m2}x_2 + \cdots + a_{mn}x_n &= b_m \end{aligned} \tag{7}$$

A *solution* to this system is any set of variables $x_1 \cdots x_n$ which simultaneously satisfies all equations. The set of all solutions to the system is called its *solution set*. The system may have one, many, or no solutions. If no solutions, the equations are said to be *inconsistent*, and their solution set is empty.

Equivalent Systems and Elementary Operations. Two systems of equations are said to be *equivalent* if they have the same solution sets. It is proved in Dantzig [2] that the following operations transform a given linear system into an equivalent system:

1. Multiplying any equation, E_t, by a constant $k \neq 0$.
2. Replacing any equation, E_t, by the equation $E_t + kE_i$, where E_i is any other equation of the system.

These operations are called elementary row operations. For example, the linear system of equations (6)

$$\begin{aligned} -x_1 + x_2 + x_3 &= 1 \\ x_1 + x_2 + x_4 &= 2 \end{aligned}$$

may be transformed into an equivalent system by multiplying the first equation by -1 and adding it to the second, yielding

$$\begin{aligned} -x_1 + x_2 + x_3 &= 1 \\ 2x_1 - x_3 + x_4 &= 1 \end{aligned}$$

Note that the solution $x_1 = 0$, $x_3 = 0$, $x_2 = 1$, $x_4 = 2$ is a solution of both systems. In fact, any solution of one system is a solution of the other.

Pivoting. A particular sequence of elementary row operations finds special application in linear programming. This sequence is called a *pivot operation*, defined as follows.

DEFINITION. A *pivot operation* consists of m elementary operations which replace a linear system by an equivalent system in which a specified variable has a coefficient of unity in one equation and zero elsewhere. The detailed steps are as follows:

1. Select a term $a_{rs}x_s$ in row (equation) r, column s, with $a_{rs} \neq 0$ called the *pivot term*.
2. Replace the rth equation by the rth equation multiplied by $1/a_{rs}$.
3. For each $i = 1, 2, \ldots, m$ except $i = r$, replace the ith equation, E_i, by $E_i - a_{is}/a_{rs}E_r$, i.e., by the sum of E_i and the replaced rth equation multiplied by $-a_{is}$.

Example. Consider the system

$$\begin{aligned} 2x_1 + 3x_2 - 4x_3 + x_4 &= 1 \qquad (E_1) \\ x_1 - x_2 + 5x_4 &= 6 \qquad (E_2) \\ 3x_1 + x_2 + x_3 &= 2 \qquad (E_3) \end{aligned}$$

Let us transform to an equivalent system in which x_1 is eliminated from all but (E_1), there having unity coefficient. Thus choose the term $2x_1$ as the pivot term. The first operation is to make the coefficient of this term unity, so we divide (E_1) by 2, yielding the equivalent system

$$\begin{aligned} x_1 + \tfrac{3}{2}x_2 - 2x_3 + \tfrac{1}{2}x_4 &= \tfrac{1}{2} && (E_1') \\ x_1 - x_2 \qquad + 5x_4 &= 6 && (E_2) \\ 3x_1 + x_2 + x_3 \qquad &= 2 && (E_3) \end{aligned}$$

The next operation eliminates x_1 from (E_2) by multiplying (E_1') by -1 and adding to (E_2), yielding

$$\begin{aligned} x_1 + \tfrac{3}{2}x_2 - 2x_3 + \tfrac{1}{2}x_4 &= \tfrac{1}{2} && (E_1') \\ - \tfrac{5}{2}x_2 + 2x_3 + \tfrac{9}{2}x_4 &= \tfrac{11}{2} && (E_2') \\ 3x_1 + x_2 + x_3 \qquad &= 2 && (E_3) \end{aligned}$$

Finally, we eliminate x_1 from (E_3) by multiplying (E_1') by -3 and adding to (E_3), yielding

$$\begin{aligned} x_1 + \tfrac{3}{2}x_2 - 2x_3 + \tfrac{1}{2}x_4 &= \tfrac{1}{2} \\ - \tfrac{5}{2}x_2 + 2x_3 + \tfrac{9}{2}x_4 &= \tfrac{11}{2} \\ - \tfrac{7}{2}x_2 + 7x_3 - \tfrac{3}{2}x_4 &= \tfrac{1}{2} \end{aligned}$$

Canonical Systems. Assume that the first m columns of the linear system (7) form a basis matrix, B. Multiplying each column of (7) by B^{-1} yields a transformed (but equivalent) system in which the coefficients of the variables $(x_1, \ldots, x_m)$ are an identity matrix. Such a system is called *canonical* and has the form shown in Table 1.

TABLE 1

Canonical System with Basic Variables $x_1, x_2, \cdots, x_m$

Dependent (basic) variables	Independent (nonbasic) variables	Constants
x_1	$+ \bar{a}_{1,m+1}x_{m+1} + \bar{a}_{1,m+2}x_{m+2} + \cdots + \bar{a}_{1n}x_n =$	$\bar{b}_1$
x_2	$+ \bar{a}_{2,m+1}x_{m+1} + \bar{a}_{2,m+2}x_{m+2} + \cdots + \bar{a}_{2n}x_n =$	$\bar{b}_2$
$\vdots$	$\vdots$	$\vdots$
x_m	$+ \bar{a}_{m,m+1}x_{m+1} + \bar{a}_{m,m+2}x_{m+2} + \cdots + \bar{a}_{mn}x_n =$	$\bar{b}_m$

The variables $x_1, \ldots, x_m$ are associated with the columns of B and are called *basic variables*. They are also called dependent, since if values are assigned to the *nonbasic* or independent variables $x_{m+1}, \ldots, x_n$, then $x_1, \ldots, x_m$ can be determined immediately. In particular, if $x_{m+1}, \ldots, x_n$ are all assigned zero values then we obtain the basic solution

$$x_1 = \bar{b}_1, \quad x_2 = \bar{b}_2, \ldots, x_m = \bar{b}_m, \quad x_{m+1} = x_{m+2} = \cdots = x_n = 0$$

If

$$\bar{b}_i \geq 0, \quad i = 1, \ldots, m$$

then this is a *basic feasible solution*. If one or more $\bar{b}_i = 0$, the basic feasible solution is *degenerate*.

Instead of actually computing B^{-1} and multiplying the linear system (7) by it, we can place (7) in canonical form by a sequence of m pivot operations. First, pivot on the term $a_{11}x_1$ if $a_{11} \neq 0$. If $a_{11} = 0$ then, since B is nonsingular, there is an element in its first row which is nonzero. Rearranging the columns makes this the (1, 1) element and allows the pivot. Repeating this procedure for the terms $a_{22}x_2, \ldots, a_{mm}x_m$ generates the canonical form. Such a form will be used to begin the simplex method.

Simplex Algorithm. The simplex method is a two-phase procedure for finding an optimal solution to linear programming problems. Phase 1 finds an initial basic feasible solution if one exists, or gives the information that one does not (in which case the constraints are inconsistent and the problem has no solution). Phase 2 uses this solution as a starting point and either (1) finds a minimizing solution or (2) yields the information that the minimum is unbounded (i.e., $-\infty$). Both phases use the simplex algorithm described here.

In initiating the simplex algorithm, we treat the objective form

$$z = c_1x_1 + c_2x_2 + \cdots + c_nx_n$$

as just another equation, i.e.,

$$-z + c_1x_1 + c_2x_2 + \cdots + c_nx_n = 0 \tag{8}$$

which we include in the set to form an augmented system of equations. The simplex algorithm is always initiated with this augmented system in canonical form. The basic variables are some m of the x's, which we renumber to make the first m, i.e., $x_1 \cdots x_m$ and $-z$. The problem can then be stated as follows.

Find values of $x_1 \geq 0, x_2 \geq 0, \ldots, x_n \geq 0$ and min z satisfying

$$\begin{array}{llllll} x_1 & & & + \bar{a}_{1,m+1}x_{m+1} + \cdots + \bar{a}_{1n}x_n & = \bar{b}_1 \\ & x_2 & & \vdots \qquad\qquad \vdots & \vdots \\ & & \ddots & & \\ & & x_m & + \bar{a}_{m,m+1}x_{m+1} + \cdots + \bar{a}_{mn}x_n & = \bar{b}_m \\ & & (-z) & + \bar{c}_{m+1}x_{m+1} + \cdots + \bar{c}_n x_n & = -\bar{z} \end{array} \tag{9}$$

In this canonical form the basic solution is

$$z = \bar{z}, \quad \bar{x}_1 = \bar{b}_1, \ldots, \bar{x}_m = \bar{b}_m, \quad x_{m+1} = x_{m+2} = \cdots = x_n = 0 \tag{10}$$

We assume that this basic solution is *feasible*, i.e.,

$$\bar{b}_1 \geq 0, \quad \bar{b}_2 \geq 0, \ldots, \bar{b}_m \geq 0 \tag{11}$$

The workings of phases 1 and 2 guarantee that this assumption is always satisfied. If (11) holds, we say that the linear programming problem is in *feasible canonical form.*

Test for Optimality. If the problem is in feasible canonical form, we have an extreme point directly at hand, represented by the basic feasible solution (10). But the form provides even more valuable information than this. By merely glancing at the numbers $\bar{c}_j$, $j = m + 1, \ldots, n$, one can tell if this extreme point is optimal or not and, further, can find a better one to go to if it is not optimal. Consider first the optimality test, given by the following theorem.

THEOREM 3. *A basic feasible solution is a minimal feasible solution with total cost $\bar{z}$ if all constants $\bar{c}_{m+1}, \bar{c}_{m+2}, \ldots, \bar{c}_n$ are nonnegative, i.e. if*

$$\bar{c}_j \geq 0 \qquad j = m + 1, \ldots, n \tag{12}$$

The $\bar{c}_j$ are called ***relative cost factors.***

PROOF. Write the last equation as

$$z = \bar{z} + \bar{c}_{m+1}x_{m+1} + \cdots + \bar{c}_n x_n$$

Since the variables $x_{m+1} \cdots x_n$ are presently zero and are constrained to be nonnegative, the only way any one of them can change is to become positive. But if $\bar{c}_j \geq 0$ for $j = m + 1, \ldots, n$, then increasing any x_j cannot decrease the objective function, z, since then $\bar{c}_j x_j \geq 0$. Since no change in the nonbasic variables can cause z to decrease, the present solution must be optimal.

The relative cost factors can also tell if there are multiple optima. Let all $\bar{c}_j \geq 0$ and let $\bar{c}_k = 0$ for some nonbasic variable, x_k. Then, if the constraints allow that variable to be made positive, no change in z results, and there are multiple optima. It is possible, however, that the variable may not be allowed by the constraints to become positive; this may occur in the case of *degenerate* solutions. We consider the effects of degeneracy later. A corollary to these results is the following:

COROLLARY. *A basic feasible solution is the unique minimal feasible solution if $\bar{c}_j > 0$ for all nonbasic variables.*

Of course, if some $\bar{c}_j < 0$, then z can be decreased by increasing the corresponding x_j, so the present solution is probably nonoptimal.[1] Thus we must consider means of improving a nonoptimal solution.

Improving a Nonoptimal Basic Feasible Solution: An Example. To illustrate, consider the problem of minimizing z, where

$$\begin{aligned} 5x_1 - 4x_2 + 13x_3 - 2x_4 + x_5 \quad &= 20 \\ x_1 - x_2 + 5x_3 - x_4 + x_5 \quad &= 8 \\ x_1 + 6x_2 - 7x_3 + x_4 + 5x_5 - z &= 0 \end{aligned} \tag{13}$$

$$x_j \geq 0, \qquad j = 1, 2, \ldots, 5 \tag{14}$$

Assume that we know that x_5, x_1, $-z$ can be used as basic variables and that the basic solution will be feasible. We can thus reduce system (13) to feasible canonical form by pivoting successively on the terms x_5 (first equation), and x_1 (second equation) ($-z$ already appears in the correct way). This yields

$$\begin{aligned} x_5 \qquad - \tfrac{1}{4}x_2 + 3x_3 - \tfrac{3}{4}x_4 &= 5 \\ x_1 \qquad - \tfrac{3}{4}x_2 + \boxed{2x_3} - \tfrac{1}{4}x_4 &= 3 \\ -z + 8x_2 - 24x_3 + 5x_4 &= -28 \end{aligned} \tag{15}$$

The circled term will be explained soon. The basic feasible solution is

$$x_5 = 5, \quad x_1 = 3, \quad x_2 = x_3 = x_4 = 0, \qquad z = 28 \tag{16}$$

Note that an arbitrary pair of variables will not necessarily yield a basic solution to (13) which is feasible. For example, had the variables x_1 and x_2 been chosen as basic variables, the basic solution would have been

$$x_1 = -12, \quad x_2 = -20, \quad x_3 = x_4 = x_5 = 0, \qquad z = -132 \tag{17}$$

which is not feasible, since x_1 and x_2 are negative.

[1] It may be that the constraints do not allow x_j to be increased, so we cannot say *definitely* that the solution is nonoptimal. If it is optimal, however, the method will soon transform all $\bar{c}_j$ to be nonnegative, and the optimality is then recognized.

For the original basic feasible solution, one relative cost factor is negative, namely $\bar{c}_3 = -24$. The optimality test of relations (12) thus fails. Further, if x_3 is increased from its present value of zero (with all other nonbasic variables remaining zero), z must decrease, since, by the third equation of (15), z is then related to x_3 by

$$z = 28 - 24x_3 \tag{18}$$

How large should x_3 become? It is reasonable to make it as large as possible, since the larger the value of x_3, the smaller the value of z. However, the constraints place a limit on the maximum value x_3 can attain. Note that, if $x_2 = x_4 = 0$, relations (15) state that the basic variables x_1, x_5 are related to x_3 by

$$\begin{aligned} x_5 &= 5 - 3x_3 \\ x_1 &= 3 - 2x_3 \end{aligned} \tag{19}$$

Thus as x_3 increases, x_5 and x_1 decrease, and they cannot be allowed to become negative. In fact, as x_3 reaches $\frac{3}{2}$, x_1 becomes zero and as x_3 reaches $\frac{5}{3}$, x_5 becomes zero. By that time, however, x_1 is already negative, so the largest value x_3 can attain is

$$x_3 = \tfrac{3}{2} \tag{20}$$

Substituting this value into (18) and (19) yields *a new basic feasible solution* with lower cost:

$$x_5 = \tfrac{1}{2}, \quad x_3 = \tfrac{3}{2}, \quad x_1 = x_2 = x_4 = 0, \qquad z = -8 \tag{21}$$

This solution reduces z from 28 to -8. The immediate objective is to see if it is optimal or not. This can be done if the system can be placed into feasible canonical form with x_5, x_3, $-z$ as basic variables. *That is, x_3 must replace x_1 as a basic variable. One of the reasons that the simplex method is efficient is that this replacement can be accomplished by doing one pivot transformation.*

Previously x_1 had a coefficient of unity in the second equation of (15) and zero elsewhere. We now wish x_3 to have this property, and this can be accomplished by pivoting on the term $2x_3$, circled in the second equation of (15). This causes x_3 to become basic and x_1 to become nonbasic, as is seen below:

$$\begin{aligned} x_5 \qquad\qquad & - \tfrac{3}{2}x_1 + \boxed{\tfrac{7}{8}x_2} - \tfrac{3}{8}x_4 = \tfrac{1}{2} \\ x_3 \qquad & + \tfrac{1}{2}x_1 - \tfrac{3}{8}x_2 - \tfrac{1}{8}x_4 = \tfrac{3}{2} \\ -z & + 12x_1 - x_2 + 2x_4 = 8 \end{aligned} \tag{22}$$

This gives the basic feasible solution (21), as predicted. It also indicates that the present solution, although better, is still not optimal, since $\bar{c}_2$, the coefficient of x_2 in the z equation, is -1. Thus we can again obtain a better solution by increasing x_2 while keeping all other nonbasic variables at zero. From (22), the current basic variables are then related to x_2 by

$$\begin{aligned} x_5 &= \tfrac{1}{2} - \tfrac{7}{8}x_2 \\ x_3 &= \tfrac{3}{2} + \tfrac{3}{8}x_2 \\ z &= -8 - x_2 \end{aligned} \tag{23}$$

Note that the second equation places no bound on the increase of x_2, but the first equation restricts x_2 to a maximum of $(\frac{1}{2})/(\frac{7}{8}) = \frac{4}{7}$, which reduces x_5 to zero. As before, we obtain a new feasible canonical form by pivoting, this time using $\frac{7}{8}x_2$ in the first equation of (22) as the pivot term. This yields the system

$$\begin{aligned} x_2 \qquad\quad - \tfrac{12}{7}x_1 - \tfrac{3}{7}x_4 + \tfrac{8}{7}x_5 &= \tfrac{4}{7} \\ x_3 \quad - \tfrac{1}{7}x_1 - \tfrac{2}{7}x_4 + \tfrac{3}{7}x_5 &= \tfrac{12}{7} \\ -z + \tfrac{72}{7}x_1 + \tfrac{11}{7}x_4 + \tfrac{8}{7}x_5 &= \tfrac{60}{7} \end{aligned} \tag{24}$$

and the basic feasible solution

$$x_2 = \tfrac{4}{7}, \quad x_3 = \tfrac{12}{7}, \quad x_1 = x_4 = x_5 = 0, \qquad z = -\tfrac{60}{7} \tag{25}$$

Since all relative cost factors for the nonbasic variables are positive, this solution is the unique minimal solution of the problem, by the corollary of the last section. The optimum has been reached in two iterations.

Degeneracy. Note that, if in the original system (15), the constant on the right-hand side of the second equation had been zero, i.e., if the basic feasible solution had been degenerate, then x_1 would have been related to x_3 by

$$x_1 = -2x_3 \tag{26}$$

and any positive change in x_3 would have caused x_1 to become negative. Thus x_3 would be forced to remain zero and z could not decrease. What is done is to go through the pivot transformation anyway and attain a new form in which the degeneracy may not be limiting. This can easily occur, for if relation (26) had been

$$x_1 = 2x_3$$

then x_3 could be made positive.

Unboundedness. If relations (23) had been

$$\begin{aligned} x_5 &= \tfrac{1}{2} + \tfrac{7}{8}x_2 \\ x_3 &= \tfrac{3}{2} + \tfrac{3}{8}x_2 \\ z &= -8 - x_2 \end{aligned}$$

then x_2 could be made as large as desired without causing x_5 and x_3 to become negative, and z could be made as small as desired. This is an *unbounded* solution. Note that it occurs whenever *all coefficients in a column with negative* $\bar{c}_j$ *are also negative* (*or zero*).

Improving a Nonoptimal Basic Feasible Solution in General. Let us now formalize the procedures of the previous section. If at least one $\bar{c}_j < 0$, then, at least if we assume nondegeneracy (all $\bar{b}_i > 0$), it is always possible to construct, by pivoting, another basic feasible solution with lower cost. If more than one $\bar{c}_j < 0$, the variable, x_s, to be increased can be chosen to be the one with the most negative $\bar{c}_j$; i.e., the one whose relative cost factor is

$$\bar{c}_s = \min_j \bar{c}_j < 0 \tag{27}$$

Although this may not lead to the greatest possible decrease in z (since it may not be possible to increase x_s very far), this is intuitively at least a good rule for choosing the variable to become basic. It is the one used in practice today because (1) it is simple and (2) it generally leads to fewer iterations than just choosing any $\bar{c}_j < 0$.

Having decided on the variable, x_s, to become basic, we increase it from zero, holding all other nonbasic variables zero, and observe the effects on the current basic variables. By (9), these are related to x_s by

$$\begin{aligned} x_1 &= \bar{b}_1 - \bar{a}_{1s}x_s \\ x_2 &= \bar{b}_2 - \bar{a}_{2s}x_s \\ &\vdots \\ x_m &= \bar{b}_m - \bar{a}_{ms}x_s \\ z &= \bar{z} + \bar{c}_s x_s, \qquad \bar{c}_s < 0 \end{aligned} \tag{28}$$

Increasing x_s decreases z, and the only factor limiting the decrease is that one of the variables $x_1 \cdots x_m$ could become negative. However, if

$$\bar{a}_{is} \leq 0, \qquad i = 1, 2, \ldots, m \tag{29}$$

then x_s can be made as large as desired. Thus we have the following theorem.

THEOREM 5 (UNBOUNDEDNESS). *If, in the canonical system for some s, all coefficients,* $\bar{a}_{is}$, *are nonpositive and* $\bar{c}_s$ *is negative, then a class of feasible solutions can be constructed where the set of z values has no lower bound.*

The class of solutions yielding unbounded z is the set

$$x_i = \bar{b}_i - \bar{a}_{is}x_s, \qquad i = 1, \ldots, m \tag{30}$$

with x_s any positive number and all other $x_i = 0$.

If, however, at least one $\bar{a}_{is}$ is positive, then x_s cannot be increased indefinitely, since eventually some basic variable will become first zero, then negative. From (28), x_i becomes zero when $\bar{a}_{is} > 0$ and when x_s attains the value

$$x_s = \frac{\bar{b}_i}{\bar{a}_{is}}, \qquad \bar{a}_{is} > 0 \tag{31}$$

The first x_i to become negative is that x_i which requires the smallest x_s to drive it to zero. This value of x_s is the greatest value for x_s permitted by the nonnegativity constraints and is given by

$$x_s^* = \frac{\bar{b}_r}{\bar{a}_{rs}} = \min_{\bar{a}_{is} > 0} \frac{\bar{b}_i}{\bar{a}_{is}} \tag{32}$$

By (28), the variable x_r then becomes nonbasic, to be replaced by x_s. We have seen from the example that a new canonical form with x_s replacing x_r as a basic variable is easily obtained by pivoting on the term $\bar{a}_{rs}x_s$. *Note that the previous operations may be viewed as simply locating that pivot term.* Finding $\bar{c}_s = \min_j \bar{c}_j < 0$ told us that the pivot term was in column s, and finding that the minimum of the ratios $\bar{b}_i/\bar{a}_{is}$ for $\bar{a}_{is} > 0$ occurred for $i = r$ told us that it was in row r.

As seen in the example, if the basic solution is degenerate, then the x_s^* given by (32) may be zero. In particular, if some $\bar{b}_i = 0$ and the corresponding $\bar{a}_{is} > 0$, then, by (32), $x_s^* = 0$. In this case the pivot operation is still carried out, but z is unchanged.

Iterative Procedure. The procedures of the previous section provide a means of going from one basic feasible solution to one whose z is at least equal to the previous z (as can occur, in the degenerate case) or lower, if there is no degeneracy. It is repeated until (1) the optimality test of relations (12) is passed or (2) the information is provided that the solution is unbounded. The following result is the main convergence theorem.

THEOREM 6. *Assuming nondegeneracy at each iteration, the simplex algorithm will terminate in a finite number of iterations.*

PROOF. Since the number of basic feasible solutions is finite, the algorithm can fail to terminate only if a basic feasible solution is repeated. Such repetition implies that the same value of z is also repeated. Under nondegeneracy, however, each value of z is lower than the previous, so no repetition can occur and the algorithm is finite.

Degenerate Case. If at some iteration the basic feasible solution is degenerate, the possibility exists that z could remain constant for some number of subsequent iterations. *It is then possible for a given set of basic variables to be repeated.* An endless loop is then set up, the optimum is never attained, and the simplex algorithm is said to have cycled. Theoretical procedures have been developed which are guaranteed to avoid cycling [2]. Such procedures are, however, of limited practical interest, for in practice *cycling has never been known to occur*, despite the many thousands of problems solved.[2] Thus, for most problems encountered in practice, no antidegeneracy procedure is necessary. Degeneracy can slow down the simplex method, however, and highly degenerate problems (i.e., problems in which many of the basic variables at a given iteration may be zero) may take quite a bit longer to solve than nondegenerate problems of the same size.

Two Phases of the Simplex Method. The simplex algorithm requires a basic feasible solution as a starting point. Such a starting point is not always easy to find and, in fact, none will exist if the constraints are inconsistent. Phase 1 of the simplex method finds an initial basic feasible solution or yields the information that none exists. Phase 2 then proceeds from this starting point to an optimal solution or yields the information that the solution is unbounded. Both phases use the simplex algorithm of the previous section.

PHASE 1. Phase 1 starts with the linear program in the general form (3)–(5) and *augments* the system to include a set of *artificial variables*, $x_{n+1}, x_{n+2}, \ldots, x_{n+m}$, so that it becomes

$$\begin{array}{lllll} a_{11}x_1 + a_{12}x_2 + \cdots + a_{1n}x_n + x_{n+1} & & & & = b_1 \\ a_{21}x_1 + a_{22}x_2 + \cdots + a_{2n}x_n & + x_{n+2} & & & = b_2 \\ \vdots & & \ddots & & \\ a_{m1}x_1 + a_{m2}x_2 + \cdots + a_{mn}x_n & & & x_{n+m} & = b_m \\ c_1x_1 + c_2x_2 + \cdots + c_nx_n & & & & - z = 0 \\ \quad x_j \geq 0, \quad j = 1, 2, \ldots, m + n & & & & \end{array} \tag{33}$$

where the b_i have all been made positive by multiplying through the original system by -1 where necessary. This system has the initial basic feasible solution

$$z = x_1 = x_2 = \cdots = x_n = 0, \quad x_{n+1} = b_1, \quad x_{n+2} = b_2, \ldots, x_{n+m} = b_m \tag{34}$$

[2] Artificial examples of cycling have been constructed, however. See Dantzig [2, chap. 10].

Any solution of (33) which has the artificial variables x_{n+i} equal to zero will have the remaining components $x_1 \cdots x_n$ comprising a solution to the original system. Thus one way to find an initial basic feasible solution to the original system is to start from (34) and use the simplex algorithm to try to drive the artificial variables to zero. This can be done by minimizing the function

$$w = x_{n+1} + x_{n+2} + \cdots + x_{n+m} \tag{35}$$

subject to (33). If min $w = 0$, then all $x_{n+i} = 0$ and conversely. Further, if min $w > 0$, then no solution to the original system exists (because then all artificial variables cannot be made zero). Since, given an initial basic feasible solution, the simplex algorithm transforms only to other basic feasible solutions, the end product of phase 1 must be a basic feasible solution to the original system, if such exists.[3]

It should be mentioned that a full set of m artificial variables may not be necessary. If the original system has some variables that may be used as part of an initial basic feasible solution, then these should be used in preference to artificial variables. The result is less work in phase 1. For example, in the system

$$\begin{aligned} 2x_1 + 3x_2 - x_3 + x_4 &= 2 \\ -x_1 + 2x_2 + 3x_3 \phantom{{}+x_4} &= 1 \end{aligned}$$

x_4 can be used as a basic variable and only one artificial variable, x_5, need be used, yielding the augmented system

$$\begin{aligned} 2x_1 + 3x_2 - x_3 + x_4 \phantom{{}+x_5} &= 2 \\ -x_1 + 2x_2 + 3x_3 \phantom{{}+x_4} + x_5 &= 1 \end{aligned}$$

The two following examples will illustrate the use of the simplex method to solve linear programming problems.

Example 1. Consider first the problem illustrated geometrically in Figure 1-10 given in relations (1), i.e.,

$$\text{maximize } z = x_1 + 3x_2$$

subject to

$$\begin{aligned} -x_1 + x_2 + x_3 \phantom{{}+x_4} &= 1 \\ x_1 + x_2 \phantom{{}+x_3} + x_4 &= 2 \\ x_i \geq 0, \quad i = 1, \ldots, 4 & \end{aligned} \tag{36}$$

[3] If the optimal phase 1 basis is degenerate and min $w = 0$, artificial variables may be basic at zero level. Procedures for dealing with this are discussed in [2].

where x_3, x_4 are slack variables. Here no phase 1 is needed, since an initial basic feasible solution is obvious. To apply directly the results of the previous sections, we rephrase the problem as

$$\text{minimize } -\textcircled{x_1} - 3x_2$$

subject to (36). The initial feasible canonical form is

$$\begin{aligned} -x_1 + x_2 + x_3 \quad &= 1 \\ x_1 + x_2 \qquad + x_4 &= 2 \\ -x_1 - 3x_2 \qquad - z &= 0 \end{aligned} \tag{37}$$

The initial basic feasible solution is

$$x_1 = x_2 = 0, \quad x_3 = 1, \quad x_4 = 2, \qquad z = 0 \tag{38}$$

This corresponds to extreme point (2) of Figure 1-10.

Iteration 1. Since $\bar{c}_2 = \min(\bar{c}_1, \bar{c}_2) = -3 < 0$, x_2 becomes basic. To see which variable becomes nonbasic, we compute the ratios $\bar{b}_i/\bar{a}_{i2}$ for all i such that $\bar{a}_{i2} > 0$. This gives

$$\frac{\bar{b}_1}{\bar{a}_{12}} = \frac{1}{1} = 1, \qquad \frac{\bar{b}_2}{\bar{a}_{22}} = \frac{2}{1} = 2$$

The minimum of these is $\bar{b}_1/\bar{a}_{12}$; thus the basic variable with unity coefficient in row 1, x_3, leaves the basis. The pivot term is $\bar{a}_{12}x_2$, i.e., the x_2 term circled in equation (37). Pivoting on this term yields

$$\begin{aligned} -x_1 + x_2 + x_3 \qquad &= 1 \\ \textcircled{2x_1} \qquad - x_3 + x_4 \quad &= 1 \\ -4x_1 \qquad + 3x_3 \qquad - z &= 3 \end{aligned} \tag{39}$$

Iteration 2. The new basic feasible solution is

$$x_1 = x_3 = 0, \quad x_2 = x_4 = 1, \qquad z \doteq -3$$

Note that z is reduced. The solution corresponds to extreme point (3) of Figure 1-10. Since $\bar{c}_1 = -4 = \min_j \bar{c}_j$, x_1 becomes basic. The only ratio $\bar{b}_i/\bar{a}_{i1}$ having $\bar{a}_{i1} > 0$ is that for $i = 2$; thus x_4 becomes nonbasic and the pivot term is $\bar{a}_{21}x_1 = 2x_1$, circled above. Pivoting yields

$$\begin{aligned} x_2 + \tfrac{1}{2}x_3 + \tfrac{1}{2}x_4 \qquad &= \tfrac{3}{2} \\ x_1 \qquad - \tfrac{1}{2}x_3 + \tfrac{1}{2}x_4 \qquad &= \tfrac{1}{2} \\ x_3 + 2x_4 - z &= 5 \end{aligned} \tag{40}$$

with basic feasible solution

$$x_1 = \tfrac{1}{2}, \quad x_2 = \tfrac{3}{2}, \quad x_3 = x_4 = 0, \quad z = -5$$

which corresponds to extreme point (4) of Figure 1-10. This is optimal, since all $\bar{c}_j > 0$. The path taken by the method is (2), (3), (4).

Example 2. The second example is that given in relations (13), which are rewritten below:

$$\begin{aligned} 5x_1 - 4x_2 + 13x_3 - 2x_4 + x_5 &= 20 \\ x_1 - x_2 + 5x_3 - x_4 + x_5 &= 8 \\ x_1 + 6x_2 - 7x_3 + x_4 + 5x_5 &= z \\ \text{all } x_i &\geq 0 \end{aligned} \tag{41}$$

Since no initial basic solution is evident, both phase 1 and phase 2 must be used. Since the constant terms are nonnegative, phase 1 is initiated with the augmented system.

$$\begin{aligned} 5x_1 - 4x_2 + 13x_3 - 2x_4 + x_5 + x_6 \qquad\qquad &= 20 \\ x_1 - x_2 + 5x_3 - x_4 + x_5 \qquad + x_7 \qquad &= 8 \\ x_1 + 6x_2 - 7x_3 + x_4 + 5x_5 \qquad\qquad - z &= 0 \\ x_6 + x_7 - w &= 0 \end{aligned} \tag{42}$$

This is reduced to canonical form by subtracting the sum of the first two equations from the last, yielding the starting system for phase 1. All computations are carried out below, with pivot terms circled. A filled or empty small circle (● or ○) under a column denotes a basic variable. The variable to become basic is denoted by *, the variable to become nonbasic by the empty small circle.

The steps for the minimization of w in phase 1 are similar to those for minimizing z. On the first cycle w is reduced from 28 to $\frac{4}{13}$, on the second cycle to zero, and a basic feasible solution $x_3 = \frac{3}{2}$, $x_5 = \frac{1}{2}$, $z = -8$ is obtained for the original system. Variables x_6 and x_7 are no longer of any use and are now dropped, as is w, and the minimization of z commences. On the third cycle z decreases from -8 to $-\frac{60}{7}$, which is minimal. The optimal solution is $x_2 = \frac{4}{7}$, $x_3 = \frac{12}{7}$, all other $x_j = 0$, $z = -\frac{60}{7}$.

CYCLE 0 (PHASE 1)

$$\begin{aligned} 5x_1 - 4x_2 + \textcircled{13x_3} - 2x_4 + x_5 + x_6 \qquad\qquad &= 20 \\ x_1 - x_2 + 5x_3 - x_4 + x_5 \qquad + x_7 \qquad &= 8 \\ x_1 - 6x_2 + 7x_3 + x_4 + 5x_5 \qquad\qquad - z \quad &= 0 \\ -6x_1 + 5x_2 - 18x_3 + 3x_4 - 2x_5 \qquad\qquad - w &= -28 \end{aligned}$$

$$\qquad\qquad * \qquad\qquad\qquad \circ \quad \bullet \quad \bullet \quad \bullet$$

CYCLE 1 (PHASE 1)

$$
\begin{array}{l}
\frac{5}{13}x_1 - \frac{4}{13}x_2 + x_3 - \frac{2}{13}x_4 + \frac{1}{13}x_5 + \frac{1}{13}x_6 = \frac{20}{13} \\
-\frac{12}{13}x_1 + \frac{7}{13}x_2 - \frac{3}{13}x_4 + \boxed{\frac{8}{13}x_5} - \frac{5}{13}x_6 + x_7 = \frac{4}{13} \\
\frac{48}{13}x_1 + \frac{50}{13}x_2 - \frac{1}{13}x_4 + \frac{72}{13}x_5 + \frac{7}{13}x_6 - z = \frac{140}{13} \\
\frac{12}{13}x_1 - \frac{7}{13}x_2 + \frac{3}{13}x_4 - \frac{8}{13}x_5 + \frac{18}{13}x_6 - w = -\frac{4}{13}
\end{array}
$$

● * ○ ● ●

CYCLE 2 (PHASE 1 AND 2)

$$
\begin{array}{l}
\frac{1}{2}x_1 - \frac{3}{8}x_2 + x_3 - \frac{1}{8}x_4 + \frac{1}{8}x_6 - \frac{1}{8}x_7 = \frac{3}{2} \\
-\frac{12}{8}x_1 + \boxed{\frac{7}{8}x_2} - \frac{3}{8}x_4 + x_5 - \frac{5}{8}x_6 + \frac{13}{8}x_7 = \frac{4}{8} \\
12x_1 - x_2 + 2x_4 + 4x_6 - 9x_7 - z = 8 \\
x_6 + x_7 - w = 0
\end{array}
$$

* ● ○ ● ●

CYCLE 3 (PHASE 2: OPTIMAL)

$$
\begin{array}{l}
-\frac{1}{7}x_1 + x_3 - \frac{2}{7}x_4 + \frac{3}{7}x_5 = \frac{12}{7} \\
-\frac{12}{7}x_1 + x_2 - \frac{3}{7}x_4 + \frac{8}{7}x_5 = \frac{4}{7} \\
\frac{72}{7}x_1 + \frac{11}{7}x_4 + \frac{8}{7}x_5 - z = \frac{60}{7}
\end{array}
$$

● ● ●

Optimal solution: $x_3 = \frac{12}{7}$, $x_2 = \frac{4}{7}$, all other $x_j = 0$, $z = -\frac{60}{7}$.

1.2.2 Revised Simplex Method. In performing a simplex iteration, much of the information contained in the tableau is not used. Only the following items are needed:

1. The relative cost factors $\bar{c}_j$. Using them, we compute

$$\bar{c}_s = \min \bar{c}_j$$

2. Assuming $\bar{c}_s < 0$, we now require the elements of the updated column

$$\bar{P}_s = (\bar{a}_{1s}, \ldots, \bar{a}_{ms})'$$

and the values of the basic variables

$$x_B = (\bar{b}_1, \ldots, \bar{b}_m)'$$

Using these, the quantity

$$\frac{\bar{b}_r}{\bar{a}_{rs}} = \min_{\bar{a}_{is} > 0} \frac{\bar{b}_i}{\bar{a}_{is}} \tag{1}$$

is calculated and a pivot operation is performed on $\bar{a}_{rs}$.

Note that only one nonbasic column of the current tableau, $\bar{P}_s$, is required. Since it is common to have many more columns than rows in a linear program, much computer time and storage can be wasted dealing with the $\bar{P}_j$ for $j \neq s$. A more efficient procedure is to generate, from the original problem data, first the $\bar{c}_j$ and then the column $\bar{P}_s$. The revised simplex method does precisely this, and the inverse of the current basis matrix is what is needed to generate the desired quantities.

The linear program to be solved is written in column form:

$$\text{minimize } z = c_1x_1 + \cdots + c_nx_n \tag{2}$$

subject to

$$P_1x_1 + \cdots + P_nx_n = b \tag{3}$$

$$x_i \geq 0, \qquad i = 1, \ldots, n \tag{4}$$

where

$$P_j = (a_{1j}, \ldots, a_{mj})' \tag{5}$$

is the jth column of the coefficient matrix, A. We assume again that this matrix has rank m and that the program is feasible. Let

$$B = [P_{j_1}, P_{j_2}, \ldots, P_{j_m}] \tag{6}$$

be a basis matrix for the program, and let

$$x_B = (x_{j_1}, x_{j_2}, \ldots, x_{j_m})' \tag{7}$$

$$c_B = (c_{j_1}, c_{j_2}, \ldots, c_{j_m}) \tag{8}$$

be the corresponding vectors of basic variables and objective coefficients. Note that c_B is a row vector. The vector x_B satisfies

$$Bx_B = b \tag{9}$$

whose solution is

$$x_B = B^{-1}b = \bar{b} \tag{10}$$

We assume that B is a feasible basis, i.e.,

$$x_B \geq 0 \tag{11}$$

As shown earlier, it is convenient to regard the z equation as equation $m + 1$, with $-z$ a permanent basic variable. This augmented system can be written in column form by defining

$$\hat{P}_j = (a_{1j}, \ldots, a_{mj}, c_j)', \qquad j = 1, \ldots, n \tag{12}$$

$$\hat{P}_{n+1} = (0, 0, \ldots, 0, 1)' \tag{13}$$

$$\hat{b} = (b_1, \ldots, b_m, 0)' \tag{14}$$

Then (2)–(3) can be written

$$\sum_{j=1}^{n} \hat{P}_j x_j + \hat{P}_{n+1}(-z) = \hat{b} \tag{15}$$

Since B is feasible, the $(m + 1)$ by $(m + 1)$ matrix

$$\hat{B} = [\hat{P}_{j_1}, \ldots, \hat{P}_{j_m}, \hat{P}_{n+1}] = \left[\begin{array}{c|c} B & 0 \\ \hline c_B & 1 \end{array}\right] \tag{16}$$

is a feasible basis for the enlarged system, (15). It is easily verified by direct matrix multiplication that the inverse of $\hat{B}$ is

$$\hat{B}^{-1} = \left[\begin{array}{c|c} B^{-1} & 0 \\ \hline -c_B B^{-1} & 1 \end{array}\right] \tag{17}$$

DEFINITION. The (row) vector

$$\pi = (\pi_1, \ldots, \pi_m) = c_B B^{-1} \tag{18}$$

is called the *vector of simplex multipliers associated with the basis B.*

Note that, if in the system (3), we multiply the first equation by $\pi_1, \ldots,$ the mth equation by π_m, sum, and subtract from the z equation, the coefficient of x_j becomes $c_j - \pi P_j$. Setting

$$c_j - \pi P_j = 0, \qquad j \text{ basic} \tag{19}$$

yields the equations

$$\pi B = c_B \tag{20}$$

whose solution is the simplex multiplier vector $c_B B^{-1}$. Thus π may also be defined as that vector which, when it multiplies the columns of the original equation system and subtracts the result from the z equation, causes the coefficients of the basic variables to vanish.

Using (17) we may write

$$\hat{B}^{-1} = \left[\begin{array}{c|c} B^{-1} & 0 \\ \hline -\pi & 1 \end{array}\right] \tag{21}$$

If each column of the system (15) is multiplied by $\hat{B}^{-1}$, the following canonical form results:

$$\begin{array}{l} x_{j_1} \\ \quad \ddots \\ \qquad x_{j_m} \end{array} + \sum_{j \text{ nonbasic}} \bar{P}_j x_j = \begin{array}{c} \bar{b}_1 \\ \vdots \\ \bar{b}_m \end{array}$$
$$-z + \sum_{j \text{ nonbasic}} \bar{c}_j x_j = -z_0 \tag{22}$$

Since

$$\begin{bmatrix} \bar{P}_j \\ \hline \bar{c}_j \end{bmatrix} = \left[\begin{array}{c|c} B^{-1} & 0 \\ \hline -\pi & 1 \end{array}\right] \begin{bmatrix} P_j \\ \hline c_j \end{bmatrix} \tag{23}$$

then the updated column is

$$\bar{P}_j = B^{-1} P_j \tag{24}$$

and the relative cost factor is

$$\bar{c}_j = c_j - \pi P_j \tag{25}$$

The two formulas above are fundamental, and will be used many times in later chapters. They show how, given the matrix $\hat{B}^{-1}$ (or, equivalently, given B^{-1} and the simplex multipliers, π) the quantities needed to perform a simplex iteration, $\bar{c}_j$ and $\bar{P}_j$, may be computed from the original problem data, c_j and P_j.

Assume now that the $\bar{c}_j$ have been computed from (25), the smallest, $\bar{c}_s$, found, and $\bar{P}_s$ generated from (24). If the values of the basic variables, $\bar{b}$, were available at the beginning of the cycle, a pivot element $\bar{a}_{rs}$ can be located in $\bar{P}_s$ by using (1). It remains to introduce P_s into the basis and remove P_{j_r}, i.e., to generate the inverse of the new basis. To see how this is done, consider the partitioned matrix

$$\left[\begin{array}{c|c|c} & & a_{1s} \\ & & \vdots \\ \hat{P}_{j_1} \cdots \hat{P}_{j_m} \hat{P}_{n+1} & u_1 \cdots u_{m+1} & a_{rs} \\ & & \vdots \\ & & a_{ms} \\ & & c_s \end{array}\right] \tag{26}$$

where u_i is the ith $(m + 1)$-dimensional unit vector. If this matrix is multiplied by $\hat{B}^{-1}$, the result is

$$\left[\begin{array}{c|c|c} & & \bar{a}_{1s} \\ & & \vdots \\ u_1 \cdots u_r \cdots u_{m+1} & \hat{B}^{-1} & \bar{a}_{rs} \\ & & \vdots \\ & & \bar{a}_{ms} \\ & & \bar{c}_s \end{array}\right] \leftarrow \text{pivot element} \tag{27}$$

Now let the above matrix be transformed by a pivot operation on $\bar{a}_{rs}$, yielding

$$[u_1 \cdots u_{r-1}\, \alpha\, u_{r+1} \cdots u_{m+1} \mid \hat{B}_{\text{new}}^{-1} \mid u_r] \tag{28}$$

where α is, in general, full of nonzero elements. The second partition now contains the desired matrix, $\hat{B}_{\text{new}}^{-1}$. To see this, compare the above with the original matrix, (26). The columns of the new basis

$$\hat{B}_{\text{new}} = [\hat{P}_{j_1} \cdots \hat{P}_{j_{r-1}}, \hat{P}_s, \hat{P}_{j_{r+1}} \cdots \hat{P}_{j_m}, \hat{P}_{n+1}] \tag{29}$$

have been reduced to a unit matrix. The pivot operations which did this must be equivalent to multiplying the original matrix (26) by $\hat{B}_{\text{new}}^{-1}$. When this is done, $\hat{B}_{\text{new}}^{-1}$ appears in the second partition. Of course, in all these operations, the first partition was carried along only for explanatory purposes, and may be dropped.

Summary of the Revised Simplex Method. At the beginning of some iteration, say iteration k, assume that the current basis inverse, $\hat{B}^{-1}$, the associated basic solution $x_B = B^{-1}b$, and the data of the original problem (A, b, and c) are available. Cycle k proceeds as follows:

1. Row $m + 1$ of $\hat{B}^{-1}$ has the form

$$\hat{B}_{m+1}^{-1} = (-\pi_1, -\pi_2, \ldots, -\pi_m, 1)$$

where the π_i are simplex multipliers. For each nonbasic variable, form the relative cost factor $\bar{c}_j$, using the equation

$$\bar{c}_j = c_j - \sum_{i=1}^{m} \pi_i a_{ij} = c_j - \pi P_j$$

This operation will often be referred to as "pricing out" the column P_j.

2. Assuming the standard column-selection rule is employed, find

$$\min_j \bar{c}_j = \bar{c}_s$$

3. If $\bar{c}_s \geq 0$, stop. The current basic solution is optimal.
4. If $\bar{c}_s < 0$, compute the transformed column

$$\begin{pmatrix} \bar{P}_s \\ \bar{c}_s \end{pmatrix} = \hat{B}^{-1} \begin{pmatrix} P_s \\ c_s \end{pmatrix}$$

5. Let

$$\bar{P}_s = (\bar{a}_{1s}, \bar{a}_{2s}, \ldots, \bar{a}_{ms})'$$

If all $\bar{a}_{is} \leq 0$, stop. The optimum is unbounded. Let

$$\hat{x} = (x_B \mid 0) + x_s(-\bar{P}_s \mid u_s)$$

where u_s is the sth unit vector. Then $\hat{x}$ is a solution, feasible for all $x_s \geq 0$, whose objective value

$$\hat{z} = c_B x_B + \bar{c}_s x_s$$

approaches $-\infty$ as $x_s \to +\infty$.

6. Otherwise, compute

$$\frac{\bar{b}_r}{\bar{a}_{rs}} = \min_{\bar{a}_{is} > 0} \frac{\bar{b}_i}{\bar{a}_{is}} = \theta$$

7. Construct the augmented matrix

$$\left[\hat{B}^{-1} \;\middle|\; \begin{matrix} \bar{P}_s \\ \bar{c}_s \end{matrix} \right]$$

and transform it by pivoting on $\bar{a}_{rs}$. The first $m + 1$ columns of the result contain the inverse of the new basis. Update the basic solution by

$$(x_B)_i \leftarrow (x_B)_i - \theta \bar{a}_{is}, \qquad i \neq r$$
$$(x_B)_r \leftarrow \theta$$

and return to step 1.

In many computer codes, especially those designed for large programs, the basis inverse is not maintained explicitly, i.e., by storing all its elements. Instead, the so-called product form is used, with the inverse stored as a product of simpler, elementary matrices. Product-form codes are described in Section 6.2, along with a comparison of the relative efficiencies of the simplex method in standard, explicit-inverse, and product-inverse form.

1.2.3 Duality in Linear Programming. Corresponding to any linear program, called the primal problem, there is another called its dual. These two problems bear interesting and useful relationships to one another. Table 1 illustrates a general primal–dual pair, with dual variables $\pi_1, \ldots, \pi_m$. Note that one can make a one-to-one correspondence between the ith dual variable

TABLE 1

Primal	Dual
minimize $z = \sum_{i=1}^{n} c_i x_i$	maximize $v = \sum_{i=1}^{m} \pi_i b_i$
subject to	subject to
$E \cup \bar{E} = \{1, 2, \ldots, m\}$: $\sum_{j=1}^{n} a_{ij} x_j = b_i, \quad i \in E$	$\sum_{i=1}^{m} \pi_i a_{ij} = c_j, \quad j \in \bar{P}$
$\sum_{j=1}^{n} a_{ij} x_j \geq b_i, \quad i \in \bar{E}$	$\sum_{i=1}^{m} \pi_i a_{ij} \leq c_j, \quad j \in P$
$P \cup \bar{P} = \{1, 2, \ldots, n\}$: $x_i \geq 0, \quad i \in P$	$\pi_i \geq 0, \quad i \in \bar{E}$
x_i unconstrained in sign, $\quad i \in \bar{P}$	π_i unconstrained in sign, $\quad i \in E$
or, in matrix form,[a] minimize $z = c'x$	maximize $v = \pi b$
subject to	subject to
$Ax \left\{ \begin{matrix} = \\ \geq \end{matrix} \right\} b$	$A'\pi' \left\{ \begin{matrix} = \\ \leq \end{matrix} \right\} c$
$x_i \geq 0, \quad i \in P$	$\pi_i \geq 0, \quad i \in \bar{E}$

[a] Note that π is a row vector.

π_i and the ith primal constraint (that with right-hand side b_i) and between the jth primal variable and the jth dual constraint (that with right-hand side c_j). Table 2 gives the primal–dual correspondence. In it, A_i is row i of the matrix

TABLE 2

Primal quantity	Corresponding dual quantity
objective $c'x \rightarrow \min$	objective $\pi b \rightarrow \max$
variable $x_j \geq 0$	constraint $\pi P_j \leq c_j$, inequality
variable x_j, unconstrained in sign	constraint $\pi P_j = c_j$, equality
constraint $A_i x = b_i$, equality	variable π_i, unconstrained in sign
constraint $A_i x \geq b_i$, inequality	variable $\pi_i \geq 0$
coefficient matrix A	coefficient matrix A'
right-hand side b	right-hand side c
cost coefficients c	cost coefficients b

A. Note that if one takes the dual of the dual, the primal is obtained. Also, by multiplying both sides of the dual constraints which are inequalities by -1 and replacing the objective max πb by $\min(-\pi b)$, the dual is transformed into a problem of primal form, but with different data. Thus any theorem beginning with an assumption about the primal and ending with an assertion regarding the dual which is true for any A, b, c also holds with the roles of primal and dual reversed.

From Table 1, the following special forms of primal–dual pairs may be obtained:

Symmetric Primal–Dual Pair ($E = \bar{P} = \varnothing$)

Primal	Dual
minimize $c'x$ subject to $Ax \geq b$ $x \geq 0$	maximize πb subject to $A'\pi' \leq c$ $\pi \geq 0$

Primal in Standard Form ($\bar{E} = \bar{P} = \varnothing$)

Primal	Dual
minimize $c'x$ subject to $Ax = b$ $x \geq 0$	maximize πb subject to $A'\pi' \leq c$ π unconstrained in sign

In symmetric form it is especially easy to see that the dual of the dual is the primal. A primal problem in any of these forms may be changed into any other by using the following devices: (1) replace an unconstrained variable by the difference of two nonnegative variables; (2) replace an equality constraint by two opposing inequalities; (3) replace an inequality constraint with an equality by adding a slack variable. Thus we may work with any of the forms when proving results. For later use, it is convenient to work with the primal in standard form.

THEOREM 1. *If $\bar{x}$ and $\bar{\pi}$ are feasible primal and dual solutions, then*

$$\bar{z} = c'\bar{x} \geq \bar{\pi} b = \bar{v} \tag{1}$$

PROOF. By the feasibility hypothesis

$$\bar{x} \geq 0, \qquad A\bar{x} = b, \qquad A'\bar{\pi}' \leq c$$

Thus

$$c'\bar{x} \geq \bar{\pi} A \bar{x} = \bar{\pi} b$$

THEOREM 2. *If both primal and dual have feasible solutions, then both have optimal solutions and*

$$\min z = \max v \tag{2}$$

PROOF. By Theorem 1, the primal objective is bounded below and the dual objective is bounded above. Thus both problems have optimal solutions. To show equality of the optimal objective values, let x^0 solve the primal. Since the primal must have a basic optimal solution, we may as well assume x^0 basic, with the optimal basis B^0, and vector of basic variables x_B^0. Thus

$$B^0 x_B^0 = b, \qquad x_B^0 \geq 0$$

The simplex multiplier vector associated with B^0 is, by equation (1.2.2–18),[4]

$$\pi^0 = c_B(B^0)^{-1}$$

with c_B the vector of cost coefficients of basic variables. Since x^0 is optimal, the relative cost factors, given by (1.2.2–25) are nonnegative:

$$\bar{c}_j = c_j - \pi^0 P_j \geq 0, \qquad \text{all } j$$

or, in matrix form,

$$A'(\pi^0)' \leq c$$

Thus π^0 satisfies the dual constraints and has objective value

$$v^0 = \pi^0 b = c_B(B^0)^{-1}b = c_B x_B^0 = z^0$$

Since this is equal to an upper bound for v, π^0 solves the dual, and the theorem is proved.

Theorems 1 and 2 have the following geometric interpretation

z, $\bar{z}$, $\bar{v}$, v $\min z = \max v$

[4] From this point on, certain cross references to equations and other elements will be in this form: The numbers preceding the hyphen refer to the section; the number after the hyphen is that of the equation, theorem, and so forth.

The objective functions of primal and dual problems are approaching the same optimal value but from different directions. Thus given both a primal and a dual feasible point, the difference between their objective values, $\bar{z} - \bar{v}$, is always greater than the difference between either value and the common optimum. The quantity $\bar{z} - \bar{v}$ can be used to formulate stop criteria in iterative procedures; e.g., stop when $\bar{z} - \bar{v}$ is less than r per cent of $\bar{z}$.

The preceding proof illustrates some important points. The dual constraints are simply the primal optimality conditions, with the relative cost factors $\bar{c}_j$ appearing as slack variables in them. Further, the simplex multiplier vector π^0 associated with any primal optimal basis solves the dual. Since, as shown in the previous section, the vector $-\pi^0$ is contained in the bottom row of $\hat{B}^{-1}$, solution of the primal automatically provides a dual solution. The same statement applies, of course, with primal and dual reversed. Thus, given any linear program, one really has a choice of two problems, primal or dual, to solve. One of these may be easier to solve than the other. For example, if the primal constraints have the form

$$\text{minimize } c_1x_1 + c_2x_2$$

subject to

$$\text{20 inequalities}\begin{cases} a_{11}x_1 + a_{12}x_2 \geq b_1 \\ \vdots \qquad\qquad\qquad \vdots \\ a_{20,1}x_1 + a_{20,2}x_2 \geq b_{20} \end{cases}$$
$$\underbrace{\qquad\qquad\qquad\qquad\qquad}_{\text{2 unknowns}}$$

then 20 slack variables are needed to convert to equalities, leading to a 22-variable, 20-equation system. At each stage we must deal with a 20×20 basis matrix. The dual constraints, however, have the structure

$$\text{2 inequalities}\begin{cases} a_{11}\pi_1 + a_{21}\pi_2 + \cdots + a_{20,1}\pi_{20} \leq c_1 \\ a_{12}\pi_2 + a_{22}\pi_2 + \cdots + a_{20,2}\pi_{20} \leq c_2 \end{cases}$$
$$\underbrace{\qquad\qquad\qquad\qquad\qquad\qquad\qquad}_{\text{20 unknowns}}$$

Here we need deal with only a 2×2 basis inverse in the revised simplex method, a great computational advantage.

The simplex multipliers π have an interpretation as prices. This is discussed in Section 3.6. They are also of great value in sensitivity analysis, as is explained in Gass [1], Dantzig [2], and in other texts.

Theorem 2 may be stated with a slightly different hypothesis.

COROLLARY. *If either the primal or dual problem has an optimal solution, then the other does also and the optimal objective values are equal.*

A proof, similar to that of Theorem 2, may be found in Gass [1].

THEOREM 3. *If either primal or dual problem has an unbounded solution, then the other problem is infeasible.*

PROOF. Assume that the primal is unbounded. Then Theorem 2 yields

$$\pi b \leq -\infty \qquad \text{for all dual feasible } \pi$$

Existence of a solution to the dual constraints would imply a finite value for the dual objective, contradicting the above. Thus the dual is infeasible. A similar argument proves the theorem with primal and dual reversed.

The relations between primal and dual given in the preceding theorems are summarized in the following table:

Primal \ Dual	Feasible	Infeasible
Feasible	$\min c'x = \max \pi b$	$c'x \to -\infty$
Infeasible	$\pi b \to +\infty$	possible

The following theorem is often useful and follows easily from Theorem 2.

THEOREM 4. *A pair* (x^0, π^0) *with* x^0 *primal feasible and* π^0 *dual feasible has* x^0 *and* π^0 *primal and dual optimal, respectively, if and only if*

$$(c' - \pi^0 A)x^0 = 0 \tag{3}$$

PROOF. For any primal feasible $\bar{x}$,

$$A\bar{x} = b \tag{4}$$

$$c'\bar{x} = \bar{z} \tag{5}$$

Multiply both sides of (4) by some dual feasible $\bar{\pi}$ and subtract the result from (5):

$$c'\bar{x} - \bar{\pi}A\bar{x} = \bar{z} - \bar{\pi}b$$

or, setting $\bar{\pi}b = \bar{v}$,

$$(c' - \bar{\pi}A)\bar{x} = \bar{z} - \bar{v} \tag{6}$$

If (x^0, π^0) are, respectively, primal and dual optimal, then, by Theorem 2

$$z^0 - v^0 = 0 = (c' - \pi^0 A)x^0$$

Conversely, if (x^0, π^0) satisfy

$$(c' - \pi^0 A)x^0 = z^0 - v^0 = 0$$

then, by Theorem 1, z attains its lower bound at x^0 and v attains its upper bound at π^0, so both x^0 and π^0 are optimal.

The condition $(c' - \pi^0 A)x^0 = 0$ is called complementary slackness. Since both x^0 and $c' - \pi^0 A$ are nonnegative vectors, each term of the scalar product sum must be zero:

$$(c' - \pi^0 A)_i x_i^0 = 0, \quad \text{all } i \tag{7}$$

Thus if the ith dual constraint is slack, i.e., $(c' - \pi^0 A)_i > 0$, $x_i^0 = 0$, while $x_i^0 > 0$ implies that the ith dual constraint holds as equality. Relation (7) may also be written in terms of the relative cost factors as

$$\bar{c}_i^0 x_i^0 = 0, \quad \text{all } i \tag{8}$$

Note that the simplex method maintains the above true at each cycle by choosing the π vector such that $\bar{c}_i = 0$ for x_i basic. The full symmetry of the relationship is best seen in the symmetric primal–dual pair. There, positivity of a variable in one problem implies that the corresponding constraint in the other holds with equality, while a slack constraint implies a zero level for the corresponding variable.

The three optimality conditions (1) primal feasibility, (2) dual feasibility, and (3) complementary slackness are useful in classifying linear programming algorithms. Most algorithms enforce two of these conditions throughout while relaxing the third, enforcing it as the iterations progress. The simplex method enforces conditions 1 and 3 while relaxing 2. The dual simplex and primal–dual methods, to be considered next, enforce the following pairs of conditions:

Algorithm	Primal feasibility	Dual feasibility	Complementary slackness
Simplex method	enforced	relaxed	enforced
Dual simplex method	relaxed	enforced	enforced
Primal–dual method	relaxed	enforced	enforced

The latter two methods use different means to reduce the amount of primal infeasibility. The dual simplex method increases the value of the dual objective at each step, while the primal–dual algorithm minimizes the infeasibility form, w, in (1.2.1–35). Other algorithms exist, corresponding to other possible table entries, the most notable being those which relax both primal and dual feasibility. These are discussed, for example, in Dantzig [2, Chap. 11].

1.2.4 Dual Simplex and Primal–Dual Algorithms. There exist a number of algorithms for linear programming which start with an infeasible solution to the primal and iteratively force it to become feasible in such a way that, when it does, it is also optimal. The most prominent of such methods are the dual simplex method and the primal–dual technique. These algorithms eliminate phase 1 of the simplex method with its attendant disadvantage—the fact that the starting point found by phase 1 may be nowhere near optimal, since the phase 1 objective ignores optimality completely. By working simultaneously toward feasibility and optimality, a smaller total number of iterations might be expected. This has been found true for network flow problems, where the out-of-kilter method [2], a generalization of the primal–dual algorithm, is very efficient. Both methods operate by maintaining dual feasibility and complementary slackness at every iteration while relaxing primal feasibility.

Both algorithms require the availability of a dual feasible solution to start with, and this limits their applicability. Still, situations often arise in which one has an infeasible basic primal solution whose associated dual solution is feasible. That is, one has a basis, B, such that $B^{-1}b$ is not a nonnegative vector, but for which the simplex multipliers

$$\pi = c_B B^{-1}$$

satisfy

$$\bar{c}_j = c_j - \pi P_j \geq 0, \qquad \text{all } j$$

with P_j the jth primal column. Examples of such situations include the following:

1. A linear program is given whose solution for a number of different right-hand-side vectors, b^j, is desired.
2. Constraints are to be added to a program whose solution is known.

In the first case, it would be inefficient to apply the two phases of the simplex method separately for each b^j. These vectors often do not differ greatly, and the solution for b^1 may, for example, be "close" to the solution for b^2. A better procedure is to solve the program for b^1, obtaining an optimal basis, B. If this basis is feasible for all other right-hand sides, i.e., if

$$B^{-1}b^j \geq 0, \qquad \text{all } j$$

then it is optimal for these as well. If not, so that

$$B^{-1}b^r \ngeq 0 \qquad \text{for some } r$$

then

$$\pi = c_B B^{-1}$$

is dual feasible, since the condition

$$\bar{c}_j = c_j - \pi P_j \geq 0$$

is independent of the right-hand side.[5] A similar situation prevails in case 2. Consider the example

$$\text{minimize } z$$

$$\text{subject to} \qquad \begin{aligned} x_1 \qquad\quad + 3x_3 + x_4 &= 1 \\ x_2 \qquad - x_3 + 5x_4 &= 2 \\ - z + 2x_3 + x_4 &= -3 \\ x_i \geq 0, \qquad \text{all } i& \end{aligned} \tag{1}$$

which is already optimized, and the additional constraint

$$x_1 + x_2 \leq 2 \tag{2}$$

which is not satisfied by the optimal solution. We add a slack variable in (2):

$$x_1 + x_2 + s = 2$$

and subtract the first two equations of (1) from the above, yielding the system

$$\begin{aligned} x_1 \qquad\qquad + 3x_3 + x_4 &= 1 \\ x_2 \qquad\quad - x_3 + 5x_4 &= 2 \\ s \quad - 2x_3 - 6x_4 &= -1 \\ - z + 2x_3 + x_4 &= -3 \end{aligned} \tag{3}$$

Again we have a basic solution which is infeasible, since $s = -1$, but whose relative cost factors are all nonnegative.

[5] An equivalent viewpoint is that changing the primal right-hand side changes the dual objective coefficients, which cannot affect dual feasibility.

Dual Simplex Method. The dual simplex algorithm is constructed to start from the kind of initial position shown in (3). We will show that it is precisely the simplex method applied to the dual but constructed so as to operate within the standard primal simplex tableau. Operationally, the algorithm still involves a sequence of pivot steps in this tableau, but with different rules for choosing the pivot element. A stepwise description follows.

The problem to be solved is initially in canonical form, with all $\bar{c}_j \geq 0$ (dual feasibility), basic $\bar{c}_j = 0$ (complementary slackness), but not all $\bar{b}_i \geq 0$ (primal feasibility relaxed).

1. Choose row r as the pivot row, where

$$\bar{b}_r = \min \bar{b}_i < 0 \tag{4}$$

2. Choose column s as the pivot column, where

$$\frac{\bar{c}_s}{-\bar{a}_{rs}} = \min_{\bar{a}_{rj}<0} \frac{\bar{c}_j}{-\bar{a}_{rj}} \tag{5}$$

If all $\bar{a}_{rj} \geq 0$, the primal is infeasible.

3. Pivot on $\bar{a}_{rs}$.
4. If

$$\bar{b}_i \geq 0, \qquad \text{all i} \tag{6}$$

stop, the current solution is optimal. Otherwise, return to step 1.

The optimality test in step 4 is valid, since nonnegativity of the $\bar{b}_i$ means that all three optimality criteria are satisfied. To understand the infeasibility test in step 2, note that if x_r is the rth basic variable, then

$$x_r = \bar{b}_r - \sum_{j=m+1}^{n} \bar{a}_{rj}x_j \tag{7}$$

If

$$\bar{b}_r < 0, \qquad \bar{a}_{rj} \geq 0 \qquad \text{for all } j$$

then no nonnegative values for the x_j can cause x_r to be nonnegative. The primal thus contains an equation that cannot be satisfied in nonnegative variables, and hence is infeasible.

Consider the application of this algorithm to the problem in (3). The pivot row is row 3, while application of step 2 yields

$$\frac{\bar{c}_s}{-\bar{a}_{3s}} = \min[\tfrac{2}{2}, \tfrac{1}{6}] = \frac{\bar{c}_4}{\bar{a}_{34}}$$

Pivoting on $\bar{a}_{34} = -6$ yields

$$\begin{array}{llllll} x_1 & & +\tfrac{1}{6}s + \tfrac{8}{3}x_3 & & = \tfrac{5}{6} \\ & x_2 & +\tfrac{5}{6}s - \tfrac{8}{3}x_3 & & = \tfrac{7}{6} \\ & & -\tfrac{1}{6}s + \tfrac{1}{3}x_3 & + x_4 & = \tfrac{1}{6} \\ & -z & +\tfrac{1}{6}s + \tfrac{5}{3}x_3 & & = -\tfrac{19}{6} \end{array} \tag{8}$$

Since all $\bar{b}_i \geq 0$, this solution is optimal.

We now show that the above algorithm is simply the simplex method applied to the dual. Let the primal be

$$\text{minimize } z = \sum_{j=1}^{n} c_j x_j$$

subject to

$$\sum_{j=1}^{n} a_{ij} x_j = b_i, \qquad i = 1, \ldots, m \tag{9}$$

$$x_i \geq 0, \qquad \text{all } i$$

and let

$$B = \begin{bmatrix} a_{11}, \ldots, a_{1m} \\ \vdots \\ a_{m1}, \ldots, a_{mm} \end{bmatrix} \tag{10}$$

be a basis for the primal. The dual is shown in detached coefficient form as follows:

	Dual variables		
	$\pi_1, \pi_2, \ldots, \pi_m$	$\bar{c}_1, \bar{c}_2, \ldots, \bar{c}_m, \bar{c}_{m+1}, \ldots, \bar{c}_n$	Constants
B'	$a_{11}, a_{21}, \ldots, a_{m1}$	1	c_1
	$\cdot \qquad \cdot$	1	$\cdot$
	$\cdot \qquad \cdot$	$\ddots$	$\cdot$
	$\cdot \qquad \cdot$		$\cdot$
	$a_{1m}, a_{2m}, \ldots, a_{mm}$	1	c_m
$\bar{B}'$	$a_{1,m+1} \quad \cdots \quad a_{m,m+1}$	1	c_{m+1}
	$\cdot \qquad \cdot$	$\ddots$	$\cdot$
	$\cdot \qquad \cdot$		$\cdot$
	$\cdot \qquad \cdot$		$\cdot$
	$a_{1n} \quad \cdots \quad a_{mn}$	1	c_n
	$b_1, \quad b_2, \ldots, b_m$	$0, 0, \ldots\ldots\ldots\ldots, 0$	
	Dual objective coefficients		

(11)

The matrix in the lower left partition is called $\bar{B}'$ and that in the upper left is B'. In equation form, the dual has n rows and $n + m$ columns, so there are n basic variables. The basis B is assumed to be dual feasible; i.e., substituting

$$\pi = c_B B^{-1}$$

into the dual constraints yields $\bar{c}_j \geq 0$. Since the above choice of π yields $\bar{c}_j = 0, j = 1, \ldots, m$, the variables $\bar{c}_j, j = m + 1, \ldots, n$, along with $\pi_1, \ldots, \pi_m$, are taken as the basic variables. The dual basis matrix corresponding to these variables is

$$B_d = \begin{array}{r} m \\ n-m \end{array}\left\{\begin{array}{c|c} \overbrace{B'}^{m} & \overbrace{0}^{n-m} \\ \hline \bar{B}' & I \end{array}\right] \tag{12}$$

Note that B and B_d are in one-to-one correspondence, each primal basis B generating a unique square block-triangular dual basis. The inverse of B_d is easily found by partitioning formulas to be

$$B_d^{-1} = \left[\begin{array}{c|c} (B^{-1})' & 0 \\ \hline -(B^{-1}\bar{B})' & I \end{array}\right] \tag{13}$$

Let us now reduce the dual to canonical form. In matrix form the dual tableau is

$$\left[\begin{array}{c|c} B' & 0 \\ \hline \bar{B}' & I \end{array}\right] \left[\begin{array}{c} \pi \\ \hline \bar{c}_{m+1} \\ \vdots \\ \bar{c}_n \end{array}\right] + \left[\begin{array}{c} I \\ \hline 0 \end{array}\right] \left[\begin{array}{c} \bar{c}_1 \\ \vdots \\ \bar{c}_m \end{array}\right] = \left[\begin{array}{c} c_1 \\ \cdot \\ \cdot \\ \cdot \\ c_n \end{array}\right] \tag{14}$$

Multiplication by B_d^{-1} yields

$$\begin{array}{l} \pi_1 \\ \quad \ddots \\ \qquad \pi_m \\ \qquad\quad \bar{c}_{m+1} \\ \qquad\qquad \ddots \\ \qquad\qquad\quad \bar{c}_n \end{array} + \left[\begin{array}{c} (B^{-1})' \\ \hline -(B^{-1}\bar{B})' \end{array}\right] \left[\begin{array}{c} \bar{c}_1 \\ \vdots \\ \bar{c}_m \end{array}\right] = \left[\begin{array}{c} (c_B B^{-1})' \\ \hline c_{m+1} - \pi P_{m+1} \\ \vdots \\ c_n - \pi P_n \end{array}\right] \tag{15}$$

$$b_1\pi_1 + \cdots + b_m\pi_m \qquad\qquad - v = 0$$

To produce the canonical form, $\pi_1, \ldots, \pi_m$ must be eliminated from the v equation. To do this, multiply the first equation by b_1, the second equation by b_2, etc., sum and subtract from the v equation. The (row) vector of objective coefficients for $\bar{c}_1, \ldots, \bar{c}_m$ now becomes

$$-b'(B^{-1})' = -(B^{-1}b)' = -\bar{b}' \tag{16}$$

Thus the final canonical form is

$$\begin{array}{cccccc} \pi_1 & & & & & \\ & \ddots & & & & \\ & & \pi_m & & & \\ & & & \bar{c}_{m+1} & & \\ & & & & \ddots & \\ & & & & & \bar{c}_n \end{array} + \begin{bmatrix} (B^{-1})' \\ \hdashline -(B^{-1}\bar{B})' \end{bmatrix} \begin{bmatrix} \bar{c}_1 \\ \vdots \\ \bar{c}_m \end{bmatrix} = \begin{bmatrix} (c_B B^{-1})' \\ \hdashline c_{m+1} - \pi P_{m+1} \\ \vdots \\ c_n - \pi P_n \end{bmatrix} \tag{17}$$

$$-\bar{b}_1\bar{c}_1 \cdots -\bar{b}_m\bar{c}_m - v = -c_B B^{-1} b$$

Note that the current value of v is

$$c_B B^{-1} b = c_B x_B$$

just the z value of the current primal basic solution.

Consider now applying the simplex method to this canonical form. Since v is to be maximized, the pivot column is chosen as column r, where

$$-\bar{b}_r = \max(-\bar{b}_i)$$

The same index r is obtained by the rule

$$\bar{b}_r = \min \bar{b}_i$$

which is step 1 of the dual simplex method. If

$$\max(-b_i) \leq 0$$

or, equivalently,

$$\min \bar{b}_i \geq 0$$

the current solution is optimal. This is the dual simplex optimality test. If

$\bar{b}_r < 0$, then, barring degeneracy,[6] the dual objective can be increased by bringing $\bar{c}_r$ into the dual basis. Since $\pi_1, \ldots, \pi_m$ are unconstrained in sign, they are not forced to leave the basis as $\bar{c}_r$ enters. Only $\bar{c}_{m+1}, \ldots, \bar{c}_n$ can leave, so the pivot row is among the last $n - m$. The general simplex rule for choosing the pivot element is

$$\frac{\bar{b}_r}{\bar{a}_{rs}} = \min_{\bar{a}_{is} > 0} \frac{\bar{b}_i}{\bar{a}_{is}}$$

where $\bar{b}_i$ is the ith right-hand-side element and $\bar{a}_{is}$ the ith element of the pivot column. From the dual tableau (17), we see that the role of $\bar{b}_i$ is played by $\bar{c}_{m+i}$. The pivot column here is that corresponding to $\bar{c}_r$, and its last $n - m$ elements are the rth column of $-(B^{-1}\bar{B})'$, which is

$$-(\bar{a}_{r,m+1}, \bar{a}_{r,m+2}, \ldots, \bar{a}_{rn})$$

The ith element of this vector corresponds to $\bar{a}_{is}$ and is $-\bar{a}_{r,m+i}$. The pivot selection rule thus becomes

$$\frac{\bar{c}_{m+k}}{-\bar{a}_{r,m+k}} = \min_{-\bar{a}_{r,m+i} > 0} \frac{\bar{c}_{m+i}}{-\bar{a}_{r,m+i}}$$

Since no $\bar{a}_{ri}$ are negative for $i \leq m$ (all are zero except $\bar{a}_{rr} = 1$), we may simplify this to

$$\frac{\bar{c}_s}{-\bar{a}_{rs}} = \min_{\bar{a}_{ri} < 0} \frac{\bar{c}_i}{-\bar{a}_{ri}}$$

This is the dual simplex column-selection rule in step 2. If all $\bar{a}_{ri} \geq 0$, the dual is unbounded which implies, by Theorem 3 of Section 1.2.3, that the primal is infeasible.

Performing the pivot increases the dual objective by an amount

$$-\bar{b}_r\left(\frac{\bar{c}_s}{-\bar{a}_{rs}}\right)$$

and leads to a new dual tableau like (17). Note that the only elements in the tableau which are ever used are the $\bar{b}_i$, $\bar{c}_i$, and the elements $\bar{a}_{ij}$. These are also contained in the primal tableau, so the transformations are done there instead. At each cycle the z value, which is equal to the value of v in the dual tableau, increases. Since the method is the simplex algorithm, the question

[6] Dual degeneracy means $\bar{c}_{m+i} = 0$ for some i; i.e., some nonbasic primal variable has a zero relative cost factor.

of convergence is easily settled. If the dual is nondegenerate at each cycle, i.e., if

$$\bar{c}_i > 0, \qquad i = m + 1, \ldots, n$$

convergence to an optimal solution is obtained in a finite number of iterations. Finite convergence even with degeneracy can be guaranteed by using any of the antidegeneracy procedures developed for the simplex method.

Primal–Dual Method. Let the problem to be solved be

$$\text{minimize } z$$

$$\text{subject to} \quad \begin{aligned} a_{11}x_1 + \cdots + a_{1n}x_n &= b_1 \\ &\vdots \\ a_{m1}x_1 + \cdots + a_{mn}x_n &= b_m \\ c_1x_1 + \cdots + c_nx_n &= z - z_0 \end{aligned} \tag{18}$$

with all variables nonnegative. It is again assumed that a feasible solution to the dual of (18) is available; i.e., there exist multipliers $\pi = (\pi_1, \ldots, \pi_m)$ such that

$$\bar{c}_j = c_j - \pi P_j \geq 0, \qquad j = 1, \ldots, n \tag{19}$$

Then, by multiplying the columns of (18) by π and subtracting from c_j, one obtains nonnegative coefficients for the x_j. We assume that this has already been done, so $c_j \geq 0, j = 1, \ldots, n$.

The initial tableau for the primal–dual method is the same as that for phase 1 of the simplex method:

$$\text{minimize } w$$

$$\text{subject to} \quad \begin{array}{llll} a_{11}x_1 + \cdots + a_{1n}x_n + x_{n+1} & & = b_1 \\ \vdots & \ddots & \vdots \\ a_{m1}x_1 + \cdots + a_{mn}x_n & + x_{n+m} & = b_m \\ \bar{d}_1x_1 + \cdots + \bar{d}_nx_n & & = w - w_0 \\ c_1x_1 + \cdots + c_nx_n & & = z - z_0 \end{array} \tag{20}$$

Assume that, at the kth iteration, this tableau has been transformed so that after relabeling the variables, it appears as follows:

Basis: $x_1 \cdots x_q \; x_{q+1} \cdots x_m$	$x_{m+1} \cdots x_{m+q}$	$\cdots x_{m+n}$	Constants	π	σ
1	$\bar{a}_{1,m+1} \cdots \bar{a}_{1,m+q}$	$\cdots \bar{a}_{1,m+n}$	$\bar{b}_1$	π_1	σ_1
$\quad 1$					
$\quad\quad \cdot$			$\cdot$	$\cdot$	$\cdot$
$\quad\quad\quad \cdot$			$\cdot$	$\cdot$	$\cdot$
$\quad\quad\quad\quad \cdot$			$\cdot$	$\cdot$	$\cdot$
$\quad\quad\quad\quad\quad 1$	$\bar{a}_{m,m+1} \cdots \bar{a}_{m,m+q} \cdots$	$\bar{a}_{m,m+n}$	$\bar{b}_m$	π_m	σ_m
$0 \cdot \cdot \cdot \cdot \cdot \cdot \cdot 0$	$\bar{d}_{m+1} \cdots \bar{d}_{m+q}$	$\bar{d}_{m+q+1} \cdots \bar{d}_{m+n}$	$w - w_0$		
$* \cdot \cdot \cdot * \; 0 \cdot \cdot \cdot 0$	$0 \cdot \cdot \cdot 0$	$\bar{c}_{m+q+1} \cdots \bar{c}_{m+n}$	$z - z_0$		
Artificial ($x_1 \cdots x_q$)		$\bar{c}_j > 0$			

Restricted primal

The following should be noted:

1. The multipliers $\pi_1, \ldots, \pi_m$ are a feasible solution to the dual of the original problem without artificial variables, (18), and generate $\bar{c}_j \geq 0$.
2. The multipliers $\sigma_1, \ldots, \sigma_m$ are the simplex multipliers of the current basis relative to the infeasibility form w and generate the $\bar{d}_j$.
3. The current basic solution is feasible for the problem with artificial variables, (20), so $\bar{b}_i \geq 0$, $i = 1, \ldots, m$.
4. The $\bar{c}_j$ for artificial variables can have either sign and may be disregarded.
5. Complementary slackness holds, since $x_j > 0$ implies $\bar{c}_j = 0$.

If $w_0 = 0$, then all artificial variables are zero. Dropping them, the remaining variables are feasible for the original problem and constitute an optimal solution, since all optimality conditions are satisfied. If $w_0 > 0$ and all $\bar{d}_j \geq 0$, then w is minimal and there is no feasible solution to (18). Otherwise, an iteration proceeds as follows:

STEP 1. The only nonbasic variables which may become positive without violating complementary slackness are those for which $\bar{c}_j = 0$. These, plus the current basic set, constitute a *restricted primal problem* over which w is now minimized, using the simplex method. At each iteration of this minimization the multipliers, σ, will change, and at its conclusion $\bar{d}_j \geq 0, j = m + 1, \ldots, m + q$.

STEP 2. If all $\bar{d}_j \geq 0$, then the current solution is either optimal ($w_0 = 0$) or no feasible solution to (18) exists ($w_0 > 0$), so terminate.

STEP 3. A new feasible solution to the dual [of (18)] is now sought, of the form

$$\pi_j^* = \pi_j + k\sigma_j, \qquad k > 0, \quad j = 1, \ldots, m \tag{21}$$

where k is yet to be selected. This solution generates new relative cost factors

$$\bar{c}_j^* = c_j - (\pi + k\sigma)P_j \tag{22}$$
$$\bar{c}_j^* = (c_j - \pi P_j) + k(-\sigma P_j) \tag{23}$$
$$\bar{c}_j^* = \bar{c}_j + k\bar{d}_j \tag{24}$$

In order that this new solution be dual feasible, we must have all $\bar{c}_j^* \geq 0$. If $\bar{d}_j \geq 0$, any $k \geq 0$ will do. For j such that $\bar{d}_j < 0$, k is limited by the relation

$$k \leq k^* = \min_{\bar{d}_j < 0} \frac{\bar{c}_j}{-\bar{d}_j} \tag{25}$$

Within this range, k is to be selected to increase as much as possible the dual objective function, given by

$$(\pi + k\sigma)b = \pi b + k\sigma b \tag{26}$$

Since σ is the vector of simplex multipliers of the current basis, B,

$$\sigma = c_B^w B^{-1}$$

where c_B^w is the vector of coefficients of the current basic variables in the w equation. Thus

$$\sigma b = c_B^w B^{-1} b = c_B^w x_B = w_0$$
$$(\pi + k\sigma)b = \pi b + kw_0$$

Since we assume $w_0 > 0$, k is selected at its upper limit, k^*:

$$k = \min_{\bar{d}_j < 0} \frac{\bar{c}_j}{-\bar{d}_j} = \frac{\bar{c}_s}{-\bar{d}_s} \tag{27}$$

Note that this is the largest k such that $z + kw$ has all coefficients nonnegative.

This choice of k forces $\bar{c}_s$, formerly positive, to become zero, and other $\bar{c}_j$ may do as well. A new restricted primal is obtained by using all nonbasic variables corresponding to $\bar{c}_j = 0$, and return is made to step 1.

Assuming nondegeneracy, each solution of the restricted primal program decreases w. Thus, no basis can be repeated and, since there are a finite number of possible bases, the procedure terminates in a finite number of iterations.

To initiate the procedure, if the initial tableau (20) has at least one $c_j = 0$ (recall, all $c_j \geq 0$), begin at step 1. Otherwise, begin at step 3 with $\sigma = (1, 1, \ldots, 1)$. Note that step 1 aims to decrease the primal infeasibility form, w, while the purpose of step 3 is to increase the dual objective, v, hence the name of the method. Also, in solving successive restricted primal programs, the optimal basis for the first is feasible for the second. Further, nonartificial variables x_j in this basis have $\bar{c}_j = \bar{d}_j = 0$, which, by (24), implies $\bar{c}_j^* = 0$. These variables thus satisfy $x_j \bar{c}_j^* = 0$, and this basis may be used as a starting point (and probably a good one) for the new restricted primal.

Example. Consider the problem given in relations (3). Since there is only one infeasibility, we need add only one artificial variable, x_a, in equation (3). Multiplying this equation by -1 and adding in x_a yields

$$x_a + 2x_3 + 6x_4 - s = 1 \tag{28}$$

Define

$$w = x_a$$

and eliminate x_a from the above by subtracting (28) from it:

$$w - 1 = -2x_3 - 6x_4 + s$$

By definition (see Section 1.2.2) the multipliers σ are

$$\sigma = (0, 0, 1)$$

while the vector π is

$$\pi = (0, 0, 0)$$

Thus the initial primal–dual tableau is as follows:

x_1	x_2	x_a	x_3	x_4	s	Constants	π	σ
1			3	1		1	0	0
	1		-1	5		2	0	0
		1	2	6	-1	1	0	1
			2	1	0	$z - 3$		
			-2	-6	1	$w - 1$		

Since, for nonbasic variables, only $\bar{c}_s = 0$, the initial restricted primal has variables x_1, x_2, x_a, and s:

Restricted Primal 1

x_1	x_2	x_a	s	$-w$	Constants
1					1
	1				2
		1			1
			1	1	-1

This is optimal as it stands, so a new dual solution is sought, of the form

$$\pi^* = \pi + k^*\sigma = (0, 0, 0) + k^*(0, 0, 1)$$

where

$$k^* = \min_{\bar{d}_j < 0} \frac{\bar{c}_j}{-\bar{d}_j} = \min[\tfrac{2}{2}, \tfrac{1}{6}] = \tfrac{1}{6}$$

Thus

$$\pi^* = (0, 0, \tfrac{1}{6}).$$

The new values of the relative cost factors are, by (24),

$$\begin{aligned} \bar{c}^* = \bar{c} + k^*\bar{d} &= (2, 1, 0) + \tfrac{1}{6}(-2, -6, 1) \\ &= (\tfrac{5}{3}, 0, \tfrac{1}{6}) \end{aligned}$$

The only zero is in the second element, $\bar{c}_4$, so x_4 plus the current basic set are the variables of the new restricted primal.

Restricted Primal 2

x_1	x_2	x_a	x_4	$-w$	Constants
1			1		1
	1		5		2
		1	6		1
			-6	1	-1

The simplex method chooses x_4 to enter the basis and x_a to leave. Performing the indicated pivot yields an optimal tableau:

x_1	x_2	x_a	x_4	$-w$	Constants
1		$-\frac{1}{6}$			$\frac{5}{6}$
	1	$-\frac{5}{6}$			$\frac{7}{6}$
		$\frac{1}{6}$	1		$\frac{1}{6}$
		1		1	0

Since $w_0 = 0$, this solution is optimal, as is verified by comparison with the solution obtained in (8).

1.3 Nonlinear Programming

Consider the problem

$$\text{minimize } z = (x_1 - 3)^2 + (x_2 - 4)^2 \tag{1}$$

subject to the linear constraints

$$\begin{aligned} x_1 &\geq 0 \\ x_2 &\geq 0 \\ 5 - x_1 - x_2 &\geq 0 \\ -2.5 + x_1 - x_2 &\leq 0 \end{aligned}$$

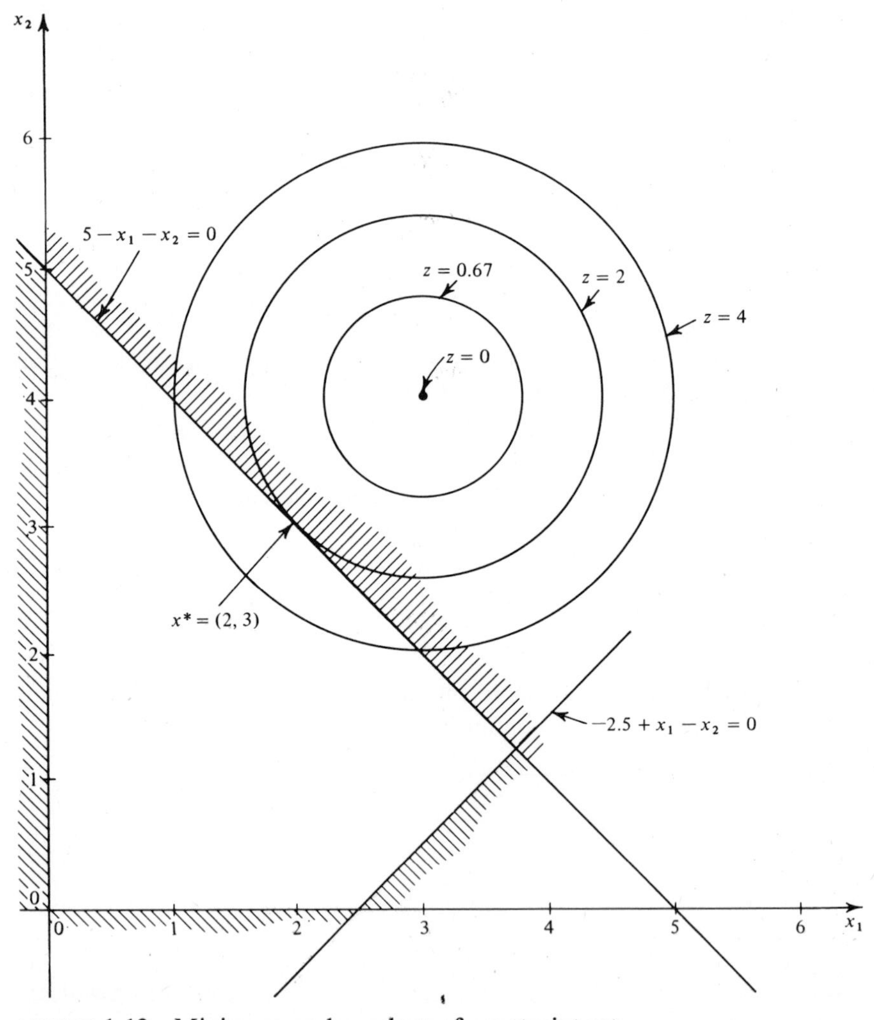

FIGURE 1-12 Minimum on boundary of constraint set.

The problem is shown graphically in Figure 1-12. The constraint set is of the linear programming type, having a finite number of corner points. The objective function, being nonlinear, has contours of constant value which are not parallel lines, as in the linear case, but concentric circles. The minimum value of z corresponds to the contour of lowest value having at least one point in common with the constraint set. This is the contour labeled $z = 2$, and the solution is at its point of tangency with the constraint set, i.e., at $x_1^* = 2$, $x_2^* = 3$. This is not an extreme point of the set, although it is a boundary point. (Recall that for linear programs the minimum is always at an extreme point.)

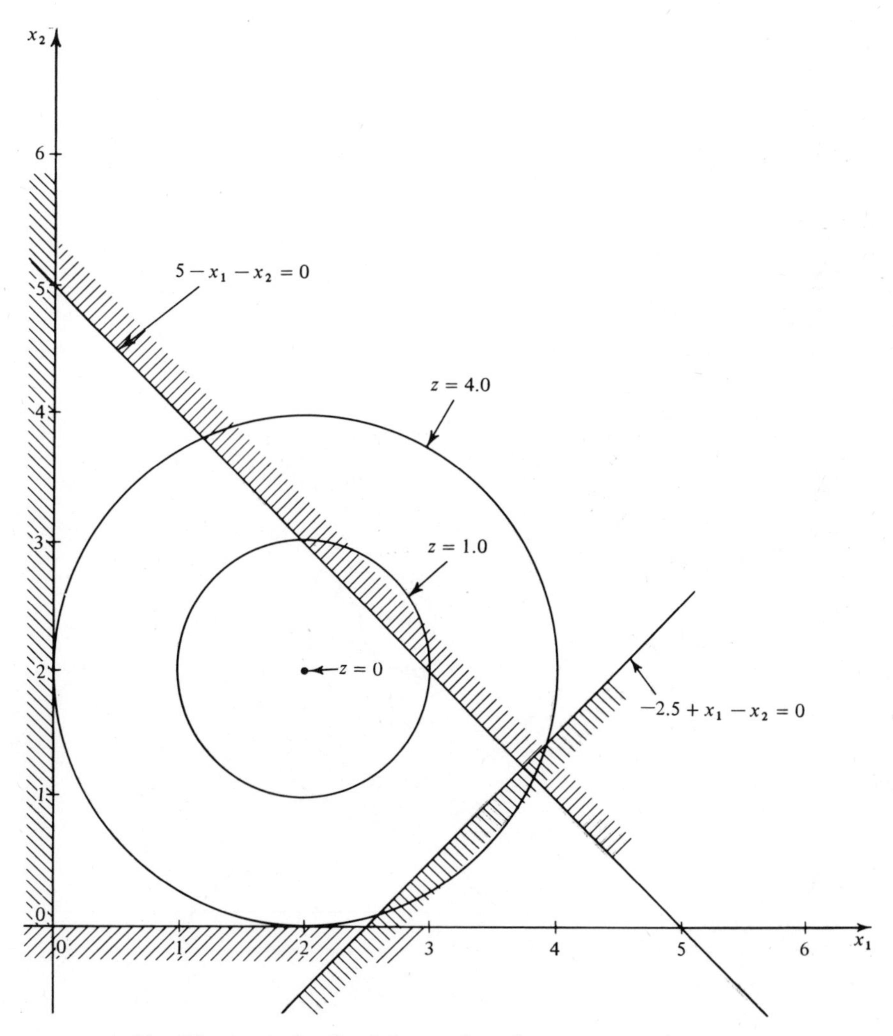

FIGURE 1-13 Unconstrained minimum interior to constraint set.

Further, if the objective function of the previous problem is changed to

$$z = (x_1 - 2)^2 + (x_2 - 2)^2 \tag{2}$$

then the situation is depicted in Figure 1-13. The minimum is now at $x_1 = 2$, $x_2 = 2$, which is not even a boundary point of the constraint set. Here the unconstrained minimum of the nonlinear function satisfies the constraints.

Neither of the previous problems had local optima. It is easy, however, to imagine nonlinear programs in which local optima occur. For example,

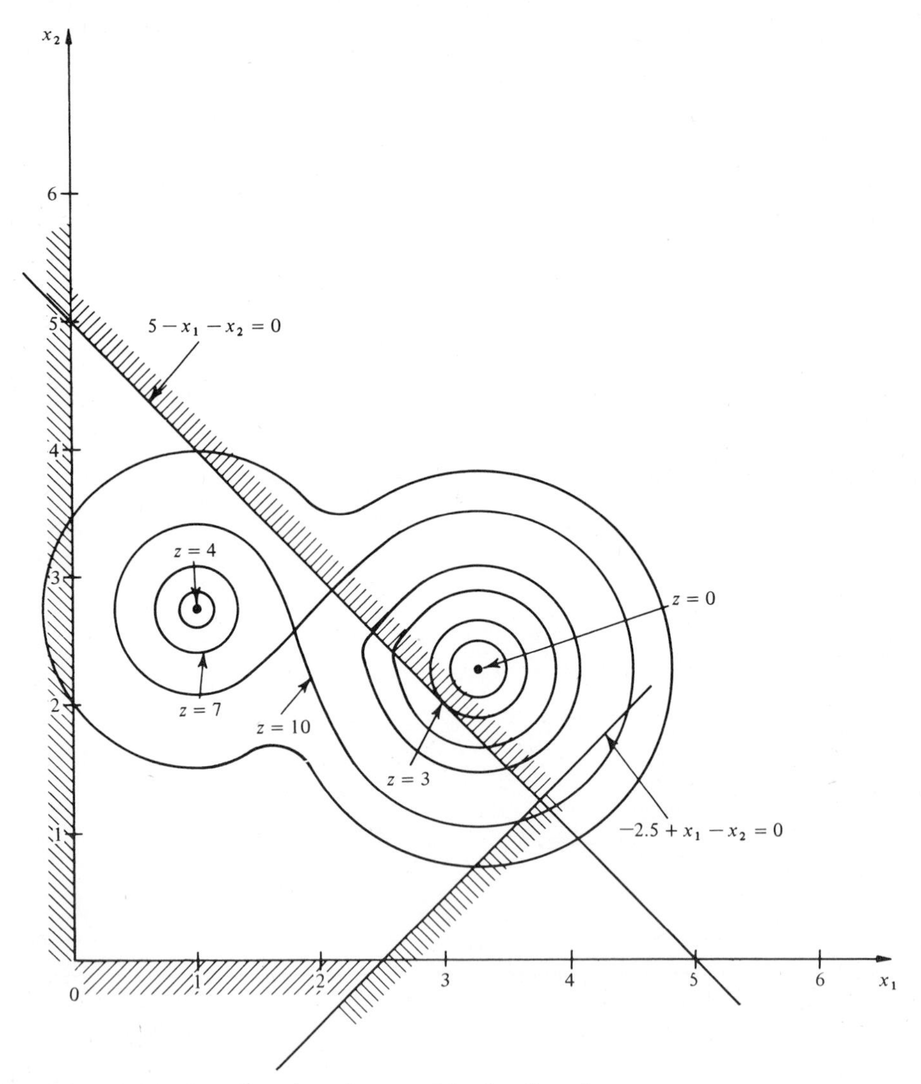

FIGURE 1-14 Local optima due to objective function.

if the objective function of the previous problem had two minima and at least one was interior to the constraint set, then the constrained problem would have two local minima. Contours of such a function are shown in Figure 1-14.

Although the examples thus far have had linear constraints, it often occurs that the chief nonlinearity of a programming problem appears in the constraints. The constraint set will then have curved boundaries. A problem with nonlinear constraints may very easily have local optima, even if the objective function has only one unconstrained minimum. Consider a problem with quadratic objective function and the constraint set shown in Figure 1-15.

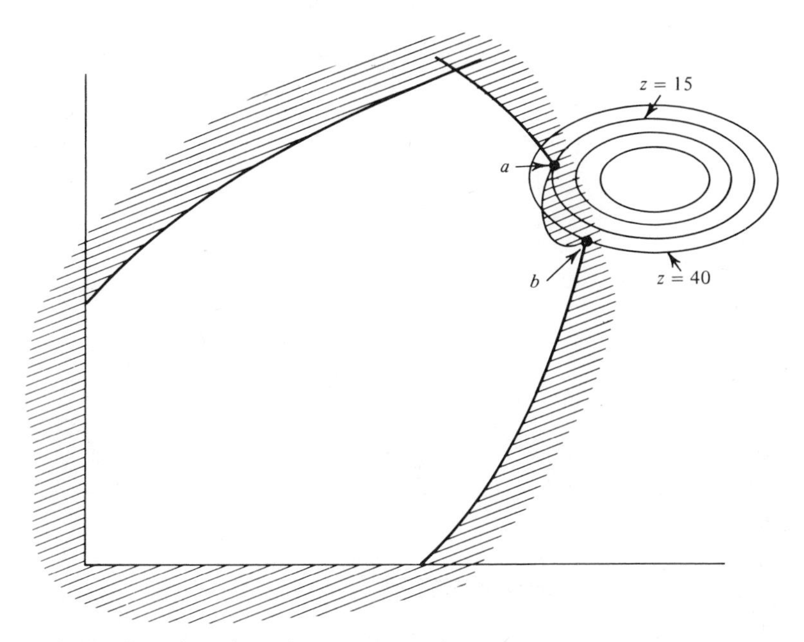

FIGURE 1-15 Local optima due to constraint set.

The problem has local optima at the two points a and b, since no point of the constraint set in the immediate vicinity of either point will yield a smaller value of z.

In summary, the optimum of a nonlinear programming problem will in general not be at an extreme point of the constraint set and may not even be on the boundary. Also, the problem may have local optima distinct from the global optimum. These properties are direct consequences of the nonlinearity. However, a class of nonlinear problems can be defined which are guaranteed to be free of distinct local optima. These are called convex programming problems and are now considered.

1.3.1 Convexity

DEFINITION 1. A set of points is said to be *convex* if, given any two points in the set, the line segment joining them is also in the set.

In the examples shown in Figure 1-16, the two sets A and B are convex, while C is not. It is easily seen that a convex set is one whose boundary does not bulge inward. Note that the constraint set of a linear programming problem is convex.

In many dimensions, these geometrical ideas must be formulated in algebraic terms. In particular, one must define what is meant by "the line segment

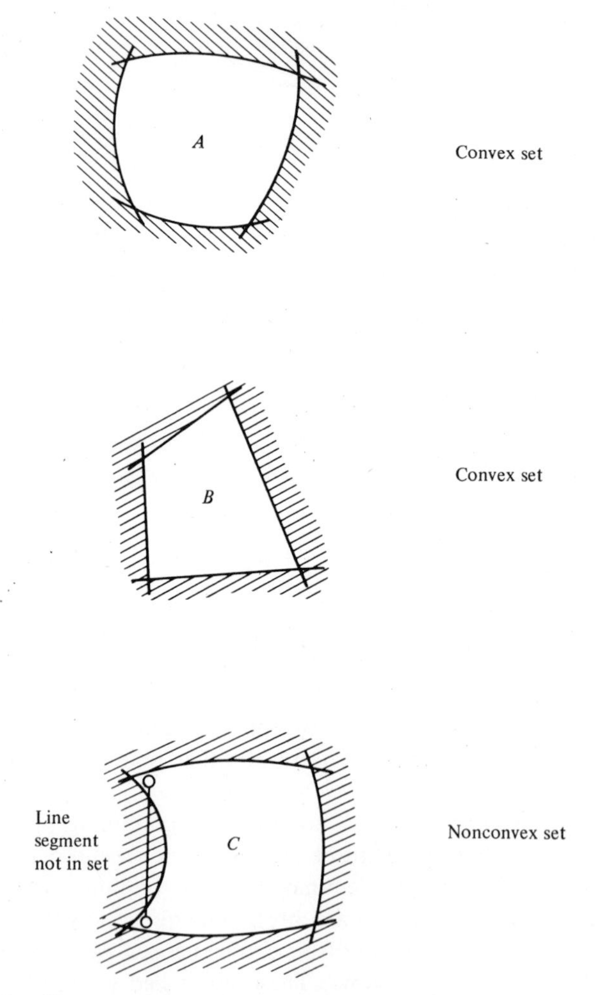

FIGURE 1-16 Convex sets.

between two points." Let the two points be x_1 and x_2 and consider the expression

$$x_3 = \lambda x_1 + (1 - \lambda)x_2 \tag{1}$$

where λ is a scalar, $0 \leq \lambda \leq 1$. Obviously, if $\lambda = 0$, $x_3 = x_2$, and if $\lambda = 1$, $x_3 = x_1$. If $0 \leq \lambda \leq 1$, then x_3 takes on all values on the line segment between x_1 and x_2. (This is easily verified in two or three dimensions.) Thus the line segment between two points is defined as the set

$$S = \{x \mid x = \lambda x_1 + (1 - \lambda)x_2, 0 \leq \lambda \leq 1\}$$

DEFINITION 2. A *function*, $f(x)$, is said to be *convex* if the line segment drawn between any two points on the graph of the function never lies *below* the graph, and concave if the line segment never lies *above* the graph.

Examples of concave and convex functions are shown in Figure 1-17. The first function is strictly convex, since the line segment is always above the function; the second function is strictly concave. Note that a linear function is both convex and concave but neither strictly convex nor strictly concave.

Algebraically, a function $f(x)$ is convex if

$$f(\lambda x_1 + (1 - \lambda)x_2) \leq \lambda f(x_1) + (1 - \lambda)f(x_2) \tag{2}$$

for all x_1, x_2 in the (convex) domain of definition of f and for all $0 \leq \lambda \leq 1$. The function is strictly convex if the strict inequality holds.

A convex programming problem is one of minimizing a convex function (or of maximizing a concave function) over a convex constraint set. The main theorem of convex programming is the following.

THEOREM 1. *Any local minimum of a convex programming problem is a global minimum.*

Note that the problem may have a number of points at which the global minimum is taken on, but the set of all such points is convex. No distinct (i.e., separated) local minima may exist with different function values. This is a nice property for a problem to have. It serves as motivation for the following results, which enable one to both characterize and recognize convex programming problems more easily. No proofs are given, but these may be found in Hadley [10] and Zoutendijk [11], among other references.

THEOREM 2. *If $f(x)$ is convex, then the set*

$$R = \{x \mid f(x) \leq k\}$$

is convex for all scalars k.

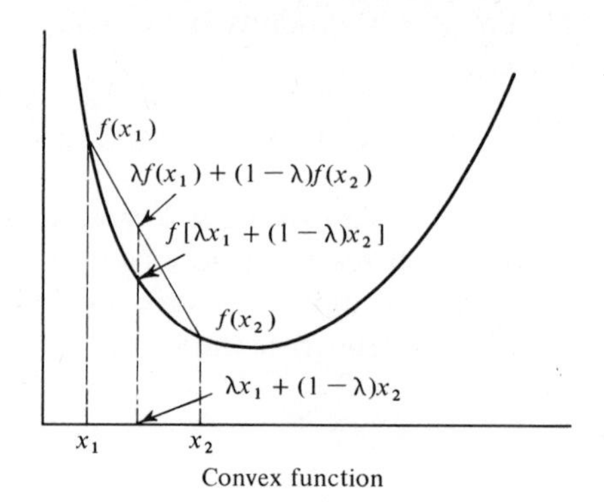

Convex function

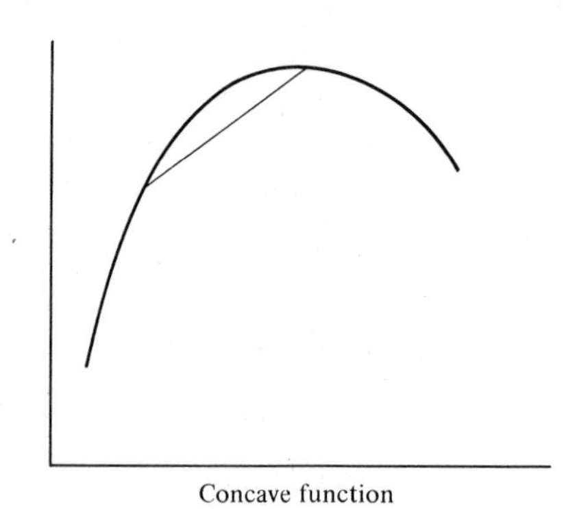

Concave function

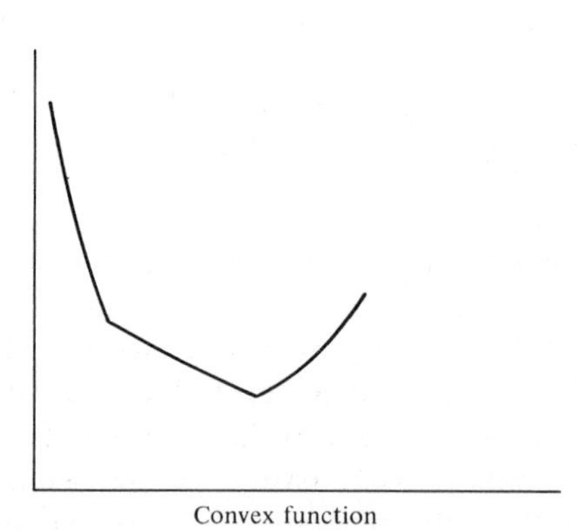

Convex function

FIGURE 1-17 Convex and concave functions.

The theorem is illustrated in Figure 1-18, where a convex quadratic function is cut by the plane $f(x) = k$. The set R above is an ellipse plus its interior, which is convex.

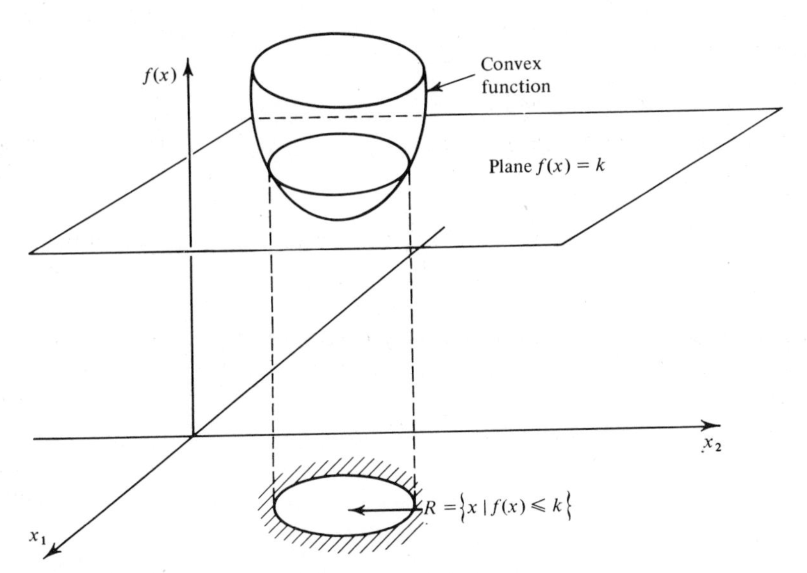

FIGURE 1-18 Illustration of Theorem 1.3.1–2.

THEOREM 3. *The intersection of any number of convex sets is convex.*

This theorem is easily verified geometrically in two dimensions. As a consequence of Theorems 2 and 3, the following holds: Given the problem

$$\min f(x)$$

subject to $$g_i(x) \leq b_i, \qquad i = 1, \ldots, m \tag{3}$$

then, if the functions f and g_i are convex, the problem is a convex programming problem. This is true because (1) each of the sets

$$R_i = \{x \mid g_i(x) \leq b_i\}$$

is convex by Theorem 2 and (2) the constraint set R is the intersection of the sets R_i and is convex by Theorem 3. Note that, since a linear function is convex, a linear programming problem is a convex programming problem. This establishes more firmly the geometrically evident fact that a linear program cannot have local optima distinct from the global optimum.

The previous result implies that convex programs can be identified by finding out if the objective and constraint functions of the problem are convex. It is thus of interest to better characterize convex functions. The main results along this line are given below.

THEOREM 4. *If $f(x)$ has continuous first and second partial derivatives, then the following statements are equivalent:*

(a) *$f(x)$ is convex.*

(b) *$f(x_1) \geq f(x_2) + \nabla f'(x_2)(x_1 - x_2)$ for any two points x_1, x_2.*

(c) *The matrix of second partial derivatives of $f(x)$ is positive semidefinite for all points x.*

Part (b) of the theorem states that the function evaluated at any point, $f(x_1)$, never lies below its tangent plane passed through any other point x_2.

THEOREM 5. *A positive semidefinite quadratic form is convex.*
This theorem is a direct consequence of part (c) of Theorem 4.

THEOREM 6. *A positive linear combination of convex functions is convex.*

THEOREM 7. *A function $f(x)$ is convex if and only if the one-dimensional function $g(\alpha) = f(x + \alpha s)$ is convex for all fixed x and s.*

Since $f(x + \alpha s)$ is the function evaluated at points along the line s passing through the point x, Theorem 7 says that a convex function is convex along any line. This affords a means of telling if a given function of n variables is not convex, for if any line in n-dimensional space can be found along which $g(\alpha)$ is not convex, $f(x)$ is not convex either.

"Mixed" Problems. Thus far, problems with both equality and inequality constraints have not been considered. The principal theoretical difficulty in dealing with such problems stems from the fact that, if $g(x)$ is nonlinear, the set

$$R = \{x | g(x) = 0\}$$

is generally not convex.[7] This is evident geometrically, since most nonlinear functions have graphs which are curved surfaces. Hence the set R will usually be a curved surface also, and the line segment joining any 2 points on this surface will generally not lie in the surface.

As a consequence of the above, the problem

$$\text{minimize } f(x)$$

$$\text{subject to} \qquad \begin{aligned} g_i(x) &\leq 0, \qquad i = 1, \ldots, m \\ h_j(x) &= 0, \qquad j = 1, \ldots, r < n \end{aligned} \tag{4}$$

may not be a convex programming problem in the variables $x_1, \ldots, x_n$ if any

[7] If $g(x)$ is both quasiconvex and quasiconcave (see section 3.11.1), then R is convex. Examples of such functions are linear functions and ratios of linear functions.

of the functions $h_j(x)$ are nonlinear. This, of course, does not preclude efficient solution of such problems, but it does make it more difficult to guarantee the absence of local optima and to generate sharp theoretical results.

In many cases the equality constraints may be used to eliminate some of the variables, leaving a problem with only inequality constraints and fewer variables. Even if the equalities are difficult to solve analytically, for example, if they are highly nonlinear, it may still be worthwhile solving them numerically. Such an approach has been successfully used for structural design [12, 13]. Here the equalities relate the stresses and deflections in the structure to the applied loads and design parameters.

Role of Convexity. Although convexity is desirable, many real-world problems turn out to be nonconvex. In addition, there is no simple way to test a nonlinear problem for convexity, since there is no simple way to test a nonlinear function for this property. Why, then, is convex programming studied? The main reasons are:

1. When convexity is assumed, many significant mathematical results have been derived in the field of mathematical programming.
2. Often results obtained under convexity assumptions can give insight into the properties of more general problems. Sometimes, such results may even be carried over to nonconvex problems but in a weaker form.

For example, it is usually impossible to prove that a given algorithm will find the global minimum of a nonlinear programming problem unless the problem is convex. For nonconvex problems, however, many such algorithms will find at least a local minimum. Here the convexity assumption has given the researcher a basis upon which to derive the method, and the method may then be applicable to more general situations. Convexity thus plays a role much the same as that of linearity in the study of dynamical systems. Many results derived from linear theory are used in the design of, for example, nonlinear control systems.

1.3.2 Kuhn–Tucker Conditions. The most important theoretical results in the field of nonlinear programming are the conditions of Kuhn and Tucker [14]. These must be satisfied at any constrained optimum, local or global, of any linear and most nonlinear programming problems. They form the basis for the development of many computational algorithms. In addition, the criteria for stopping many algorithms, i.e., for recognizing when a local constrained optimum has been achieved, are derived directly from them. The geometrical significance of these results is illustrated below and their analytical relation to the classical Lagrange multiplier conditions is noted.

Cones. The idea of a cone aids the understanding of the Kuhn–Tucker conditions. A cone is a set of points, R, such that, if x is in R, λx is also in R for $\lambda \geq 0$. A convex cone is a cone which is a convex set. An example of a

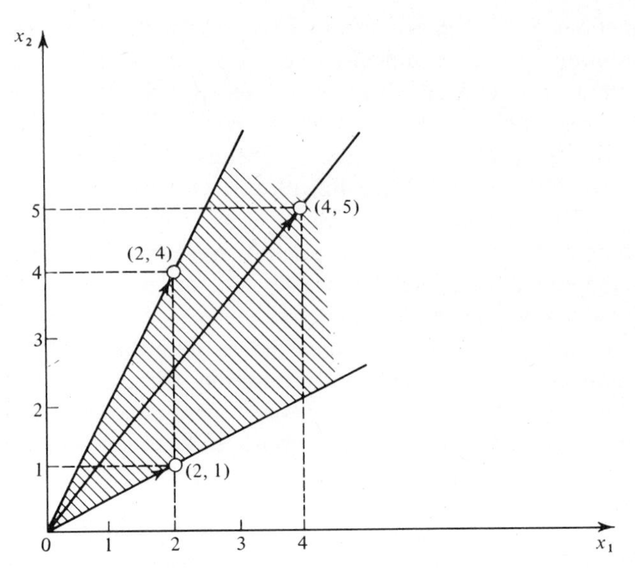

FIGURE 1-19 Convex cone.

convex cone in two dimensions is shown in Figure 1-19. In two and three dimensions, the definition of a convex cone coincides with the usual meaning of the word.

It is easily shown from the above definitions that the set of all nonnegative linear combinations of a finite set of vectors is a convex cone, i.e., that the set

$$R = \{x \mid x = \lambda_1 x_1 + \lambda_2 x_2 + \cdots + \lambda_m x_m, \lambda_i \geq 0, i = 1, \ldots, m\}$$

is a convex cone. The vectors $x_1, x_2, \ldots, x_m$ are called the generators of the cone. For example, the cone of Figure 1-19 is generated by the vectors (2, 1) and (2, 4). Thus any vector which can be expressed as a nonnegative linear combination of some other vectors lies in the cone generated by these vectors. For example, in Figure 1-19 the vector (4, 5) in the cone is given by

$$(4, 5) = 1 \times (2, 1) + 1 \times (2, 4)$$

Kuhn–Tucker Conditions—Geometrical Interpretation. The Kuhn–Tucker conditions are predicated on this fact: At any constrained optimum, no (small) allowable change in the problem variables can improve the objective function. To illustrate this consider the nonlinear programming problem

$$\text{minimize } f(x, y) = (x - 2)^2 + (y - 1)^2$$

subject to

$$\begin{aligned} g_1(x, y) &= -y + x^2 \leq 0 \\ g_2(x, y) &= x + y - 2 \leq 0 \end{aligned} \tag{1}$$

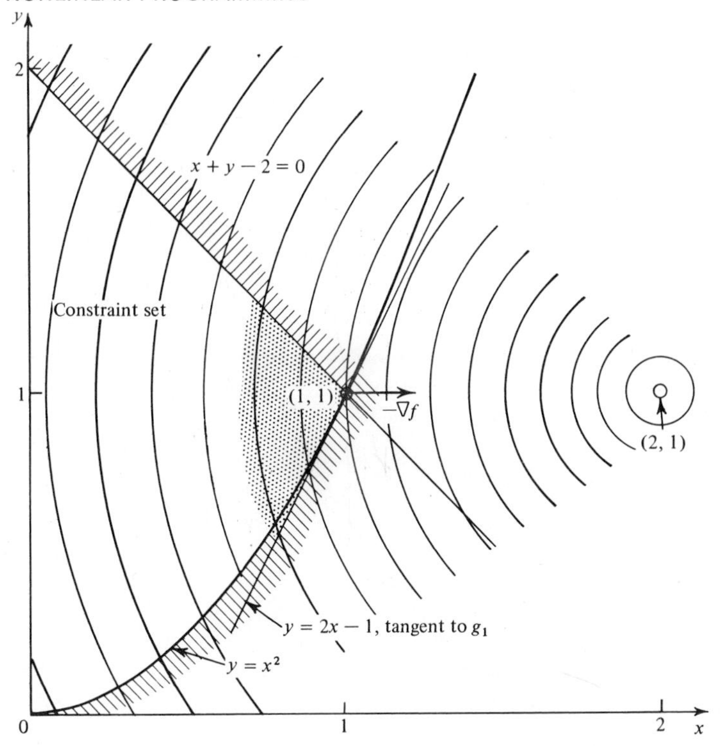

FIGURE 1-20 Geometry of constrained optimization problem.

The problem is shown geometrically in Figure 1-20. It is evident that the optimum is at the intersection of the two constraints at (1, 1). Define a feasible direction as a vector such that a small move along that vector violates no constraints. At (1, 1), the set of all feasible directions lies between the line $x + y - 2 = 0$ and the tangent line to $y = x^2$ at (1, 1), i.e., the line $y = 2x - 1$. In other words, this set is the cone generated by these lines. The vector $-\nabla f$ points in the direction of maximum rate of decrease of f and a small move along any direction making an angle of less than 90° with $-\nabla f$ will decrease f. Thus, at the optimum, no feasible direction can have an angle of less than 90° between it and $-\nabla f$.

Consider Figure 1-21, in which the gradient vectors ∇g_1 and ∇g_2 are drawn. Note that $-\nabla f$ is contained in the cone generated by ∇g_1 and ∇g_2. What if this were not so? If $-\nabla f$ were slightly above ∇g_2, it would make an angle of less than 90° with a feasible direction just below the line $x + y - 2 = 0$. If $-\nabla f$ were slightly below ∇g_1, it would make an angle of less than 90° with a feasible direction just above the line $y = 2x - 1$. Neither case can occur at an optimal point, and both cases are excluded if and only if $-\nabla f$ lies within the cone generated by ∇g_1 and ∇g_2. Of course, this is the same as requiring that ∇f lie within the cone generated by $-\nabla g_1$ and $-\nabla g_2$. This is

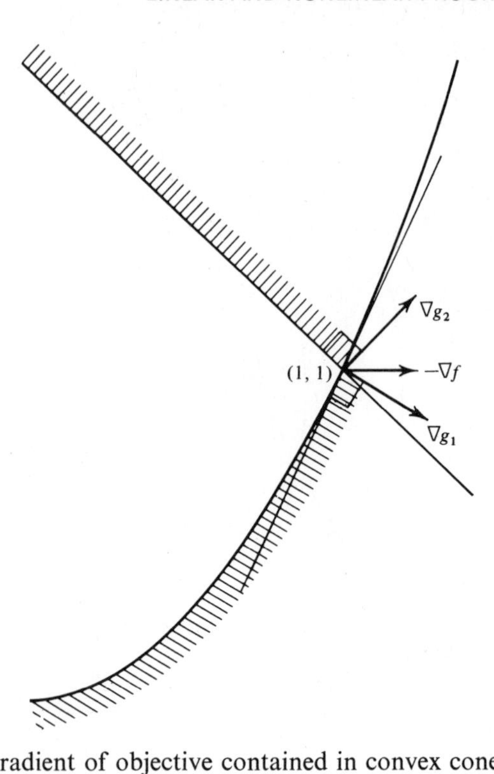

FIGURE 1-21 Gradient of objective contained in convex cone.

the usual statement of the Kuhn–Tucker conditions, i.e., if f and all g_i are differentiable, a necessary condition[8] for a point x° to be a constrained minimum of the problem

$$\text{minimize } f(x)$$

$$\text{subject to} \qquad g_i(x) \leq 0, \qquad i = 1, \ldots, m \tag{2}$$

is that, at x^0, ∇f lie within the cone generated by the negative gradients of the binding constraints.[9]

Algebraic Statement of the Kuhn–Tucker Conditions. The above results may be stated in algebraic terms. For ∇f to lie within the cone described above, it must be a nonnegative linear combination of the negative gradients of the binding constraints; i.e., there must exist numbers u_i^0 such that

$$\nabla f(x^0) = \sum_{i \in I} u_i^0(-\nabla g_i(x^0)) \tag{3}$$

[8] For this condition to be necessary, a constraint qualification must be satisfied; see the following derivation.
[9] Binding constraints are those which hold as equalities at x^0.

where

$$u_i^0 \geq 0, \qquad i \in I \tag{4}$$

and I is the set of indices of the binding constraints.

These results may be restated to include all constraints by defining the coefficient u_i^0 to be zero if $g_i(x^0) < 0$. Note that if this is done, then $u_i^0 \geq 0$ if $g_i(x^0) = 0$, and $u_i^0 = 0$ if $g_i(x^0) < 0$; i.e., the product $u_i^0 g_i(x^0)$ is zero for all i. Conditions (3)–(4) then become

$$\nabla f(x^0) = \sum_{i=1}^{m} u_i^0(-\nabla g_i(x^0)) \tag{5}$$

$$u_i^0 \geq 0, \quad u_i^0 g_i(x^0) = 0, \quad g_i(x^0) \leq 0, \qquad i = 1, \ldots, m \tag{6}$$

Relations (5)–(6) are the form in which the Kuhn–Tucker conditions are usually stated.

Lagrange Multipliers. The Kuhn–Tucker conditions are closely related to the classical Lagrange multiplier results for equality constrained problems. Form the Lagrangian

$$L(x, u) = f(x) + \sum_{i=1}^{m} u_i g_i(x) \tag{7}$$

where the u_i are viewed as Lagrange multipliers for the inequality constraints $g_i(x) \leq 0$. Then (5)–(6) state that $L(x, u)$ must be stationary in x at (x^0, u^0) with the multipliers, u^0, satisfying (6). The stationarity of L is the same condition as in the equality constrained case—the additional conditions in (6) arise because the constraints here are inequalities.

Derivation of the Kuhn–Tucker Conditions. As mentioned at the beginning of this section, the Kuhn–Tucker conditions are necessary for a local optimum of "most" nonlinear programs. The following material, based on that in Fiacco and McCormick [22], derives these conditions and specifies the class of programs for which they hold.

Consider again a program of the form (2):

$$\text{minimize } f(x)$$

subject to

$$g_i(x) \leq 0, \qquad i = 1, \ldots, m \tag{8}$$

where all functions f and g_i are assumed differentiable. Let x^0 be a local minimum for this program, and consider small perturbations, $x^0 + y$, about x^0. Let

$$B^0 = \{i \mid g_i(x^0) = 0\} \tag{9}$$

be the index set of the binding constraints at x^0. Since all others must be negative, sufficiently small perturbations y cannot cause them to be violated. Thus, to specify the set of all admissible perturbations, i.e., the set of all y such that $x^0 + \alpha y$ is feasible for some range of positive α, we need consider only those constraints with indices in B^0. For these, y must satisfy

$$g_i(x^0 + y) \leq 0, \qquad i \in B^0 \tag{10}$$

Since each g_i is differentiable, we may expand in Taylor series about x^0:

$$g_i(x^0) + \nabla g_i'(x^0)y + 0(y) \leq 0, \qquad i \in B^0 \tag{11}$$

where $0(y)$ denotes terms of second order. For y sufficiently small, the sign of the left hand side of (11) is determined by the zero and first order terms, so we can neglect $0(y)$. Then, since $g_i(x^0) = 0$, (11) yields

$$\nabla g_i'(x^0)y \leq 0, \qquad i \in B^0 \tag{12}$$

as conditions specifying admissible perturbations. Certainly, if y is admissible it must satisfy the above. We can show, however, that the converse is not true without some additional conditions. Consider the constraints posed by Kuhn and Tucker [14]:

$$\begin{aligned} g_1(x) &= -x_1 \leq 0 \\ g_2(x) &= -x_2 \leq 0 \\ g_3(x) &= -(1 - x_1)^3 + x_2 \leq 0 \end{aligned}$$

The constraint set is shown in Figure 1-22. At $x_1 = 1$, $x_2 = 0$ the constraint

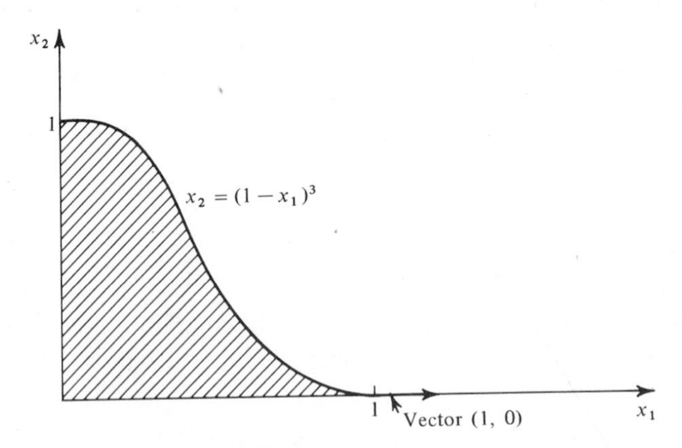

FIGURE 1-22 Constraint set with cusp.

set has an outward pointing "cusp" where the value and the first two derivatives of the curve $(1 - x_1)^3$ are zero. Let

$$x^0 = (1, 0)$$

Then the set B^0 is

$$B^0 = \{2, 3\}$$

and the inequalities (12) are

$$\nabla g_2'(x^0)y = -y_2 \leq 0$$
$$\nabla g_3'(x^0)y = y_2 \leq 0$$

The vector $y_1 = 1$, $y_2 = 0$ satisfies these inequalities, yet points out of the constraint set. Thus there may be inadmissible perturbations satisfying (12). A constraint qualification which rules out such occurrences will be stated later.

It is convenient to define the following sets:

$$Z_1^0 = \{y \mid \nabla g_i'(x^0)y \leq 0,\ i \in B^0,\ \nabla f'(x^0)y \geq 0\} \tag{13}$$
$$Z_2^0 = \{y \mid \nabla g_i'(x^0)y \leq 0,\ i \in B^0,\ \nabla f'(x^0)y < 0\} \tag{14}$$
$$Z_3^0 = \{y \mid \nabla g_i'(x^0)y > 0 \text{ for some } i \in B^0\} \tag{15}$$

These sets are disjoint, and any vector $y \in E^n$ must be in one of them. The motivation for defining them is that if x^0 is a local minimum, then, as noted earlier, no admissible perturbation from x^0 can decrease the objective, f. Assume for the moment that a vector y is admissible if and only if it satisfies (12). Then, if x^0 is locally optimal, all y satisfying (12) must satisfy

$$\nabla f'(x^0)y \geq 0 \tag{16}$$

or, stated in terms of the sets Z_i^0,

$$Z_2^0 = \varnothing \tag{17}$$

where $\varnothing$ is the empty set. In fact, we will show that (17) is necessary and sufficient for the multipliers u^0 in (5) to exist. The proof requires the following lemma.

FARKAS' LEMMA. *Let $\{P_0, P_1, \ldots, P_r\}$ be an arbitrary set of vectors. There exist $\beta_i \geq 0$ such that*

$$P_0 = \sum_{i=1}^{r} \beta_i P_i \tag{18}$$

if and only if

$$P_0' y \geq 0 \tag{19}$$

for all y satisfying

$$P_i' y \geq 0, \qquad i = 1, \ldots, r \tag{20}$$

PROOF. $\Rightarrow$: Assume that (18) holds. Then

$$P_0' y = \sum_{i=1}^{r} \beta_i P_i' y, \qquad \beta_i \geq 0$$

and for all y satisfying (20), $P_0' y \geq 0$,
$\Leftarrow$: If (19) and (20) hold, then the linear program

$$\text{minimize } P_0' y$$

subject to $$P_i' y \geq 0, \qquad i = 1, \ldots, r$$

has an optimal solution $y = 0$. By the Corollary to Theorem 1.2.3-2, its dual is feasible and has a finite optimal solution. This dual, with variables $\beta_1, \ldots, \beta_r$, has constraints

$$\sum_{i=1}^{r} \beta_i P_i = P_0, \qquad \beta_i \geq 0$$

Since these are feasible, the theorem is proved.

THEOREM 1. *Assume* (a) f *and all* g_i *in* (8) *are differentiable, and* (b) x^0 *is a local minimum for* (8). *Then there exist multipliers* $u_i^0 \geq 0$ *such that*

$$\nabla f(x^0) + \sum_{i=1}^{m} u_i^0 \nabla g_i(x^0) = 0 \tag{21}$$

$$u_i^0 g_i(x^0) = 0, \qquad i = 1, \ldots, m \tag{22}$$

if and only if

$$Z_2^0 = \varnothing \tag{23}$$

PROOF. Choose the vectors $\nabla f(x^0)$ and $-\nabla g_i(x^0)$, $i \in B^0$, as the vectors $\{P_0, \ldots, P_r\}$ in Farkas' lemma. Then there exist $u_i^0 \geq 0$, $i \in B^0$, such that

$$\nabla f(x^0) = \sum_{i \in B^0} u_i^0(-\nabla g_i(x^0))$$

if and only if

$$\nabla f'(x^0)y \geq 0$$

for all y satisfying (12), i.e., if and only if $Z_2^0 = \varnothing$. Now define $u_i^0 = 0$ for $i \notin B^0$. Then (21) holds, and since at least one of each pair $(u_i^0, g_i(x^0))$ is zero, (22) holds as well.

The above theorem completely defines the class of programs such that the Kuhn–Tucker conditions are necessary for a local minimum. This is the class for which $Z_2^0 = \varnothing$ at each such minimum. Of course, $Z_2^0 = \varnothing$ is not an easy condition to check. The previous example suggests that $Z_2^0 \neq \varnothing$ only in exceptional cases, i.e., when there are singularities or cusps on the boundaries of the constraint set. Still, it is instructive to pose sufficient conditions guaranteeing $Z_2^0 = \varnothing$. Some are:

1. All constraint functions are linear (Karlin [23], p. 203).
   All constraint functions are convex and the constraint set has a nonempty interior (Karlin [23], p. 239, problem 12; See also Theorem 3, Section 1.3.3).
3. The gradients of all binding constraints are linearly independent (Fiacco and McCormick [22], p. 37).
4. The Kuhn–Tucker constraint qualification (Kuhn and Tucker [14]).

The last of these will be dealt with in detail. To state it, we first define an arc $\alpha(\theta)$ in E^n as a vector of n functions

$$\alpha(\theta) = (\alpha_1(\theta), \ldots, \alpha_n(\theta))$$

defined for all θ in an interval $[0, \delta]$ with $\delta > 0$. The arc is differentiable if each $\alpha_i(\theta)$ is differentiable, and its tangent vector is then given by

$$D\alpha(\theta) = \left(\frac{d\alpha_1}{d\theta}, \ldots, \frac{d\alpha_n}{d\theta}\right)$$

The arc is said to emanate from a point x^0 if $\alpha(0) = x^0$ and is tangent to a vector y at a point θ if $D\alpha(\theta) = y$. It is contained in a set $R \subseteq E^n$ if $\alpha(\theta) \in R$ for all $0 \leq \theta \leq \delta$.

Kuhn–Tucker Constraint Qualification. Let x^0 satisfy the constraints (8) and assume that all constraint functions g_i are differentiable. Then the Kuhn–Tucker constraint qualification holds at x^0 if every y satisfying the inequalities (12) is tangent to a differentiable arc emanating from x^0 and contained in the constraint set.

THEOREM 2 (KUHN–TUCKER THEOREM). *Assume*

(a) *All functions f and g_i in* (8) *are differentiable.*

(b) *x^0 is a local minimum for the problem* (8).

(c) *The Kuhn–Tucker constraint qualification holds at x^0.*

Then there exists a vector $u^0 = (u_1^0, \ldots, u_m^0) \geq 0$ such that the Kuhn–Tucker conditions, (21)–(22), *hold at x^0.*

PROOF. By Theorem 1, we need only show $Z_2^0 = \varnothing$. Let y satisfy the inequalities (12). If there is no such y, $Z_2^0 = \varnothing$ and the theorem is proved. By the constraint qualification, y is tangent to an arc, $\alpha(\theta)$, at x^0, i.e.,

$$D\alpha(0) = y$$

where, since the arc emanates from x^0,

$$\alpha(0) = x^0$$

Consider the behavior of the objective, f, along the arc, given by the function $f(\alpha(\theta))$. Since x^0 is optimal and $\alpha(\theta)$ is feasible for all $\theta \in [0, \delta]$, then, for all $\theta \in [0, \delta]$,

$$f(\alpha(\theta)) \geq f(\alpha(0))$$

Since f and α are differentiable, the above can hold if and only if

$$\left.\frac{df(\alpha(\theta))}{d\theta}\right|_{\theta=0} = \nabla f'(\alpha(0))D\alpha(0) \geq 0$$

But $\alpha(0) = x^0$ and $D\alpha(0) = y$, so

$$\nabla f'(x^0)y \geq 0$$

Thus $y \in Z_1^0$ and $Z_2^0 = \varnothing$.

Note that the Kuhn–Tucker constraint qualification does not hold at the point (1, 0) of our previous example, since the vector $y = (1, 0)$ is not tangent to any arc emanating from (1, 0) and contained in the constraint set.

1.3.3 Saddle Points and Sufficiency Conditions. Thus far the results stated, the Kuhn–Tucker conditions, have been shown only to be necessary, but hold for both convex and nonconvex problems. The main restriction on their application is that the objective and constraint functions must be differentiable. There are other conditions, again centering around a Lagrangian function, which hold even in the absence of differentiability. These are the saddle-point criteria and are sufficient for a given point to be optimal for almost any mathematical program. If the program is convex and satisfies a constraint qualification, the saddle-point conditions are both necessary and sufficient. This result can then be used to show that, if the problem is convex and all functions are differentiable, the Kuhn–Tucker conditions are also necessary and sufficient.

The problem under consideration is stated in slightly more general form than in the previous section:

$$\text{minimize } f(x) \tag{1}$$

subject to

$$g_i(x) \leq 0, \qquad i = 1, \ldots, m \tag{2}$$

$$x \in S \tag{3}$$

This is called the primal problem. The quantity x is an n-vector, S is an arbitrary subset of E^n, and f and the g_i are arbitrary real-valued functions, defined on S.

The set S may be used to impose any kind of additional restrictions. For example, S may be the set of all n vectors with integral-valued components. For purposes of decomposition, Chapter 8 shows how the constraints of the problem should be partitioned into the functions g_i and into those determining S.

The Lagrangian function associated with this problem is

$$L(x, u) = f(x) + \sum_{i=1}^{m} u_i g_i(x), \qquad u_i \geq 0 \tag{4}$$

Definition. A point (x^0, u^0) with $u^0 \geq 0$ and $x^0 \in S$ is said to be a *saddle point* for $L(x, u)$ if it satisfies

$$L(x^0, u^0) \leq L(x, u^0) \qquad \text{for all } x \in S \tag{5}$$

$$L(x^0, u^0) \geq L(x^0, u) \qquad \text{for all } u \geq 0 \tag{6}$$

That is, x^0 minimizes $L(x, u^0)$ over S and u^0 maximizes $L(x^0, u)$ over all $u \geq 0$. In two variables, $L(x, u)$ would have the saddle shape shown in Figure 1-23, with the saddle point yielding simultaneously a minimum in one variable and a maximum in the other.

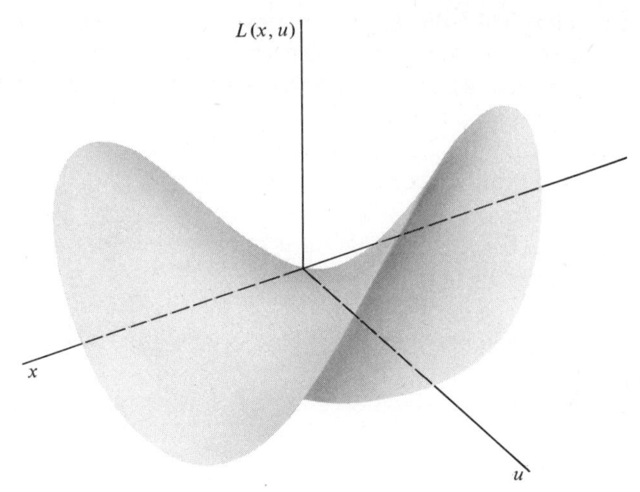

FIGURE 1-23 Saddle-shaped function.

The following theorem gives necessary and sufficient conditions for a saddle point of $L(x, u)$.

THEOREM 1. *Let $u^0 \geq 0$ and $x^0 \in S$. Then (x^0, u^0) is a saddle point for $L(x, u)$ if and only if*

(a) *x^0 minimizes $L(x, u^0)$ over S* (7)

(b) $g_i(x^0) \leq 0, \quad i = 1, \ldots, m$ (8)

(c) $u_i^0 g_i(x^0) = 0, \quad i = 1, \ldots, m$ (9)

PROOF. ⇒: Relation (5) is equivalent to condition (a). By (6),

$$f(x^0) + \sum_i u_i^0 g_i(x^0) \geq f(x^0) + \sum_i u_i g_i(x^0) \qquad \text{for all } u_i \geq 0 \tag{10}$$

or

$$\sum_i (u_i - u_i^0) g_i(x^0) \leq 0 \qquad \text{for all } u_i \geq 0 \tag{11}$$

If (b) is violated for some i, u_i may be chosen sufficiently large so that (11) is violated. Thus (b) must hold. If all $u_i = 0$, (11) yields

$$\sum_i u_i^0 g_i(x^0) \geq 0 \tag{12}$$

But $u_i^0 \geq 0$ and $g_i(x^0) \leq 0$ imply

$$\sum_i u_i^0 g_i(x^0) \leq 0 \tag{13}$$

Thus the above sum is equal to zero. Since each term has the same sign, (c) must hold as well.

$\Leftarrow$: Condition (a) is identical with (5). By (c),

$$L(x^0, u^0) = f(x^0) \tag{14}$$

By definition

$$L(x^0, u) = f(x^0) + \sum_i u_i g_i(x^0) \tag{15}$$

Since $g_i(x^0) \leq 0$, the term $u_i g_i(x^0)$ is nonpositive for all $u_i \geq 0$. Removing it from the right-hand side of (15) yields

$$L(x^0, u) \leq f(x^0) = L(x^0, u^0) \qquad \text{for all } u \geq 0 \tag{16}$$

which is (6). Thus the theorem is proved.

Similarity of the saddle-point conditions (a)–(c) and the Kuhn–Tucker conditions should be noted. (They will be proved equivalent for convex differentiable programs.) Conditions (b) and (c) are common to both. Condition (a) replaces the stationarity of the Lagrangian function by its minimization. As mentioned earlier, formulation in terms of minimization has certain advantages. In Chapter 8 it guarantees that the subproblems which arise are minimization problems. It also permits treatment of programs with nondifferentiable functions, discrete constraint sets, etc.

The usefulness of a saddle point is brought out in the following theorem.

THEOREM 2 (SUFFICIENCY OF SADDLE POINT). *If (x^0, u^0) is a saddle point for $L(x, u)$, then x^0 solves the primal problem* (1)–(3).

PROOF. Since (x^0, u^0) is a saddle point, conditions (a)–(c) of Theorem 1 must hold. Setting

$$g(x) = (g_1(x), \ldots, g_m(x))' \tag{17}$$

(a) becomes

$$f(x^0) + u^0 g(x^0) \leq f(x) + u^0 g(x) \qquad \text{for all } x \in S \tag{18}$$

By condition (c), $u^0 g(x^0) = 0$, so the above becomes

$$f(x^0) \leq f(x) + u^0 g(x) \qquad \text{for all } x \in S \tag{19}$$

For all points $x \in S$ which satisfy $g(x) \leq 0$; i.e., for all x feasible for the primal, the term $u^0 g(x)$ is nonpositive, so (19) becomes

$$f(x^0) \leq f(x) \qquad \text{for all } x \text{ satisfying (2)–(3)} \tag{20}$$

Thus x^0 solves the primal.

Theorem 2 requires some comment. The theorem applies to any mathematical program, including programs where S is a finite set, f or the g_i are not convex, etc. Of course, a saddle point may fail to exist for such problems. Existence can be guaranteed, at present, only for convex programs. For other problems, certain attractive features of solving the primal problem by minimizing $L(x, u)$, to be considered in Chapter 8, may motivate the analyst to search for a saddle point anyway.

It is not difficult to construct programs whose Lagrangian functions have no saddle point.

Example 1. Consider the problem [24]

$$\text{minimize } \{-x^2 \mid 1 - 2x = 0, 0 \leq x \leq 1\} \tag{21}$$

taking

$$S = \{x \mid 0 \leq x \leq 1\} \tag{22}$$

the problem of minimizing the Lagrangian is

$$\underset{0 \leq x \leq 1}{\text{minimize}} \{L(x, u) = -x^2 + u(1 - 2x)\} \tag{23}$$

where u may have any sign, since it corresponds to an equality constraint. Since $L(x, u)$ is concave in x for all u, its minimum over the interval $[0, 1]$ occurs at either zero or one. No value of u can shift this minimum to the optimal solution at $x^0 = \frac{1}{2}$, so there is no saddle point.

Example 2

$$\text{minimize } \{x_1^2 - 2x_2 \mid x_2 \leq 0\} \tag{24}$$

The problem of minimizing $L(x, u)$ is

$$\text{minimize } (x_1^2 + (u - 2)x_2) \tag{25}$$

No finite minimum exists for $u \neq 2$. If $u = 2$ the minimizing values of x are $x_1(2) = 0$ and $x_2(2)$ arbitrary. If $x_2(2)$ is chosen to be zero, the point $(0, 0; 2)$ is a saddle point. However, since any value of x_2 minimizes $L(x, 2)$, finding this minimum yields little information as to an optimal value of x_2. Of course, in this example, x_2 could be chosen to have the unique value satisfying primal feasibility and complementary slackness. However, in more complex problems, these conditions may still not be enough to determine unique values for all variables. Such situations can arise whenever a variable appears linearly in f and all g_i and is not bounded above and below by the constraints determining S.

Existence of Saddle Points. The following theorem deals with existence of a saddle point for problems with convex objective and constraint functions.

THEOREM 3 (KARLIN [23]). *Let S be a convex subset of E^n, f a convex function defined on S, and $g(x) = (g_1(x), \ldots, g_m(x))'$ a vector of convex functions defined on S. Assume that there exists a point $x \in S$ such that $g(x) < 0$. If x^0 is a point at which $f(x)$ assumes its minimum subject to $g(x) \leq 0$, $x \in S$, then there is a vector of multipliers $u^0 \geq 0$ such that (x^0, u^0) is a saddle point of*

$$L(x, u) = f(x) + ug(x) \tag{26}$$

Conversely, if (x^0, u^0) is a saddle point of $L(x, u)$, then x^0 minimizes $f(x)$ subject to $g(x) \leq 0$, $x \in S$.

The proof requires the following lemma, proved in [23].

LEMMA 1 (SEPARATION THEOREM). *Let A and B be convex subsets of E^n. Assume that A contains an interior point, and that no interior point of A is contained in B. Then there is a hyperplane separating A and B, i.e., there is a nonzero vector c and constant α such that*

$$cx \leq \alpha \quad \text{for all } x \in A \tag{27}$$

$$cx \geq \alpha \quad \text{for all } x \in B \tag{28}$$

The geometric interpretation is shown in Figure 1-24.

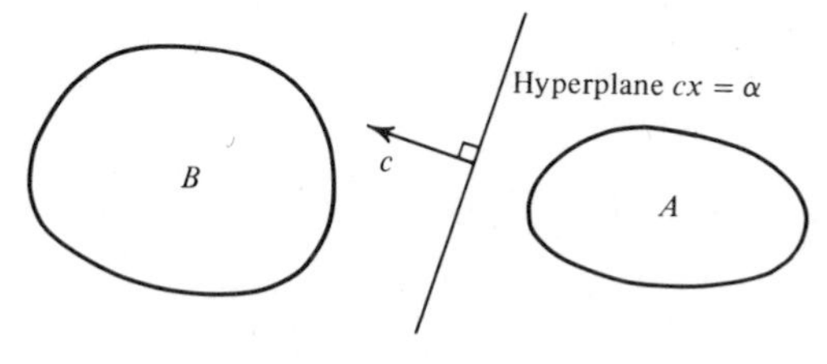

FIGURE 1-24 Separating hyperplane with normal vector oriented toward B.

PROOF OF THEOREM 3. The "conversely" part of the theorem has been proved in Theorem 2, with no convexity or regularity assumptions attached. To prove the other portion, the plan of attack is to construct two convex sets and, using Lemma 1, separate them by a hyperplane. The inequalities implied by the separation are then manipulated to yield conditions (a)–(c) of Theorem 1, which are necessary and sufficient for a saddle point.

Define two sets, A and B, subsets of E^{m+1}, by

$$A = \{(y_0, y) \mid y_0 \geq f(x),\ y \geq g(x) \text{ for some } x \in S\} \tag{29}$$

and

$$B = \{(y_0, y) \mid y_0 \leq f(x^0),\ y_i \leq 0,\ i = 1, \ldots, m\} \tag{30}$$

These two sets in E^2 might appear as shown in Figure 1-25. The set B is an

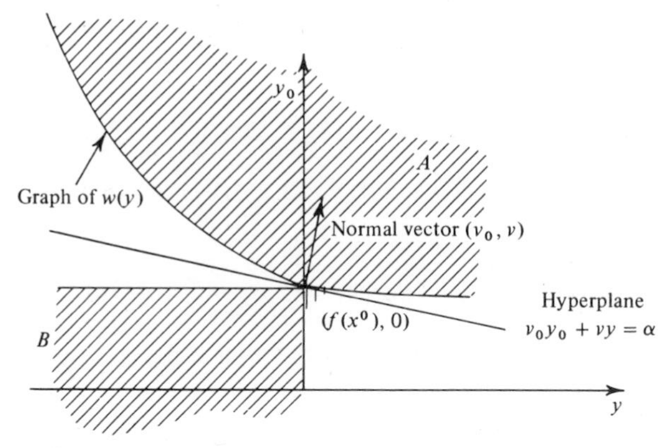

FIGURE 1-25 Separation of A and B.

orthant centered at $(f(x^0), 0)$. The set A consists of all points on and above the graph of the function

$$w(y) = \inf\{f(x) \mid g(x) \leq y,\ x \in S\} \tag{31}$$

The function $w(y)$ is nonincreasing; i.e.,

$$y_2 \geq y_1 \Rightarrow w(y_2) \leq w(y_1) \tag{32}$$

In addition, under the hypotheses of this theorem, $w(y)$ is easily shown to be convex. Thus A is a convex set and, since $f(x^0)$ is minimal, A and B have no interior points in common. Since B has an interior point, we may apply

Lemma 1 and separate these sets by a hyperplane, as shown in Figure 1-25, with the normal vector oriented toward A. This yields the inequality

$$v_0 y_0 + vy \geq v_0 z_0 + vz \qquad \text{for all } \begin{pmatrix} y_0 \\ y \end{pmatrix} \in A, \text{ all } \begin{pmatrix} z_0 \\ z \end{pmatrix} \in B \tag{33}$$

If any component of (v_0, v) were negative, then the corresponding component of (z_0, z) could be made sufficiently negative so that (33) would be violated. Thus (v_0, v) has all nonnegative components with at least one positive.

We can show that $v_0 > 0$. Since $(f(x^0), 0) \in B$, and $(f(x), g(x)) \in A$ for any $x \in S$, (33) yields

$$v_0 f(x) + vg(x) \geq v_0 f(x^0) \qquad \text{for all } x \in S \tag{34}$$

If $v_0 = 0$, then (34) yields

$$vg(x) \geq 0 \qquad \text{for all } x \in S \tag{35}$$

with v having at least one positive component. But this contradicts the hypothesis that there is an element $x \in S$ such that $g(x) < 0$, which would imply $vg(x) < 0$. Thus $v_0 > 0$.

Let

$$u^0 = \frac{v}{v_0} \tag{36}$$

Then (34) becomes

$$f(x) + u^0 g(x) \geq f(x^0) \qquad \text{for all } x \in S \tag{37}$$

If $x = x^0$ in (37) we obtain

$$u^0 g(x^0) \geq 0 \tag{38}$$

But

$$u^0 \geq 0, \qquad g(x^0) \leq 0 \Rightarrow u^0 g(x^0) \leq 0 \tag{39}$$

By (38) and (39),

$$u^0 g(x^0) = 0 \tag{40}$$

Adding (40) to (37) yields

$$L(x^0, u^0) = f(x^0) + u^0 g(x^0) \leq L(x, u^0) = f(x) + u^0 g(x) \tag{41}$$

for all $x \in S$. That is, x^0 minimizes $L(x, u^0)$ over S. Relations (40)–(41) and $g(x^0) \leq 0$ are then conditions (a)–(c) of Theorem 1, which are necessary and sufficient that (x^0, u^0) be a saddle point. Thus the theorem is proved.

Some of the sets and functions used in this proof play an important role in the developments of Chapter 8. The separating hyperplane with normal vector $(1, u^0)$ is a supporting hyperplane to the set of points on and above the graph of $w(y)$. The fact that no supporting hyperplane to this set has an intercept with the y_0 axis larger than the value $f(x^0)$ yields a number of duality theorems, stated more fully in Chapter 8.

Sufficiency of the Kuhn–Tucker Conditions. If the assumption that the objective and constraint functions are differentiable is added to the hypotheses of Theorem 3, the Kuhn–Tucker conditions can be shown to be both necessary and sufficient. Let the problem be as considered in Section 1.3.2:

$$\text{minimize } f(x)$$

subject to $$g_i(x) \leq 0, \qquad i = 1, \ldots, m$$

This is obtained by setting $S = E^n$ in relation (3) of this section. We then have the following.

THEOREM 4. *Assume the following in the program above:*

(a) *f and all g_i are convex and differentiable.*

(b) *There exists a point x such that*

$$g(x) = (g_1(x), \ldots, g_m(x)) < 0$$

Then x^0 solves the program if and only if there exists an m vector $u^0 \geq 0$ such that the Kuhn–Tucker conditions are satisfied at (x^0, u^0), i.e., if and only if

$$\nabla_x L(x^0, u^0) = 0$$
$$u^0 g(x^0) = 0$$

where

$$L(x, u) = f(x) + ug(x)$$

PROOF. By Theorem 3, x^0 is optimal if and only if there exists $u^0 \geq 0$ such that (x^0, u^0) is a saddle point for $L(x, u)$. By Theorem 1, (x^0, u^0) is a saddle point if and only if

(a) x^0 minimizes $L(x, u^0)$ over E^n.

(b) $g_i(x^0) \leq 0$, $i = 1, \ldots, m$.

(c) $u^0 g(x^0) = 0$.

Since $u^0 \geq 0$, $L(x, u^0)$ is a positive linear combination of convex functions, hence is convex (Theorem 1.3.1-6). Since f and all g_i are differentiable, $L(x, u^0)$ is also differentiable. Thus $L(x, u^0)$ assumes its unconstrained minimum at x^0 if and only if

$$\nabla_x L(x^0, u^0) = 0$$

The above, combined with condition (c), proves the theorem.

According to the above result, the Kuhn–Tucker conditions are an infallible test for an optimal point for differentiable convex programs satisfying a constraint qualification. For example, the gradient projection algorithm of Rosen [15], to be discussed shortly, determines if the current x is optimal by constructing, when possible, the relations $\nabla_x L(x, u) = 0$, $ug(x) = 0$, and seeing if the multipliers u_i are nonnegative.

1.3.4 Methods of Nonlinear Programming. Many, if not most, existing methods of nonlinear programming fall roughly into two main categories: (1) methods of feasible directions, and (2) penalty function techniques. The meaning of these terms will be made explicit in the following sections, where two methods from category 1 and one from category 2 will be presented. An effort will be made to point out the relations between working features of the methods and the theory of previous sections.

Constrained optimization problems are generally more difficult to solve than those without constraints. However, it is sometimes possible to eliminate inequality constraints by appropriate transformations. For example, if a constraint is of the form

$$l_i \leq x_i \leq u_i$$

then the transformation

$$x_i = l_i + (u_i - l_i) \sin^2 y_i$$

replaces x_i by the unconstrained variable y_i. Other transformations, as well as sequences of transformations, are discussed in [8].

Methods of Feasible Directions. Methods of feasible directions embody the same philosophy as the techniques of unconstrained minimization discussed previously but are constructed to deal with inequality restrictions. Briefly, the idea is to pick a starting point satisfying the constraints and to find a direction such that (1) a small move in that direction violates no constraint, and (2) the objective function improves in that direction. One then moves some distance in this direction, obtaining a new and better point, and repeats the procedure until a point is obtained such that no direction satisfying both (1) and (2) can be found. In general, such a point is a constrained local minimum of the problem. This local minimum need not be global, however, as will be illustrated later. A direction satisfying (1) is called feasible while a direction satisfying both (1) and (2) is called a usable feasible direction. There are many ways of choosing such directions, thus many different methods of feasible directions.

An iterative procedure of this type is shown geometrically in Figure 1-26.

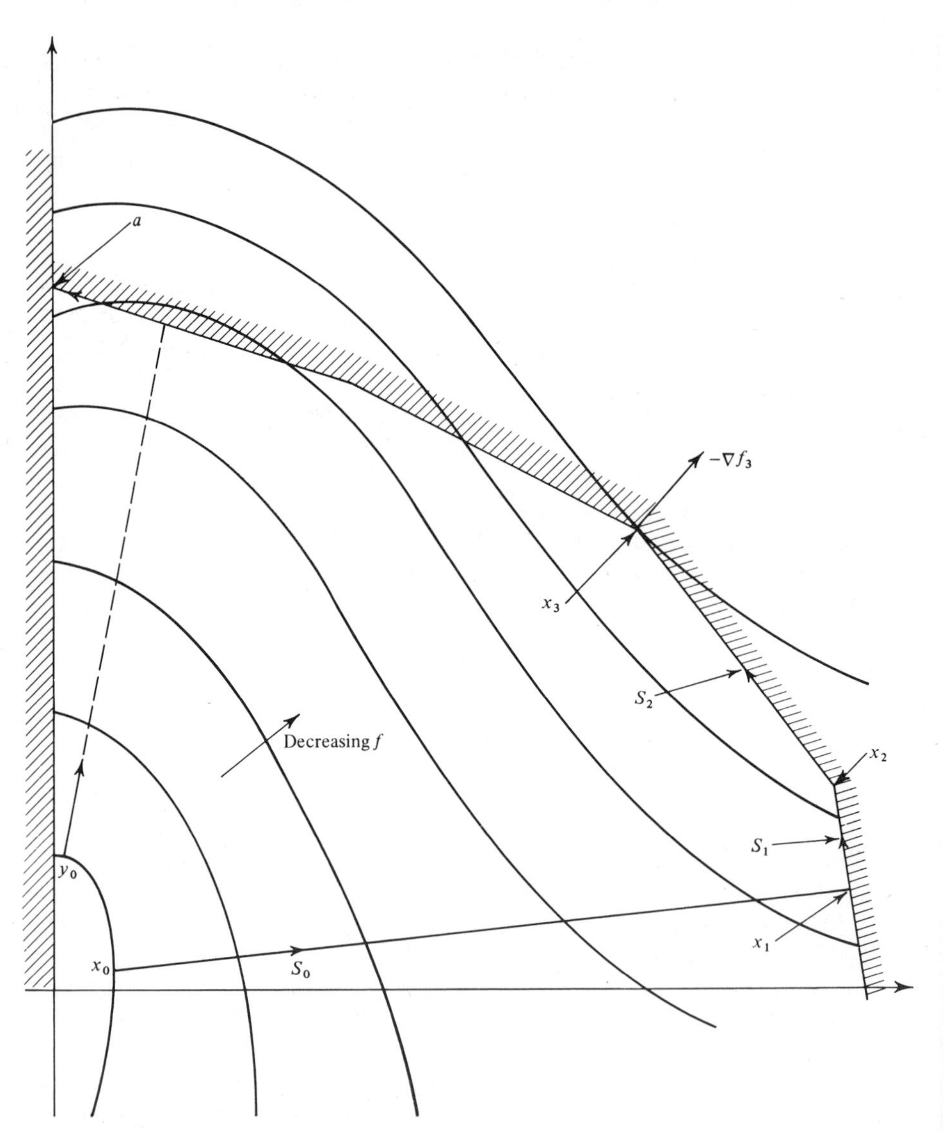

FIGURE 1-26 Method of feasible directions.

Let the starting point be x_0 and let the usable feasible direction chosen be $s_0 = -\nabla f(x_0)$. If the procedure of choosing the distance moved along s_0 to minimize f along s_0 is adopted, then the new point is x_1. Proceeding in the negative gradient direction at x_1 violates the constraints. There are many feasible directions in which one could move at this point, i.e., any direction pointing into the constraint set or along a constraint boundary. The locally "best" direction, s, is that feasible direction along which f decreases most rapidly, i.e., along which $-s' \nabla f(x)$ is maximized. This is the feasible direction making the smallest angle with $-\nabla f(x)$ and is the projection of $-\nabla f$ onto the constraint boundary. At the point x_1, this is the vector s_1. The farthest one can move along s_1 is the point x_2. Repeating the procedure leads to x_3 with negative gradient $-\nabla f_3$. At this point there is no usable feasible direction. This is equivalent to the statement that no feasible direction at x_3 makes an angle of less than 90° with $-\nabla f(x_3)$. It is easily seen that, in this case, x_3 is the global minimum of f over the constraint set.

The global minimum is not, however, always reached by this procedure. In the example shown in Figure 1-26, the same procedure, starting with the y_0 shown, leads to a local minimum at the point a which is distinct from the global minimum at x_3. This example illustrates the difficulties such procedures may encounter with local optima. These difficulties are common to all methods and one can be sure of avoiding them only if the problem is a convex programming problem.

These geometrical ideas are now examined algebraically in terms of two methods of feasible directions: a method of Zoutendijk [11] and the gradient projection method of Rosen [15].

Zoutendijk's Procedure. Consider the problem

$$\text{minimize } f(x) \tag{1}$$

subject to

$$g_i(x) \geq 0, \qquad i = 1, \ldots, m$$

The constraint set is

$$R = \{x \mid g_i(x) \geq 0, i = 1, \ldots, m\}$$

Assume that the starting point x_0 satisfies the constraints. The problem is to choose a vector s whose direction is both usable and feasible. Let

$$g_i(x_0) = 0, \qquad i \in I$$

where the indices of the binding constraints at x_0 form the set I. For feasible vectors, s, a small move along the vector from x_0 makes no binding constraint negative, i.e.,

$$\frac{d}{d\alpha} g_i(x_0 + \alpha s)\bigg|_{\alpha=0} = \nabla g_i'(x_0)s \geq 0, \qquad i \in I$$

A usable feasible vector has the additional property that

$$\frac{d}{d\alpha} f(x_0 + \alpha s)\bigg|_{\alpha=0} = \nabla f'(x_0)s < 0$$

Therefore the function initially decreases along the vector. In searching for a "best" vector s along which to move, one could choose that feasible vector minimizing $\nabla f'(x_0)s$. However, if some of the binding constraints were non-linear, this could lead to certain difficulties, as shown in Figure 1-27. Starting

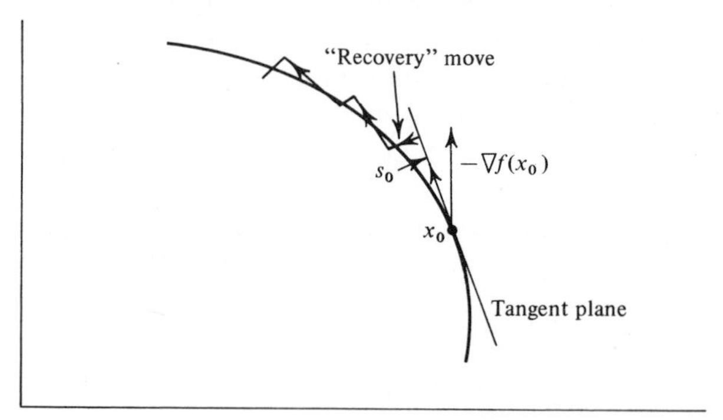

FIGURE 1-27 Zigzagging.

at x_0 the feasible direction, s_0, that minimizes $\nabla f'(x_0)s$ is the projection of $-\nabla f(x_0)$ on the tangent plane through x_0. Since the constraint surface is curved, movement along s_0 for any finite distance violates the constraint. Thus a recovery move must be made to come back inside the constraint set. Repetitions of the procedure lead to the inefficient zigzagging shown. The moral of this story is that in looking for a locally "best" direction it is wise to choose one that, in addition to decreasing f, also moves away from the boundaries of the nonlinear constraints. Hopefully, this will avoid zigzagging. Such a direction is the solution of the following problem:

$$\text{minimize } \xi \tag{2}$$

subject to

$$\nabla g_i'(x_0)s + \theta_i \xi \geq 0, \qquad i \in I, \quad 0 \leq \theta_i \leq 1 \tag{3}$$

$$\nabla f'(x_0)s - \xi \leq 0 \tag{4}$$

$$s's = 1 \tag{5}$$

If all $\theta_i = 1$, then any vector (s, ξ) satisfying (3)–(4) with $\xi < 0$ is a usable feasible direction. That with minimum ξ is a best direction which simultaneously makes $\nabla f'(x_0)s$ as negative and $\nabla g_i'(x_0)s$ as positive as possible, i.e., steers away from nonlinear constraint boundaries. Other values of θ_i enable one to emphasize certain constraint boundaries relative to others. The requirement (5) is a normalization requirement ensuring that min ξ be finite. If it were not included and a vector (s, ξ) existed satisfying (3)–(4) with ξ negative, then ξ could be made to approach $-\infty$, since (3)–(4) are homogeneous. Other normalizations are also possible.

Since the vectors ∇f and ∇g_i are evaluated at a fixed point x_0, the above "direction-finding" problem is almost linear, the only nonlinearity being (5). It is shown by Zoutendijk that the problem may be handled by a modified version of linear programming. Thus the direction-finding problem may be solved with reasonable efficiency.

Of course, once a direction has been selected, the problem of selecting the step size still remains. This problem may be dealt with almost as in the unconstrained case. It is still desirable to minimize the objective function along the vector s, but now no constraint may be violated. Thus α is determined to minimize $f(x_k + \alpha s)$ subject to the constraint $x_k + \alpha s \in R$. Any of the techniques discussed in Section 1.1 can be used. A new point is thus determined, and the direction-finding problem is repeated. If at some point the minimum $\xi \geq 0$, then there is no feasible direction s satisfying $\nabla f'(x)s < 0$ and the procedure terminates. The final point will generally be a local minimum of the problem. Zoutendijk shows that for convex programs, the procedure converges to the global minimum.

Rosen's Gradient Projection Method. At each iteration of Zoutendijk's method, an optimization problem must be solved to find a direction in which to move. Although the direction found was in some sense "best," the procedure can be time consuming. An alternative is provided by the gradient projection method of Rosen. Here a usable feasible direction is found without solving an optimization problem. The direction, however, may not be locally best in any sense. The method is efficient when all constraints are linear,[10] although less so in the nonlinear case. It is of particular interest here because of the direct way in which it uses the Kuhn–Tucker conditions both to generate new directions and as a stop criterion.

Only the procedure for linear constraints will be discussed. The problem is of the form (1), but the constraints are linear and are written

$$a_i'x \geq b_i, \qquad i = 1, \ldots, m$$

The equation

$$a_i'x = b_i$$

[10] The algorithm described in reference [19], an extension of Rosen's method, appears to be the best currently available for linearly constrained problems.

represents a hyperplane and generally forms one of the boundaries of the constraint set, as in Figure 1-28. The procedure starts at some point, x_0,

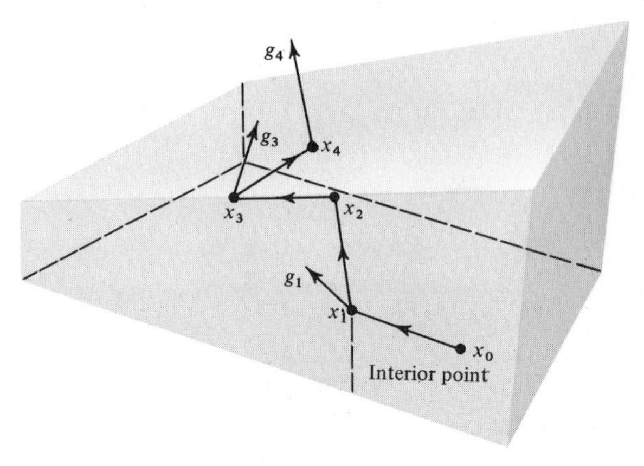

FIGURE 1-28 Gradient projection method.

satisfying the constraints and the kth iteration proceeds as follows:

1. Calculate $\nabla f(x_k)$.
2. Associate with x_k those constraint planes passing through x_k.
3. Find the projection of $-\nabla f(x_k)$ onto the intersection of the planes associated with x_k. In case there are no such planes (as when x_k is interior to the constraint set), this intersection is the whole space, and thus the projection is $-\nabla f(x_k)$ itself.
4. If the projection is not the zero vector, then minimize along the projection vector subject to the constraints, as described previously, replace k by $k + 1$, and return to step 1.
5. If the projection vector is zero, then $\nabla f(x_k)$ may be written

$$\nabla f(x_k) = \sum_i u_i a_i \tag{6}$$

i.e., as a linear combination of normals a_i to the planes associated with x_k (see below).

(a) If all $u_i \geq 0$, then x_k is the solution of the problem, for the Kuhn–Tucker conditions are satisfied.

(b) Otherwise, define a new set of planes to be associated with x_k by deleting from the present set some one plane for which $u_i < 0$, and return to step 3.

In the example of Figure 1-28 the points x_3 and x_4 have been obtained as the result of minimizing on the segments (x_2, x_3) and (x_3, x_4).

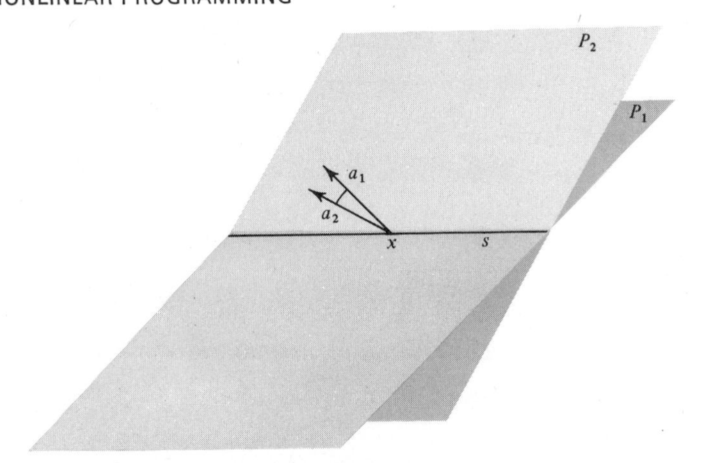

FIGURE 1-29 Two planes intersecting in a line.

Relation (6) in step 5 can be explained geometrically. Consider the intersection of two planes in a line s as shown in Figure 1-29. Let x be a point on the intersection. The normals to these two planes, a_1 and a_2, are the gradients of the corresponding constraint functions. Since a_1 and a_2 are normal to their respective planes, they are normal to the intersection s. Moreover, if a_1 and a_2 are linearly independent (i.e., nonparallel in two dimensions), then any vector normal to the intersection can be written as a linear combination of a_1 and a_2. If the projection of $\nabla f(x)$ on s at x is 0, then $\nabla f(x)$ is perpendicular to s at x. Thus

$$\nabla f(x) = u_1 a_1 + u_2 a_2$$

which is (6) for two intersecting planes.

To explain step 5(a), note that for linear constraints the vectors a_i are the gradients $\nabla g_i(x)$ in the Kuhn–Tucker conditions. Since $u_i > 0$ only for those constraints for which $a_i'x = 0$ and $u_i = 0$ otherwise, $u_i g_i(x) = 0$ is satisfied for all i. Thus the Kuhn–Tucker conditions are satisfied. For step 5(b), it is shown by Rosen [15] and Hadley [10] that the procedure of dropping a plane corresponding to $u_k < 0$ and projecting on the remaining constraints leads to a usable feasible direction, as does the procedure for projecting ∇f upon the intersection of all binding constraints.

Most of the computational labor in Rosen's method is in computing the various projections. The projection of a vector a onto a given vector space S is another vector b, which can be obtained from a by multiplying it by a projection matrix P. Rosen shows that to project on the intersection of the constraints $a_i'x = b_i$ for $i \in I$, the appropriate projection matrix is

$$P = I - A(A'A)^{-1}A'$$

where A is a matrix whose columns are the vectors a_i. Of course, it is necessary to recompute P when the set of constraints changes, and this can be cumbersome if one set differs radically from the next. The procedure of dropping one constraint at a time allows the formation of successive projection matrices one from the other by partitioning [15], since successive A matrices differ in only one column.

It is not difficult to see why the procedure is not as efficient when the constraints are nonlinear. Then the projection is made onto the tangent planes to the constraints. Movement along these planes may then, unlike the linear case, violate the constraints. Thus a recovery step is needed and zigzagging can easily occur.

Penalty Function Techniques. The problem

$$\text{minimize } f(x) \tag{7}$$

subject to

$$g_i(x) \geq 0, \qquad i = 1, \ldots, m \tag{8}$$

can be attacked from an entirely different viewpoint. Define the "penalty" function

$$\phi(y) = \begin{cases} 0, & y \geq 0 \\ \infty, & y < 0 \end{cases} \tag{9}$$

and consider the problem of minimizing

$$\psi(x) = f(x) + \sum_{i=1}^{m} \phi(g_i(x)) \tag{10}$$

subject to no constraints. Obviously, if all constraints $g_i(x) \geq 0$ are satisfied, then the summation term contributes nothing and minimizing ψ is equivalent to minimizing f. If any $g_i < 0$, then $\phi(g_i) = \infty$, which will certainly not be anywhere near the minimum of $\psi(x)$; i.e., the summation term penalizes any violation of the constraints. Thus any procedure which minimizes ψ will never select a point outside the constraint set and will, in fact, select that point of the constraint set which minimizes $f(x)$. In this way a constrained problem is solved using unconstrained minimization. Since powerful methods of unconstrained minimization are available, such an approach is quite attractive.

There are, however, difficulties which give even the most powerful minimizer trouble and which must be overcome. To illustrate these, consider the problem

$$\text{minimize } x_1^2 + x_2^2$$

subject to $x_1 \geq 3$

whose solution is $x_1 = 3$, $x_2 = 0$. The function $\psi(x)$ is

$$\psi(x) = x_1^2 + x_2^2 + \phi(x_1 - 3)$$

Contours of ψ in the feasible region (i.e., to the right of the line $x_1 = 3$) are circles with center at the origin. Here the penalty term $\phi(x_1 - 3)$ has no effect. Just to the left of the line $x_1 = 3$, ψ becomes unbounded, so that moving to the left from that line one traverses all contours of constant value at once. A gradient minimizer starting at x_0 would move to the boundary at x_1 and could proceed no further. In fact, the function ψ is discontinuous and has no derivative along $x_1 = 3$, so minimization is almost hopeless.

These difficulties may be relieved by defining other, less "harsh" penalty functions having the necessary smoothness properties. For example, the function

$$\langle y \rangle^2 = \begin{cases} 0, & y \geq 0 \\ y^2, & y < 0 \end{cases} \tag{11}$$

is suitable. This is continuous and has continuous first derivatives for all y. Of course, since the penalty for constraint violations is not infinite, some violations are possible. To reduce these violations, multiply the penalty function (11) by a large positive constant k. Then the unconstrained problem to be solved is

$$\text{minimize } f(x) + k \sum_{i=1}^{m} \langle g_i(x) \rangle^2 \tag{12}$$

The difficulty of "bunched" or eccentric contours still remains. Consider applying this new penalty function to the previous problem by minimizing

$$\psi(x) = x_1^2 + x_2^2 + k\langle x_1 - 3 \rangle^2$$

The contours of the function to the right of $x_1 = 3$ are circular, but to the left they are elongated ellipses, showing the same bunching effect as before. Even the most efficient minimizer may converge slowly.

Recent work has shown that a more relaxed approach is better. Rather than solving only one unconstrained problem, a sequence of such problems is solved, each bringing us closer to the final solution. The overall effect is a more efficient procedure for solving constrained optimization problems. For example, in (12) one would select an initial $k_0 > 0$, but not too large, and minimize. At the jth step, select $k_j > k_{j-1}$ and repeat the minimization, with $k_j \rightarrow \infty$ as $j \rightarrow \infty$. In general, the sequence of unconstrained minima approaches the solution of the original constrained problem.

With the penalty function (11) intermediate solutions will violate the constraints. Thus the method approaches the constrained minimum from outside the constraint set. In many cases, this may be unsatisfactory. If small violations of the constraints are not permitted, then intermediate solutions often cannot be used. The method will be inefficient if the objective or constraint functions are ill-behaved exterior to the constraint set. Moreover, the approach cannot be used at all when any of these functions are not defined outside the constraint set.

Fiacco–McCormick Method. A method which avoids these difficulties by approaching the optimum from inside the constraint set is now widely used [16, 17, 22]. Define the function

$$P(x, r) = f(x) + r \sum_{i=1}^{m} \frac{1}{g_i(x)} \tag{13}$$

where $r > 0$. Choose $r_1 > 0$, and x_0 strictly inside the constraint set and consider the problem of minimizing $P(x, r_1)$ starting from x_0 subject to no constraints. Intuitively one would expect that a minimum of $P(x, r_1)$ exists inside the constraint set, since on the boundary of this set, i.e., for some $g_i(x) = 0$, $P(x, r_1) \to +\infty$. Thus the trajectory of steepest descent leading from the point x_0, a path on which $P(x, r_1)$ is strictly decreasing, cannot penetrate the boundary of the constraint set. The minimizing point depends, of course, on the choice of r_1, and is denoted by $x(r_1)$. By the above arguments, $x(r_1)$ is inside the constraint set.

Consider repeating this minimization process for a sequence of r values $r_1 > r_2 > \cdots > r_k > 0$. Each minimizing point $x(r_k)$ will be strictly inside the constraint set. Further, by reducing r the influence of the term $r \sum_{i=1}^{s} 1/g_i(x)$, which "penalizes" closeness to the constraint boundaries is reduced and, in minimizing P, more effort is concentrated upon reducing $f(x)$. Thus the sequence of points $x(r_1), x(r_2), \ldots, x(r_k)$ can come closer and closer to the boundary of the set R if it is profitable, in terms of reducing f, to do so. One would thus expect that as r approaches zero, the minimizing point $x(r)$ approaches the solution of the original problem (7)–(8). Such a method is particularly attractive when dealing with problems with markedly nonlinear constraints, since it approaches the solution value from inside the constraint set. Motion along the boundaries of this set, which can be very cumbersome when the boundaries have large curvature, is completely avoided.

Fiacco and McCormick prove that, under certain assumptions, the above conjectures are true. In particular, it is shown that if

1. the interior of the constraint set is nonempty
2. the functions f and g_i are twice continuously differentiable

3. the set of points in the constraint set for which $f(x) \leq k$ is bounded for all $k < \infty$
4. the function $f(x)$ is bounded below for x in the constraint set

then at least one finite local minimum of $P(x, r)$ exists in the interior of the constraint set for any $r > 0$. To prove more, the usual convexity assumptions must be invoked; i.e.,

5. $f(x)$ is convex
6. the $g_i(x)$ are concave functions.
7. $P(x, r)$ is strictly convex in the interior of the constraint set for any $r > 0$.

Conditions (5)–(6) guarantee that the problem (7)–(8) is a convex programming problem. For such a problem, any local minimum is global. Condition (7) is not implied by (5)–(6), but only small additional requirements on f and the g_i are needed for it to hold [16]. Under conditions (1)–(7) it is shown [16] that the values $P(x(r_1), r_1), \ldots, P(x(r_k), r_k), \ldots$ approach v_0, the constrained minimum value of $f(x)$, as $r_k \to 0$. In addition, it is shown that

$$f(x(r_{k+1})) \leq f(x(r_k)) \tag{14}$$

i.e., that f is monotonically nonincreasing as r is reduced.

Duality. The above are the main convergence results. Two other features of the method remain. The first involves the dual nonlinear program of Wolfe [18] (see Section 8.4) and has direct application in bounding the problem minimum and terminating the method. Wolfe defines as a dual problem for (7)–(8) the problem

$$\text{maximize } L(x, u) = f(x) - \sum_{i=1}^{m} u_i g_i(x) \tag{15}$$

subject to

$$\nabla f(x) = \sum_{i=1}^{m} u_i \nabla g_i(x) \tag{16}$$

$$u_i \geq 0, \qquad i = 1, \ldots, m \tag{17}$$

He shows that under the differentiability and convexity assumptions (2) and (5)–(6) the following results hold:

1. If the primal problem (7)–(8) has a solution at x, then there exists u such that (x, u) solves the dual problem and the extreme values of the problems are equal.

2. For any point x satisfying the primal constraints (8) and (x, u) satisfying the dual constraints (16)–(17)

$$f(x) \geq L(x, u)$$

i.e., $f(x)$ is an upper bound for the problem minimum and $L(x, u)$ is a lower bound.

It is easily shown that this method provides a point (x, u) satisfying the dual constraints at each P-minimum. At such a minimum the condition

$$\nabla P(x(r_k), r_k) = \nabla f(x(r_k)) - \sum_{i=1}^{m} \frac{r_k}{g_i^2(x(r_k))} \nabla g_i(x(r_k)) = 0 \tag{18}$$

must hold. If the correspondence

$$u_i(r_k) = \frac{r_k}{g_i^2(x(r_k))} > 0 \tag{19}$$

is made, then (16)–(17) are satisfied. Fiacco and McCormick show that, under assumptions (1)–(7) the sequence of points $\{(x(r_k), u(r_k))\}$ converges to the dual solution (x, u). Thus, at each P-minimum v_0 is bounded by the inequalities

$$L(x, u) = f(x(r_k)) - r_k \sum_{i=1}^{m} \frac{1}{g_i(x(r_k))} \leq v_0 \leq f(x(r_k))$$

These allow termination of the method as soon as the difference between upper and lower bounds becomes less than some given tolerance.

References

1. S. I. Gass, *Linear Programming: Methods and Applications*, McGraw-Hill, Inc., New York, 1958.
2. G. B. Dantzig, *Linear Programming and Extensions*, Princeton University Press, Princeton, N.J., 1963.
3. F. B. Hildebrand, *Methods of Applied Mathematics*, Prentice-Hall, Inc., Englewood Cliffs, N.J., 1952, pp. 120–121.
4. H. Curry, "Methods of Steepest Descent for Non-linear Minimization Problems," *Quart. Appl. Math.*, **2**, 1954, pp. 258–261.
5. R. Fletcher and M. J. D. Powell, "A Rapidly Convergent Descent Method for Minimization," *Brit. Computer J.*, **6**, 1963, pp. 163–168.
6. R. Fletcher and C. M. Reeves, "Function Minimization by Conjugate Gradients," *Brit. Computer J.*, **7**, 1964, pp. 149–154.

7. M. J. D. Powell, "An Efficient Method for Finding the Minimum of a Function of Several Variables without Using Derivatives," *Brit. Computer J.*, 7, 1964, pp. 155–162.
8. M. J. Box, "A Comparison of Several Current Optimization Methods and the Use of Transformations in Constrained Problems," *Brit. Computer J.*, **9**, 1966, pp. 67–78.
9. H. H. Rosenbrock, "An Automatic Method for Finding the Greatest or Least Value of a Function," *Brit. Computer J.*, **3**, 1960, p. 175.
10. G. Hadley, *Nonlinear and Dynamic Programming*, Addison-Wesley Publishing Company, Inc., Reading, Mass., 1964.
11. G. Zoutendijk, *Methods of Feasible Directions*, American Elsevier Publishing Company, Inc., New York, 1960.
12. L. A. Schmit, "Structural Design by Systematic Synthesis," *Proceedings of the Second National Conference on Electronic Computation, Structural Division, A.S.C.E., Pittsburgh, Pa.*, September 1960, pp. 105–132.
13. C. Y. Sheu and W. Prager, "Recent Developments in Optimal Structural Design," *Applied Mechanics Reviews*, **21**, No. 10, 1968, pp. 985–992.
14. H. W. Kuhn and A. W. Tucker, "Nonlinear Programming," *Proceedings of the Second Berkeley Symposium on Mathematical Statistics and Probability, Berkeley, Calif., 1950*, pp. 481–492.
15. J. B. Rosen, "The Gradient Projection Method for Nonlinear Programming: Part I—Linear Constraints," *SIAM J. Appl. Math.*, No. 8, 1960, pp. 181–217.
16. A. V. Fiacco and G. P. McCormick, "The Sequential Unconstrained Minimization Technique for Nonlinear Programming, A Primal Dual Method," *Management Sci.*, **10**, No. 2, 1964, pp. 360–366.
17. A. V. Fiacco and G. P. McCormick, "Computational Algorithm for the Sequential Unconstrained Minimization Technique for Nonlinear Programming," *Management Sci.*, **10**, No. 2, 1964, pp. 601–617.
18. P. Wolfe, "A Duality Theorem for Non-linear Programming," *Quart. Appl. Math.*, **19**, No. 3, 1961, pp. 239–244.
19. D. Goldfarb, "Extension of Davidon's Variable Metric Method to Maximization Under Linear Inequality and Equality Constraints," *SIAM J. Appl. Math.*, **17**, No. 4, 1969, pp. 739–764.
20. D. J. Wilde, *Optimum Seeking Methods*, Prentice-Hall, Inc., Englewood Cliffs, N.J., 1964, chap. 2.
21. J. D. Pearson, "Variable Metric Methods of Minimisation," *Brit. Computer J.*, **12**, 1969, pp. 171–178.
22. A. V. Fiacco and G. P. McCormick, *Sequential Unconstrained Minimization Techniques for Nonlinear Programming*, John Wiley & Sons, Inc., New York, 1968.
23. S. Karlin, *Mathematical Methods and Theory in Games, Programming and Economics*, Vol. 1, Addison-Wesley Publishing Company, Inc., Reading, Mass, 1959.
24. J. E. Falk, "Lagrange Multipliers and Nonlinear Programming," *J. Math. Anal. Appl.*, **19**, No. 1, 1967.

2

Large Mathematical Programs with Special Structure

2.1 Introduction

The solution of problems involving real systems is often hampered by size—the problem is simply too big. In mathematical programming, size is determined by the number of variables, the number and complexity of the constraint functions, and the complexity of the system objective. What "large" means depends on the capabilities of solution algorithms, the speed and capacity of available computing equipment, etc. However, as mathematical programming becomes more widely used, problems are formulated which tax even the best algorithms and the largest computers. For example, in linear programming, problems of up to 4,095 rows and many more variables may be solved using the revised simplex method, as implemented in the IBM MPS/360 programming system. This requires about 248,000 32-bit words of core memory. However, according to Dantzig [1], problems having 10^4 equations and 10^6 variables have been solved. To do this requires the construction of special algorithms, which, in turn, demands an exploitation of system structure.

Fortunately, large programs almost always have special structure. For linear programs, the structure of the system being studied is reflected in the pattern of zero and nonzero elements in the constraint coefficient matrix. When the program is large, the density of nonzero matrix elements is generally less than a few per cent. Moreover, these are often arranged in an ordered way, for example, in blocks along the main diagonal except for a few rows or columns. Alternatively, some linear programs have many columns or rows (but not both) and the set of columns (rows) is defined by some explicit relations, e.g., linear inequalities. This enables a search for a particular row or column to be conducted efficiently, with the desired element generated upon request. It is worth noting that, if a truly large linear program did not have special structure, the task of gathering data for it would be almost hopeless. For example, 10^{10} numbers are required to specify a 10^4 by 10^6 linear program with full constraint matrix.

In the nonlinear case, problems are harder to classify. One can define a matrix whose (i, j) element is unity if variable i appears in constraint j and zero otherwise, and this matrix reflects problem structure. Also important are which variables appear linearly, and if any nonlinear functions are separable. Hybrid problems, e.g., problems which are linear with special structure when a relatively small number of nonlinear variables are fixed, are common.

Approaches to solving large mathematical programs may be divided into two classes: direct methods, and decomposition or partitioning techniques. Direct methods specialize an existing algorithm to a particular class of problems. These are most common in linear programming, where the basic tool is the simplex method. Here the main computational problem is the manipulation of the basis inverse. In the case of special matrix structure, the necessary operations may often be performed while storing a matrix of reduced size. Proposals for doing this include generalized upper bounding procedures [2], and compact basis triangularization [3].

Indirect methods are characterized by a decomposition of the original system into subsystems, each with a smaller, independent subproblem. Since the subsystems interact, solving the subproblems will not, in general, yield the correct solution. The multilevel approach [4] proposes that one account for the interactions by defining one or more "second-level" subsystems which influence, in some way, the original subsystems, defined to be on the first level. This influence may take many forms, depending on the original problem, the type of first level decomposition, etc., and must clearly be allowed for when the system is initially decomposed. The goal of the second level is to coordinate the actions of first-level units so that the solution of the original problem is obtained. All large organizations operate in this way.

One can extend this idea by defining third-level subsystems, each of which coordinates a number of second-level units, etc. The resulting structure has the form of a pyramid of decision making units, as is shown in Figure 2-1.

All existing decomposition algorithms are two level in structure. The coordinator influences the subproblems by varying parameters in their objective functions, adding constraints, etc. Advantages of this approach include the freedom to solve each subproblem using any algorithm desired, the ability to write and debug smaller independent computer programs, and the reduced amount of fast computer memory required. The chief disadvantage is that the subproblems must be solved a number of times, so much computer time may be used.

The main portion of this book is concerned with algorithms for solving various classes of large optimization problems. However, it is best to see first what types of problems are to be solved, since problem structure deter-

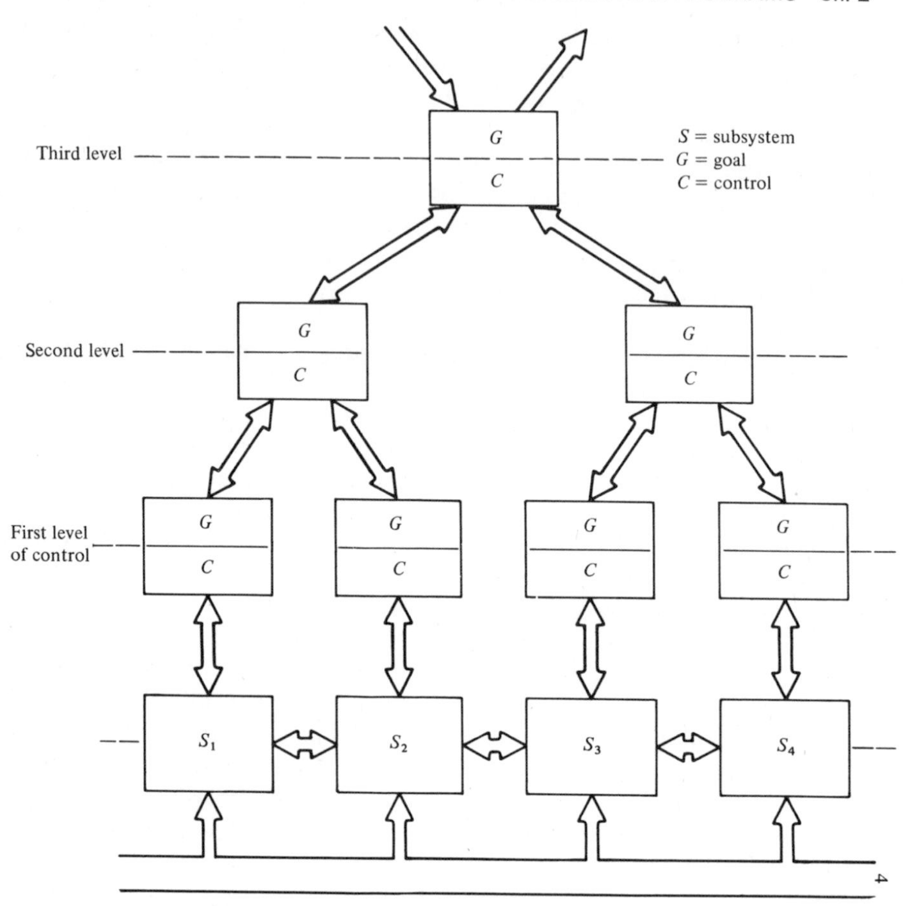

FIGURE 2-1 Multilevel control structure.

mines the form of the solution algorithm. Thus we consider, in the following sections, a number of examples. It will often be useful to abstract these problems from their physical settings and discuss them all using the same terminology. The concepts of activity analysis, which provide a systematic basis for constructing mathematical programming models, permit this to be done.

2.2 Activity Analysis

The mathematical programmer, in considering a system whose operation he is to improve, views it as being composed of a number of fundamental activities. These are the separate, detailed actions which must be performed for the system to function. Each activity is considered to be a "black box," into which flow certain inputs (labor, raw materials, money) and out of which flow outputs (finished goods, profit, etc.). These inputs and outputs will be called items and the quantity of each activity, a numerical measure of how much of it is performed, is called the activity level.

Consider for example, a transportation problem, where there are quantities of a single commodity stored at various warehouses, and demands to be met at certain destinations. The problem is to determine a shipping schedule which meets all demands at minimal transportation cost. Here the basic activity involved is "shipping some quantity from warehouse i to destination j." The activity level, x_{ij}, could be the number of units shipped. The input items to this activity are the goods to be shipped and money, the cost of shipment, while the item output is some quantity of the commodity at destination j. A block diagram of this activity is shown in Figure 2.2.

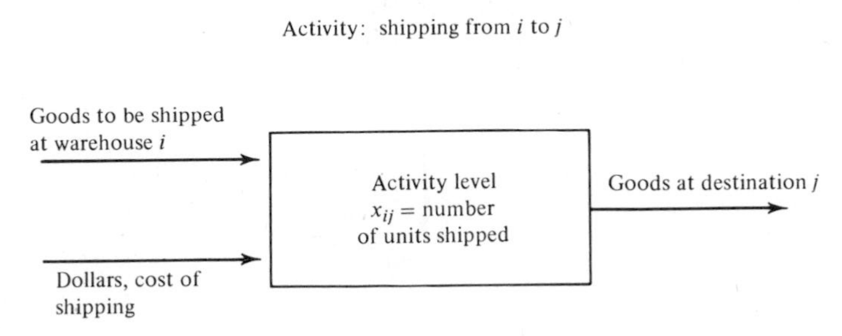

FIGURE 2-2 Transportation activity.

There are as many activities for this problem as there are warehouse-destination pairs. They are related by the fact that the input items are in limited supply so if one activity ships a large quantity, the amounts shipped by other activities are limited. Also, the output items must together equal or exceed customer demands. Let us regard one of the items, here money, as having particular value. Then the problem of determining all activity levels such that (1) all item constraints are met and (2) the total amount of the special item either required or produced is maximized or minimized is a mathematical program.

Under certain hypotheses on the activities, this program will be linear. Those given below are due to Dantzig [5].

Assumption 1. (Proportionality). The quantity of each of the items input to or output by an activity is proportional to the activity level.

Assumption 2. (Nonnegativity). Levels of all activities are nonnegative. One can choose only to operate an activity (at a positive level) or not operate it (the level is zero).

Assumption 3. (Additivity). The total quantity of any item either consumed or produced by all activities is the sum of the amounts consumed or produced by each activity.

Assumption 4. (*Linear Objective Function*). One of the items is singled out as being particularly valuable. The quantity of this item used or produced by the system is to be either maximized or minimized. By assumptions 1 and 3, this quantity is a linear function of the activity levels, called the objective function.

By assumption 1, if the level of activity j is x_j, then the amount of item i flowing into or out of the activity is $a_{ij}x_j$. The constant a_{ij} is the amount of item i required to operate activity j at unit level. The sign convention adopted here is that a_{ij} is positive for inputs and negative for output items. An activity is then completely characterized, in an input–output sense, by the activity vector,

$$p_j = \begin{pmatrix} a_{1j} \\ a_{2j} \\ \vdots \\ a_{mj} \end{pmatrix} \tag{1}$$

Assumption 2 implies

$$x_j \geq 0, \qquad j = 1, \ldots, n \tag{2}$$

In any program, a number of the items, typically resources, will be in short supply, while others, usually some outputs, will need to meet requirements imposed from outside the system. Thus, for each item, there is a material-balance equation, stating that the total flow of the item to or from the system is constrained. By assumption 3, this can be written

$$\sum_{j=1}^{n} a_{ij}x_j = b_i \tag{3}$$

where b_i is the quantity of flow required, again positive for input items and negative for outputs. If the constraint on item flow is an inequality, then by including a slack activity, i.e., "not using item i," an equation such as (3) results. Finally, if c_j is the flow of the item to be optimized into activity j when $x_j = 1$, then the objective form is

$$z = \sum_{j=1}^{n} c_j x_j \tag{4}$$

which we assume is to be minimized.

Relaxing assumptions 1–4 leads to nonlinear programs. If 1 is relaxed, separable programs result;

$$\text{minimize} \sum_j f_j(x_j) \tag{5}$$

subject to

$$\sum_j g_{ij}(x_j) = b_i, \qquad i = 1, \ldots, m \tag{6}$$

$$x_j \geq 0 \tag{7}$$

where f_j and g_{ij} may be nonlinear. Deleting assumptions 1 and 3 admits general nonlinear objective and constraint functions.

2.3 Production and Inventory Problem

A manufacturer must schedule the production of two different products quarterly over a year's time to meet given demand forecasts for each product in each quarter. Production may be stored for later sale, and scarce items are storage capacity and labor. All demands must be met in the time period when they occur, and there are time-varying costs for labor and storage.

This system has three input items, labor hours, storage capacity, and operating costs, and two output items, finished products 1 and 2, destined for customers. In time period t, six activities, including slacks, may be identified:

Activity	Level	Units of activity level
Producing product 1	$x_{1t} \geq 0$	one unit of product 1
Producing product 2	$x_{2t} \geq 0$	one unit of product 2
Storing product 1	$y_{1t} \geq 0$	one unit of product 1
Storing product 2	$y_{2t} \geq 0$	one unit of product 2
Not using storage capacity	$s_{1t} \geq 0$	one unity of capacity
Not using available labor	$s_{2t} \geq 0$	one hour of labor

Let

a_i = number of units of capacity required to store one unit of product i

v_{it} = dollar cost of storing one unit of product i in period t

b_i = number of labor hours required to produce one unit of product i

c_{it} = dollar cost of 1 labor hour used to produce item i in period t

l_t = labor hour availability in period t

f_t = availability of storage capacity in period t

d_{it} = forecasted demand for item i in period t

Then the constraints of the problem may be written as follows:

The number of labor hours available is limited:

$$b_1x_{1t} + b_2x_{2t} + s_{2t} = l_t \tag{1}$$

Storage capacity is limited:

$$a_1y_{1t} + a_2y_{2t} + s_{1t} = f_t \tag{2}$$

Material balance for inventories:

$$y_{it} = y_{i,t-1} + x_{it} - d_{it}, \qquad i = 1, 2 \tag{3}$$

Cost equation:

$$z_t = v_{1t}y_{1t} + v_{2t}y_{2t} + c_{1t}b_1x_{1t} + c_{2t}b_2x_{2t} \tag{4}$$

These may be displayed in detached coefficient form as shown in Table 1. Structurally, the important thing to note is that the only activities which enter the material-balance equations of more than one period are the storage activities, which have inputs in t and outputs in $t + 1$. The effect of this on the coefficient matrix for a multiperiod problem is shown in Table 2. This coefficient matrix is of the "staircase" form, with all elements zero except for blocks on and below the main diagonal. The diagonal blocks arise from activities carried on within a time period, while the nonzero entries in the off-diagonal coupling blocks are provided by the storage activities. Both this structure and the role played by the storage activities are typical of dynamic situations, where the goal is optimal allocation of resources over time.

2.4 Dynamic Leontief Model

The methods of input–output analysis, first put forth by Leontief, have found wide use in economic planning and forecasting. The simplest static model views an economic system as composed of n activities, typically industries or groups of industries, obeying assumptions 1–3 of Section 2.2 with each producing a single output item. The output of any activity flows either to other activities (including itself) or to the outside world, where we assume there are given requirements, called final demands, for the output. These may include public and private consumption, as well as exports. Let

x_i = level of activity i = number of units of industry i's output produced

d_i = final demand for the output of i

a_{ij} = amount of i's output required to produce one unit of j's output

TABLE 1

Items	Activities								Constants
	Storing 1 in $t-1$: $y_{1,t-1}$	Storing 2 in $t-1$: $y_{2,t-1}$	Storing 1 in t: y_{1t}	Storing 2 in t: y_{2t}	Producing 1 in t: x_{1t}	Producing 2 in t: x_{2t}	Not using storage in t: s_{1t}	Not using labor in t: s_{2t}	
Labor hours					b_1	b_2		1	l_t
Storage capacity			a_1	a_2			1		f_t
Finished product 1	-1		1		-1				$-d_{1t}$
Finished product 2		-1		1		-1			$-d_{2t}$
Cost			v_{1t}	v_{2t}	$c_{1t}b_1$	$c_{2t}b_2$			0

TABLE 2

Constraint Equations for Multiperiod Production and Inventory Problem

y_{11}	y_{21}	x_{11}	x_{21}	s_{11}	s_{21}	y_{12}	y_{22}	x_{12}	x_{22}	s_{12}	s_{22}	y_{13}	y_{23}	x_{13}	x_{23}	s_{13}	s_{23}	y_{14}	y_{24}	x_{14}	x_{24}	s_{14}	s_{24}	Exogenous flows
		b_1	b_2		1																			l_1 (labour hours)
a_1	a_2			1																				f_1 (storage capacity)
1		-1																						$y_{10} - d_{11}$ (product 1)
	1		-1																					$y_{20} - d_{21}$ (product 2)
								b_1	b_2		1													l_2
						a_1	a_2			1														f_2
-1						1		-1																$-d_{12}$
	-1						1		-1															$-d_{22}$
														b_1	b_2		1							l_3
												a_1	a_2			1								f_3
						-1						1		-1										$-d_{13}$
							-1						1		-1									$-d_{23}$
																				b_1	b_2		1	l_4
																		a_1	a_2			1		f_4
												-1						1		-1				$-d_{14}$
													-1						1		-1			$-d_{24}$
v_{11}	v_{21}	$c_{11}b_1$	$c_{21}b_2$			v_{12}	v_{22}	$c_{12}b_1$	$c_{22}b_2$			v_{13}	v_{23}	$c_{13}b_1$	$c_{23}b_2$			v_{14}	v_{24}	$c_{14}b_1$	$c_{24}b_2$			z(min) cost

The material-balance equations state that industry i's output either goes to other industries or to final demand:

$$x_i - \sum_{j=1}^{n} a_{ij}x_j = d_i, \qquad i = 1, \ldots, n \tag{1}$$

or, in matrix form,

$$(I - A)x = d \tag{2}$$

where A is the matrix of coefficients (a_{ij}). Both A and d have nonnegative elements, and, if a solution to this $n \times n$ linear system exists, it is meaningful only if it has nonnegative components.

As stated above, the model has either a unique solution or no solutions. Wagner [6] has considered problems involving a choice among alternative solutions by introducing a time variable, yielding a dynamic Leontief model. Here, in each of a number of time periods, the economic system must sustain its own internal operation through exchange of items and meet final demands. It also has the opporunity to build new productive capacity for use in the future, and to store items for future use. Clearly, there could be many ways in which such a system might operate. In each period one must decide whether productive capacity is to be used to produce for final demand, to build new capacity, or to produce for stockpiling. In addition, there is a decision as to how present stockpiles are to be used. If final demands are not too high there are many alternatives in one time period, while otherwise most production must go to final demand. Of course, it may be that all demands cannot be met. In any case, the existing alternatives compound themselves over time. The introduction of an objective function permits selection of one (or more) of the possible programs as optimal.

As in the static case, an $n \times n$ matrix of input–output coefficients, A, is assumed known, perhaps empirically. In addition, there is available an $n \times n$ matrix of capital coefficients, B, where b_{ij} is the quantity of industry i's output required to build one unit of capacity for industry j. Both A and B may vary with time, indicating technological changes in the industries. Let

d_t = final demand in period t

c_t = production capacity existing at the end of period t, measured in units of x_t

x_t = production levels, period t

n_t = new capacity built in period t

s_t = level of stockpiles at the end of period t, measured in units of x_t

u_t = unused or surplus capacity, period t

The material-balance equation for period t expresses the fact that the production in that period, x_t, goes either to other production activities, to final demand, to build new capacity, or to change the levels of stockpiles:

$$x_t = Ax_t + d_t + Bn_t + s_t - s_{t-1}, \qquad t = 1, \ldots, T \tag{3}$$

The capacity existing at the end of t is that existing at the end of $t - 1$ plus new capacity added in t;

$$c_t = c_{t-1} + n_t \tag{4}$$

or

$$c_t = c_0 + \sum_{\tau=1}^{t} n_\tau \tag{5}$$

Production in t must not exceed capacity to produce in t,

$$x_t \leq c_t$$

or, introducing the slack vector u_t,

$$x_t + u_t = c_t \tag{6}$$

Finally, all quantities, by their nature, must be nonnegative,

$$d_t, c_t, x_t, n_t, s_t, u_t \geq 0 \tag{7}$$

To examine the structure of these relations, let the following quantities be given; initial capacity, c_0, the initial level of stockpiles, s_0, and a set of final demands, $d_1, d_2, \ldots, d_T$. Then, using (5) to eliminate the variables c_t, (3) and (6) may be written in detached coefficient form as shown in Table 1. This matrix has a block-triangular structure, similar to the staircase form discussed in Section 2.3 but more complex. There are still diagonal blocks coupled by storage activities. However, since capacity built in period t has implications for all future time periods, the column corresponding to n_t has nonzero entries in the constraints of periods $t, t + 1, \ldots, T$. If the number of stockpiling and capacity-building activities is small relative to the number associated with production and unused capacity, then, by fixing the levels of these activities, all coupling through off-diagonal blocks vanishes, and the coefficient matrix becomes block diagonal. This is called a "weakly coupled" system, with few nonzero off-diagonal entries relative to diagonal elements.

Within this model, a number of interesting questions may be posed. One set deals with feasibility, i.e., the existence of solutions satisfying (3)–(6).

TABLE 1

Activities												
x_1	u_1	n_1	s_1	x_2	u_2	n_2	s_2	x_3	u_3	n_3	s_3	Items
$(I - A)$		$-B$	$-I$									$d_1 - s_0$
I	I	$-I$										c_0
			I	$(I - A)$		$-B$	$-I$					d_2
		$-I$		I	I	$-I$						c_0
							I	$(I - A)$		$-B$	$-I$	d_3
		$-I$				$-I$		I	I	$-I$		c_0
											↘	↓
		$-I$				$-I$				$-I$		
		↓				↓				↓		

The simplest of these asks whether, given initial capacity c_0 and initial stockpiles s_0, a certain set of final demands $d_1, \ldots, d_T$ can be met. The answer is certainly affirmative if initial capacity suffices to meet all future demands. Then no new capacity or stockpiles are needed, and setting $n_t = s_t = 0$, $t = 1, \ldots, T$ in (3) yields

$$x_1 = (I - A)^{-1}(d_1 - s_0) \tag{8}$$

$$x_t = (I - A)^{-1}d_t, \qquad t = 2, \ldots, T \tag{9}$$

which is feasible if u_t in (6) is nonnegative, i.e., if

$$c_0 \geq (I - A)^{-1}(d_1 - s_0) \tag{10}$$

$$c_0 \geq (I - A)^{-1}d_t, \qquad t = 2, \ldots, T \tag{11}$$

Final demands cannot be met if initial capacity and stockpiles cannot satisfy even the first period requirements, i.e., if

$$c_0 \ngeq (I - A)^{-1}(d_1 - s_0) \tag{12}$$

The most interesting and difficult case is when (10) is satisfied but (11) is not for some $t \geq 2$. Then the economy must build capacity and/or stockpiles to meet future demands. Phase 1 of the simplex method may be used here to find a feasible schedule of actions or to determine that none exists.

Other feasibility problems deal with the maximum level of final demands in one or more periods that can be satisfied. Assume, for example, that one wishes to increase satisfied final demands for the outputs of industries 1 and 3 in period t and that increases shall be in the ratio of 2:1. Adding a new activity vector to the model, with a minus two in position 1, a minus one in position 3, and zeros elsewhere and maximizing the level of this activity answers this question. If various increases are desired in different time periods, and if the relative values of meeting demands in these time periods are assigned, then the problem is solved by adding new activity vectors in each period, using the relative values as cost coefficients and maximizing the summed objective function over time. A similar problem might arise in defense planning. Suppose that minimum levels of civilian final demand are set, and the economic planners wish to allocate the remaining economic resources to maximize, say, munitions production. First a new activity vector is formed, giving the amounts of each industry's output required to produce one unit of munitions. This activity is introduced on the right-hand side of (3). Then, if the total amount of munitions produced over T periods is to be maximized, a price of unity is associated with each activity, and the resulting objective form maximized. If varying priorities were placed on obtaining munitions in different time periods, prices other than unity could be used.

Other similar problems may be posed, a number of which are discussed in [6]. In any of these, if a feasible solution exists it may not be unique, since the model has more unknowns than equations. In these cases one of the solutions may be selected as optimal by defining an objective function. Suppose, for example, that it is desired to minimize the number of labor hours required to satisfy a given final demand schedule. Let h_{1t} be a vector whose components give the number of labor hours required to produce one unit of each product in period t and h_{2t} a similar vector giving the number of labor hours needed to build one unit of capacity for the various industries in t. Then the objective function to be minimized would be

$$z = \sum_t (h_{1t}x_t + h_{2t}n_t) \tag{13}$$

Alternatively, in the period prior to initiation of the plan, the initial capacity, c_0, might be controllable. Assume that it is known that the present initial capacity is such that there is no feasible program which will meet all demands. The problem then might be to minimize the cost of initial capacity needed for feasibility. This may be dealt with by moving c_0 over to the left side of Table 1 and letting the objective function be

$$z = \nu_0 c_0 \tag{14}$$

where ν_0 is a vector of cost coefficients relative to c_0. Such a problem could arise during a prewar mobilization period, where the government desires to know the amount of new capacity which should be built in order to support future production requirements.

2.5 Angular and Dual-Angular Structures

Consider a company having two divisions, producing two and three products, respectively. Each has its own internal resources for production, e.g., labor and machines. The divisions are coupled by the fact that there is a shared resource which both use, for example, a raw material of limited availability. Let division 1 have limited amounts b_1 and b_2 of its internal resources, and division 2 have similar availability b_3. If the limitation on the common resource is b_0, then the problem is as follows:

$$\text{minimize } z = c_1x_1 + c_2x_2 + d_1y_1 + d_2y_2 + d_3y_3$$

$$\begin{aligned}
\text{subject to} \quad & \boxed{a_{11}x_1 + a_{12}x_2 + a_{13}y_1 + a_{14}y_2 + a_{15}y_3} & \leq b_0 \\
& \boxed{\begin{matrix} a_{21}x_1 + a_{22}x_2 \\ a_{31}x_1 + a_{32}x_2 \end{matrix}} & \begin{matrix} \leq b_1 \\ \leq b_2 \end{matrix} \\
& \qquad\qquad \boxed{a_{41}y_1 + a_{42}y_2 + a_{43}y_3} & \leq b_3 \\
& x_i \geq 0, \qquad y_i \geq 0
\end{aligned} \tag{1}$$

where x_i and y_i are the quantities of the various products produced by the divisions and the a_{ij} are the quantities of resource i required to produce one unit of product j.

These constraints consist of two independent subsets of inequalities, each pertaining to internal operation of one division, and one coupling constraint, introduced by the common resource. This situation generalizes to programs having the structure

$$\text{minimize } z = c_1x_1 + c_2x_2 + \cdots + c_px_p$$

$$\begin{aligned}
\text{subject to} \quad & A_1x_1 + A_2x_2 + \cdots + A_px_p = b_0 \\
& B_1x_1 \qquad\qquad\qquad\qquad\quad = b_1 \\
& \qquad\quad B_2x_2 \qquad\qquad\qquad = b_2 \\
& \qquad\qquad\qquad \ddots \\
& \qquad\qquad\qquad\qquad B_px_p = b_p \\
& x_1 \geq 0, x_2 \geq 0, \ldots, x_p \geq 0
\end{aligned} \tag{2}$$

where the x_i and b_i are now column vectors. The matrices B_i arise from constraints on items input and output only by the activities x_i, while the items b_0 are shared by all activities, with input–output matrices A_i. Such a structure is called angular. If the number of common constraints is small relative to the total, the system can be said to be weakly coupled, and a decomposition approach would appear to be attractive.

Often a linear program will be large only because it deals with similar subsystems, repeated either in time or in location, with the subsystems having relatively few items in common. Thus angular structures occur frequently in large problems. Examples include multiplant production and distribution problems, where the coupling constraints may arise from common budget limitations on the plants, from common capital used for expansion, or from demands for products whose production involves more than one plant.

It is interesting to note that the constraint matrix of the Hitchcock–Koopmans transportation problem may be partitioned into angular form as shown below:

$$
\begin{array}{cccccccccccccc}
x_{11} & \cdots & & x_{12} & \cdots & & x_{13} & \cdots & & x_{14} & \cdots & & & \\
\hline
1 & & & 1 & & & 1 & & & 1 & & & & b_1 \\
& 1 & & & 1 & & & 1 & & & 1 & & = & b_2 \\
& & 1 & & & 1 & & & 1 & & & 1 & & b_3 \\
\hline
1 & 1 & 1 & & & & & & & & & & = & a_1 \\
& & & 1 & 1 & 1 & & & & & & & = & a_2 \\
& & & & & & 1 & 1 & 1 & & & & = & a_3 \\
& & & & & & & & & 1 & 1 & 1 & = & a_4
\end{array}
\qquad (3)
$$

Here the origin availabilities play the role of the divisional constraints while the destination requirements couple these together. (This may be reversed by rearranging the columns.) Techniques for solving problems with angular structure are considered in Chapter 3.

The dual of the angular problem (2) is

$$
\text{maximize } b_1'\pi_1 + b_2'\pi_2 + \cdots + b_p'\pi_p + b_0'\pi
$$

$$
\begin{array}{llllll}
\text{subject to} & B_1'\pi_1 & & & + A_1'\pi \le c_1' \\
& & B_2'\pi_2 & & + A_2'\pi \le c_2' \\
& & & \ddots & \qquad \vdots \\
& & & B_p'\pi_p & + A_p'\pi \le c_p'
\end{array}
\qquad (4)
$$

This is a block-diagonal problem with coupling variables, π, replacing the coupling constraints of (2). Such structures, called dual angular, are weakly coupled if the number of coupling variables is substantially less than the total number of variables associated with the diagonal blocks. Since a solution to the dual of a linear program is readily obtained if a (basic) solution to the primal is available, methods for solving dual-angular problems can be applied to angular systems and vice versa.

Dual-angular problems arise from situations in which the bulk of the activities may be partitioned into subsets which input or output disjoint subsets of items. The remaining activity levels comprise the coupling variables. An important example is the production and inventory problem of Section 2.2, structured there in staircase form. The only coupling variables are the levels of the storage activities, and the system may be restructured in dual-angular form by moving the columns corresponding to these activities to the right, as shown in Table 1. Any staircase structure may be so rearranged. A similar comment applies to general block-triangular systems, such as the dynamic Leontief model of Section 2.4. Thus procedures for dual-angular structures are immediately applicable to a wide variety of problems involving planning over time.

Large linear programs may also result when uncertainty is introduced in a decision problem. The following case gives rise to a program of dual-angular structure.

Consider a problem of stochastic linear programming with recourse. That is, we have a set of linear constraints

$$Ax + By = b \tag{5}$$

$$x \geq 0, \qquad y \geq 0 \tag{6}$$

in which b is a random vector. The following sequence of events is assumed: First x is chosen, then "nature" chooses b according to a known probability distribution, and finally the decision maker chooses $y \geq 0$ to satisfy (5). The vector y is a "recourse vector" which the decision maker uses to eliminate infeasibilities in the relations

$$Ax = b, \qquad x \geq 0 \tag{7}$$

caused by the random choice of b. It is assumed that the x chosen always permits a feasible choice of y. That is, the set

$$F = \{x \mid \text{the relations } By = b - Ax,\ y \geq 0 \text{ are feasible for all possible } b\} \tag{8}$$

is nonempty, and the choice of x must be from F. Let $c'x$ and $d'y$ be the cost of choosing x and y, respectively. For given x, and after b has been observed, the decision regarding y is a deterministic problem, and y is obviously chosen to solve the problem

$$\text{minimize } \{d'y \mid By = b - Ax,\ y \geq 0\}$$

TABLE 1

Production and Inventory Problem in Dual-Angular Form

x_{11}	x_{21}	s_{11}	s_{21}	x_{12}	x_{22}	s_{12}	s_{22}	x_{13}	x_{23}	s_{13}	s_{23}	x_{14}	x_{24}	s_{14}	s_{24}	y_{11}	y_{21}	y_{12}	y_{22}	y_{13}	y_{23}	y_{14}	y_{24}	Item flows
b_1	b_2		1																					l_1
		1														a_1	a_2							f_1
−1																1								$y_{10} - d_{11}$
	−1																1							$y_{20} - d_{21}$
				b_1	b_2		1																	l_2
						1												a_1	a_2					f_2
				−1												−1		1						$-d_{12}$
					−1												−1		1					$-d_{22}$
								b_1	b_2		1													l_3
										1										a_1	a_2			f_3
								−1										−1		1				$-d_{13}$
									−1										−1		1			$-d_{23}$
												b_1	b_2		1									l_4
														1								a_1	a_2	f_4
												−1								−1		1		$-d_{14}$
													−1								−1		1	$-d_{24}$
$c_{11}b_1$	$c_{21}b_2$			$c_{12}b_1$	$c_{22}b_2$			$c_{13}b_1$	$c_{23}b_2$			$c_{14}b_1$	$c_{24}b_2$			v_{11}	v_{21}	v_{12}	v_{22}	v_{13}	v_{23}	v_{14}	v_{24}	z (min)

The above minimum depends on b and x and is denoted by $\phi(x, b)$. Since, in the first stage, b is random, $\phi(x, b)$ is a random variable for each x. The choice of x is made to minimize $c'x$ plus the expected value of $\phi(x, b)$:

$$\underset{x \in F}{\text{minimize}} \left\{ c'x + E_b \min_{y} \{d'y \mid By = b - Ax, y \geq 0\} \right\} \tag{9}$$

The function $E_b\phi(x, b)$ is generally nonlinear. The two-stage problem (9) is considered in Dantzig [7], where $\phi(x, b)$ and $E_b\phi(x, b)$ are shown to be convex.

A certain special case, however, gives rise to a linear program. Let b be drawn from a finite set $\{b_1, \ldots, b_n\}$ with probabilities $p_1, \ldots, p_n$. If we let y_i represent the second-stage variable when $b = b_i$, then $x \in F$ if and only if the following system is feasible:

$$\begin{aligned} Ax + By_1 \qquad\qquad\qquad &= b_1 \\ Ax \qquad + By_2 \qquad\quad &= b_2 \\ \vdots \qquad\qquad\qquad\qquad\quad &\;\;\vdots \\ Ax \qquad\qquad\quad + By_n &= b_n \\ x \geq 0, \quad y_i \geq 0, \quad &\text{all } i \end{aligned} \tag{10}$$

The first and second stage variables may be selected simultaneously by minimizing

$$z = c'x + p_1 d'y_1 + \cdots + p_n d'y_n \tag{11}$$

subject to (10). This is a dual–angular problem, which can be quite large if A has more than a few rows and n is large. In [7] and [8], Dantzig suggests that the problem be handled by taking its dual and applying the Dantzig–Wolfe decomposition procedure (see Chapter 3). Rosens partitioning method of Section 5.4 may also be used. Whatever the choice of algorithm, additional efficiency may be obtained by taking advantage of the repetitive appearance of A and B.

As an example, consider the following inventory problem. A known quantity of goods, q, is stored at a warehouse. An amount s is to be shipped to a destination, with shipping cost $c_s s$, where a demand, d, then occurs. If $d > s$, a penalty cost $p(d - s)$ is incurred, which may come about, for example, by outside purchase of the goods at price p. For convenience, we let

$$u = d - s, \qquad d > s \tag{12}$$

If $s > d$, a quantity

$$e = s - d, \qquad s > d \tag{13}$$

must be stored at the destination, at a cost $c_{st}e$. For known demand, d, the problem may be formulated as a linear program, of the form

$$\begin{aligned} s + y \qquad\qquad &= q \\ s \qquad + u - e \quad &= d \\ z = c_s s \qquad + pu + c_{st}e &\to \min \end{aligned} \tag{14}$$

All variables must be nonnegative, and the condition $ue = 0$ must be guaranteed, in order that u and e satisfy their defining relations (12)–(13). Since any basic solution has $ue = 0$, this is no problem.

If d is random, the problem is of the form discussed previously, where not all components of b are random. Since demand can occur only in integral units and very large demands are unlikely, it is reasonable to assume that d is drawn from a finite set $\{d_1, \ldots, d_n\}$ with probabilities $p_1, \ldots, p_n$. Then the two-stage problem is equivalent to the following dual-angular program:

$$\text{minimize } c_s s + p_1(pu_1 + c_{st}e_1) + \cdots + p_n(pu_n + c_{st}e_n)$$

$$\text{subject to} \quad \begin{array}{llllll} s + y & & & & = q \\ s & + u_1 - e_1 & & & = d_1 \\ s & & + u_2 - e_2 & & = d_2 \\ & & & & \vdots \\ s & & & + u_n - e_n & = d_n \\ \end{array} \tag{15}$$

$$s \geq 0, \qquad u_i \geq 0, \qquad v_i \geq 0, \qquad \text{all } i$$

Solution procedures for dual-angular programs are discussed in Chapter 5.

A union of angular and dual-angular structure leads to problems with both coupling constraints and coupling variables:

$$\text{minimize } z = c_1x_1 + c_2x_2 + \cdots + c_px_p + cy$$

$$\text{subject to} \quad \begin{array}{lllll} A_1x_1 + A_2x_2 + & \cdots + A_px_p & + D_0y & = b_0 \\ B_1x_1 & & + D_1y & = b_1 \\ & B_2x_2 & + D_2y & = b_2 \\ & \ddots & \vdots & \\ & & B_px_p + D_py & = b_p \end{array} \tag{16}$$

$$x_i \geq 0, \qquad y \geq 0$$

A solution procedure for such problems is considered in Section 5.3.

2.6 Linear Programs with Many Rows or Columns

A linear program may be large due to its constraint coefficient matrix having either many rows or many columns or both. In many cases, only one of these is large enough to cause real difficulty. Further, if both rows and columns are numerous, the Dantzig–Wolfe decomposition principle (see Chapter 3) may be used to reduce the number of rows at the expense of many more columns. In almost all cases, the set of all rows or columns has a great deal of structure. For example, in the decomposition principle, the jth column of the master program has the form Ax^j, with x^j the jth extreme point of a bounded convex polyhedron. Other examples are considered in detail in Chapter 4.

One wishes to solve such problems without having to examine all rows or columns. To see how this may be done, consider the operations involved in either the simplex or dual simplex methods. The primal–dual pair of problems to be solved is

Primal		*Dual*	
	minimize $c'x$		maximize $b'\pi$
subject to	$Ax = b$	subject to	$A'\pi \leq c$
	$x \geq 0$		

In either the primal or dual simplex methods, the following three steps are performed:

1. Select a column of A, column s.
2. Select a row of A, row r.
3. Perform a pivot operation, using a_{rs} as pivot element, and return to steps 1 and 2.

The pivot operation need not be performed on the entire matrix A, which is $m \times n$ with $n > m$. As in the revised simplex method, only a nonsingular $m \times m$ matrix, B, need be manipulated, and it is B^{-1} that is updated by the pivot operations.

Let

$$\begin{aligned} \pi = c_B B^{-1} &= \text{simplex multipliers associated with } B \\ \bar{b} = B^{-1}b &= \text{values of primal basic variables} \\ p^j &= \text{column } j \text{ of } A \\ \bar{c}_j = c_j - \pi p^j &= \text{relative cost factor for } x_j \\ \bar{a}_{ij} &= (i, j) \text{ element of } B^{-1}A \end{aligned}$$

Then operations 1 and 2 in primal and dual simplex methods are performed as indicated in Table 1.

TABLE 1

	Primal simplex method	Dual simplex method
(1) Optimality criterion	$\bar{c}_j \geq 0$, all j	$\bar{b}_i \geq 0$, all i
(2) Rule for selecting variable to enter basic set	If $\bar{c}_s = \min \bar{c}_j < 0$, x_s enters (pivot in column s)	If $\bar{b}_r = \min \bar{b}_i < 0$, $\bar{c}_r$ enters in the dual and x_r leaves in the primal (pivot in row r)
(3) Rule for selecting variable to leave basic set	If $\bar{b}_r/\bar{a}_{rs} = \min_{\bar{a}_{is}} > 0\ \bar{b}_i/\bar{a}_{is}$, x_r leaves (pivot in row r) If all $\bar{a}_{is} \leq 0$, primal is unbounded	If $-\bar{c}_s/\bar{a}_{rs} = \min_{\bar{a}_{ri}} < 0\ \bar{c}_i/-\bar{a}_{ri}$, $\bar{c}_s$ leaves in the dual and x_s enters in the primal (pivot in column s) If all $\bar{a}_{ri} \geq 0$, primal is infeasible
(4) Pivot element	$\bar{a}_{rs}$	$\bar{a}_{rs}$

In the primal simplex method the order of operations is 1, 2, 3. The minimization in row (3) of Table 1 is necessary to maintain primal feasibility. That in row (2) is not necessary, only desirable (see Section 4.1 for evidence of its desirability), since introducing any x_j for which $\bar{c}_j < 0$ will, barring degeneracy, cause the primal objective to decrease. In the dual simplex method, the order of operations is 2, 1, 3, the minimization in row (3) is necessary to maintain dual feasibility, and that in row (2) is desirable but not necessary.

In the standard application of either method, these minimization operations are performed by computing all the numbers involved and comparing. For problems with many rows or columns, this is not practical. Instead, as pointed out by Gomory [9], one of the minimizations is performed using a minimization algorithm, e.g., linear programming and dynamic programming. By solving a suitably constructed subproblem, the algorithm generates that column or row of the tableau required by the simplex or dual simplex method. Since generating a row of the dual tableau is equivalent to generating a column of the primal, such procedures will be called column generation. Depending on which minimization is done by the algorithm, the quantity to be minimized is either linear in the problem data or is a ratio of linear functions. Although the latter problem is not as well known as the former, it is easily solved by a slight modification of the simplex method if the feasible set is polyhedral (see Section 3.11.1).

If the problem to be solved has many columns, it can be regarded as the primal of the pair (1)–(2) and attacked by the primal or dual simplex methods, using linear or ratio minimization, respectively. If it has many rows, it is regarded as the dual, (2), and similar comments apply. In what follows, we outline briefly a number of examples, which serve both to illustrate the general procedure and have interest in themselves.

Solution of Linear Optimal Control Problems Using Column Generation. Consider a linear, continuous time, dynamic system whose state at time t is described by a vector $x(t) = (x_0(t), \ldots, x_n(t))$. The system evolves according to the time-invariant differential equations

$$\frac{dx}{dt} = Ax + Bu, \qquad 0 \le t \le T \tag{1}$$

$$x(0) = (0, x_1^0, \ldots, x_n^0) \text{ given} \tag{2}$$

where A and B are $(n + 1) \times (n + 1)$ and $(n + 1) \times m$ constant matrices, respectively. The vector $u = (u_1, \ldots, u_m)$ is composed of control functions defined over $[0, T]$, which can be chosen freely, subject to the restriction that the value of u at time t, $u(t)$, must lie in a compact convex set, $S(t)$:

$$u(t) \in S(t), \qquad 0 \le t \le T \tag{3}$$

Assume for simplicity that the last n components of the state at $t = T$ are prescribed

$$x_i(T) = x_i^T, \qquad i = 1, \ldots, n \tag{4}$$

The problem of (open-loop) optimal control of the system is to choose u to minimize $x_0(T)$. If, for example,

$$\frac{dx_0}{dt} = 1, \qquad x_0(0) = 0 \tag{5}$$

then $x_0(T) = T$ and the problem is to transfer the system between two given points in minimal time.

Dantzig [10] has shown that linear programming can be used to solve such problems. The differential equations (1) may be solved in closed form, yielding

$$x(T) = e^{AT}x(0) + \int_0^T e^{A(T-t)}Bu(t)\,dt \tag{8}$$

Let

$$p = \int_0^T e^{A(T-t)}Bu(t)\,dt \tag{9}$$

If

$$R = \left\{p \mid p = \int_0^T e^{A(T-t)}Bu(t)\,dt,\ u \in S\right\} \tag{10}$$

then, since S is convex and (9) involves only linear operations, it is easily shown that R is convex. Assume that we have available a set of control vectors, $u^i \in S$, $i = 1, \ldots, q$, $q \geq n + 1$, generating a set of vectors $\{p^i\}$. By convexity, the convex hull of $\{u^i\}$ is contained in S, and the convex hull of $\{p^i\}$ is in R. Assume further that some element in the convex hull of $\{u^i\}$, u_f, generates a vector p_f satisfying (8); i.e., we have an initial feasible control. Then it is easy to construct a feasible linear program which finds that control in the convex hull of the set $\{u^i\}$ which minimizes $x_0(T)$:

$$\text{minimize } x_0(T)$$

subject to

$$e^{AT}x(0) + \sum_i \lambda_i p^i = x(T) \tag{11}$$

$$\sum_i \lambda_i = 1 \tag{12}$$

$$\lambda_i \geq 0$$

Assume that, at the kth iteration, this program has been solved, using the revised simplex method, yielding an optimal basis, B^k, an optimal solution $x_0^k(T)$, $\{\lambda_i^k\}$, and a vector of simplex multipliers (π^k, π_0^k), where π^k is an $n + 1$ vector and π_0^k is a scalar. The control yielding the value $x_0^k(T)$ is

$$u^k = \sum_i \lambda_i^k u^i \tag{13}$$

This is an approximation to the optimal control. Suppose we wish to refine this approximation by adding another point to the initial "grid" of points, $\{u^i\}$. To see which new point is best, imagine that a very large collection of points is selected, and the vectors p^i calculated for each. The constraints (11) now have many columns, p^i, with a variable λ_i for each. Pricing out a generic column using the multipliers (π^k, π_0^k) yields a relative cost factor

$$\bar{c}_i = -\pi^k p^i - \pi_0^k \tag{14}$$

which is to be minimized. Since p^i was a general column generated by any $u^i \in S$, minimizing $\bar{c}_i$ is equivalent to maximizing

$$\int_0^T \pi^k e^{A(T-t)} Bu(t)\, dt \tag{15}$$

subject to

$$u \in S \tag{16}$$

Clearly this is accomplished by maximizing the integrand for each t:

$$\left.\begin{array}{ll} & \text{maximize } \lambda^k(t)Bu(t) \\ \text{subject to} & u(t) \in S(t) \end{array}\right\} \begin{array}{l}\text{subproblem} \\ \text{(one for each } t)\end{array} \qquad \begin{array}{r}(17)\\(18)\end{array}$$

where

$$\lambda^k(t) = \pi^k e^{A(T-t)} \tag{19}$$

These subproblems replace step 1 in the primal simplex method. If $S(t)$ is polyhedral, the subproblem is a linear program for each t, and if $S(t)$ is independent of t, then only the objectives of these programs vary. If $S(t)$ is determined by upper and lower bounds on each component of u, then a "bang-bang" solution is obtained, with each u_i on one bound or the other, switching when the ith component of $\lambda^k(t)B$ changes sign.

Let $\hat{u}^{k+1}$ solve the subproblems. The condition min $\bar{c}_i < 0$ here becomes

$$-\int_0^T \pi^k e^{A(T-t)} B\hat{u}^{k+1}(t) - \pi_0^k < 0 \tag{20}$$

If the above is satisfied, then, barring degeneracy, introduction of the new column $\hat{p}^{k+1}$ will decrease $x_0(T)$. If not, the current solution can be proved optimal (see Section 4.4.1). In [11], Dantzig has given a convergence proof for this procedure as the number of iterations tends to infinity.

A number of points should be noted. There are an uncountable infinity of possible columns, yet column generation permits a computationally feasible (although not finite) algorithm to be developed. If, at optimality, $\lambda_{j_1}, \ldots, \lambda_{j_{n+1}}$ are basic with values $\lambda_{j_i}^*$, the optimal solution, u^*, is a convex combination of $n+1$ basic proposals:

$$u^* = \sum_{i=1}^{n+1} \lambda_{j_i}^* u^{j_i} \tag{21}$$

Since the column generated by each u^{j_i} is in the final basis, these columns price out to zero:

$$\int_0^T \pi^* e^{A(T-t)} B u^{j_i}\, dt = -\pi_0^*, \qquad i = 1, \ldots, n+1 \tag{22}$$

If

$$\lambda^* = \pi^* e^{A(T-t)} \tag{23}$$

then, by (21) and (22),

$$\int_0^T \lambda^* B u^*\, dt = -\pi_0^* \tag{24}$$

Since (π^*, π_0^*) are the simplex multipliers at optimality, they must cause all possible columns to price out nonnegative. In terms of the subproblem (15)–(16) this is equivalent to

$$\max_{u \in S} \int_0^T \lambda^* Bu \, dt \leq -\pi_0^* \tag{25}$$

Thus, by (24) and (25),

$$\max_{u \in S} \int_0^T \lambda^* Bu \, dt = \int_0^T \lambda^* Bu^* \, dt \tag{26}$$

or

$$\lambda^*(t) Bu^*(t) = \max_{u(t) \in S(t)} \lambda^*(t) Bu(t) \tag{27}$$

almost everywhere in the interval $[0, T]$. This is the statement of Pontryagin's maximum principle [12] for this problem. We see here that it can be derived using concepts from linear programming.

The procedures discussed here are considered in more detail in Sections 4.3 and 4.4.

Program with Many Rows. Consider a linear program whose constraints are partitioned into two subsets

$$\text{minimize } cx \tag{28}$$

subject to

$$A_1 x = b_1 \} m_1 \text{ rows} \tag{29}$$

$$A_2 x = b_2 \} m_2 \text{ rows} \tag{30}$$

$$x \geq 0$$

Assume that the polyhedron

$$S = \{x \mid A_2 x = b_2, \, x \geq 0\} \tag{31}$$

is bounded, and let x^j be the jth extreme point of S. If an arbitrary element of S, x, is expressed as

$$x = \sum_j \lambda_j x^j, \; \sum_j \lambda_j = 1, \qquad \lambda_j \geq 0 \tag{32}$$

(see Theorem 3.2-2) and (32) is substituted into (28) and (29), the following "master program" results:

$$\text{minimize} \sum_j (cx^j) \lambda_j \tag{33}$$

subject to
$$\sum_j (A_1 x^j)\lambda_j = b_1 \tag{34}$$
$$\sum_j \lambda_j = 1 \tag{35}$$
$$\lambda_j \geq 0 \tag{36}$$

This master program is equivalent to the original; i.e., if $\{\lambda_j^0\}$ solves the above, then

$$x^0 = \sum_j \lambda_j^0 x^j \tag{37}$$

solves (28)–(30). It has many columns but only $m_1 + 1$ rows, as compared with $m_1 + m_2$ rows of the original problem.

In Chapter 3 we show how to attack this program by column generation, using a primal method. Here we use a dual method. The dual of the master program is

$$\text{maximize } \pi b_1 + \pi_0 \tag{38}$$

subject to
$$\pi A_1 x^j + \pi_0 \leq (cx^j), \qquad \text{all } j \tag{39}$$

This has only $m_1 + 1$ variables but many inequalities. Abadie and Williams [13] have shown how to solve the pair (33)–(36) and (38)–(39) by using the dual simplex method on (33)–(36). Assume that an initial basic and complementary slack solution to the dual is available. This would be the case in a problem already solved whose solution is desired for another right hand side. The only difficulty occurs in step 1 mentioned earlier, where a pivot column must be chosen. Assume that the pivot element is in row r, let B be the current (infeasible) basis for the master program, and let b_r be the rth row of B^{-1}. The dual simplex method requires that we find

$$\min_{\bar{a}_{rj} < 0} \frac{-\bar{c}_j}{\bar{a}_{rj}} \tag{40}$$

Let the vector b_r be partitioned

$$b_r = (\hat{b}_r, b_r^0)$$

with $\hat{b}_r$ an m_1 vector and b_r^0 a scalar. Then, for the master program,

$$\bar{a}_{rj} = (\hat{b}_r A_1)x^j + b_r^0 \tag{41}$$
$$-\bar{c}_j = (\pi A_1 - c)x^j + \pi_0 \tag{42}$$

so we must minimize the ratio

$$\frac{(\pi A_1 - c)x^j + \pi_0}{(\hat{b}_r A_1)x^j + b_r^0} \tag{43}$$

over all extreme points, x^j, of S for which the denominator is negative. It is shown in Section 3.11.1 that if a ratio of linear functions attains a finite minimum over a convex polyhedron, there is an extreme point which is optimal. Thus the desired extreme point solves the subproblem

$$\text{minimize } \frac{(\pi A_1 - c)x + \pi_0}{(\hat{b}_r A_1)x + b_r^0} \tag{44}$$

subject to

$$A_2 x = b_2 \tag{45}$$

$$x \geq 0 \tag{46}$$

$$(\hat{b}_r A_1)x + b_r^0 < 0 \tag{47}$$

Solution of this subproblem replaces step 1 of the dual simplex method, with all other computations proceeding as usual. If x^{j*} solves (44)–(47), the primal column (dual row) generated to enter the current basis is

$$\begin{bmatrix} A_1 x^{j*} \\ 1 \end{bmatrix} \tag{48}$$

with cost coefficient cx^{j*}. Finite algorithms for solving (44)–(47) are presented in Section 3.11.1. Other problems with many rows are discussed in Sections 4.5 and 7.3.

2.7 Nonlinear Programs with Coupling Variables

An important class of nonlinear programs has the form

$$\text{minimize } c_1(y)x_1 + c_2(y)x_2 + \cdots + c_p(y)x_p + c_0(y)$$

subject to

$$\begin{array}{llllll} A_1(y)x_1 & & & & \geq b_1(y) \\ & A_2(y)x_2 & & & \geq b_2(y) \\ & & \ddots & & \vdots \\ & & & A_p(y)x_p & \geq b_p(y) \\ & & & & Dy \geq \beta \end{array} \tag{1}$$

Here the cost coefficients c_i, constraint matrices A_i, and right-hand-side vectors b_i are all functions of a vector of coupling variables, y. If values are assigned to these coupling variables, the problem is linear in the vectors x_i

and has a block-diagonal constraint matrix, so it decomposes into p independent subproblems. Problems of this type are nonlinear generalizations of linear programs with dual-angular structure, reducing to that case when the A_i and c_i are independent of y and $c_0(y)$ and the $b_i(y)$ are linear. As in the linear case, such problems lend themselves to solution by partitioning methods, which vary first y, then solve linear subproblems to obtain new x_i, etc. Rosen has detailed a method of this type [14], which is discussed in Section 7.2.

Such problems arise frequently in scheduling the production and distribution of products in the petroleum industry, when overall optimization of a multirefinery complex is considered. An example is shown in Figure 2-3.

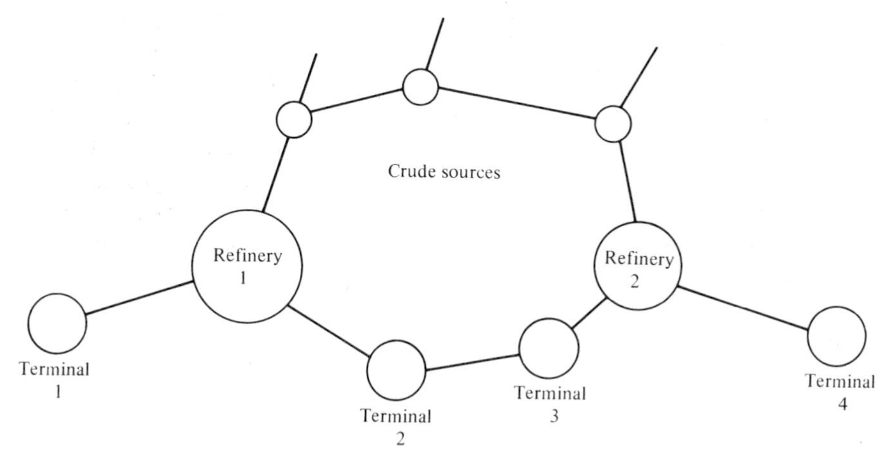

FIGURE 2-3 Multirefinery complex.

Each refinery contains a number of processing units (distillation, cracking, reforming, etc.) which transform the crude inputs into finished products (motor fuel, jet fuel, fuel oil, etc.). These are then sent to shipping terminals for pickup by customers. Typical decisions which must be made are:

1. How shall the various raw materials be allocated to each refinery, and to certain processing units within each refinery?
2. How shall the streams within each refinery be allocated to process units? This decision is complicated by the fact that certain intermediate streams in one refinery may be shipped to other refineries for further use.
3. What operating conditions shall be maintained in the various processing units? These include temperature, pressure, recycle rates, and catalyst activities. The dependence of stream properties on these variables is almost always highly nonlinear.
4. Which materials should be blended into each product in order that specifications on product properties are met? An example is the requirement

of a minimal octane number for motor fuels. Again, components to be blended may be shipped between refineries.

5. Which terminal product demands should be met from which refineries?

These decisions are generally interrelated. The transportation decisions in 5 depend not only on shipping costs, but also on the availabilities of the various products at each refinery (4) and on the relative efficiencies with which the products may be produced (1–3). Since the refineries often differ significantly in age, capacity, equipment, etc., the problem can be quite complex. The time span involved is assumed to be relatively long, e.g., a quarter or a half year, so that the intricacies of daily operations need not be considered.

To see how such a problem can lead to a nonlinear program such as (1), consider the following simplified example [15]. Two refineries, R_1 and R_2, produce products P_1 and P_2 for delivery to terminals T_1 and T_2. The cumulative demands for these products at each terminal over the planning interval are assumed known, the demand for P_i at T_j denoted by $D(P_i, T_j)$. At refinery R_1 three materials, M_1, M_2, and M_3, are available for blending into P_1 and P_2. In addition, M_1 may be transshipped from R_1 to R_2, where it is blended with M_4 and M_5. The system is shown schematically in Figure 2-4. Consider

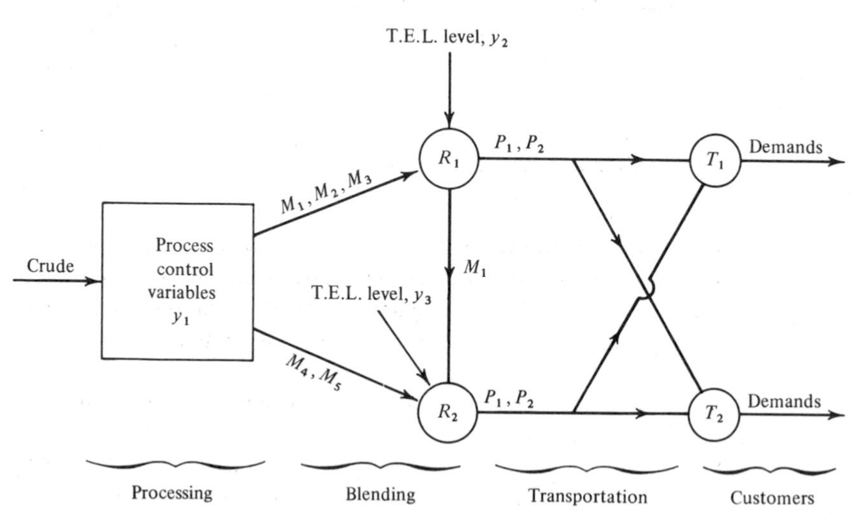

FIGURE 2-4 Processing and distribution system.

first the blending operation in refinery R_1. Products P_1 and P_2 must meet minimal octane number specifications s_1 and s_2. The octane number depends both on the volumes of each material present in the blend and (nonlinearly) on the concentration of tetraethyl lead (TEL) added, y_2. Let $f_i(y)$ be the octane number of material M_i when the TEL concentration is y cc/bbl, and let x_{ij} be the volume of material i blended to produce product j. It is usually assumed [16] that the contribution of M_i to the octane number of product j is the product of $f_i(y)$ with the fractional volume of M_i, $f_i(y)(x_{ij}/\sum_i x_{ij})$, and that the contributions of the various materials are additive. Thus, if s_j is the minimal octane number of product j, the octane constraint becomes

$$\sum_{i=1}^{3} f_i(y)\left(x_{ij}\Big/\sum_{i=1}^{3} x_{ij}\right) \geq s_j \tag{2}$$

which can be rewritten

$$\sum_{i=1}^{3} (f_i(y) - s_j)x_{ij} \geq 0 \tag{3}$$

The blending variables x_{ij} must also be constrained so that no more of a material is used than is available. The amounts available are determined by the processing units feeding the blending operation, thus by the control variables, y_1. If $b_i(y_1)$ is the availability of material i, then the constraints that no more of each material is used than is available are

$$x_{11} + x_{12} + (M_1 \text{ shipped to } R_2) \leq b_1(y_1) \tag{4}$$

$$x_{21} + x_{22} \leq b_2(y_1) \tag{5}$$

$$x_{31} + x_{32} \leq b_3(y_1) \tag{6}$$

In matrix form these restrictions, (3)–(6), become

P_1			P_2				
M_1	M_2	M_3	M_1	M_2	M_3		
1			1			$+ (M_1 \text{ to } R_2) \leq b_1(y_1)$	
	1			1		$\leq b_2(y_1)$	
		1			1	$\leq b_3(y_1)$	
$a_{11}(y_2)$	$a_{21}(y_2)$	$a_{31}(y_2)$				≥ 0	
			$a_{12}(y_2)$	$a_{22}(y_2)$	$a_{32}(y_2)$	≥ 0	(7)

where

$$a_{ij}(y) = f_i(y) - s_j \tag{8}$$

A similar set of constraints holds for blending in refinery 2:

P_1			P_2			
M_1	M_4	M_5	M_1	M_4	M_5	
1			1			$-(M_1 \text{ to } R_2) = 0$
	1			1		$\le b_4(y_1)$
		1			1	$\le b_5(y_1)$
$a_{11}(y_3)$	$a_{41}(y_3)$	$a_{51}(y_3)$				≥ 0
			$a_{12}(y_3)$	$a_{42}(y_3)$	$a_{52}(y_3)$	≥ 0 (9)

where y_3 is the concentration of TEL used at R_3. Note that the operation of shipping M_1 from R_1 to R_2 has introduced a variable common to the blending problems of these refineries, which is treated as a coupling variable. If three or four materials had been transshipped, it might be preferable to combine these two blocks into one, rather than introducing additional coupling variables. This is because the computing effort with Rosen's partitioning algorithm increases rapidly with the number of coupling variables [17]. The proper compromise in such situations is discussed in [15].

The constraints describing the transportation operation express the facts that demands must be met and that no more may be shipped than is available. In matrix form

(10)

T_1				T_2					
R_1		R_2		R_1		R_2			
P_1	P_2	P_1	P_2	P_1	P_2	P_1	P_2		
1				1				$-(P_1 \text{ from } R_1) = 0$	no more can
	1				1			$-(P_2 \text{ from } R_1) = 0$	be shipped
		1				1		$-(P_1 \text{ from } R_2) = 0$	than is
			1				1	$-(P_2 \text{ from } R_2) = 0$	available
1		1						$= D(P_1, T_1)$	demands
	1		1					$= D(P_2, T_1)$	must
				1		1		$= D(P_1, T_2)$	be met
					1		1	$= D(P_2, T_2)$	

To obtain a complete set of constraints, material-balance equations must be written expressing the facts that the volume of P_i sent from R_j to the terminals is equal to the total volume of the materials from which P_i was blended. Thus, for example,

$$(P_1 \text{ from } R_1) = x_{11} + x_{21} + x_{31} \tag{11}$$

These constraints serve to further couple the system together—the variables

(P_i from R_j) are additional coupling variables. When they are included in the blending constraints of refineries 1 and 2, the overall system appears as shown in Table 1. This has the form of (1), with five coupling variables and three diagonal blocks, each representing a distinct phase of system operation.

The objective function for the system is revenue minus cost. The revenue is fixed if all demands must be met exactly, so the problem is one of cost minimization. The costs are those of shipping, blending, and processing. If c_{ij} is the unit cost of shipping from R_i to T_j, total shipping costs are

$$c_s = \sum_{i,j} c_{ij}[(P_1R_iT_j) + (P_2R_iT_j)] \tag{12}$$

where $(P_iR_jT_k)$ represents the variable in that cell of the transportation array, (10). As an example of blending costs, if c is the cost per volume of TEL, then the total cost of TEL used is

$$c_b = \sum_{i,j} [c(\$/\text{cc}) \cdot y(\text{cc}/\text{bbl})]x_{ij}(\text{bbl}) \tag{13}$$

so the cost coefficient of the blending variable depends on y. We may lump all the processing costs together by defining functions $\phi_i(y_1)$ = cost of producing a volume $b_i(y_1)$ of material M_i when the control variable settings are y_1. The total processing cost then becomes

$$c_p = \sum_{i=1}^{5} \phi_i(y_1) \tag{14}$$

Adding (12)–(14) yields the total cost.

A realistic model would require a much more detailed representation of all operations, particularly the processing prior to blending. Problems of this form with up to 60 coupling variables, 6 blocks, and 500–600 inequality restrictions have been solved in the petroleum industry [15].

2.8 Mixed-Variable Programs and a Location Problem

The previous section dealt with problems which are linear and block diagonal when certain variables are fixed. A related class of problems includes some of these but also admits a number of important new problems. Elements of this class have the form

$$\text{minimize } c'x + f(y) \tag{1}$$

subject to

$$Ax + F(y) \geq b \tag{2}$$

$$x \geq 0, \qquad y \in S \tag{3}$$

TABLE 1

P_1			P_2		
M_1	M_2	M_3	M_1	M_2	M_3
1			1		
	1			1	
		1			1
1	1	1			
			1	1	1
$a_{11}(y_2)$	$a_{21}(y_2)$	$a_{31}(y_2)$			
			$a_{12}(y_2)$	$a_{22}(y_2)$	$a_{32}(y_2)$

Gasoline Blending at Refinery 1

P_1			P_2		
M_1	M_4	M_5	M_1	M_4	M_5
1			1		
	1			1	
		1			1
1	1	1			
			1	1	1
$a_{11}(y_3)$	$a_{41}(y_3)$	$a_{51}(y_3)$			
			$a_{12}(y_3)$	$a_4(y_3)$	$a_{52}(y_3)$

Gasoline Blending at Refinery 2

M_1 to R_2	P_1 from R_1	P_2 from R_1	P_1 from R_2	P_2 from R_2		*Description*
1					$\leq b_1(y_1)$	avail. of M_1
					$\leq b_2(y_1)$	avail. of M_2
					$\leq b_3(y_1)$	avail. of M_3
	-1				$= 0$	mat. bal., P_1
		-1			$= 0$	mat. bal., P_2
					≥ 0	spec. on P_1
					≥ 0	spec. on P_2
-1					$= 0$	M_1 avail. at R_2
					$\leq b_4(y_1)$	avail. of M_4
					$\leq b_5(y_1)$	avail. of M_5
			-1		$= 0$	mat. bal., P_1
				-1	$= 0$	mat. bal., P_2
					≥ 0	spec. on P_1
					≥ 0	spec. on P_2

T_1				T_2										
R_1		R_2		R_1		R_2								
P_1	P_2	P_1	P_2	P_1	P_2	P_1	P_2							
1				1					-1				$= 0$	P_1 from R_1
	1				1					-1			$= 0$	P_2 from R_1
		1				1					-1		$= 0$	P_1 from R_2
			1				1					-1	$= 0$	P_2 from R_2
1		1											$= D(P_1, T_1)$	demands
	1		1										$= D(P_2, T_1)$	demands
				1		1							$= D(P_1, T_2)$	demands
					1		1						$= D(P_2, T_2)$	demands

Shipping of Gasoline from Refineries to Terminals

The vector x has n components, A is an $m \times n$ constant matrix, and c is an n vector of constants, so if y is fixed, the problem is linear. The quantity $f(y)$ is a (possibly nonlinear) function of a p vector y, $F(y)$ is an m vector of such functions, and S is an arbitrary subset of E^p. By appropriately specifying these quantities, a number of important problems are obtained. If S is the set of all p vectors with nonnegative integral-valued components and F and f are linear, the problem is the well-known mixed continuous–discrete variable linear program. If $S = E^p$ or a convex polyhedron in E^p and f and F are allowed to be nonlinear, the problem is of the class studied in Section 7.2, where c and A are not allowed to depend on y. If A is block diagonal, f and F linear, and $S = E^p$ or $(E^p)^+$, the problem is a dual-angular linear program, as in Section 2.5. In all cases, it is important to note that the terms in y must appear additively.

The partitioning algorithm of Benders [18], described in Section 7.3, is specially constructed to deal with such problems. It solves (1)–(3) by choosing $y^* \in S$, solving the linear program

$$\text{minimize } c'x \tag{4}$$

subject to

$$Ax \geq b - F(y^*)$$

and using information from the dual solution of (4) to select a better y vector, or to determine that no better choice exists. By operating in this way, full advantage is taken of the partial linearity of the problem. This is particularly important if the matrix A has special structure, e.g., is block diagonal or is of the transportation type. Then the linear program (4) is relatively easy to solve. Such structure is usually lost in an algorithm which varies x and y simultaneously, since the structures of F and A generally get mixed together, and any special properties of A are lost.

As an example, consider a location problem. A finite number of sites are being considered by a firm for locating new plants and warehouses. These are to mesh in with the existing warehouse-plant complex in such a way that the total cost of transporting goods to customers (which can occur directly from plants or via warehouses) and of storage at warehouses is minimized. Such problems involve both discrete activities (whether or not to construct a facility) and continuous ones (how much to produce and ship).

A particular network is shown [19] in Figure 2-5. An older plant, P_0, serves customers c_1 through c_4. Construction of a new plant, P_n, and a new warehouse, W, is being considered. The numbers in the boxes and on the links of the network are unit transportation costs, capacities of the facilities, and requirements of the customers. The cost functions for the new plant and warehouse include a capital outlay for construction, represented by the discontinuity at the origin. Table 1 shows the costs of the various alternatives. There are four of these, corresponding to constructing (1) or not constructing

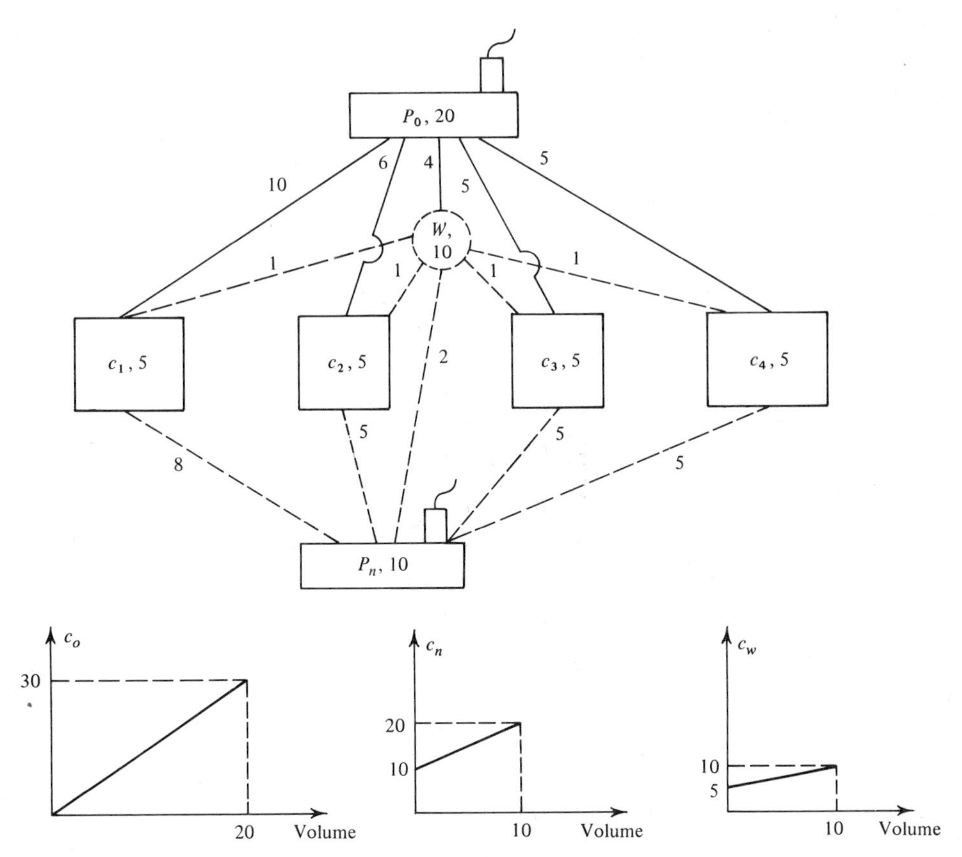

FIGURE 2-5 Production and distribution network.

(0) the two new facilities. To find the cost of each alternative, the minimal cost pattern of production and shipment is computed. For example, if only the new plant is constructed, it is used to capacity (cost = 20), the older less efficient plant produces only 10 units (cost = 15), the new plant satisfies c_1 and c_2 ($f_{n1} = f_{n2} = 5$, cost = 65), and the older plant satisfies c_3 and c_4 ($f_{03} = f_{04} = 5$, cost = 50), for a total cost of 150.

If there are only a few alternatives, this procedure of evaluating each in turn is probably best. However, the number of alternatives can become large very rapidly, owing either to more new facilities to be considered or to a more realistic representation of the cost of a facility, as illustrated in Figure 2-6. The discontinuities represent the capital investment required to progress from a smaller facility to the next size larger, while the decreasing slopes indicate economies of scale. The above facility operates in one of three distinct states, one for each of the linear segments. In problems like this, enumeration usually takes too long. Formulation as a mixed-variable problem of the form

TABLE 1. COSTS OF VARIOUS ALTERNATIVES

New plant	New warehouse	Volume P_0	Volume P_n	Volume W	Flows	Cost
0	0	20	0	0	$f_{01} = f_{02} = f_{03} = f_{04} = 5$	160
1	0	10	10	0	$f_{n1} = f_{n2} = f_{03} = f_{04} = 5$	150
0	1	20	0	10	$f_{0w} = 10, f_{w1} = f_{w2} = f_{03} = f_{04} = 5$	140
1	1	10	10	10	$f_{nw} = 10, f_{w1} = f_{w2} = f_{03} = f_{04} = 5$	125

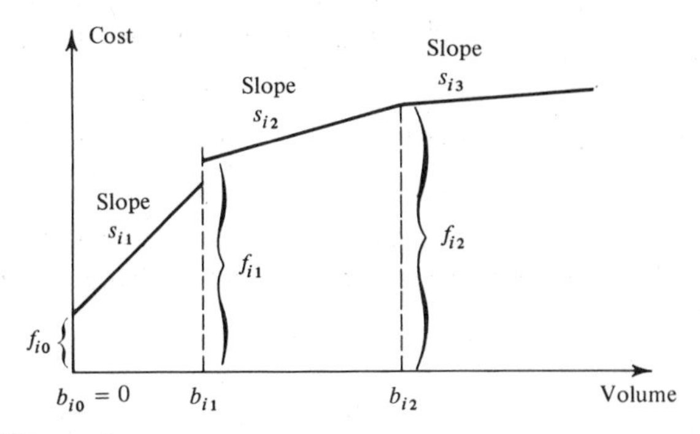

FIGURE 2-6 Discontinuous cost.

(1)–(3) permits an optimal solution to be obtained while (usually) examining only a small fraction of the alternatives.

To obtain this formulation, let

r_j = requirement of customer j
c_{ij} = per unit cost of shipping from location i to customer j
x_{ij} = number of units shipped from i to j
m_{ik} = number of units produced (stored) by plant (warehouse) i at marginal cost s_{ik}
I_p, I_w, I_c = sets of indices of plants, warehouses, and customers, respectively

The variables m_{ik} must be in the correct piecewise linear ranges and at most one can be positive for each i. To ensure this we define zero–one variables y_{ik} such that

$$y_{ik} = 0 \Rightarrow m_{ik} = 0 \tag{5}$$

$$y_{ik} = 1 \Rightarrow b_{i,k-1} \leq m_{ik} \leq b_{ik} \tag{6}$$

and at most one $y_{ik} = 1$ for each i. The following system of linear constraints accomplishes this:

$$y_{ik}b_{i,k-1} \leq m_{ik} \leq y_{ik}b_{ik} \tag{7}$$

$$\sum_k y_{ik} = 1, \qquad y_{ik} = 0 \text{ or } 1 \tag{8}$$

Since customer demands must be met,

$$\sum_i x_{ij} = d_j, \qquad j \in I_c \tag{9}$$

The variables x_{ij} and m_{ik} are related by the fact that we can ship only what is produced,

$$\sum_{k} m_{ik} = \sum_{j \in I_c \cup I_w} x_{ij}, \qquad i \in I_p \tag{10}$$

The warehouses are subject to conservation equations, stating that inflow equals outflow in any period (no storage allowed from one period to the next):

$$\sum_{i \in I_p} x_{ij} = \sum_{k \in I_c} x_{jk}, \qquad j \in I_w \tag{11}$$

Finally, flows into warehouses must not exceed the capacity of the warehouse,

$$\sum_{i \in I_p} x_{ij} \le \alpha_j, \qquad j \in I_w \tag{12}$$

The cost to be minimized is

$$z = \sum_{i,j} c_{ij} x_{ij} + \sum_{i,k} s_{ik} m_{ik} + \sum_{i,k} y_{ik}(f_{i,k-1} - s_{ik} b_{i,k-1}) \tag{13}$$

The first term above is the transportation cost; the second and third give the cost of constructing and operating the facilities.

The problem (7)–(13) is of the form (1)–(3), with F and f linear and S the set of all zero–one alternatives for the y_{ik}. When applied to this problem, Benders' algorithm of Section 7.3 iteratively solves first a linear program with fixed y_{ik}, then a linear integer program to obtain new y_{ik}. The dual solution of the linear program is used to add a new constraint to the integer program, leaving a smaller set of alternatives for future examination. The procedure thus can be viewed as an automated version of the case-study approach, in which information from previous cases is used in deciding which alternative to examine next.

Problems

1. Give an expression for e^{At} when A has distinct eigenvalues.
2. Prove that, if p is given by (2.4-9), then the set of all vectors p that can be generated by $u \in S$ is convex.
3. Develop a procedure for generating an initial set of vectors $\{p^i\}$, satisfying (2.4-11).
4. Consider parametric linear programs, with a single parameter in the objective or in the right-hand side. Show that, if the program has many columns, the portion of the standard parametric algorithm which chooses a new column to enter the basis can be replaced by a subproblem involving minimizing the ratio of linear functions over a convex polyhedron.

References

1. G. B. Dantzig, "Large-Scale System Optimization: A Review," Operations Research Center, University of California at Berkeley, Rept. *ORC 65–9*, 1965.
2. G. B. Dantzig and R. M. Van Slyke, "Generalized Upper Bounding Techniques," *J. Computer System Sci.*, **1**, 1967, pp. 213–226.
3. G. B. Dantzig, "Compact Basis Triangularization for the Simplex Method," in *Recent Advances in Mathematical Programming*, R. L. Graves and P. Wolfe, eds., McGraw-Hill, Inc., New York, 1963, pp. 125–132.
4. M. D. Mesarovic, D. Macko, and Y. Takahara, "Theory of Multi-Level Hierarchical Systems," Academic Press, 1970.
5. G. B. Dantzig, *Linear Programming and Extensions*, Princeton University Press, Princeton, N.J., 1963, chap. 3.
6. H. M. Wagner, "A Linear Programming Solution to Dynamic Leontief Type Models," *Management Sci.*, **3**, No. 3, 1957, pp. 234–254.
7. G. B. Dantzig, *Linear Programming and Extensions*, Princeton University Press, Princeton, N.J., 1963, chap. 25, sec. 3.
8. G. B. Dantzig and A. Madansky, "On the Solution of Two-Stage Linear Programs under Uncertainty," The RAND Corporation, *Paper P–2039*, 1960.
9. R. Gomory, "Large and Nonconvex Problems in Linear Programming," *Proc. Symp. Appl. Math.*, No. 15, 1963, pp. 125–139, also *IBM Rept. RC–765.*
10. G. B. Dantzig, "Linear Control Processes and Mathematical Programming," *J. SIAM Control*, **4**, No. 1, 1966, pp. 56–60.
11. G. B. Dantzig, *Linear Programming and Extensions*, Princeton University Press, Princeton, N.J., 1963, chaps. 22 and 24.
12. L. S. Pontryagin, et al., *The Mathematical Theory of Optimum Processes*, Wiley-Interscience, New York, 1962.
13. J. M. Abadie and A. C. Williams, "Dual and Parametric Methods in Decomposition," in *Recent Advances in Mathematical Programming*, R. L. Graves and P. Wolfe, eds., McGraw-Hill, Inc., New York, 1963.
14. J. B. Rosen, "Convex Partition Programming," in *Recent Adances in Mathematical Programming*, R. L. Graves and P. Wolfe, eds., McGraw-Hill, Inc., New York, 1963, pp. 159–176.
15. J. B. Rosen and J. C. Ornea, "Solution of Nonlinear Programming Problems by Partitioning," *Management Sci.*, **10**, No. 1, 1963, pp. 160–173.
16. G. Hadley, *Linear Programming*, Addison-Wesley Publishing Company, Inc., Reading, Mass., 1962, pp. 458–463.
17. J. C. Ornea and G. G. Eldredge, "Nonlinear Partitioned Models for Plant Scheduling and Economic Evaluation," *Proceedings of the A.I.Ch.E.–Inst. Chem. Eng. Joint Meeting, London, June 13–17, 1965*, Symposium on Application of Mathematical Models in Chemical Engineering Research, Design, and Production.
18. J. F. Benders, "Partitioning Procedures for Solving Mixed-Variables Programming Problems," *Numerische Mathematik*, **4**, 1962, pp. 238–252.
19. B. R. Buzby, "Nonlinear Distribution Problems," private communication, Dec. 1965.

3

The Dantzig-Wolfe Decomposition Principle

3.1 Introduction

The publication of the Dantzig–Wolfe decomposition principle in 1960 [1] initiated the extensive work on large-scale mathematical programming which has followed. The procedure is most efficient when applied to linear programs whose coefficient matrices have angular structure, i.e., one or more independent blocks linked by coupling equations. It operates by forming an equivalent "master program," with only a few more rows than there are coupling equations in the original problem but with very many columns. This program is solved without tabulating all these columns by generating them when the simplex method needs them, a technique we call "column generation." The resulting algorithm involves iteration between a set of independent subproblems whose objective functions contain variable parameters, and the master program. The subproblems receive a set of parameters (simplex multipliers or prices) from the master program. They send their solutions to the master program, which combines these with previous solutions in an optimal way and computes new prices. These are again sent to the subproblems, and the iteration proceeds until an optimality test is passed. The procedure has an elegant economic interpretation, in which the master program coordinates the actions of the subproblems by setting prices on resources used by these problems.

The chapter begins with brief sections on a theorem to be used frequently later and on column generation. The basic procedure and some of its variants are then developed, and a numerical example is given. The economic interpretation of the decomposition principle is described, and a lower bound for the minimal cost is developed. A number of applications follow, to transportation and generalized transportation problems and to a class of resource allocation problems. Finally, an alternative technique for solving the master program is described, involving application of the primal–dual method of linear programming.

3.2 A Theorem on Convex Combinations

The development of the Dantzig–Wolfe decomposition principle depends principally on two notions. The first of these, column generation, is considered in the following section. The second is a theorem stating that a point is in a closed bounded convex polyhedron if and only if it can be written as a convex combination of extreme points of the polyhedron. Its proof requires the following results concerning convex sets, proofs of which may be found in [2]:

1. Let X be a convex set and y an element not contained in X. Then there is a hyperplane which separates y and X.
2. If X is a compact convex set, then any supporting hyperplane[1] to X contains an extreme point of X.

The theorem of interest follows:

THEOREM 1. *Let X be a compact convex set in E^n, $E(X)$ the set of its extreme points, and $C[E(X)]$ the convex hull[2] of $E(X)$. Then $C[E(X)] = X$.*

PROOF

(a) To show that $X \supseteq C[E(X)]$, write any element, y, of $C[E(X)]$ as

$$y = \sum_i \lambda_i x^i, \quad \lambda_i \geq 0, \qquad \sum_i \lambda_i = 1, \quad x^i \in E(X) \tag{1}$$

Then, since $x^i \in X$ and X is convex, $y \in X$.

(b) To show that $C[E(X)] \supseteq X$, let $x^* \in X$, but $x^* \notin C[E(X)]$. Then x^* can be separated from $C[E(X)]$ by a hyperplane; i.e., there exists (c, α) such that

$$c'x^* = \alpha \tag{2}$$

$$c'y < \alpha \qquad \text{for all } y \in C[E(X)] \tag{3}$$

Let $\alpha^0 = \max\{c'x \mid x \in X\}$. The number α^0 exists since X is compact. Then (c, α^0) defines a supporting hyperplane to X, since

$$c'x \leq \alpha^0 \qquad \text{for all } x \in X$$

[1] This is a hyperplane such that all of X is on one side of it and at least one point of X is contained in the hyperplane.

[2] The smallest convex set containing $E(X)$.

As such, it must contain an extreme point of X, x^i, so

$$c'x^i = \alpha^0 \tag{4}$$

But this contradicts the fact that all points of $C[E(X)]$ (thus all extreme points of X) satisfy (3), since $\alpha^0 \geq \alpha$.

Theorem 1 may be rewritten in a form more immediately useful for later developments by using (1) and specializing to the case where X has a finite number of extreme points.

THEOREM 2. *Let $X = \{x \mid Ax = b, x \geq 0\}$ be nonempty and bounded, and let x^i, $i = 1, \ldots, r$ be its extreme points. Then any element $x \in X$ may be written*

$$x = \sum_{i=1}^{r} \lambda_i x^i, \quad \lambda_i \geq 0, \quad i = 1, \ldots, r, \qquad \sum_{i=1}^{r} \lambda_i = 1 \tag{5}$$

The extension of this result when X is not bounded is as follows.

THEOREM 3. *Let $X = \{x \mid Ax = b, x \geq 0\}$ be nonempty. Then a point, x, is in X if and only if it can be written as a convex combination of extreme points of X plus a nonnegative linear combination of extreme rays (homogeneous solutions) of X; i.e.,*

$$x = \sum_i \lambda_i x^i$$

where

$$\sum_i \lambda_i \delta_i = 1, \qquad \lambda_i \geq 0$$

and

$$\delta_i = \begin{Bmatrix} 1 \\ 0 \end{Bmatrix} \text{ depending on whether } x^i \text{ is an } \begin{Bmatrix} \text{extreme point} \\ \text{extreme ray} \end{Bmatrix} \text{ of } X$$

This result is used to extend the decomposition principle to the case of unbounded subsystem constraint sets in Section 3.7.

3.3 Column Generation[3]

Consider a linear program

$$\text{minimize } z = \sum_{j=1}^{n} c_j x_j$$

[3] This material is covered in more detail in Section 2.6.

subject to $$\sum_{j=1}^{n} p_j x_j = b$$

$$x_j \geq 0, \qquad j = 1, \ldots, n$$

where the p_j and b are m-component vectors, $m < n$. Assume that an initial basic feasible solution, x_B, is available, with associated basis matrix B, and cost coefficients c_B. Such a solution, if it exists, may be found by using a phase 1 procedure. The simplex multipliers associated with this basis are

$$\pi = c_B B^{-1}$$

and are always made available by the simplex method. To improve the basic feasible solution we "price out" all columns corresponding to nonbasic variables by forming their relative cost coefficients

$$\bar{c}_j = c_j - \pi p_j$$

If

$$\min_j \bar{c}_j = \bar{c}_s < 0$$

then, barring degeneracy, the current solution may be improved by introducing x_s into the basis via a pivot transformation.

If there are many columns, i.e., from several thousands to several millions, then finding $\min \bar{c}_j$ by computing each $\bar{c}_j$ and comparing may be tedious, if not impossible. Fortunately, in such large problems, the set of all columns generally has a well-defined structure, owing to the structure of the real-world situation giving rise to the problem. This is almost always true, and fortunately so, since otherwise the task of gathering unordered numerical data to specify all the columns would be hopeless. An example is the cutting-stock problem (Section 4.1), in which the admissible columns have components, a_i, which are nonnegative integers and satisfy the linear inequality $\sum_i l_i a_i \leq l$ for given numbers l_i, l. Another example is the subsequent decomposition principle, in which the columns, p_{ij}, all have the form $A_i x_i^j$ with A_i the ith partition of the coefficient matrix of the coupling constraints and x_i^j the jth extreme point of the ith angular block. Thus these columns are also defined by a system of linear equations and inequalities.

In general, then, let us assume that all columns, p_j, are drawn from a set, S, which typically is the set of all m vectors satisfying some system of equations or equalities. The column to enter the basis may then be chosen by solving the subproblem

$$\underset{p_j \in S}{\text{minimize}}\ c(p_j) - \pi p_j$$

where $c(p_j) \equiv c_j$ is a given function of p_j. Depending on the structure of S and the form of $c(\cdot)$, a variety of techniques may be used to solve this subproblem, e.g., linear programming in the decomposition principle, dynamic programming in the cutting-stock case. This approach is called column generation because, in solving the subproblem, only a small subset of columns in S are typically examined, and these are generated when needed. Thus none of the columns need be kept in computer storage, a definite advantage for large problems.

Of course, linear programs may be large not only in their number of variables but in their number of constraints as well. One way to deal with this difficulty is to transform the problem into an equivalent linear program with fewer constraints but many more variables. Column-generation methods are then applied to the equivalent problem. This is the basis of the Dantzig–Wolfe decomposition principle for linear programs, whose description follows.

3.4 Development of the Decomposition Principle

Consider a linear program whose constraint matrix has a p-block angular structure, with $p \geq 1$:

$$A = \begin{bmatrix} A_1 & A_2 & \cdots & A_p \\ B_1 & & & \\ & B_2 & & \\ & & \ddots & \\ & & & B_p \end{bmatrix}$$

Any linear program may be considered to have this form with $p = 1$ by partitioning its constraints into two subsets:

$$\text{minimize } z = cx \tag{1}$$

subject to

$$\hat{A}_1 x = b_1 \qquad (m_1 \text{ constraints}) \tag{2}$$

$$A_2 x = b_2 \qquad (m_2 \text{ constraints}) \tag{3}$$

$$x \geq 0 \tag{4}$$

Assume that the convex polyhedron

$$S_2 = \{x \mid A_2 x = b_2, x \geq 0\} \tag{5}$$

is bounded; this assumption will be relaxed later. Then, by Theorem 3.2-2, any element of S_2 may be written

$$x = \sum_j \lambda_j x^j \tag{6}$$

where

$$\sum_j \lambda_j = 1, \qquad \lambda_j \geq 0 \tag{7}$$

and the x^j are extreme points of the polyhedron S_2.

The original problem (1)–(4) may be viewed as follows: Choose, from all solutions of (3)–(4), those which satisfy (2) and minimize z. To enforce satisfaction of (2), substitute (6) into (2), obtaining

$$\sum_j (\hat{A}_1 x^j)\lambda_j = b_1 \tag{8}$$

Substituting (6) into (1) gives an expression for z in terms of the variables λ_j:

$$z = \sum_j (cx^j)\lambda_j \tag{9}$$

Defining

$$\hat{A}_1 x^j = p_j \tag{10}$$

$$cx^j = f_j \tag{11}$$

relations (7)–(9) are seen to comprise a linear program in the λ_j:

$$\text{minimize} \sum_j f_j \lambda_j \tag{12}$$

subject to

$$\sum_j p_j \lambda_j = b \tag{13}$$

$$\sum_j \lambda_j = 1 \tag{14}$$

$$\lambda_j \geq 0 \tag{15}$$

This program, called the master program, is completely equivalent to the original. It has only $m_1 + 1$ rows, compared to the $m_1 + m_2$ rows of the original problem, a sizable saving if m_2 is large. It also has as many columns as the polyhedron S_2 has extreme points, which may be many thousands if m_2 is large.

Rather than tabulating all these columns, we use a column-generation technique, creating columns to enter the basis as they are needed. To see how this is done, consider the relative cost coefficient for the variable λ_j:

$$\bar{f}_j = f_j - \pi \begin{bmatrix} p_j \\ --- \\ 1 \end{bmatrix} \tag{16}$$

Partition π as

$$\pi = (\pi_1, \pi_0)$$

where π_1 corresponds to the constraints (13) and the scalar π_0 to the single constraint (14). Then, using the definitions of f_j and p_j in (10) and (11), $\bar{f}_j$ may be written

$$\bar{f}_j = (c - \pi_1 \hat{A}_1)x^j - \pi_0 \tag{17}$$

The usual simplex criterion asks that we find

$$\min_j \bar{f}_j = \bar{f}_s = (c - \pi_1 \hat{A}_1)x^s - \pi_0 \tag{18}$$

in order to choose a variable, λ_s, to enter the basis. Recall that x^j is an extreme point of S_2, and note that $\bar{f}_j$ is linear in x^j. Since an optimal solution of a linear program whose constraint set is bounded always occurs at an extreme point of that set, (18) is equivalent to the subproblem

$$\text{minimize } (c - \pi_1 \hat{A}_1)x \tag{19}$$

subject to

$$A_2 x = b_2, \qquad x \geq 0 \tag{20}$$

To find a column to enter the basis of the master program, solve this subproblem, obtaining a solution, x^s. Then the column to enter the basis is

$$p_s = \begin{bmatrix} \hat{A}_1 x^s \\ \hline 1 \end{bmatrix} \tag{21}$$

with cost coefficient

$$f_s = cx^s \tag{22}$$

This approach becomes particularly attractive if p, the number of independent blocks in the angular structure, is greater than one, i.e., if the problem to be solved has the form

$$\text{minimize } z = c_1 x_1 + c_2 x_2 + \cdots + c_p x_p \tag{23}$$

subject to

$$\begin{aligned} A_1 x_1 + A_2 x_2 + \cdots + A_p x_p &= b_0 \\ B_1 x_1 \qquad\qquad\qquad &= b_1 \\ B_2 x_2 \qquad\qquad &= b_2 \\ \ddots \qquad\qquad &\;\;\vdots \\ B_p x_p &= b_p \end{aligned} \tag{24}$$

$$x_1 \geq 0, \qquad x_2 \geq 0 \cdots x_p \geq 0, \qquad p > 1 \tag{25}$$

then the subproblem (19)–(20) becomes

$$\text{minimize} \sum_{i=1}^{p} (c_i - \pi_1 A_i)x_i \tag{26}$$

subject to $$B_i x_i = b_i, \quad x_i \geq 0, \quad i = 1, \ldots, p \tag{27}$$

Since the objective, (26), is additively separable in the x_i and the constraints on the x_i, (27), are independent, this problem reduces to the p independent subproblems

$$\text{minimize } (c_i - \pi_1 A_i)x_i \tag{28}$$

subject to $$B_i x_i = b_i, \quad x_i \geq 0 \tag{29}$$

Decomposition Algorithm. A two-level algorithm for the solution of (23)–(25) may now be formulated, with the master program on the second level and the subproblems (28)–(29) on the first. Assume that an initial basic feasible solution for the master program, (12)–(15), is available, with basis matrix B and simplex multipliers (π_1, π_0).

STEP 1. Using the simplex multipliers π_1, solve the subproblems (28)–(29) obtaining solutions $x_i(\pi_1)$ and optimal objective values z_i^0. Let $x(\pi_1) = (x_1(\pi_1), \ldots, x_p(\pi_1))'$.

STEP 2. Compute $\min \bar{f}_j$, which, by (17), is

$$\min \bar{f}_j = \sum_{i=1}^{p} z_i^0 - \pi_0 \tag{30}$$

If

$$\min \bar{f}_j \geq 0 \tag{31}$$

stop. The optimal solution to (23)–(25) is

$$x^0 = \sum_{j \text{ basic}} \lambda_j x^j \tag{32}$$

where the x^j are the extreme points of S_2 corresponding to basic λ_j.

STEP 3. If $\min \bar{f}_j < 0$, form the column

$$p = \begin{pmatrix} \sum_i A_i x_i(\pi_1) \\ 1 \end{pmatrix} \tag{33}$$

transform it by multiplication by B^{-1}, and, using the usual simplex pivot operation, obtain a new basis inverse and a new vector of simplex multipliers. Go back to step 1 and repeat.

If the master program is nondegenerate (or ϵ-perturbed), then each iteration decreases z by a nonzero amount. Since there are only a finite number of possible bases, and none is repeated, the decomposition principle will find the optimal solution in a finite number of iterations.

Note that the optimal solution, x^0, is *not necessarily* any one of the subproblem proposals. By (32), x^0 is some convex combination of a number of such solutions. Thus the master program cannot obtain the overall optimum simply by sending appropriate prices, π_1, to the subproblems; it must have the freedom to combine subproblem solutions into an overall plan. In this sense, the decomposition obtained here cannot be viewed as complete decentralization of the decision-making process. A better term would be "centralized planning without complete information at the center," as in Dantzig [8].

Restricted Master Program. As previously formulated, the Dantzig–Wolfe decomposition principle solves optimization problems on the first level, but not at the second, where only a single pivot operation is performed. By defining a restricted master program a more symmetric formulation is possible in which both levels solve linear programs. This is simply the master program (12)–(15) with all columns dropped but those in the current basis and that column about to enter. Then the restricted master program is

$$\text{minimize } f_1\lambda_1 + \cdots + f_m\lambda_m + f\lambda \tag{34}$$

subject to

$$p_1\lambda_1 + \cdots + p_m\lambda_m + p\lambda = b_0 \tag{35}$$

$$\lambda_1 + \cdots + \lambda_m + \lambda = 1 \tag{36}$$

$$\lambda_i \geq 0, \qquad i = 1, \ldots, m, \qquad \lambda \geq 0 \tag{37}$$

where $m = m_1 + 1$, the λ_i are the current basic variables and λ is the variable entering. Assuming that the variable λ had an $\bar{f} < 0$, it enters the basis. If the current basis is nondegenerate, the variable leaving will have a positive $\bar{f}$, so the new solution will be optimal. The only case in which more than one iteration is needed to pass the optimality test is when the current basis is degenerate and the pivot element in the entering column is negative. Thus, little is gained by solving (34)–(37). However, as we now show, other formulations of the master program can make the restricted master much more worthwhile.

Alternative Strategies for Decomposing the Primal. There are a number of ways in which the primal may be decomposed, each of which affects the form of the master and restricted master programs. Specifically, if solutions to the constraints

$$B_i x_i = b_i, \qquad x_i \geq 0 \tag{38}$$

are written

$$x_i = \sum_j \lambda_{ij} x_i^j \tag{39}$$

with the x_i^j extreme points of $B_i x_i = b_i$, $x_i \geq 0$, then it is easily verified that the following master program results:

$$\text{minimize} \sum_{i,j} f_{ij} \lambda_{ij} \tag{40}$$

subject to

$$\sum_{i,j} p_{ij} \lambda_{ij} = b_0 \tag{41}$$

$$\sum_j \lambda_{ij} = 1, \qquad i = 1, \ldots, p, \qquad \lambda_{ij} \geq 0 \tag{42}$$

where

$$f_{ij} = c_i x_i^j \tag{43}$$

$$p_{ij} = A_i x_i^j \tag{44}$$

This program differs from the master program (12)–(14) obtained previously in two respects. It has p convexity rows rather than one, and each subsystem $B_i x_i = b_i$, $x_i \geq 0$, is represented separately by a set of variables λ_{ij} (previously, all subsystems were aggregated together). The latter feature opens up some interesting computational possibilities. To see these, consider the solution of (40)–(42) by the simplex method using column generation. Let B be a feasible $(m_1 + p) \times (m_1 + p)$ basis matrix, and let $(\pi, \pi_{01}, \ldots, \pi_{0p})$ be the simplex multipliers for this basis, with π associated with (41) and the π_{0i} with (42). Pricing out the column associated with λ_{ij} yields

$$\bar{f}_{ij} = (c_i - \pi A_i) x_i^j - \pi_{0i} \tag{45}$$

The problem of finding, for fixed i, $\min_j \bar{f}_{ij}$ is equivalent to solving the ith subproblem

$$\text{minimize } (c_i - \pi A_i) x_i$$

subject to

$$B_i x_i = b_i, \qquad x_i \geq 0$$

These are the same subproblems as were obtained previously. Let the minimal objective value for the above be z_i^0. If

$$\min_i \min_j \bar{f}_{ij} = \min_i (z_i^0 - \pi_{0i}) \geq 0 \tag{46}$$

the current solution is optimal. If not, the column to enter the basis is that with

$$\min_{1 \leq i \leq p} (z_i^0 - \pi_{0i}) \tag{47}$$

If the minimum above occurs for $i = s$ and $x_s(\pi_1)$ solves subproblem s, the column entering is given by

$$\begin{bmatrix} A_s x_s(\pi_1) \\ \hline u_s \end{bmatrix} \tag{48}$$

where u_s is a p-component vector with a one in position s and zeros elsewhere. Although this column now enters the basis yielding new multipliers (π_1, π_0), similar columns could have been generated from the subproblem solutions of other subsystems. A number of these probably had negative relative costs using the old multipliers. If any such column prices out negative using the new multipliers (and this is again probable since successive sets of multipliers should not differ greatly), it may again be used to reduce the objective. This is strong motivation for forming a new restricted master program, with a "new" column for each subsystem. Let $m = m_1 + p$, define $\{p_{ij}\}$ as the columns of the current basis, and let $p_1^*, \ldots, p_p^*$ be the columns generated from the latest subproblem solutions. The new restricted master program is

$$\text{minimize} \sum_{i=1}^{p} \sum_{J\ basic} f_{ij}\lambda_{ij} + \sum_{i=1}^{p} f_i^* \lambda_i^* \tag{49}$$

$$\text{subject to} \quad \sum_{i=1}^{p} \sum_{J\ basic} p_{ij}\lambda_{ij} + \sum_{i=1}^{p} p_i^* \lambda_i^* = b_0 \tag{50}$$

$$\sum_{J\ basic} \lambda_{ij} + \qquad \lambda_i^* = 1, \qquad i = 1, \ldots, p \tag{51}$$

$$\lambda_i \geq 0, \qquad \lambda_i^* \geq 0 \tag{52}$$

This program has p more variables than constraints, rather than one more as in the previous case, (34)–(37). One would thus expect that a greater decrease in z would result. Of course, the size of the basis has been increased from $m_1 + 1$ to $m_1 + p$. Most of the added labor may be eliminated, however, by using a variant of the simplex method, the generalized upper bounding algorithm (see Section 6.4) to solve each restricted master program. An instance in which the advantage of having many columns in the restricted master program was found, computationally, to outweigh the disadvantage of increased basis size is discussed in Section 4.2.

Thus far master programs with one and with p convexity rows have been discussed, the former obtained by grouping all diagonal blocks together, the latter by treating them all separately. Obviously, other groupings of these blocks lead to master programs with any number of convexity rows between one and p. Although the extreme case of p seems of greatest computational value, some intermediate program may have special advantages in a particular application.

3.5 Example of the Decomposition Principle

To illustrate the decomposition principle, consider the problem

$$\text{minimize } z = -x_1 - x_2 - 2y_1 - y_2$$

subject to

$$\begin{aligned}
x_1 + 2x_2 + 2y_1 + y_2 &\leq 40 \\
x_1 + 3x_2 \qquad\qquad &\leq 30 \\
2x_1 + x_2 \qquad\qquad &\leq 20 \\
y_1 &\leq 10 \\
y_2 &\leq 10 \\
y_1 + y_2 &\leq 15
\end{aligned}$$

and all variables nonnegative. This problem has one coupling constraint and two independent blocks. The feasible regions for these blocks are shown shaded in Figure 3-1. As mentioned in Section 3.4, this problem can be

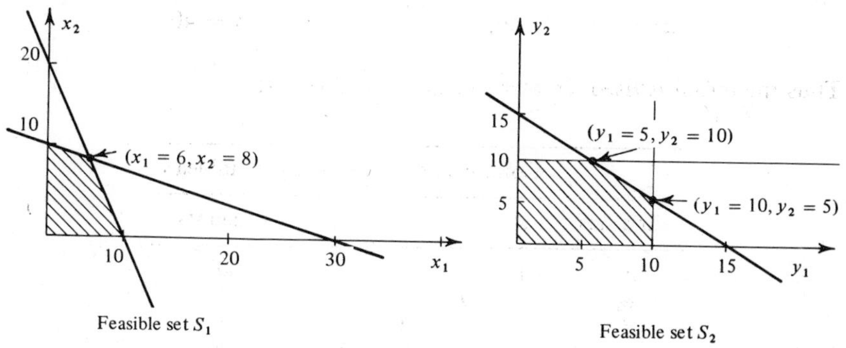

FIGURE 3-1 Feasible regions of subproblems.

decomposed in a number of ways, each leading to a different number of convexity constraints, $\sum_j \lambda_{ij} = 1$, in the master program. To make fullest use of the restricted master program, two convexity constraints will be included. Let x and y be feasible solutions for blocks 1 and 2, respectively, and write

$$x = \sum_i \alpha_i x^i, \qquad y = \sum_i \beta_i y^i$$

with the x^i, y^i extreme points of these blocks. The coupling constraint and objective function are written in vector notation as

$$z = c_1 x + c_2 y$$
$$a_1 x + a_2 y + s = 40$$

where $c_1 = (-1, -1)$, $c_2 = (-2, -1)$, $a_1 = (1, 2)$, $a_2 = (2, 1)$, and $s \geq 0$ is a slack variable. The master program then becomes

$$\text{minimize } z = \sum_i (c_1 x^i)\alpha_i + \sum_i (c_2 y^i)\beta_i$$

subject to

$$\sum_i (a_1 x^i)\alpha_i + \sum_i (a_2 y^i)\beta_i + s = 40$$

$$\sum_i \alpha_i \qquad\qquad = 1$$

$$\sum_i \beta_i \qquad = 1$$

$$\alpha_i \geq 0, \qquad \beta_i \geq 0$$

An initial basic feasible solution may be obtained by choosing

$$x^1 = y^1 = (0, 0), \qquad \alpha_1 = \beta_1 = 1, \qquad s = 40$$

Thus the initial revised simplex tableau is as follows:

Basic variables	Columns of master program tableau				
	s	α_1	β_1	$-z$	Constants
s	1				40
α_1		1			1
β_1			1		1
$-z$				1	0

Iteration 1. The negatives of the simplex multipliers for the master program are contained in the bottom row of the above tableau and are currently all zero. Thus the first set of subproblems is

$$\text{minimize } z_1 = c_1 x = -x_1 - x_2 \qquad \text{minimize } z_2 = c_2 y = -2y_1 - y_2$$

$$\text{subject to} \quad x \in S_1 \qquad \text{subject to} \quad y \in S_2$$

with optimal solutions obtained by inspection:

$$x^2 = (6, 8), \quad z_1^0 = -14, \quad y^2 = (10, 5), \quad z_2^0 = -25$$

The minimal relative cost factors are

$$\min \bar{f}_x = z_1^0 - \pi_{01} = -14$$
$$\min \bar{f}_y = z_2^0 - \pi_{02} = -25$$

Associated with each subproblem solution is a column of the master program. The columns corresponding to the current solutions are

$$\begin{pmatrix} a_1 x^2 \\ 1 \\ 0 \\ c_1 x^2 \end{pmatrix} = \begin{pmatrix} 22 \\ 1 \\ 0 \\ -14 \end{pmatrix}, \quad \begin{pmatrix} a_2 y^2 \\ 0 \\ 1 \\ c_2 y^2 \end{pmatrix} = \begin{pmatrix} 25 \\ 0 \\ 1 \\ -25 \end{pmatrix}$$

A restricted master program is now formed using the current master program basis plus these new columns:

$$\text{minimize } -14\alpha_2 - 25\beta_2$$

$$\begin{array}{llllll} \text{subject to} & s & & & + 22\alpha_2 + 25\beta_2 & = 40 \\ & & \alpha_1 & & + \alpha_2 & = 1 \\ & & & \beta_2 & + \beta_2 & = 1 \end{array}$$

with all variables nonnegative. This is solved by the revised simplex method, yielding the following sequence of tableaus:

	Relative cost factors				
Cycle	s	α_1	β_1	α_2	β_2
0	0	0	0	-14	-25
1	0	0	25	-14	0
2	$\frac{14}{22}$	0	$\frac{200}{22}$	0	0

Basic variables	Columns of master program tableau					Transformed entering column
	s	α_1	β_1	$-z$	Constants	
	Cycle 0: β_2 enters, β_1 leaves					
s	1				40	25
α_1		1			1	0
β_1			1		1	1
$-z$				1	0	-25
	Cycle 1: α_2 enters, s leaves					
s	1		-25		15	22
α_1		1			1	1
β_2			1		1	0
$-z$			25	1	25	-14
	Cycle 2: optimal					
α_2	$\frac{1}{22}$		$-\frac{25}{22}$		$\frac{15}{22}$	
α_1	$-\frac{1}{22}$	1	$\frac{25}{22}$		$\frac{7}{22}$	
β_2			1		1	
$-z$	$\frac{14}{22}$		$\frac{200}{22}$	1	$34\frac{6}{11}$	

The new solution to the primal problem is

$$\begin{aligned} x &= \alpha_1 x^1 + \alpha_2 x^2 = \tfrac{7}{22}(0, 0) + \tfrac{15}{22}(6, 8) \\ y &= \beta_2 y^2 \qquad\qquad\;\; = (10, 5) \end{aligned}$$

with objective value $-34\frac{6}{11}$.

Iteration 2. The new simplex multipliers are found in the bottom row of the final tableau

$$(\pi, \pi_{01}, \pi_{02}) = (-\tfrac{14}{22}, 0, -\tfrac{200}{22})$$

These are used to form new subproblem objectives

$$\begin{aligned} z_1 &= (c_1 - \pi a_1)x = [(-1, -1) + \tfrac{14}{22}(1, 2)]x \\ &= -\tfrac{8}{22}x_1 + \tfrac{6}{22}x_2 \\ z_2 &= (c_2 - \pi a_2)y = [(-2, -1) + \tfrac{14}{22}(2, 1)]y \\ &= -\tfrac{16}{22}y_1 - \tfrac{8}{22}y_2 \end{aligned}$$

Minimizing these over the subsystem constraint sets yields optimal solutions

$$x^3 = (10, 0), \qquad z_1^0 = -\tfrac{80}{22}$$
$$y^3 = (10, 5), \qquad z_2^0 = -\tfrac{200}{22}$$

with minimal relative cost factors

$$\min \bar{f}_x = z_1^0 - \pi_{01} = -\tfrac{80}{22}$$
$$\min \bar{f}_y = z_2^0 - \pi_{02} = 0$$

and associated columns

$$\begin{pmatrix} a_1 x^3 \\ 1 \\ 0 \\ c_1 x^3 \end{pmatrix} = \begin{pmatrix} 10 \\ 1 \\ 0 \\ -10 \end{pmatrix}, \qquad \begin{pmatrix} a_2 y^3 \\ 0 \\ 1 \\ c_2 y^3 \end{pmatrix} = \begin{pmatrix} 25 \\ 0 \\ 1 \\ -25 \end{pmatrix}$$

The new restricted master program consists of the optimal basic columns of the previous master program (all other columns are dropped) plus these new columns:

$$\text{minimize } -14\alpha_2 - 25\beta_2 - 10\alpha_3 - 25\beta_3$$

subject to

$$\begin{aligned} 22\alpha_2 + 25\beta_2 + 10\alpha_3 + 25\beta_3 &= 40 \\ \alpha_1 + \alpha_2 \qquad + \alpha_3 \qquad &= 1 \\ \beta_2 \qquad + \beta_3 &= 1 \end{aligned}$$

The optimal solution for the previous restricted master is an initial basic feasible solution for this program, with basis inverse given in the final tableau of iteration 1. Solution by the revised simplex method follows

Relative cost factors						
Cycle	α_1	α_2	α_3	β_2	β_3
0	0	0	$-\frac{80}{22}$	0	0
1	$\frac{80}{12}$	0	0	0	0

Basic variables	Columns of master program tableau					Transformed entering column
	s	α_1	β_1	$-z$	Constants	
Cycle 0: α_3 enters, α_1 leaves						
α_2	$\frac{1}{22}$		$-\frac{25}{22}$		$\frac{15}{22}$	$\frac{10}{22}$
α_1	$-\frac{1}{22}$	1	$\frac{25}{22}$		$\frac{7}{22}$	$\frac{12}{22}$
β_2			1		1	0
$-z$	$\frac{14}{22}$		$\frac{200}{22}$	1	$34\frac{6}{11}$	$-\frac{80}{22}$

Cycle 1: optimal

α_2	$\frac{1}{12}$	$-\frac{10}{12}$	$-\frac{25}{12}$		$\frac{5}{12}$
α_3	$-\frac{1}{12}$	$\frac{22}{12}$	$\frac{25}{12}$		$\frac{7}{12}$
β_2			1		1
$-z$	$\frac{4}{12}$	$\frac{80}{12}$	$\frac{200}{12}$	1	$\frac{110}{3}$

The current feasible solution to the primal problem is

$$\begin{aligned} x &= \alpha_2 x^2 + \alpha_3 x^3 = \tfrac{5}{12}(6, 8) + \tfrac{7}{12}(10, 0) \\ y &= \beta_2 y^2 \qquad\qquad\; = (10, 5) \end{aligned}$$

with objective value $z = -\frac{110}{3}$

Iteration 3. The new simplex multipliers and subsystem objectives are

$$\begin{aligned} (\pi, \pi_{01}, \pi_{02}) &= -(\tfrac{4}{12}, \tfrac{80}{12}, \tfrac{200}{12}) \\ z_1 &= [(-1, -1) + \tfrac{4}{12}(1, 2)]x = -\tfrac{2}{3}x_1 - \tfrac{1}{3}x_2 \\ z_2 &= [(-2, -1) + \tfrac{4}{12}(2, 1)]y = -\tfrac{4}{3}y_1 - \tfrac{2}{3}y_2 \end{aligned}$$

The optimal subproblem solutions are

$$\begin{aligned} &x^4 = (6, 8) \text{ or } x^5 = (10, 0), \qquad z_1^0 = -\tfrac{20}{3} \\ &y^4 = (10, 5), \qquad z_2^0 = -\tfrac{50}{3} \end{aligned}$$

with minimal relative cost factors

$$\begin{aligned} \min \bar{f}_x &= z_1^0 - \pi_{01} = 0 \\ \min \bar{f}_y &= z_1^0 - \pi_{02} = 0 \\ \min \bar{c}_s &= -\pi_{01} = \tfrac{80}{12} \end{aligned}$$

Since these are nonnegative, the current primal solution is optimal.

3.6 Economic Interpretation of the Decomposition Principle[4]

The Dantzig–Wolfe decomposition principle has an interesting economic interpretation, based upon viewing the simplex multipliers π as prices. Consider a given basis B, with associated basic feasible solution

$$x_B = B^{-1}b \tag{1}$$

If differential changes in the item availabilities, db, are made, what is the

[4] See Baumol and Fabian [7] and Dantzig [8] for further reading on this material.

effect on the objective, z, assuming the current basis is maintained? In particular, what is $\partial z/\partial b_i$? Of course, for the question to be meaningful, the new solution

$$x_B^n = B^{-1}(b + db) \tag{2}$$

must be nonnegative for arbitrary db. This will be true if we assume that x_B has all positive components, i.e., is nondegenerate. Then by definition

$$\pi = c_B B^{-1} \tag{3}$$

$$z = c_B x_B = c_B(B^{-1}b) = (c_B B^{-1})b \tag{4}$$

Thus

$$z = \pi b = \sum_i \pi_i b_i \tag{5}$$

Under the nondegeneracy assumption, π does not change if small changes db are made, since B need not change. Thus, by (5),

$$\frac{\partial z}{\partial b_i} = \pi_i, \qquad i = 1, \ldots, m \tag{6}$$

If z has units of dollars, then π_i has units of (dollars per unit of item b_i). The multiplier π_i is called a "shadow price" for item i, since it represents the amount of objective value which may be gained by a small increase in item i's availability, assuming that all items are transformed into activity levels as dictated by (1). The adjective "shadow" is attached since π_i need not be the true market price of the item. Note that B need not be an optimal basis, and the π_i may be either positive or negative.

Returning to the decomposition principle, consider the actions of the ith subsystem and the way they affect the overall objective, z, viewed here as a total cost to be minimized. If subsystem i chooses an activity vector x_i, it incurs a direct cost $c_i x_i$. It also uses an amount, $A_i x_i$, of the shared items, thus denying them to other subsystems, and possibly increasing their costs. Subsystem i must be made to take this indirect, but important, contribution to cost into account. The simplest way to do this is to announce a set of shadow prices, π, for the shared items, and force the subsystems to pay for whatever quantities of the resources they use. If a particularly valuable item is assigned a high price, then this should discourage subsystems from using excessive quantities of it, as they might if no penalty were imposed.

It is easy to see that the decomposition principle coordinates the actions of the subsystems in precisely this way. The objective function of subproblem i is, by (3.4–28),

$$z_i = c_i x_i - \pi A_i x_i \tag{7}$$

The term $c_i x_i$ is the direct contribution to overall cost, z, incurred in operation at level x_i, while $\pi A_i x_i$, being the product of the per-unit price vector π of the shared items times the amount of these items used, is a cost charged to the subsystem for using shared resources. Thus the subsystem objective, z_i, is a measure of the total cost, both direct and indirect, incurred by the subsystem. The minus sign in front of $\pi A_i x_i$ is justified as follows. Using a positive amount of item $j(\{A_i x_i\}_j > 0)$ is equivalent to decreasing its availability. If item j is valuable, a decrease in its availability should increase the cost; hence, by (6), $\pi_j < 0$. Then $-\pi_j$, its price, is positive, as is the term $-\pi_j(A_i x_i)_j$, which increases the subsystem cost, as it should.

A similar interpretation may be placed on the optimality criterion, which is

$$\min_j \bar{f}_{ij} = z_i^0 - \pi_{0i} \geq 0 \tag{8}$$

or

$$z_i^0 \geq \pi_{0i}, \qquad i = 1, \ldots, p \tag{9}$$

where z_i^0 is the optimal value of the objective of subproblem i and π_{0i} is the simplex multiplier corresponding to the ith convexity constraint,

$$\sum_j \lambda_{ij} = 1 \tag{10}$$

By (6), π_{0i} is the change in z caused by a small change in the right-hand side of (10). Increasing this quantity from its current value of unity implies that the weights λ_{ij} assigned to the various proposals already submitted by subsystem i may sum to more than one; i.e., these proposals may play a greater role in the overall plan. Thus π_{0i} is a measure of the marginal cost of the current proposals. As discussed previously, z_i^0 is a measure of the total (direct and indirect) cost of a new proposal, x_i^0. Thus the optimality test (9) says that if, for all subsystems, introducing the "best" new proposal will increase costs more than using the current proposals, the current proposals, suitably weighted, are optimal. If (9) does not hold, then the new proposal replaces one of the old ones.

Related to this pricing interpretation is the following question. The reader will recall that the decision-making scheme presented here is not truly decentralized, since the central agency actually makes the final decisions by assigning optimal weights to subsystem proposals. Why cannot a system of prices be invented which induces the subsystems to choose optimal decision levels themselves? Curiously the reason this is, in general, not possible is that the problem is completely linear. This implies that, no matter what their objective functions, the subsystems can only arrive at activity levels x_i which are extreme points of their constraint sets. However, the optimal solution of the original problem, $(x_1^*, \ldots, x_p^*)$, need not be such that any x_i^* is an extreme point of the corresponding subsystem constraints, since an extreme point solution of the angular system (3.4-23)–(3.4-25) need not have components which are extreme points of the diagonal blocks. The optimal activity levels x_i^* may well be interior to the subsystem constraints, and only by weighting the various extreme points can interior points be generated.

Of course, if the subsystem objective were nonlinear, then nonextreme point solutions could be obtained directly. Thus complete decentralization might be possible in the case of suitable nonlinearities. That this is true is discussed in Chapter 8.

3.7 Lower Bound for the Minimal Cost

Consider again the master program (3.4-40)–(3.4-42), rewritten here as

$$\text{minimize } z = \sum_{i,j} f_{ij}\lambda_{ij} \tag{1}$$

subject to

$$\sum_{i,j} p_{ij}\lambda_{ij} = b_0 \tag{2}$$

$$\sum_{j} \lambda_{ij} = 1, \qquad i = 1, \ldots, p \tag{3}$$

$$\lambda_{ij} \geq 0 \tag{4}$$

Multiply (2) by the simplex multipliers π_1 and (3) by the multipliers π_{0i}, sum, and subtract from the cost equation, yielding

$$z - \pi_1 b_0 - \sum_i \pi_{0i} = \sum_{i,j} \lambda_{ij}(f_{ij} - \pi_1 p_{ij} - \pi_{0i}) \tag{5}$$

The quantity in parentheses on the right of (5) is simply the relative cost factor for λ_{ij}, $\bar{f}_{ij}$. By solving the subproblems (3.4-28)–(3.4-29), we compute $\min_j \bar{f}_{ij}$ for each i. Since λ_{ij} is nonnegative, $\bar{f}_{ij}$ may be replaced by its minimal value and the right-hand side of (5) cannot increase, so

$$z - \pi_1 b_0 - \sum_i \pi_{0i} \geq \sum_{i,j} \lambda_{ij}\left(\min_j \bar{f}_{ij}\right) \tag{6}$$

$$z - \pi_1 b_0 - \sum_i \pi_{0i} \geq \sum_i \left(\min_j \bar{f}_{ij}\right) \sum_j \lambda_{ij} \tag{7}$$

Using (3), (7) becomes

$$z \geq \sum_i \left(\min_j \bar{f}_{ij}\right) + \pi_1 b_0 + \sum_i \pi_{0i} \tag{8}$$

Since (8) holds for all values of z obtainable from (1)–(4), it holds for the minimal value, so

$$\min z \geq \sum_i \left(\min_j \bar{f}_{ij}\right) + \pi_1 b_0 + \pi_0 e_p \tag{9}$$

where e_p is the p-dimensional sum vector. Relation (9) may be placed in a more compact form by writing the last two terms on the right as

$$(\pi_1, \pi_0)\begin{pmatrix} b_0 \\ e_p \end{pmatrix} = c_B B^{-1}\begin{pmatrix} b_0 \\ e_p \end{pmatrix} = c_B x_B = z_B \tag{10}$$

where z_B is the value of z associated with the current basis, B. Thus (9) becomes

$$\min z \geq z_B + \sum_{i=1}^{p} \left(\min_j \bar{f}_{ij}\right) \tag{11}$$

which is the desired lower bound.

Note that the quantities $\min_j \bar{f}_{ij}$ must be computed in performing the simplex iterations, so little additional effort is required to obtain this bound. Further, it is easy to see that relation (11) becomes an equality at the optimum. At optimality

$$\min_j \bar{f}_{ij} \geq 0 \tag{12}$$

But j is an index denoting extreme points of the subsystem constraints, and for extreme points corresponding to basic variables λ_{ij}, $\bar{f}_{ij} = 0$. Thus, when $z_B = \min z$,

$$\min_j \bar{f}_{ij} = 0 \tag{13}$$

and (11) is an equality. If the right-hand side of (11) is computed at each iteration, the sequence of numbers so obtained converges to $\min z$ from below (but not necessarily monotonically). This sequence may be used with the sequence of z values obtained to bracket the minimum and terminate computations when, for example, the current z value is predicted to be within α per cent of the minimum.

Modifications for Unbounded Subsystem Constraint Sets. Previous developments have assumed that the polyhedra

$$S_i = \{x \mid B_i x = b_i,\ x \geq 0\}, \qquad i = 1, \ldots, p \tag{14}$$

were bounded. If this is not true, then the theorem that any point in S_i can be represented as a convex combination of extreme points of S_i does not hold, and this is the result upon which our development of the decomposition principle has been based. A slightly modified principle may be derived, however, by applying Theorem 3.2-3, which states: Any point in the set S_i may be written as a convex combination of extreme points of S_i plus a nonnegative linear combination of extreme rays of S_i. These extreme rays are solutions to the homogeneous system

$$B_i x = 0, \qquad x \geq 0 \tag{15}$$

and are finite in number. An unbounded solution to a linear program can be approached by moving outward along an extreme ray, and, if such a situation exists, it will be detected by the simplex method.

Applying this generalized result, any element, x, of S_i is written

$$x = \sum_i \lambda_i x^i \tag{16}$$

where

$$\sum_i \lambda_i \delta_i = 1, \qquad \lambda_i \geq 0 \tag{17}$$

$$\delta_i = \begin{Bmatrix} 1 \\ 0 \end{Bmatrix} \text{ depending on whether } x^i \text{ is an } \begin{Bmatrix} \text{extreme point} \\ \text{extreme ray} \end{Bmatrix} \text{ of } S_i \tag{18}$$

The decomposition principle results from substituting (16) into the coupling constraints and objective of the original angular problem. The only difference in the master program is the change in the convexity constraint, $\sum_i \lambda_i = 1$, to the form given in (17) and (18).

3.8 Application to Transportation Problems

The Hitchcock–Koopmans transportation problem is: Given m origins each with stocks a_i of a single good, and n destinations with requirements b_j, find a least-cost shipping schedule which meets all requirements. Letting

x_{ij} = number of units shipped from origin i to destination j

c_{ij} = per unit cost of shipping from i to j

the problem becomes

$$\text{minimize} \sum_{i,j} c_{ij} x_{ij} \tag{1}$$

subject to
$$\sum_j x_{ij} = a_i, \qquad i = 1, \ldots, m \tag{2}$$

$$\sum_i x_{ij} = b_j, \qquad j = 1, \ldots, n \tag{3}$$

$$x_{ij} \geq 0 \tag{4}$$

where the a_i and b_j are positive and $\sum_i a_i = \sum_j b_j$ to guarantee (2) and (3) consistent.

Since this problem has the property that integral a_i and b_j imply an integral optimal solution, it is essentially combinatorial. A number of very efficient methods of solution, almost all combinatorial, exist, and problems with as many as 5000 constraint equations may be solved relatively easily. Still, even larger problems do arise in practice and solution eventually becomes difficult due either to memory or time limitations.

If the number of destinations is much greater than the number of origins ($n \gg m$), the problem may be considered to have special structure. To take advantage of this, Williams [3] has applied the decomposition principle to such problems. The larger subset of equations, (3), is taken to be the subsystem constraints, with (2) the coupling relations. The result is a master program with only m constraints (the convexity condition may be dropped, as will be shown shortly). Since each variable appears in exactly one equation of (3), the subproblems will be solvable by inspection, requiring only a scanning of the cost matrix, (c_{ij}), by columns.

The reader should note that the comments made would still apply if the roles of (2) and (3) were reversed and $m \gg n$. An example of such a situation is given in the next section.

Let x_{ij}^k be components of the kth extreme point of (3)–(4) and write any solution to these equations as

$$x_{ij} = \sum_k \lambda_k x_{ij}^k, \qquad \sum_k \lambda_k = 1, \qquad \lambda_k \geq 0 \tag{5}$$

Substituting (5) into (2) and (1) yields the master program

$$\text{minimize} \sum_k f_k \lambda_k \tag{6}$$

subject to
$$\lambda_k \geq 0, \qquad \sum_k v_{ik} \lambda_k = a_i, \qquad i = 1, \ldots, m \tag{7}$$

where

$$f_k = \sum_{i,j} c_{ij} x_{ij}^k \tag{8}$$

$$v_{ik} = \sum_j x_{ij}^k \tag{9}$$

The convexity condition, $\sum_k \lambda_k = 1$, is implied by (7), since

$$\sum_i a_i = \sum_{i,k} \lambda_k v_{ik} = \sum_k \lambda_k \sum_i \sum_j x_{ij}^k \tag{10}$$

$$\sum_i a_i = \sum_{i,k} \lambda_k v_{ik} = \sum_k \lambda_k \sum_j \sum_i x_{ij}^k = \sum_k \lambda_k \left(\sum_j b_j\right) \tag{11}$$

with $\sum_i a_i = \sum_j b_j > 0$.

To obtain the subproblems, price out a column of (7), obtaining the relative cost coefficient,

$$\bar{f}_k = f_k - \sum_i \pi_i v_{ik} \tag{12}$$

$$\bar{f}_k = \sum_{i,j} (c_{ij} - \pi_i) x_{ij}^k \tag{13}$$

Since the x_{ij}^k are extreme points of (3)–(4), minimizing $\bar{f}_k$ is equivalent to solving:

$$\text{minimize} \sum_{i,j} (c_{ij} - \pi_i) x_{ij} \tag{14}$$

subject to $$x_{ij} \geq 0, \quad \sum_i x_{ij} = b_j, \quad j = 1, \ldots, n \tag{15}$$

The objective function of this problem is separable and each x_{ij} appears in only one equation of (15). Thus (14)–(15) is equivalent to solving, for each j,

$$\text{minimize} \sum_i (c_{ij} - \pi_i) x_{ij} \tag{16}$$

subject to (15). The optimal solution is to find the smallest coefficient in (16) and set its x_{ij} equal to b_j; i.e., find

$$\min_i (c_{ij} - \pi_i) = c_{s_j j} - \pi_{s_j} \tag{17}$$

and set

$$x_{ij}^k = \begin{cases} b_j, & i = s_j \\ 0, & i \neq s_j \end{cases} \tag{18}$$

The minimal relative cost factor is, by (13),

$$\min \bar{f}_k = \sum_j b_j \left[\min_i (c_{ij} - \pi_i)\right] \tag{19}$$

If this is negative, the coefficients v_{ik} are formed using (18) and a new column enters the basis.

The minimization in (17) requires only the scanning by columns of a cost matrix which has π_i subtracted from all elements in row i. The associated solution, (18), ships all requirements of destination j from the origin having the smallest per unit (modified) cost relative to j.

3.9 Generalized Transportation Problems and a Forestry-Cutting Example

The generalized transportation problem has the form: Find nonnegative y_{ij} which minimize

$$z = \sum_{i,j} \hat{c}_{ij} y_{ij} \tag{1}$$

subject to

$$\sum_j a_{ij} y_{ij} = a_i, \qquad i = 1, \ldots, m \tag{2}$$

$$\sum_i b_{ij} y_{ij} = b_j, \qquad j = 1, \ldots, n \tag{3}$$

where the a_i and b_j are nonnegative and the a_{ij} are positive. Here, as in the Hitchcock–Koopmans transportation problem, each variable y_{ij} has only two nonzero coefficients, but they need not be unity. This problem may be placed in standard form by the transformation

$$x_{ij} = a_{ij} y_{ij}$$

Since the a_{ij} are positive, $x_{ij} \geq 0 \Leftrightarrow y_{ij} \geq 0$, and the problem becomes

$$\text{minimize } z = \sum_{i,j} c_{ij} x_{ij} \tag{4}$$

subject to

$$\sum_j x_{ij} = a_i, \qquad i = 1, \ldots, m \tag{5}$$

$$\sum_i d_{ij} x_{ij} = b_j, \qquad j = 1, \ldots, n \tag{6}$$

$$x_{ij} \geq 0$$

where

$$d_{ij} = \frac{b_{ij}}{a_{ij}}, \qquad c_{ij} = \frac{\hat{c}_{ij}}{a_{ij}}$$

Such problems are fairly common in applications [4]–[5]. One example is a forestry-cutting problem formulated by Tcheng [6]. A large number of woodlands (in the thousands) are to be scheduled for cutting over an n-year planning horizon. The expected yield from woodland i if cut in year j is known, and there are limitations on the total minimum and maximum acreage that may be cut in any year. A cutting schedule is to be determined such that the total yield over the n years is maximized. Let

r_i = number of acres in woodland i

$b_j^{\min}, b_j^{\max}$ = minimum and maximum number of acres that may be cut in year j

c_{ij} = yield from woodland i if cut in year j

x_{ij} = the fraction of woodland i to be cut in year j

Then the problem is

$$\text{maximize } z = \sum_{i,j} c_{ij}x_{ij} \tag{7}$$

subject to

$$\sum_{i=1}^{m} r_i x_{ij} + u_j = b_j^{\max} \qquad j = 1, \ldots, n \tag{8}$$

$$\sum_{i=1}^{m} r_i x_{ij} - v_j = b_j^{\min}, \qquad j = 1, \ldots, n \tag{9}$$

$$\sum_{j=1}^{n} x_{ij} = 1, \qquad i = 1, \ldots, m \tag{10}$$

$$x_{ij}, u_j, v_j \geq 0$$

where u_j and v_j are slack and surplus variables. Subtracting (9) from (8) yields

$$\text{maximize } z = \sum_{i,j} c_{ij}x_{ij} \tag{11}$$

subject to

$$\sum_{j=1}^{n} x_{ij} = 1, \qquad i = 1, \ldots, m \tag{12}$$

$$u_j + v_j = b_j^{\max} - b_j^{\min}, \qquad j = 1, \ldots, n \tag{13}$$

$$\sum_{i=1}^{m} r_i x_{ij} + u_j = b_j^{\max}, \qquad j = 1, \ldots, n \tag{14}$$

which is a generalized transportation problem with upper bounded slack variables, u_j.

The difficulty in solving (11)–(14) is due to the multiplicity of equations corresponding to each woodland, (12). Thus, in contrast to the previous section, here is a (generalized) transportation problem with many more "origins" than "destinations," i.e., $m \gg n$. We therefore reverse the procedure of Section 3.8, viewing (13)–(14) as coupling constraints and (12) as subsystem relations. Writing (11)–(14) in matrix form,

$$\text{maximize } z = cx \tag{11}$$

subject to

$$A_1 x + u = b^{\max} \tag{14}$$

$$u \leq b^{\max} - b^{\min}, \qquad u \geq 0 \tag{13}$$

$$\sum_{j=1}^{n} x_{ij} = 1, \qquad i = 1, \ldots, m, \quad x_{ij} \geq 0 \tag{12}$$

Let x^k be the kth extreme point of (12) above. Then the master program becomes

$$\text{maximize } z = \sum_k (cx^k)\lambda_k \tag{15}$$

subject to

$$\sum_k (A_1 x^k)\lambda_k + u = b^{\max} \tag{16}$$

$$\sum_k \lambda_k = 1 \tag{17}$$

$$u \leq b^{\max} - b^{\min} \tag{18}$$

$$u \geq 0, \qquad \lambda_k \geq 0 \tag{19}$$

The subproblems are developed as in Section 3.8. If π_j is the simplex multiplier corresponding to the jth constraint of (16), subproblem solution requires only scanning of a modified cost matrix, with elements $(c_{ij} - a_i\pi_j)$. The upper bounds (18) may be dealt with either by upper bounding techniques or by direct incorporation into the master-program basis.

This technique was applied by Tcheng [6] to a forestry-cutting problem with 1,166 woodlands to be cut over a 24-year period, with $b_j^{\max} = 9{,}700$ acres, $b_j^{\min} = 8{,}900$ acres, woodland acreages varying from 30 to 1,000 acres, and expected wood volume yield from each woodland increasing linearly with time. Solution was on an IBM 7044 computer with $32k$ core-storage capacity, using double-precision arithmetic. A summary of computations is given in Table 1. The objective increases rapidly during early iterations, but convergence becomes much slower as the computations progress. This phenomenon has occurred in other linear programs with a very large number of columns and will be discussed in subsequent sections.

TABLE 1

No. of iterations	Total yield in cords	Increase per 90 iterations	Cumulative computation time (min)
90	1,498,481	1,498,481	75
180	1,504,541	6,060	150
270	1,505,922	1,381	225
360	1,506,556	634	300
450	1,506,908	352	375
540	1,507,022	114	450
630	1,507,171	149	525
720	1,507,242	71	600
810	1,507,289	47	675
900	1,507,321	32	750

Solution of large transportation and generalized transportation problems is also possible using generalized upper bounding procedures. These are discussed in Chapter 6.

3.10 Optimal Allocation of Limited Resources

3.10.1 General Formulation. A variety of resource-allocation problems, including multi-item production scheduling, may be formulated as follows. Let there be I activities and a single resource, to be allocated to these activities in each of T time periods; let x_{it} be the level of activity i in time period t and define

$$x_i = (x_{i1}, \ldots, x_{iT}) \tag{1}$$

The set of allowable activity levels is in part determined by the necessity of satisfying external demands and by the internal characteristics of the activity. For example, machine characteristics may limit the rate of production of a commodity. All such constraints, whether referring to demand or production, are called technological constraints and are symbolized by defining a constraint set S_i such that

$$x_i \in S_i \tag{2}$$

Operating activity i incurs a cost and uses up the resource. Let

$c_i(x_i)$ = cost of operating activity i at levels x_i

$y_{it}(x_i)$ = amount of the resource used by activity i in time period t

b_t = resource availability in period t

$$Y_i(x_i) = (y_{i1}(x_i), \ldots, y_{iT}(x_i)) \tag{3}$$

$$b = (b_1, \ldots, b_T) \tag{4}$$

The problem of optimal resource allocation is to choose $x_1, \ldots, x_I$ to minimize the cost of operating all activities over time:

$$\text{minimize } z = \sum_{i=1}^{I} c_i(x_i) \tag{5}$$

subject to the resource limitations in all time periods:

$$\sum_{i=1}^{I} Y_i(x_i) \leq b \tag{6}$$

and the technological constraints

$$x_i \in S_i, \qquad i = 1, \ldots, I \tag{7}$$

The resource constraints (6) couple the activities together, forcing one to consider them all simultaneously when minimizing z. It is not difficult to extend this formulation to the case of more than one resource. An example of such a problem is given later in this section.

The problem of scheduling the production of I items over time may be included in this formulation by letting the activity level, x_{it}, be the quantity of item i produced in time period t. The resource could be labor or machinery, and $c_i(x_i)$ a production cost. Assuming that customer demands, r_{it}, are known for all items in all time periods, the constraints $x_i \in S_i$ could arise from upper and lower bounds on inventory levels, η_{it}, and production quantities, x_{it}; i.e.,

$$S_i = \{x_i \mid (x_{it})_{\min} \leq x_{it} \leq (x_{it})_{\max}, (\eta_{it})_{\min} \leq \eta_{it} \leq (\eta_{it})_{\max}, t = 1, \ldots, T\} \tag{8}$$

where the inventory levels are given by

$$\eta_{it} = \eta_{i,t-1} + x_{it} - r_{it}, \qquad t = 1, \ldots, T, \quad \eta_{i0} \text{ given} \tag{9}$$

Approximate Solution by Linear Programming. The problem (5)–(7) may be difficult to solve if I is large, e.g., many items, or if the x_{it} are discrete, as when production levels can be varied only by allocating a limited number of machines. Manne [9] has shown that an approximate solution may be obtained for the case $I \gg T$ by recasting the problem as a large linear program as follows. Assume that, for each i, the set of all x_i that need be considered as candidates for optimality in (5)–(7) is finite. This is surely the case if S_i is finite, as when the x_{it} are discrete, and is shown true in other situations later.

Let the jth such schedule from S_i be x_i^j. Define

$$c_{ij} \equiv c_i(x_i^j) \tag{10}$$

$$Y_{ij} \equiv Y_i(x_i^j) \tag{11}$$

and consider the following integer program:

$$\text{minimize} \sum_{i,j} c_{ij}\theta_{ij} \tag{12}$$

subject to

$$\sum_j \theta_{ij} = 1, \qquad i = 1, \ldots, I \tag{13}$$

$$\sum_{i,j} Y_{ij}\theta_{ij} \leq b \tag{14}$$

$$\theta_{ij} \geq 0 \tag{15}$$

$$\theta_{ij} = \text{integer} \tag{16}$$

Constraints (13), (15), and (16) require that, for each i, exactly one $\theta_{ij} = 1$, with all others zero. Thus θ_{ij} selects a particular schedule for item i, and (12) and (14) cause the schedules selected to satisfy the resource constraints and minimize the cost.

Solving (12)–(15) as an integer program would, however, be very difficult, owing to the potentially vast number of columns (one for each admissible schedule) and the many constraints if I or T is large. Fortunately, if $I \gg T$, the constraint θ_{ij} = integer may be dropped, and most of the variables in an optimal solution of the linear program (12)–(15) will still be integral. The precise result is this:

THEOREM. *If $I > T$, then any basic feasible solution of the linear program* (12)–(15) *has the property that at least $I - T$ of the indices i have precisely one θ_{ij} positive* (*and hence unity*).

PROOF. Any basic solution has at most $I + T$ variables positive. Each of the I constraints (13) requires at least one positive variable to satisfy it. The remaining T basic variables are then allocated to at most T different constraints of (13), leaving at least $(I - T)$ of these with exactly one positive variable.

There are many practical problems where $I \gg T$, for example in scheduling production of hundreds of items over a 12-month horizon. However, for such cases, solution of (12)–(15) is difficult, owing to the multiplicity of columns and many rows. Dzielinski and Gomory [10] have considered the application of the Dantzig–Wolfe decomposition principle to this problem.

This leads to a master program with only $T + 1$ rows and even more columns which, however, may be dealt with by column generation. Thus an efficient solution is possible.

The constraint matrix of (12)–(15) has the following structure:

$$
\begin{array}{cccc}
\theta_{11} \cdots \theta_{1j_1} & \theta_{21} \cdots \theta_{2j_2} & \cdots & \theta_{I1} \cdots \theta_{Ij_I} \\
\boxed{Y_{11} \cdots Y_{1j_1}} & \boxed{Y_{21} \cdots Y_{2j_2}} & \cdots & \boxed{Y_{I1} \cdots Y_{Ij_I}} \\
\boxed{1\ 1 \cdots \ 1} & & & \\
 & \boxed{1\ 1 \cdots \ 1} & & \\
 & & \ddots & \\
 & & & \boxed{1\ 1 \cdots \ 1}
\end{array}
$$

The constraints $\sum_j \theta_{ij} = 1$ obviously play the role of the subsystem constraints, while (14) couples the system together.

Defining,

$$e_i = (1, 1, \ldots, 1),\ j_i \text{ components} \tag{17}$$

$$\theta_i = (\theta_{i1}, \ldots, \theta_{ij_i}) \tag{18}$$

$$c_i = (c_{i1}, \ldots, c_{ij_i}) \tag{19}$$

$$\theta = (\theta_1, \ldots, \theta_I) \tag{20}$$

$$c = (c_1, \ldots, c_I) \tag{21}$$

$$L_i = [Y_{i1} \cdots Y_{ij_i}] \tag{22}$$

$$L = [L_1 L_2 \cdots L_I] \tag{23}$$

$$A = \begin{bmatrix} e_1 & & & \\ & e_2 & & \\ & & \ddots & \\ & & & e_I \end{bmatrix} \tag{24}$$

$$e = (1, 1, \ldots, 1),\ I \text{ components}$$

The linear program (12)–(15) may be more compactly written

$$\text{minimize } c\theta \tag{25}$$

subject to

$$L\theta + s = b \tag{26}$$

$$A\theta = e \tag{27}$$

$$\theta \geq 0,\ s \geq 0 \tag{28}$$

Since the set of all solutions to (27)–(28) is bounded, any solution may be written as a convex combination of extreme points, d_q:

$$\theta = \sum_q \lambda_q d_q, \qquad \lambda_q \geq 0, \qquad \sum_q \lambda_q = 1 \tag{29}$$

substituting (29) into (25)–(26) yields the following master program:

$$\text{minimize } z = \sum_q (cd_q)\lambda_q \tag{30}$$

subject to

$$\sum_q (Ld_q)\lambda_q + s = b \tag{31}$$

$$\sum_q \lambda_q = 1, \qquad \lambda_q \geq 0, s \geq 0 \tag{32}$$

This program has $T + 1$ constraints and $\prod_i(j_i)$ columns, one for each extreme point of (27)–(28). To develop the subproblems, let (π, π_0) be a vector of simplex multipliers for (31)–(32). The relative cost coefficient for column q is

$$\bar{c}_q = cd_q - \pi L d_q - \pi_0 \tag{33}$$

To minimize $\bar{c}_q$, recall that d_q is an extreme point of the constraints

$$\left.\begin{aligned} e_i\theta_i &= 1 \\ \theta_i &\geq 0 \end{aligned}\right\} i = 1, 2, \ldots, I \tag{34}$$

If d_q is partitioned as

$$d_q = (d_{q1}, \ldots, d_{qI}) \tag{35}$$

then it is easily verified that each d_{qi} has the form

$$d_{qi} = (0, 0, \ldots, 0, 1, 0, \ldots, 0) \tag{36}$$

The quantities cd_q and Ld_q may be similarly partitioned:

$$cd_q = \sum_{i=1}^{I} c_i d_{qi} \tag{37}$$

$$Ld_q = [L_1 \cdots L_I] \begin{bmatrix} d_{q1} \\ \vdots \\ d_{qI} \end{bmatrix} = \sum_{i=1}^{I} L_i d_{qi} \tag{38}$$

Thus the relative cost coefficient is

$$\bar{c}_q = \sum_{i=1}^{I} (c_i - \pi L_i)d_{qi} - \pi_0 \tag{39}$$

Since (39) is separable, minimizing $\bar{c}_q$ is equivalent to minimizing, for each i,

$$(c_i - \pi L_i)d_{qi} \tag{40}$$

over all d_{qi} of the form (36). The solution is obviously to put the "1" in the position where the minimal component of the vector $(c_i - \pi L_i)$ occurs. Using the definitions in (19), (10)–(11), and (22), this minimum is found by solving the subproblem

$$\underset{j}{\text{minimize}} \; (c_i(x_i^j) - \pi Y_i(x_i^j)) \tag{41}$$

Since $x_i^j \in S_i$, this is equivalent to

$$\text{minimize } c_i(x_i) - \pi Y_i(x_i) \tag{42}$$

subject to

$$x_i \in S_i \tag{43}$$

Note that this is a single item scheduling and inventory problem with a penalty term, πY_i, for using the common resource. The quantity $-\pi_t$ may be viewed as the price of this resource in period t.

Let x_i^0 solve subproblem i. By (39),

$$\min \bar{c}_q = \sum_{i=1}^{I} (c_i(x_i^0) - \pi Y_i(x_i^0)) - \pi_0 \tag{44}$$

Pricing out the slack columns, the relative cost coefficient is

$$\bar{c}_t = -\pi_t \tag{45}$$

If

$$\min \bar{c}_q < 0 \tag{46}$$

$$\min \bar{c}_q < \min_t \bar{c}_t \tag{47}$$

Then the column

$$\begin{bmatrix} Ld_q \\ 1 \end{bmatrix} = \begin{bmatrix} \sum_i Y_i(x_i^0) \\ 1 \end{bmatrix}$$

is formed and enters the current basis. If (47) does not hold, but $\min_t \bar{c}_t < 0$, then a slack column enters, while if all columns price out nonnegative the current basis is optimal.

3.10.2 Specializing the Model—Lot Sizes and Labor Allocations. As shown by Dzielinski and Gomory [10], under certain assumptions on the production process and the costs involved, the subproblems (42)–(43) take on an especially convenient form. Let production be of the batch type, with a given item incurring a setup in any time period in which its production is nonzero. This might occur in the machine shop of a metal-working company producing many different parts [9]. The common resource is labor, and each setup requires some number of labor hours to accomplish it. There are different types and pay classes of labor and the labor force may be increased or decreased in each time period—but with associated costs of hiring and layoff. This is a case where the resource levels b_t may be varied, and the problem is to choose optimal resource availabilities and optimal resource allocations simultaneously.

It is assumed that there is no backlogging of requirements for any item, initial and final inventories are zero, and production has no upper bound. Thus the constraints $x_i \in S_i$ take the form

$$\eta_{it} = \sum_{\tau=1}^{t} (x_{i\tau} - r_{i\tau}) \geq 0, \qquad t = 1, \ldots, T-1 \tag{1}$$

$$\eta_{iT} = \sum_{\tau=1}^{T} (x_{i\tau} - r_{i\tau}) = 0 \tag{2}$$

$$x_{it} \geq 0, \qquad \text{all } i, t \tag{3}$$

The cost $c_i(x_i)$ is a linear production cost, which may be viewed as a cost of materials, and which may vary with item and time:

$$c_i(x_i) = \sum_{t=1}^{T} c_{it} x_{it} \tag{4}$$

The vector of resource consumptions, $Y_i(x_i)$, gives the number of hours of labor required to implement the schedule x_i. Each of a number of different kinds of labor is required, indexed by type $k = 1, \ldots, K$ and payment class $r = 1, \ldots, R$, e.g., $r = 1 \Rightarrow$ straight time, $r = 2 \Rightarrow$ straight time plus overtime, etc.

The number of labor hours used includes time to set up the production facilities (required whenever $x_{it} > 0$) and time to actually perform the production. Let

a_{ik} = number of hours required to set up production for item i using labor of type k

b_{ik} = number of hours required by labor of type k to produce one unit of item i

Then the amount of labor of type k required to produce an amount x_{it} is

$$y_{ikt}(x_{it}) = a_{ik}\delta(x_{it}) + b_{ik}x_{it} \tag{5}$$

where

$$\delta(x) = \begin{cases} 1, & x > 0 \\ 0, & \text{otherwise} \end{cases} \tag{6}$$

Let

h_k^r = number of hours a (k, r) worker will work in a time period, assumed fixed

w_{kt}^r = number of workers of (k, r) type working in time period t, a decision variable

Then the constraints on resource availabilities become

$$\sum_i y_{ikt}(x_{it}) \leq \sum_r h_k^r w_{kt}^r, \qquad t = 1, \ldots, T, \quad k = 1, \ldots, K \tag{7}$$

Defining

$$Y_{ik}(x_i) = (y_{ik1}, y_{ik2}, \ldots, y_{ikT}) \tag{8}$$

$$b_k = \left(\sum_r h_k^r w_{k1}^r, \ldots, \sum_r h_k^r w_{kT}^r\right) \tag{9}$$

the resource constraints (7) become

$$\sum_i Y_{ik}(x_i) \leq b_k, \qquad k = 1, \ldots, K \tag{10}$$

We see that this problem represents an extension of that in (3.10.1-5)–(3.10.1-7) since there are K different resources rather than just one.

Equations must be written governing the change in labor force over time. Let

w_{kt}^+ = increase in labor force, type k, period t

w_{kt}^- = decrease in labor force, type k, period t

Then

$$\sum_r w_{kt}^r = \sum_r w_{k,t-1}^r + w_{kt}^+ - w_{kt}^-, \qquad w_{k0}^r \text{ given} \tag{11}$$

The objective function now includes costs of increasing and decreasing the labor force and of paying the workers

$$z = \sum_i c_i(x_i) + \sum_{k,t,r} c_{kt}^r w_{kt}^r + \sum_{k,t} (c_{kt}^+ w_{kt}^+ + c_{kt}^- w_{kt}^-) \tag{12}$$

where c_{kt}^r = wages paid to (k, r) worker in time period t

c_{kt}^+ = cost of increasing labor force of type k by one unit in period t

c_{kt}^- = similar cost of decrease

The production scheduling problem is to choose the schedules $x_1, \ldots, x_I$ to minimize (12) subject to (10), (11), and (1)–(3).

Here the sets S_i defined by (1)–(3) are not finite. We will show, however, that only a finite number of elements from each S_i can ever enter an optimal solution. Let us defer this question for a moment and write the approximating linear program (3.10.1-12)–(3.10.1-15) in terms of some finite but arbitrary set of schedules $x_i^j \in S_i$. As before, let

$$c_{ij} \equiv c_i(x_i^j), \qquad Y_{ik}(x_i^j) \equiv Y_{ijk} \tag{13}$$

Then the approximating linear program is

$$\text{minimize} \sum_{i,j} c_{ij}\theta_{ij} + \sum_{k,t,r} c_{kt}^r w_{kt}^r + \sum_{k,t} (c_{kt}^+ w_{kt}^+ + c_{kt}^- w_{kt}^-) \tag{14}$$

subject to

$$\sum_{i,j} Y_{ijk}\theta_{ij} \le b_k, \qquad k = 1, \ldots, K \tag{15}$$

$$\sum_r w_{kt}^r = \sum_r w_{k,t-1}^r + w_{kt}^+ - w_{kt}^-, \qquad k = 1, \ldots, K, \quad t = 1, \ldots, T \tag{16}$$

$$\sum_j \theta_{ij} = 1, \qquad i = 1, \ldots, I \tag{17}$$

and all variables nonnegative.

As shown previously, if $I > 2KT$ [the number of constraints in (15)–(16)] then $I - 2KT$ of the items will have a single schedule in any basic solution of this program.

Application of the decomposition principle proceeds as before, with each of the K sets of resource constraints (15) handled as the single set was in Section 3.10.1. Let

$$L_{ik} = [Y_{i1k}, Y_{i2k}, \ldots, Y_{ij_ik}] \tag{18}$$

$$L_k = [L_{1k} \mid L_{2k} \mid, \ldots, \mid L_{Ik}] \tag{19}$$

If d_q again represents an extreme point of the constraints (3.10.1-34), the master program is

$$\text{minimize} \sum_q (cd_q)\lambda_q + \sum_{k,t,r} c_{kt}^r w_{kt}^r + \sum_{k,t} (c_{kt}^+ w_{kt}^+ + c_{kt}^- w_{kt}^-) \tag{20}$$

$$\text{subject to} \quad \sum_q (L_k d_q)\lambda_q \le b_k, \qquad k = 1, \ldots, K \tag{21}$$

$$\sum_q \lambda_q = 1 \tag{22}$$

$$\sum_r w_{kt}^r = \sum_r w_{k,t-1}^r + w_{kt}^+ - w_{kt}^-, \qquad \text{all } k, t \tag{23}$$

and all variables nonnegative. Note that the vectors b_k involve the decision variables w_{kt}^r, as shown by (9). The columns corresponding to the variables w_{kt}^r, w_{kt}^+, and w_{kt}^-, and the various slacks must be priced out separately in the decomposition algorithm. The following discussion shows how to deal with the columns corresponding to the variables λ_q.

Let B be a feasible basis for the master program, let π_k be the (row) vector of simplex multipliers associated with the kth set of resource constraints, (21), and let π_0 be the multiplier for (22). Pricing out the column corresponding to λ_q yields the relative cost coefficient

$$\bar{c}_q = cd_q - \sum_k \pi_k L_k d_q - \pi_0 \tag{24}$$

As in Section 3.10.1, d_q can be partitioned

$$d_q = (d_{q1}, \ldots, d_{qI}) \tag{25}$$

with each d_{qi} a unit vector, so (24) can be rewritten

$$\bar{c}_q = \sum_i \left(c_i - \sum_k \pi_k L_{ik} \right) d_{qi} - \pi_0 \tag{26}$$

To minimize $\bar{c}_q$, each term

$$\left(c_i - \sum_k \pi_k L_{ik} \right) d_{qi} \tag{27}$$

must be minimized. Since each d_{qi} is a unit vector, the smallest component of the vector in parentheses in (27) must be located:

$$\text{minimize } c_i(x_i^j) - \sum_k \pi_k Y_{ik}(x_i^j) \tag{28}$$

Since $x_i^j \in S_i$, this is equivalent to

$$\text{minimize} \sum_t c_{it} x_{it} - \sum_{k,t} \pi_{kt}(a_{ik}\delta(x_{it}) + b_{ik}x_{it}) \tag{29}$$

or

$$\text{minimize} \sum_t \left\{ c_{it} - \sum_k \pi_{kt} b_{ik} \right\} x_{it} - \sum_t \left\{ \sum_k \pi_{kt} a_{ik} \right\} \delta(x_{it}) \tag{30}$$

subject to the single item constraints (1)–(3). If the π_{kt} could be guaranteed to be nonpositive, then this would be a single-item, deterministic inventory problem of the Wagner–Whitin type [11], since then all coefficients of x_{it} and $\delta(x_{it})$ would be nonnegative. Solution is then possible by a very efficient dynamic programming algorithm. To guarantee $\pi_{kt} \leq 0$, note that the relative cost coefficient for the slack variable in (21) is

$$\bar{c}_{kt} = -\pi_{kt}$$

and the simplex condition that no slack variable may enter the current basis and reduce the cost is

$$\pi_{kt} \leq 0 \tag{31}$$

Thus, to ensure (31), simply bring into the current basis all slack vectors which will reduce costs until (31) holds, whereupon the subproblems (30) may be solved as Wagner–Whitin problems. Note that, before a final decision as to the column to enter the basis can be made, the columns corresponding to the variables w_{kt}^r, w_{kt}^+, and w_{kt}^- must be priced out also.

Returning to the question of a finite number of candidates for optimality from the S_i, the fact that the subproblem (30) is of the Wagner–Whitin form resolves this issue. Wagner and Whitin show that optimal solutions to such problems have the property: $x_{it} > 0$ implies $\eta_{i,t-1} = 0$. Thus the only possible optimal production levels are partial sums of the demands, r_{it}, a finite set.

3.10.3 Computational Experience. Dzielinski and Gomory have solved a number of scheduling problems using the decomposition principle. The master program for these problems has the form (20)–(23) with additional inequalities restricting the size of the work force in each of S shifts. Their program used the revised simplex method with product form of the basis inverse. The computation was initiated in phase 1 with no production plans and a basis of slack and artificial variables. A summary of the results obtained in solving 10 problems is given in Table 1.

TABLE 1

	Problem									
	(1)	(2)	(3)	(4)	(5)	(6)	(7)	(8)	(9)	(10)
I: Parameters That Determine Size of Problems										
(1) I: number of items	35	963	428	428	428	428	100	100	100	100
(2) K: number of facilities	2	2	2	2	2	2	1	1	1	1
(3) T: number of periods	3	3	3	5	7	8	12	12	12	12
(4) S: number of shifts	2	2	2	2	2	2	2	2	2	2
(5) $(2 \cdot K \cdot T) \cdot (S + 1)$: work-force columns	36	36	36	60	84	96	72	72	72	72
(6) $(K \cdot T) \cdot 3$: slack columns	18	18	18	30	42	48	36	36	36	36
(7) $(S + 2) \cdot (K \cdot T) + 2$ number interacting constraint rows	26	26	26	42	58	66	50	50	50	50
(8) $(2^{T-1}) \cdot I$: item production schedules	140	3,847	1,712	6,784	27,392	54,434	204,800	204,800	204,800	204,800
II: Size of Problems as Ordinary Linear Programming Problems										
(9) (1) + (7): number of rows	61	989	454	466	486	494	150	150	150	150
(10) (5) + (6) + (8): number of columns	194	3,901	1,766	6,874	27,518	54,928	204,908	204,908	204,908	204,908

III: Size of Problems with Decomposition and Dynamic Programming

(11) (7): number of rows	27	27	27	43	59	67	51	51	51	51
(12) (5) + (6): number of columns	54	54	54	90	126	144	108	108	108	108
(13) $(I) \cdot (T(T+1)/2)$: number of elementary steps in dynamic programming algorithm	210	5,778	2,568	6,420	11,984	15,408	7,800	7,800	7,800	7,800

IV: Computations of Problems with Decomposition and Dynamic Programming

(14) Total time to compute optimal solution (min of 7090 time)	3.36[a,b]	24.5[a,b]	5.66[a]	9.0[a]	10.70[a]	17.92[a]	5.90[a]	8.87[a]	8.07[a]	9.00[a]
(15) Total phase II iterations	48	34	22	35	52	65	45	49	46	51
(16) Average time to create a production plan (min)	0.07	0.72[b]	0.264	0.345	0.483	0.510	0.183	0.225	0.223	0.215
(17) Number of production plans created	16	12	19	24	20	29	26	34	31	36

[a] This does not include 7090 time to set up the problem and output the results.

[b] This problem was run using a version of the dynamic programming subroutine that was later improved. The improved routine, used in the remaining problems, requires about one third less time than the original routine.

In part I of the table, (5) + (6) gives the number of columns corresponding to variables w_{kt}^r, w_{kt}^+, w_{kt}^-, and all slack variables, while number (7) gives the number of rows not of the form $\sum_j \theta_{ij} = 1$. It is clear from part II that a number of these problems—certainly the last four—would have been very difficult, if not impossible, to run as standard linear programs, owing to the very large number of columns. These would require a vast amount of memory for storage; enough core would seldom be available, and storage and retrieval from disk or tape at every simplex iteration could require prohibitive amounts of time. Decomposition or some similar column-generation approach, which creates the columns as needed, is mandatory.

The number of phase 2 iterations required is quite small—usually somewhere near the number of rows, m, of the program. This is about the lowest that can be expected, m to $3m$ being a commonly used rule of thumb, and is very encouraging in light of the experience cited in Section 3.9 for the forestry problem. One might expect that, owing to the vast number of columns in the master program, a large number of near-optimal solutions would occur prior to optimality. This would be especially serious here for the following reason. The linear program being decomposed is only valuable when its optimal solution possesses a majority of integer variables. All basic solutions of this program have this property (if $I > 2KT$), but nonoptimal basic solutions of the master program do not, since only at optimality need they correspond to basic solutions of the original problem. Thus intermediate solutions of the master program could have an arbitrary number of items with fractional schedules, in which case early termination would not yield a meaningful solution.

Figure 3-2 shows the percentage of items with more than one schedule in the basis versus iteration count for the last four problems of Table 1. This

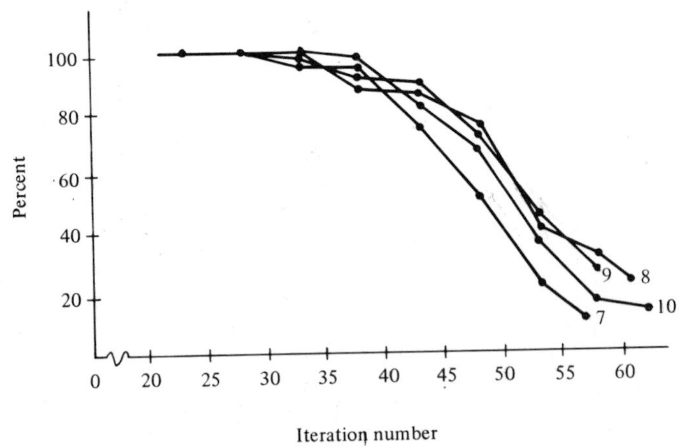

FIGURE 3-2 Percentage of items with split lots in the solution to problems 7, 8, 9, and 10.

percentage decreases as the iterations progress, indicating that, for some problems, termination of the decomposition principle short of optimality still yields useful results.

An alternative approach for solving (3.10.1–12)(3.10.1–15) is given in Section 4.2. A direct attack on the problem (3.10.1–5)(3.10.1–7) is discussed in Section 8.8.3.

3.11 Primal–Dual Approach to the Master Program

The primal–dual method [12], discussed in Section 1.2.4, is a variant of the simplex method, which has shown itself to be particularly efficient for a number of problems having special structure, especially transportation problems. Our eventual goal is the application of this technique to the master program of the decomposition principle As will be shown, this leads to subproblems involving the minimization of a *ratio* of linear functions subject to linear equalities and inequalities. In this preliminary section, we discuss the problem of minimizing ratios of linear functions, a topic of interest in itself. Application to the master program is considered in Section 3.11.2.

3.11.1 Linear Fractional Programming. The problem of concern in this section is

$$\text{minimize } \frac{px + \alpha}{qx + \beta} \tag{1}$$

$$\text{subject to} \qquad a_i x \leq b_i, \qquad i = 1, \ldots, m \tag{2}$$

This problem is of interest for a variety of reasons, [13]–[14]. It is included here because the subproblems arising from the decomposition approach in the next section are of this form, and we wish to show that they may be solved by the simplex method.

A ratio of linear functions is neither convex nor concave. However, its contours of constant value, the sets

$$S_b = \{x \mid px + \alpha/qx + \beta = b\} \tag{3}$$

are hyperplanes [although replacing the equality by inequality in (3) does not yield half-spaces]. Dorn [13] has shown that any local minimum of a linear fractional program is global and, if a finite optimal solution is attained at a finite point, there is an extreme point which is optimal. Defining

$$S = \{x \mid a_i x \leq b_i,\ i = 1, \ldots, m\} \tag{4}$$

these results hold assuming only that the numerator and denominator of (1) do not vanish simultaneously for any $x \in S$. However, the fractional programs of the next section include the restriction that the denominator of the ratio is positive over S. Since this property simplifies the following proofs, we assume it holds.

Assumption 1

$$qx + \beta > 0 \qquad \text{for any } x \in S$$

The notation is simplified by adding a component to x:

$$x = (x_1, \ldots, x_n, x_{n+1}) \tag{5}$$

and adding the constraints

$$x_{n+1} \geq 1, \qquad x_{n+1} \leq 1 \tag{6}$$

so the ratio may be written

$$f(x) = \frac{px}{qx} \tag{7}$$

DEFINITION 1. A continuous function $f(x)$ is *quasiconvex* over a convex set, S, if either of the following equivalent statements holds:

(a) $\{x \mid f(x) \leq c, x \in S\}$ is convex for all c (8)

(b) $x_1, x_2 \in S, f(x_2) < f(x_1) \Rightarrow f(\lambda x_2 + (1 - \lambda)x_1) \leq f(x_1), \quad 0 < \lambda < 1$ (9)

DEFINITION 2. A *function* $f(x)$ is *strictly quasiconvex* over a convex set, S, if the strict inequality holds in (9).

The following theorem is due to Ponstein [17].

THEOREM 1. *Any local minimum of a strictly quasiconvex function is a global minimum.*

The proof is found in [17]. Counterexamples can be constructed to show that the theorem is not true when the strictness is dropped.

The importance of this theorem for our purposes is that, under assumption 1, $f(x) = px/qx$ is strictly quasiconvex over the set S given in (4).[5]

This leads immediately to

THEOREM 2. *Under assumption* 1, *any local minimum of the fractional program* (1)–(2) *is global.*

To prove the next result, the following definition is needed.

[5] Problem 12 asks that this be proved.

DEFINITION 3. A *function* $f(x)$ is *quasiconcave* over a convex set, S, if $-f(x)$ is quasiconvex.

Since, under assumption 1, the set

$$\{x \mid -px/qx \leq c, x \in S\}$$

is convex for all c, $f(x) = px|qx$ is quasiconcave over the set S in (4). For continuous quasiconcave functions, the defining relation (9) with f replaced by $-f$ yields

$$f(x_2) > f(x_1) \Rightarrow f(\lambda x_1 + (1 - \lambda)x_2) \geq f(x_1) \tag{10}$$

or, equivalently,

$$f(\lambda x_1 + (1 - \lambda)x_2) \geq \min[f(x_1), f(x_2)] \tag{11}$$

for any $x_1,x_2 \in S$ and any $0 < \lambda < 1$. Relation (11) will be used as a defining relation for quasiconcavity in what follows.

Since we assume, in the angular problem of the next section, that the subsystem constraints define a bounded set, the following theorem is stated and proved under that assumption.

THEOREM 3. *Assume that the set S in* (4) *is bounded and that assumption* 1 *holds. Then $f(x) = px/qx$ takes on its minimum value over S at an extreme point of S.*

PROOF. Let x^i, $i = 1, \ldots, s$, be the extreme points of S. Then, by Theorem 3.2-2, any $x \in S$ may be written

$$x = \sum_{i=1}^{s} \alpha_i x^i, \qquad \sum_{i=1}^{s} \alpha_i = 1, \qquad \alpha_i \geq 0 \tag{12}$$

Since f is quasiconcave, (11) yields

$$f(\alpha x^1 + (1 - \alpha)x^2) \geq \min[f(x^1), f(x^2)], \qquad \text{all } 0 \leq \alpha \leq 1 \tag{13}$$

Let

$$x_3 = \alpha x^1 + (1 - \alpha)x^2 \tag{14}$$

Then, again using (11),

$$f(\beta x_3 + (1 - \beta)x^3) \geq \min[f(x_3), f(x^3)], \qquad 0 \leq \beta \leq 1 \tag{15}$$

Since, by (13), $f(x_3) \geq \min[f(x^1), f(x^2)]$

$$\min[f(x_3), f(x^3)] \geq \min[f(x^1), f(x^2), f(x^3)] \tag{16}$$

Also

$$\beta x_3 + (1 - \beta)x^3 = \beta\alpha x^1 + \beta(1 - \alpha)x^2 + (1 - \beta)x^3 \tag{17}$$

The coefficients of the above linear combination sum to one and are nonnegative. Thus, as both α and β range between zero and one, $\beta x_3 + (1 - \beta)x^3$ assumes all values of the form $\alpha_1 x^1 + \alpha_2 x^2 + \alpha_3 x^3$, where the α's are nonnegative and sum to one.

Using this fact and (16), (15) becomes

$$f(\alpha_1 x^1 + \alpha_2 x^2 + \alpha_3 x^3) \geq \min[f(x^1), f(x^2), f(x^3)] \tag{18}$$

Repeated applications of this procedure yield

$$f(x) = f\left\{\sum_{i=1}^{s} \alpha_i x^i\right\} \geq \min_i\{f(x^i)\} \tag{19}$$

which is the desired result.

This proof obviously holds for any quasiconcave function. Martos [18] shows that the theorem holds as an if and only if. That is, let L be a polyhedral subset of a bounded convex polyhedron, S. Then the following statement holds for all L if and only if $f(x)$ is quasiconcave over S: $f(x)$ assumes its minimum over L at an extreme point of L.

An extension to Theorem 3 may be proved, stating that if a linear fractional program with positive denominator has an unbounded solution, then the objective goes to $-\infty$ along an extreme ray (homogeneous solution) of the set S. This makes the parallel between linear fractional programs and linear programs fairly complete.

Algorithms for Linear Fractional Programming. A number of computational algorithms have been proposed for linear fractional programs (see [13] of Section 2.6 and [13]–[15] of this section). Those of greatest interest here are due to Charnes and Cooper [15] and Gilmore and Gomory [14]. The Charnes–Cooper method assumes only that the constraint set is bounded. It searches out optima for which the denominator of the ratio is restricted to either positive or negative values, as desired. Since the subproblems arising in the next section include the constraint that the denominator be positive, this algorithm seems appropriate for their solution. The procedure of Gilmore and Gomory is a direct extension of the simplex method, requiring only small modifications of existing simplex codes. To apply the method, the fractional program must have the essential properties of a linear program. As seen previously, this is true only if the denominator of the ratio has one sign over the constraint set; otherwise it vanishes and the program may have an infinite solution not attained at an extreme point or approached along an extreme ray. The algorithm allows linear fractional programs with many columns to be attacked by column generation procedures; an application of this type is considered in Section 4.1.

Charnes–Cooper Algorithm. Let the linear fractional program be written in the form

$$\text{minimize } \frac{px + \alpha}{qx + \beta} \tag{20}$$

subject to

$$Ax \leq b \tag{21}$$

$$x \geq 0 \tag{22}$$

where the set

$$S = \{x \mid Ax \leq b, x \geq 0\} \tag{23}$$

is assumed to be nonempty and bounded. We make the transformation of variables

$$y = tx, \qquad t \geq 0 \tag{24}$$

and consider the following linear program:

$$\text{minimize } py + \alpha t \tag{25}$$

subject to

$$Ay - bt \leq 0 \tag{26}$$

$$qy + \beta t = \gamma, \qquad \gamma \neq 0 \tag{27}$$

$$y \geq 0, \qquad t \geq 0 \tag{28}$$

which is obtained from the original by multiplying through by t and setting the denominator to a specific value, γ. This linear program has the following property: Every (y, t) satisfying (26)–(28) has $t > 0$. For if not, there exists some y satisfying $Ay \leq 0$, $y \geq 0$. By (27), this y is not the zero vector. Then, if x satisfies (21)–(22), $x + \alpha y$ satisfies these also for all $\alpha \geq 0$, contradicting the boundedness of S.

The following relationship holds between this linear program and the fractional program:

THEOREM 4. *If*

(a) *both* γ *and* $qx^0 + \beta$ *are positive, for* x^0 *some optimal solution of* (20)–(22),

(b) (y^*, t^*) *solves the linear program,* (25)–(28),

then y^*/t^* *solves the fractional program* (20)–(22).

PROOF. Clearly y^*/t^* is feasible for the fractional program. Assume that y^*/t^* is not optimal. Then x^0 yields a smaller objective value:

$$\frac{px^0 + \alpha}{qx^0 + \beta} < \frac{p(y^*/t^*) + \alpha}{q(y^*/t^*) + \beta} \tag{29}$$

Since γ and $qx^0 + \beta$ have the same sign, there is a $\theta > 0$ such that

$$qx^0 + \beta = \theta\gamma \tag{30}$$

Consider the pair

$$\hat{y} = \frac{x^0}{\theta}, \qquad \hat{t} = \frac{1}{\theta} \tag{31}$$

It is easily verified that $(\hat{y}, \hat{t})$ is feasible for the linear program (25)–(28), and dividing the numerator and denominator of the left-hand side of (29) by θ yields

$$\frac{p\hat{y} + \alpha\hat{t}}{\gamma} < \frac{py^* + \alpha t^*}{\gamma} \tag{32}$$

which contradicts the optimality of (y^*, t^*). Thus the theorem is true.

This theorem enables us to solve a fractional program by solving the linear program (25)–(28) if the sign of the denominator at the optimum is positive. If it is negative, replacing (p, α) and (q, β) by their negatives makes it positive and leaves the ratio unaltered. Thus any finite minimum to the fractional program may be located by solving the following pair of linear programs:

Denominator positive	Denominator negative
minimize $py + \alpha t$ subject to $Ay - bt \le 0$ $qy + \beta t = 1$ $y \ge 0, \quad t \ge 0$	minimize $-py - \alpha t$ subject to $Ay - bt \le 0$ $-qy - \beta t = 1$ $y \ge 0, \quad t \ge 0$

Of course, only one of the pair need be solved if the sign of the denominator at the optimum is known.

Adaptation of the Simplex Method. Let us again assume that the denominator of the ratio is positive on the constraint set S. Then, as shown earlier, any finite minimum for the fractional program is attained at an extreme point, and unbounded solutions are approached along extreme rays. The problem can thus be solved by moving from extreme point to extreme point, always choosing a new extreme point with lower objective value. The following result shows how a better extreme point may be chosen.

THEOREM 5. *Let $f(x) = z_1(x)/z_2(x)$ be a ratio of linear functions defined on a set S, let $y \in S$, and let s be a given direction. Assume that*

(a) $z_2(x) > 0$ *for all* $x \in S$.

(b) $y + \alpha s \in S$ *for some range* $0 < \alpha \leq \delta$, $\delta > 0$.

(c) $g(0) = \left.\dfrac{df(y + \alpha s)}{d\alpha}\right|_{\alpha = 0} < 0$

Then

$$h(\alpha) = f(y + \alpha s)$$

is monotone decreasing for all $0 < \alpha \leq \delta$.

PROOF. By assumptions (a) and (b), $h(\alpha)$ is continuous and differentiable in the range $[0, \delta]$. Thus the theorem is proved if we can show that

$$g(\alpha) = \frac{dh}{d\alpha}$$

is negative on $[0, \delta]$. Differentiating,

$$\begin{aligned} g(\alpha) &= \frac{d}{d\alpha} \frac{z_1(y + \alpha s)}{z_2(y + \alpha s)} \\ &= \frac{z_2(y + \alpha s)z_1(s) - z_1(y + \alpha s)z_2(s)}{z_2^2(y + \alpha s)} \\ &= \frac{(z_2(y) + \alpha z_2(s))z_1(s) - (z_1(y) + \alpha z_1(s))z_2(s)}{z_2^2(y + \alpha s)} \\ &= \frac{z_2(y)z_1(s) - z_1(y)z_2(s)}{z_2^2(y + \alpha s)} \end{aligned} \tag{33}$$

By evaluating $g(0)$ from (33) it is easily seen that

$$g(\alpha) = \frac{z_2^2(y)}{z_2^2(y + \alpha s)} g(0) \tag{34}$$

Since neither $z_2(y)$ nor $z_2(y + \alpha s)$ is zero and $g(0) < 0$, $g(\alpha) < 0$ for all α in $[0, \delta]$.

It is clear from the above that if $g(0) > 0$, $h(\alpha)$ is increasing. Thus if the directional derivative of the ratio along an edge leading to a new extreme point is negative, the value at the new point is lower. If, at some extreme point, all such directional derivatives are nonnegative, the point is optimal. All that is needed is an efficient way to calculate the directional derivatives along all edges. Let the linear constraints be placed in canonic form relative to some feasible basis, B, and add the two equations

$$\sum_j c_{1j}x_j - z_1 = 0 \tag{35}$$

$$\sum_j c_{2j}x_j - z_2 = 0 \tag{36}$$

to the system, with $-z_1$ and $-z_2$ permanent basic variables. If the basic x_j are eliminated from (35) and (36), the equations take the form

$$\begin{array}{lllllll} x_1 & & & + \bar{a}_{1,m+1}x_{m+1} & + \cdots + & \bar{a}_{1n}x_n & = \bar{b}_1 \\ \quad\ddots & & & \vdots & & \vdots & \vdots \\ \quad\quad x_m & & & + \bar{a}_{m,m+1}x_{m+1} & + \cdots + & \bar{a}_{mn}x_n & = \bar{b}_m \\ & -z_1 & & + \bar{c}_{1,m+1}x_{m+1} & + \cdots + & \bar{c}_{1n}x_n & = -\bar{z}_1 \\ & & -z_2 & + \bar{c}_{2,m+1}x_{m+1} & + \cdots + & \bar{c}_{2n}x_n & = -\bar{z}_2 \end{array} \tag{37}$$

To decide which nonbasic variable will enter the basis, we require the directional derivatives

$$\bar{c}_j = \frac{\partial(z_1/z_2)}{\partial x_j} = \frac{\bar{z}_2\,\partial z_1/\partial x_j - \bar{z}_1\,\partial z_2/\partial x_j}{\bar{z}_2^2}, \qquad j = m+1, \cdots, n \tag{38}$$

$$\bar{c}_j = \frac{\partial(z_1/z_2)}{\partial x_j} = \frac{\bar{z}_2\bar{c}_{1j} - \bar{z}_1\bar{c}_{2j}}{\bar{z}_2^2} \tag{39}$$

As seen from (39), these are easily calculated if the relative cost factors $\bar{c}_{1j}$ and $\bar{c}_{2j}$ are known. These, in turn, may be computed in revised simplex style from the original problem data and certain simplex multipliers. Let the current $(m + 2) \times (m + 2)$ basis be

$$\hat{B} = \left[\begin{array}{c|c|c} B & 0 & 0 \\ \hline c_{1B} & 1 & 0 \\ \hline c_{2B} & 0 & 1 \end{array}\right] \tag{40}$$

where c_{1B} and c_{2B} are the vectors of objective coefficients of basic variables in z_1 and z_2, respectively. It is easily verified that the inverse of $\hat{B}$ is

$$\hat{B}^{-1} = \left[\begin{array}{c|c|c} B^{-1} & 0 & 0 \\ \hline -c_{1B}B^{-1} & 1 & 0 \\ \hline -c_{2B}B^{-1} & 0 & 1 \end{array}\right] \tag{41}$$

Let

$$\pi_1 = c_{1B}B^{-1}$$
$$\pi_2 = c_{2B}B^{-1}$$

be the simplex multipliers of the current basis relative to z_1 and z_2, respectively. Then the nonbasic columns in (37) are

$$\left[\begin{array}{c} \bar{p}_j \\ \hline \bar{c}_{1j} \\ \bar{c}_{2j} \end{array}\right] = \hat{B}^{-1}\left[\begin{array}{c} p_j \\ \hline c_{1j} \\ c_{2j} \end{array}\right] = \left[\begin{array}{c} B^{-1}p_j \\ \hline c_{1j} - \pi_1 p_j \\ c_{2j} - \pi_2 p_j \end{array}\right] \tag{42}$$

Thus the relative cost factors needed to evaluate $\bar{c}_j$ in (39) can be computed using the multipliers π_1, π_2, and the original problem data from the expressions

$$\bar{c}_{1j} = c_{1j} - \pi_1 p_j \tag{43}$$
$$\bar{c}_{2j} = c_{2j} - \pi_2 p_j \tag{44}$$

The algorithm proceeds exactly as the revised simplex method, with the directional derivative $\bar{c}_j$ playing the role of relative cost factor. The only additional work required is in maintaining rows in $\hat{B}^{-1}$ for the objective functions, computing $\bar{c}_{1j}$ and $\bar{c}_{2j}$ from (43)–(44), and finally computing $\bar{c}_j$ from (39). Barring degeneracy, the rational objective decreases at each step, no basis is repeated, and an optimal solution is found in a finite number of iterations.

3.11.2 Application of the Primal–Dual Method to the Master Program. The problem of interest here is again the p-block angular structure

$$\text{minimize } z = \sum_i c_i x_i \tag{1}$$

subject to

$$\sum_i A_i x_i = b_0 \tag{2}$$

$$B_i x_i = b_i, \qquad x_i \geq 0, \quad i = 1, \ldots, p \tag{3}$$

If we assume that the sets

$$S_i = \{x_i \mid B_i x_i = b_i,\ x_i \geq 0\} \tag{4}$$

are bounded, then, as shown in Section 3.4, the master program is

$$\text{minimize } z = \sum_{i,j} f_{ij}\lambda_{ij} \tag{5}$$

subject to

$$\sum_{i,j} p_{ij}\lambda_{ij} = b_0 \tag{6}$$

$$\sum_j \lambda_{ij} = 1, \qquad i = 1, \ldots, p \tag{7}$$

where

$$f_{ij} = c_i x_i^j \tag{8}$$

$$p_{ij} = A_i x_i^j \tag{9}$$

with x_i^j the jth extreme point of S_i. Bell [16] has considered the application of the primal–dual method to this master program. To initiate the process, a solution to the dual must be available. The dual of (5)–(7) is

$$\text{maximize } v = b_0'\pi + \sum_{i=1}^{p} \pi_{0i} \tag{10}$$

subject to

$$\bar{f}_{ij} = f_{ij} - \pi p_{ij} - \pi_{0i} \geq 0, \qquad \text{all } i, j \tag{11}$$

Using (8) and (9), the dual constraints may be rewritten

$$\pi_{0i} \leq (c_i - \pi A_i)x_i^j \tag{12}$$

These may be satisfied by setting

$$\pi = 0 \tag{13}$$

$$\pi_{0i} = \min\{c_i x_i \mid B_i x_i = b_i,\ x_i \geq 0\} \tag{14}$$

The algorithm proceeds as indicated in the description of the primal–dual method, Section 1.2.4, modified only when necessary to allow for the fact that the columns of the master program are very numerous and not explicitly available. The initial primal–dual tableau is

$$\sum_{i,j} p_{ij}\lambda_{ij} + x_1 = b_0 \tag{15}$$

$$\sum_j \lambda_{ij} + x_{2i} = 1, \qquad i = 1, \ldots, p \tag{16}$$

$$\sum_{i,j} \bar{d}_{ij}\lambda_{ij} = w - w_0 \tag{17}$$

$$\sum_{i,j} \bar{f}_{ij}\lambda_{ij} = z - z_0 \tag{18}$$

where x_1 and x_{2i} are artificial variables, and the infeasibility form, w, has been placed in canonical form by subtracting (15) and (16) from it. The simplex multipliers of the initial basis relative to (17) are

$$\hat{\sigma} = (\sigma, \sigma_{01}, \ldots, \sigma_{0p}) = (1, 1, \ldots, 1) \tag{19}$$

and the initial dual-feasible solution is given by (13)–(14):

$$\hat{\pi} = (0, \pi_{01}, \ldots, \pi_{0p}) \tag{20}$$

Beginning at step 1 of the primal–dual algorithm, the variables λ_{ij} of the first restricted primal program are those for which x_i^j was optimal in (14), since then, by (11), $\bar{f}_{ij} = 0$. For these, form the columns

$$p_{ij} = \begin{pmatrix} A_i x_i^j \\ u_j \end{pmatrix} \tag{21}$$

These are the nonbasic columns of the first restricted primal, over which w is minimized using the revised simplex method.

In checking for termination at step 2 of the algorithm, the only way to test directly for all $\bar{d}_{ij} \geq 0$ is to solve a set of subproblems to compute $\min_j \bar{d}_{ij}$ for all i and test these. Fortunately, the tests of step 2 may be accomplished indirectly, via the dual. If $w_0 = 0$, terminate; the current solution is optimal. If $w_0 > 0$, then either there is no feasible solution, i.e., min $w > 0$, or we simply have not yet reached $w = 0$. To tell which is correct, use the result that an infeasible primal and feasible dual imply an unbounded dual optimum. Thus, if min $w > 0$, an unbounded solution will eventually appear in step 3 of the algorithm, signaled by $k^* = +\infty$.

Moving, then, to step 3, a new feasible solution, $\hat{\pi}^*$ to the dual is found, where

$$\hat{\pi}^* = \hat{\pi} + k^*\hat{\sigma} \tag{22}$$

where, by (1.2.4-27),

$$k^* = \min_{\bar{d}_{ij} < 0} \frac{\bar{f}_{ij}}{-\bar{d}_{ij}} \tag{23}$$

As in computing min $\bar{f}_{ij}$, the quantities $\bar{f}_{ij}$, $\bar{d}_{ij}$ are very numerous and not explicitly available. Thus subproblems are used to compute the minima in (23). By definition

$$\bar{d}_{ij} = -\sigma p_{ij} - \sigma_{0i} = -\sigma A_i x_i^j - \sigma_{0i} \tag{24}$$

$$\bar{f}_{ij} = f_{ij} - \pi p_{ij} - \pi_{0i} = (c_i - \pi A_i)x_i^j - \pi_{0i} \tag{25}$$

so (23) becomes

$$k^* = \min_{-\sigma A_i x_i^j - \sigma_{0i} < 0} \frac{(c_i - \pi A_i)x_i^j - \pi_{0i}}{\sigma A_i x_i^j + \sigma_{0i}} \tag{26}$$

The x_i^j are extreme points of the bounded convex polyhedra, S_i, and, by Theorem 3.11.1-3, the minimum of a ratio like (26) over such sets occurs at an extreme point. Thus, for each i, the minima in (26) may be found by solving the linear fractional programs:

$$\text{minimize} \frac{(c_i - \pi A_i)x_i - \pi_{0i}}{\sigma A_i x_i + \sigma_{0i}} \tag{27}$$

subject to

$$B_i x_i = b_i, \qquad x_i \geq 0 \tag{28}$$

$$\sigma A_i x_i + \sigma_{0i} > 0 \tag{29}$$

Each of these must be solved, with the smallest of the objective values taken as k^*. If, for each i, there are no points x_i^j for which $\bar{d}_{ij} < 0$ [i.e., such that (29) is satisfied], then, by the discussion following (21), $k^* = +\infty$ and the master program has no feasible solution.

A new restricted master program may now be formed, and a new primal–dual iteration begins. This program includes all existing columns for which $\bar{f}_{ij} = 0$, plus some new columns for which this condition holds. These are columns of the form (21), where x_i^j solves any of the subproblems (27)–(29) for which the optimal objective value is k^*. Note that all alternative optima for such subproblems may be used. In fact, if one is omitted, then its $\bar{f}_{ij}^* = 0$ and, on the next cycle, if the corresponding $\bar{d}_{ij}$ is negative, the new k^* will be zero. Thus an iteration will be wasted, in the sense that no increase will be made in the dual objective function. To ease this difficulty without searching out all alternative subproblem optima, Bell [16] suggests solving first those subproblems whose optimal objective values equaled k^* at the previous cycle. If the new optimal values are zero, then no other subproblems need be solved, and a new restricted master can be formed with the new subproblem solutions.

The Charnes–Cooper algorithm, Section 3.11.1, may be used to solve the subproblems. Only one of the linear programs, that for positive denominator, need be solved. If there are no points for which the denominator in (27) is positive, this linear program will be infeasible, since there are no points x_i

satisfying (3) for which the constraints $qy_i + \beta t = t(qx_i + \beta) = 1,\ t \geq 0$, hold. As remarked previously, this implies that this subproblem imposes no upper bound on k.

3.11.3 Example of the Primal–Dual Method. Consider the problem of Section 3.5, whose master program is

$$\text{minimize } z = \sum_i (c_1 x^i)\alpha_i + \sum_i (c_2 y^i)\beta_i$$

$$\text{subject to} \quad \begin{aligned} \sum_i (a_1 x^i)\alpha_i + \sum_i (a_2 y^i)\beta_i + s &= 40 \\ \sum_i \alpha_i &= 1 \\ \sum_i \beta_i &= 1 \\ \alpha_i \geq 0, \quad \beta_i \geq 0, \quad s &\geq 0 \end{aligned}$$

where $c_1 = (-1, -1)$, $c_2 = (-2, -1)$, $a_1 = (1, 2)$, $a_2 = (2, 1)$. The x^i and y^i are extreme points of the sets S_1 and S_2 given in Section 3.5 and are listed here for convenience:

$$\begin{aligned} x^1 &= (6, 8) & y^1 &= (10, 5) \\ x^2 &= (10, 0) & y^2 &= (5, 10) \\ x^3 &= (0, 10) & y^3 &= (10, 0) \\ x^4 &= (0, 0) & y^4 &= (0, 10) \\ & & y^5 &= (0, 0) \end{aligned}$$

To initiate the primal–dual procedure, a feasible solution for the dual of the master program must be available. This dual has constraints

$$\begin{aligned} (a_1 x^i)\pi_1 + \pi_{01} &\leq c_1 x^i, \quad i = 1, \ldots, 4 \\ (a_2 y^i)\pi_1 + \pi_{02} &\leq c_2 y^i, \quad i = 1, \ldots, 5 \\ \pi_1 &\leq 0 \end{aligned} \tag{30}$$

These can be satisfied by setting

$$\pi_1 = 0$$

$$\pi_{01} = \min\{c_1 x^i \mid x^i \in S_1\} = -14, \text{ attained at } x^1 = (6, 8) \tag{31}$$

$$\pi_{02} = \min\{c_2 y^i \mid y^i \in S_2\} = -25, \text{ attained at } y^1 = (10, 5) \tag{32}$$

Note that most of the terms $(c_1 x^i)$ and $(c_2 y^i)$ are negative, whereas the primal–dual method requires them all to be nonnegative to ensure dual feasibility.

To achieve nonnegativity, we multiply the constraints of the master program by $(\pi_1, \pi_{01}, \pi_{02})$ and subtract from the z equation. Then, adding artificial variables and the w equation, the initial primal–dual tableau is

$$
\begin{array}{llllll}
\sum_i (a_1x^i)\alpha_i + \sum_i (a_2y^i)\beta_i + s + x_a^1 & & & & = 40 & \\
\sum_i \alpha_i & + x_a^2 & & & = 1 & \\
\sum_i \beta_i & & + x_a^3 & & = 1 & (33) \\
\sum_i (\bar{c}_i^x)\alpha_i + \sum_i (\bar{c}_i^y)\beta_i - \pi_1 s & & & - z & = 39 & \\
x_a^1 + x_a^2 + x_a^3 & & & & - w = 0 &
\end{array}
$$

where

$$
\begin{aligned}
\bar{c}_i^x &= c_1x^i - (a_1x^i)\pi_1 - \pi_{01} \\
\bar{c}_i^y &= c_2y^i - (a_2y^i)\pi_1 - \pi_{02}
\end{aligned}
$$

are the slack variables in the dual constraints, (30).

Iteration 1

First Restricted Primal. The first restricted primal problem has columns corresponding to the variables x_a^1, x_a^2, x_a^3, $-z$, $-w$ in the tableau (33), plus whatever other columns of that tableau have zero coefficients in the z equation. These correspond to extreme points at which the minima in (31) and (32) were attained, and the slack column:

$$
\begin{bmatrix} a_1x^1 \\ 1 \\ 0 \\ 0 \\ 0 \end{bmatrix} = \begin{bmatrix} 22 \\ 1 \\ 0 \\ 0 \\ 0 \end{bmatrix}, \quad \begin{bmatrix} a_2y^1 \\ 0 \\ 1 \\ 0 \\ 0 \end{bmatrix} = \begin{bmatrix} 25 \\ 0 \\ 1 \\ 0 \\ 0 \end{bmatrix}, \quad \begin{bmatrix} 1 \\ 0 \\ 0 \\ 0 \\ 0 \end{bmatrix}
$$

We note that, since all columns of the restricted primal have zero coefficients in the z equation, this equation plays no role, and may be dropped. The initial basis for the restricted primal and its inverse are

$$
\begin{array}{cccc} x_a^1 & x_a^2 & x_a^3 & -w \end{array}
$$

$$
B = \begin{bmatrix} 1 & & & \\ & 1 & & \\ & & 1 & \\ 1 & 1 & 1 & 1 \end{bmatrix}, \quad B^{-1} = \begin{bmatrix} 1 & & & \\ & 1 & & \\ & & 1 & \\ -1 & -1 & -1 & 1 \end{bmatrix}
$$

Solution of this restricted primal by the revised simplex method is as follows:

Relative cost factors			
cycle	α_1	β_1	s
0	-23	-26	0
1	-23	0	-1
2	0	0	$\frac{1}{22}$

Basic variables	Columns of master-program tableau					Transformed entering column
	x_a^1	x_a^2	x_a^3	$-w$	Constraints	
Cycle 0: β_1 enters, x_a^3 leaves						
x_a^1	1				40	25
x_a^2		1			1	0
x_a^3			1		1	(1)
$-w$	-1	-1	-1	1	-42	-26
Cycle 1: α_1 enters, x_a^1 leaves						
x_a^1	1		-25		15	(22)
x_a^2		1			1	1
β_1			1		1	0
$-w$	-1	-1	25	1	-16	-23
Cycle 2: optimal						
α_1	$\frac{1}{22}$		$-\frac{25}{22}$		$\frac{15}{22}$	
x_a^2	$-\frac{1}{22}$	1	$\frac{25}{22}$		$\frac{7}{22}$	
β_1			1		1	
$-w$	$\frac{1}{22}$	-1	$-\frac{25}{22}$	1	$-\frac{7}{22}$	

Optimality Test. Since $w_0 = \frac{7}{22} > 0$, the current solution is not optimal. To see if the primal is infeasible, rather than checking the condition $\bar{d}_j \geq 0$, we wait until the next step. There primal infeasibility is signaled by an unbounded dual solution; i.e., $k^* \to +\infty$. This, in turn, is detected by infeasibility of all subproblems.

New Dual Solution. The simplex multipliers relative to the infeasibility form are found in the bottom row of the final tableau above:

$$\sigma = (\sigma_1, \sigma_{01}, \sigma_{02}) = (-\tfrac{1}{22}, 1, \tfrac{25}{22})$$

The new dual solution has the form

$$(\pi)_{\text{new}} = (\pi)_{\text{old}} + k\sigma$$
$$= (0, -14, -25) + k(-\tfrac{1}{22}, 1, \tfrac{25}{22})$$

where $k = k^*$, given in (23). This requires solving the linear fractional subproblems, (27)–(29), which are

Subproblem 1

$$\text{minimize } \frac{(c_1 - \pi_1 a_1)x - \pi_{01}}{\sigma_1 a_1 x + \sigma_{01}} = 22\left[\frac{-x_1 - x_2 + 14}{-x_1 - 2x_2 + 22}\right]$$

subject to

$$(x_1, x_2) \in S_1$$
$$-x_1 - 2x_2 + 22 > 0$$

Subproblem 2

$$\text{minimize } \frac{(c_2 - \pi_1 a_2)y - \pi_{02}}{\sigma_1 a_2 y + \sigma_{02}} = 22\left[\frac{-2y_1 - y_2 + 25}{-2y_1 - y_2 + 25}\right] = 22$$

subject to

$$(y_1, y_2) \in S_2$$
$$-2y_1 - y_2 + 25 > 0$$

Subproblem Solutions

$$x = x^2 = (10, 0), \qquad \text{objective value} = \tfrac{22}{3}$$
$$y = y^2 \text{ or } y^3 \text{ or } y^4 \text{ or } y^5, \qquad \text{objective value} = 22$$

The value of k^* is the smallest of these optimal objective values:

$$k^* = \min(\tfrac{22}{3}, 22) = \tfrac{22}{3}$$

The new dual solution thus is

$$(\pi_{\text{new}}) = (0, -14, -25) + \tfrac{22}{3}(-\tfrac{1}{22}, 1, \tfrac{25}{22})$$
$$= (-\tfrac{1}{3}, -\tfrac{20}{3}, -\tfrac{50}{3})$$

Comparison with the final restricted master tableau of Section 3.5 shows that this dual solution is optimal.

Iteration 2

Second Restricted Primal. This new restricted primal has as its columns, the columns in the optimal basis of the first restricted primal, plus the column corresponding to the optimal solution to subproblem 1:

$$\begin{bmatrix} a_1 x^2 \\ 1 \\ 0 \\ 0 \end{bmatrix} = \begin{bmatrix} 10 \\ 1 \\ 0 \\ 0 \end{bmatrix}$$

Solution by the revised simplex method is shown below:

Cycle	Relative cost factors σ_2
0	$-\frac{12}{22}$
1	0

Basic variables	Columns of master-program tableau x_a^1	x_a^2	x_a^3	$-w$	Constraints	Transformed entering column
		Cycle 0: α_2 enters, x_a^2 leaves				
α_1	$\frac{1}{22}$		$-\frac{25}{22}$		$\frac{15}{22}$	$\frac{10}{22}$
x_a^2	$-\frac{1}{22}$	1	$\frac{25}{22}$		$\frac{7}{22}$	$\frac{12}{22}$
β_1			1		1	0
$-w$	$\frac{1}{22}$	-1	$-\frac{25}{22}$	1	$-\frac{7}{22}$	$-\frac{12}{22}$
		Cycle 1: optimal				
α_1	$\frac{1}{12}$	$-\frac{10}{12}$	$-\frac{25}{12}$		$\frac{5}{12}$	
α_2	$-\frac{1}{12}$	$\frac{22}{12}$	$\frac{25}{12}$		$\frac{7}{12}$	
β_1			1		1	
$-w$				1	0	

Since $w_0 = 0$, the current solution is optimal. The corresponding solution to the original angular problem is

$$x = \alpha_1 x^1 + \alpha_2 x^2 = \tfrac{5}{12}(6, 8) + \tfrac{7}{12}(10, 0)$$
$$y = \beta_1 y^1 = (10, 5)$$

which checks with the solution obtained in Section 3.5.

3.12 Three Algorithms for Solving the Master Program —A Comparison[6]

In Chapters 2 and 3, the three best known linear programming algorithms have been applied to the Dantzig–Wolfe master program. These are the

[6] This section is based in part on the material in Jacobsen [19].

primal simplex method (Section 3.4), the dual simplex method (Section 2.6), and the primal–dual algorithm of the previous section. Although the relative efficiencies of these in solving the master program can only be determined empirically, some preliminary judgments can be made. Comparison is based on three factors:

1. The advantages of maintaining primal feasibility at each cycle.
2. The ease with which information generated while solving the subproblems can be used.
3. The number of times the subproblems are re-solved and the ease of solution.

Note that, in general, fewer subproblem solutions occur as better use is made of generated information.

Considering first the primal simplex method, it is obviously the only one of the three which maintains primal feasibility. This is important, in that the algorithm can be terminated short of optimality while still yielding a "useful" solution. The lower bound of Section 3.7 can be used to formulate a rational stop criterion. According to Jacobsen [19], use of this bound has eased the problem of slow convergence of the primal method near the optimum, a difficulty which motivated, in large part, the development of the other approaches. Regarding points 2 and 3, the subproblems are usually solved using the simplex method, for which efficient computer codes are available. The dual algorithms must use algorithms for linear fractional programming, for which good codes are currently scarcer. Since the constraint sets of the subproblems do not change from iteration to iteration, the previous optimal basis may be used to initiate the new solution. Of course, special algorithms may be used if the subproblems have special structure, e.g., if they are network flow problems. Finally, any extreme point generated while solving a subproblem whose associated column prices out negative may be used in the restricted master program of Section 3.4. Thus all subproblems need not be solved at each cycle (although then the lower bound cannot be computed), and those solved need not be pursued to optimality (although more iterations may then be required—see Section 4.1). In short, the procedure is quite flexible, and excellent use can be made of generated information.

In both the dual and primal–dual approaches, the main disadvantage is that primal feasibility is not attained until optimality, so early termination is not possible. Regarding the linear fractional subproblems, the ease with which they can be solved depends, of course, on which algorithm is used. The Charnes–Cooper method of Section 3.11.1 permits solution by solving a single linear program. This program is formed by adding a variable (t in 3.11.1-25) and a constraint (3.11.1-27) to the subproblem constraints $B_i x_i = b_i$, $x_i \geq 0$. Since the added constraint involves the denominator of the ratio, and this depends on the multipliers, $\hat{\sigma}$, of the master program, this constraint changes at each cycle. Thus any special structure of the subproblem is lost, and a phase 1 procedure is required, making re-solution more difficult. Because of this, Jacobsen [19] recommends the algorithm of Abadie and Williams (see [13] of Section 2.6) instead. This solves the fractional program by solving a sequence of linear programs, each with the constraints $B_i x_i = b_i$, $x_i \geq 0$, but with a different objective. The structure of B_i is thus preserved, and the old optimal basis can be used to initiate each new solution. The disadvantage, of course, is that more than one linear programming solution may be needed. Whatever algorithm is used, all subproblems must be completely solved at each cycle, or dual feasibility may not be maintained.

In using generated information, the dual methods seem not to fare as well as the primal. This is especially true for the dual simplex approach, where no method has yet been devised for using nonoptimal solutions. The primal–dual method is better in this respect, as alternative optima for some of the subproblems may be used to form columns for the new restricted master. However, near-optimal solutions are not useful, as they are in the primal method. Further, as pointed out in the previous section, if all alternative optima are not used, the dual objective may not increase in the next iteration. Thus the flexibility inherent in the primal method is not available.

Some additional comments are in order regarding unbounded subsystem constraint sets. This poses little added difficulty for the primal approach, but the dual methods have some trouble getting started. To see this, note that the initial dual solution given in (3.11.2-13) and (3.11.2-14) can be used only if each objective $c_i x_i$ has a finite minimum subject to the constraints $B_i x_i = b_i$, $x_i \geq 0$. If not, adding a constraint of the form $\sum_j x_{ij} \leq m$, m large and positive, bounds the subproblem. However, at optimality, if this constraint is binding, a check must be made to see whether or not the master program has an unbounded solution. In applying the primal simplex method, unboundedness is revealed prior to termination by the usual simplex test.

In order that this discussion should not be overly biased toward the primal method, we add a few words of caution. The primal approach was the first to be presented and so has received much more attention. Future work on the dual algorithms may reverse many of their apparent disadvantages. Also, despite all the above arguments, solution by some dual method may, in many cases, simply require fewer cycles and less work. Only numerical experiments can reveal this.

Problems

1. Prove Theorem 3.2-3.
2. Solve the problem of Section 3.5 using a master program with only one convexity constraint and no restricted master program. Compare the iteration by iteration results with those in Section 3.5.

3. Consider the problem [20], p. 407

$$\text{minimize } z = -x_1 - 8x_2 - 5x_3 - 6x_4$$

subject to

$$\begin{aligned} x_1 + 4x_2 + 5x_3 + 2x_4 &\leq 7 \\ 2x_1 + 3x_2 \qquad\qquad &\leq 6 \\ 5x_1 + x_2 \qquad\qquad &\leq 5 \\ 3x_3 + 4x_4 &\geq 12 \\ x_3 \qquad &\leq 4 \\ x_4 &\leq 3 \\ x_i \geq 0, \qquad i = 1, \ldots, 4 \end{aligned}$$

Solve this problem using the Dantzig–Wolfe decomposition principle, with a master program having two convexity constraints and a restricted master with two new columns at each cycle. Compare with the solution given in [20].

4. Given a linear program of angular form with five blocks, group together the first two and last three, apply the Dantzig–Wolfe decomposition principle, and obtain a master program with two convexity constraints. Write the subproblems and the optimality test. Sketch briefly the iterative procedure.
5. Consider a linear program with some of the variables having upper bounds. Use the decomposition principle and obtain a master program with no upper bounds. Write an explicit solution for the subproblems and outline the iterative solution procedure.
6. Given a linear program with staircase structure, apply the decomposition principle to this problem by regarding every other stage as coupling constraints and the remainder as subsystem relations. Use this idea recursively and discuss the structure of the algorithm obtained (see Dantzig [8] for an approach of this type).
7. Give a step-by-step description of the Dantzig–Wolfe decomposition algorithm for the case of unbounded subsystem constraint sets, justifying all steps. Also show how to construct an extreme ray for a subproblem when the simplex method shows it to be unbounded.
8. Consider the dual-angular program (2.5-10)–(2.5-11). Show that its dual can be written

$$\text{maximize} \sum_{i=1}^{n} p_i b_1' \pi_i$$

subject to

$$\sum_{i=1}^{n} p_i A' \pi_i \leq c$$

$$B' \pi_i \leq d, \qquad i = 1, \ldots, n$$

Apply the Dantzig–Wolfe decomposition procedure to this problem, and obtain a master program with one convexity row. Outline the iterative procedure, and discuss how the repetitive appearance of A and B may be used to advantage. Show that, at optimality, the vector of simplex multipliers for the rows of the master program which are not convexity rows are a set of optimal x values for the original dual-angular program.

9. Obtain a lower bound on the minimal cost for transportation and generalized transportation problems, similar to that obtained for the master program in Section 3.7. What is the most general class of linear programs for which such a bound may be derived? Does the bound hold as an equality at optimality? If so, does it converge monotonically to the optimal value?
10. Construct a transportation problem with four origins and two destinations, solve it by the decomposition principle and by the simplex method, and compare the results.
11. Show that solving the subproblems of Section 3.9 requires only scanning a cost matrix with elements $(c_{ij} - a_i\pi_j)$.
12. Prove that, under assumption 1 of Section 3.11.1, $f(x) = px/qx$ is strictly quasiconvex over S.
13. Prove that any local minimum of a strictly quasiconvex function is global.
14. Present a counterexample to show that the assertion of Problem 13 is false if the strictness is removed.
15. Prove Theorem 3.11.1-2 without assumption 1, assuming only that both px and qx do not vanish simultaneously on S. (See [13], but note that the proof given there is in error.)
16. Is the following statement true? $f(x)$ attains its minimum over a bounded convex polyhedron, S, at an extreme point of S only if $f(x)$ is quasiconcave over S. Give an example to justify your answer.
17. Consider the ratio

$$f(x) = \frac{2x_1 + x_2 - 3}{x_1 - x_2 + 1}$$

Sketch in the x_1,x_2 plane the set

$$S = \{(x_1, x_2) \mid f(x) \leq 1\}$$

Also sketch the sets

$$S^+ = S \cap \{(x_1, x_2) \mid x_1 - x_2 + 1 > 0\}$$
$$S^- = S \cap \{(x_1, x_2) \mid x_1 - x_2 + 1 < 0\}$$

Note that S is not convex, so $f(x)$ cannot be a convex function. The sets S^+ and S^- are convex, as must be the case if $f(x)$ is to be both quasiconvex and quasiconcave over the region for which its denominator is positive.

18. Solve the master program corresponding to Problem 3 using the primal–dual method.
19. Extend the primal–dual decomposition of Section 3.11.2 to the case of unbounded subsystem constraint sets.

References

1. G. B. Dantzig and P. Wolfe, "The Decomposition Algorithm for Linear Programming," *Econometrica*, **9**, No. 4, 1961; also *Operations Res.*, **8**, Jan., Feb., Oct. 1960.

2. S. Karlin, *Mathematical Methods and Theory in Games, Programming and Economics*, Vol. 1, Addison-Wesley Publishing Company, Inc., Reading, Mass., 1961.
3. A. C. Williams, "A Treatment of Transportation Problems by Decomposition," *J. Soc. Ind. Appl. Math.*, **10**, No. 1, 1962, pp. 35–48.
4. W. W. Garvin, *Introduction to Linear Programming*, McGraw-Hill, Inc., New York, 1960.
5. S. I. Gass, *Linear Programming, Methods and Applications*, McGraw-Hill, Inc., New York, 2nd ed., 1964.
6. Tse-Hao Tcheng, "Scheduling of a Large Forestry-Cutting Problem by Linear Programming Decomposition," Ph.D. Thesis, University of Iowa, 1966.
7. W. J. Baumol and T. Fabian, "Decomposition, Pricing for Decentralization, and External Economies," *Management Sci.*, **11**, No. 1, 1964, pp. 1–32.
8. G. B. Dantzig, *Linear Programming and Extensions*, Princeton University Press, Princeton, N.J., 1963, chap. 23.
9. A. S. Manne, "Programming of Economic Lot Sizes," *Management Sci.*, **14**, 1958, pp. 115–135.
10. B. P. Dzielinski and R. E. Gomory, "Optimal Programming of Lot Sizes, Inventory and Labor Allocations," *Management Sci.*, **11**, No. 9, 1965, pp. 874–890.
11. H. M. Wagner and T. M. Whitin, "A Dynamic Version of the Economic Lot Size Model," *Management Sci.*, **5**, 1958, pp. 89–96.
12. L. R. Ford and D. R. Fulkerson, "A Primal–Dual Algorithm for the Capacitated Hitchcock Problem," *Naval Res. Logistics Quart.*, **4**, No. 1, 1957, pp. 47–54.
13. W. S. Dorn, "Linear Fractional Programming," *IBM Res. Rept. RC–830*, Nov. 1962.
14. P. C. Gilmore and R. E. Gomory, "A Linear Programming Approach to the Cutting Stock Problem—Part II," *Operations Res.*, **11**, 1963, pp. 863–888.
15. A. Charnes and W. W. Cooper, "Programming with Linear Fractional Functionals," *Naval Res. Logistics Quart.*, **9**, Sept.–Dec. 1962, pp. 181–186.
16. E. J. Bell, "Primal–Dual Decomposition Programming," Ph.D. Thesis. Operations Research Center, University of California at Berkeley, 1965; *Rept. ORC 65–23.*
17. J. Ponstein, "Seven Kinds of Convexity," *SIAM Rev.*, **9**, No. 1, 1967.
18. B. Martos, "The Direct Power of Adjacent Vertex Programming Methods," *Management Sci.*, **12**, No. 3, 1965, pp. 241–252.
19. S. Jacobsen, "Comparison of the Primal, Dual, and Primal-Dual Decomposition Algorithms" (Notes on Operations Research 6), Operations Research Center, University of California at Berkeley, 1967, *Rept. ORC 67–17.*
20. G. Hadley, "*Linear Programming*", Addison-Wesley Publishing Company, Inc., Reading, Mass., 1962.

4

Solution of Linear Programs with Many Columns by Column-Generation Procedures

The development of the Dantzig–Wolfe decomposition principle consists of two distinct phases. First, a problem with many rows is transformed into an equivalent master program with fewer rows but many more columns. Then column-generation methods are applied to this problem. The two phases may be separated and the second, column generation, may be applied to problems other than the master program. Such problems may arise from a wide variety of situations and usually could not be solved without such special techniques, since the number of possible columns may be many millions (see the cutting-stock problem, Section 4.1). This chapter examines a number of such applications. Particular attention should be paid to the different subproblems which arise and to the variety of methods used to solve them. These include dynamic programming (Sections 4.1 and 4.2), linear programming (Section 4.3), nonlinear programming (Section 4.4), and network-flow algorithms (Section 4.5).

4.1 Cutting-Stock Problem

Consider the problem of cutting an unlimited number of pieces of stock of various lengths (for example, rolls of paper) so that at least n_i pieces of length l_i are furnished, $i = 1, \ldots, I$. The objective is to meet the demands, n_i, while minimizing the total number of rolls that must be cut. Since cutting each roll involves some waste, this keeps total waste at a low level.

Assume for simplicity that all the pieces available for cutting have length l. If $l_i \leq l$, $i = 1, \ldots, I$, the problem has a solution. Define a cutting pattern as a particular way of cutting a length of stock, e.g., a_1 pieces of length l_1, a_2 pieces of length l_2, etc. A cutting pattern is thus uniquely determined by the vector $a = (a_1, a_2, \ldots, a_I)$. There are a finite number of cutting patterns, although this number may be very large. The problem may be stated: Specify the number of times each cutting pattern should be used so that all demands are met and the total number of rolls used is minimized.

Let j index the various cutting patterns and define

$$a_{ij} = \text{number of pieces of length } l_i \text{ cut by pattern } j$$

$$p_j = (a_{1j}, \ldots, a_{Ij}), \text{ a vector describing the } j\text{th cutting pattern}$$

$$x_j = \text{number of times pattern } j \text{ is used}$$

$$n = (n_1, \ldots, n_I), \text{ a vector of demands for lengths } l_1 \cdots l_I.$$

Then the constraints that all demands must be met become

$$\sum_j p_j x_j \geq n \tag{1}$$

$$x_j \geq 0, \qquad x_j \text{ integer} \tag{2}$$

The objective function is

$$\text{minimize } z = \sum_j x_j \tag{3}$$

which minimizes the total number of pieces of length l cut.

The problem is very difficult to solve if the x_j must be integers, so this constraint is dropped. In many problems, especially those with high demands, the optimal fractional values of x_j are large enough so that little is lost in rounding to integers. However, solution is still difficult, since the number of columns p_j may be very large. For example, if the stock to be cut has length $l = 200$ inches and there are demands for 40 different lengths from 20 to 80 inches, the number of possible patterns could easily exceed 10 to 100 million [2]. This, however, may be dealt with as before, by pricing out the columns and defining suitable subproblems.

The relative cost coefficient for a nonbasic variable x_j is

$$\bar{c}_j = 1 - \sum_i \pi_i a_{ij} \tag{4}$$

with the π_i simplex multipliers for the constraints (1). In order that a vector a describe a legitimate cutting pattern its coordinates must satisfy

$$a_i = \text{nonnegative integer} \tag{5}$$

$$\sum_{i=1}^{I} l_i a_i \leq l \tag{6}$$

Thus the problem of minimizing $\bar{c}_j$ over all possible columns becomes that of maximizing

$$\sum_i \pi_i a_i \tag{7}$$

subject to (5) and (6). Problems of this form are called knapsack problems, because of the following interpretation: a_i is the number of pieces of equipment of weight l_i and relative usefulness π_i to be carried on a camping trip in a knapsack which will hold at most a total weight l. Such problems may be solved efficiently by a number of algorithms, e.g., dynamic programming [3]. Gilmore and Gomory describe an algorithm in [2] reported to be approximately five times faster than the standard dynamic programming recursion.

If the optimal objective value in (7) is v, then the simplex criterion for possible improvement, min $\bar{c}_j < 0$, becomes, by (4),

$$v > 1 \tag{8}$$

and the corresponding pattern vector is formed and enters the basis. If (8) does not hold, and if the reduced costs of the slack variables, $\bar{c}_s$, satisfy

$$\bar{c}_s = \pi_s \geq 0$$

the current solution is optimal.

Gilmore and Gomory [1, 2] have successfully applied this formulation to a number of practical problems in the paper industry. Four series of test problems were solved, A1 through A4, each containing one 40-length, one 35-length, one 30-length, one 25-length, and one 20-length problem. Problems from different series having the same number of ordered lengths differ in their right-hand-side vectors, n. A typical 30-length test problem, problem A2-3, and its solution are shown in Table 1. Demands are quite high, as are the number of times the various cutting patterns are used. Thus rounding off to integers has little effect. The patterns chosen involve little trim, as the total length cut is quite close to $l = 218$. A cutting-knife limit was employed, restricting the total number of cuts in a pattern, $\sum_i a_i$, to be 7, 5, 5, 9, and 7 for series A1–A5, respectively. This arises due to a limitation on the number of knives available in paper plants to actually perform the cutting. Such restrictions are easily incorporated into the knapsack algorithm [2].

In solving these problems, advantage was taken of the fact that the simplex multipliers for different rows of (1) were often identical. This can be used to simplify the knapsack calculation, since if, in (5)–(7), say π_1 and π_2 are equal and $l_2 > l_1$, then there is an optimal solution in which a_2 is zero. This is because, if a_2 were positive, its value could be assigned to a_1 without violating (6) or changing the value of the objective, (7). Thus l_2 and its associated variable a_2 may be dropped, yielding a smaller knapsack problem to be solved.

TABLE 1. TYPICAL TEST PROBLEM

Stock length, 218.00; *cutting-knife limit*, 5

Order length, l_i	Demands, n_i
81.00	4415
70.00	291
68.25	4765
67.50	4827
66.75	90
66.00	691
64.00	263
63.75	141
63.00	133
60.00	390
56.25	459
56.00	343
52.50	766
52.00	58
51.75	2736
51.00	212
50.00	720
49.50	133
46.50	529
45.50	185
44.50	94
41.25	393
38.50	47
38.00	95
35.00	411
33.50	36
33.00	273
32.00	56
31.50	171
21.50	140

No. of times cutting pattern used	No. of lengths to be cut	Ordered lengths
2382.50000	1	81.00
	2	68.25
302.55552	2	81.00
	1	56.00
36.00000	1	81.00
	2	51.75
	1	33.50
1194.38860	1	81.00
	2	67.50
94.99999	1	67.50
	1	60.00
	1	52.50
	1	38.00

No. of times cutting pattern used	No. of lengths to be cut	Ordered lengths
141.00000	1	67.50
	1	63.75
	1	51.75
	1	35.00
86.40738	1	67.50
	1	66.00
	1	63.00
	1	21.50
77.48144	1	60.00
	3	52.50
26.40740	2	67.50
	1	51.00
	1	31.50
529.00000	1	67.50
	2	51.75
	1	46.50
133.00000	1	67.50
	1	66.00
	1	49.50
	1	35.00
19.59259	1	63.00
	2	51.75
	1	51.00
56.00000	2	67.50
	1	51.00
	1	32.00
144.24070	2	56.25
	2	52.50
145.25922	1	67.50
	2	51.75
	1	45.50
5.50000	1	67.50
	2	52.00
	1	45.50
272.99999	2	67.50
	1	50.00
	1	33.00
94.00000	1	70.00
	2	51.75
	1	44.50
53.59262	3	51.75
	1	41.25
	1	21.50
34.24078	1	67.50
	2	52.50
	1	45.50
170.51850	1	60.00
	1	56.25
	1	51.75

No. of times cutting pattern used	No. of lengths to be cut	Ordered lengths
	1	50.00
169.70368	2	67.50
	2	41.25
109.99999	2	66.00
	1	51.00
	1	35.00
27.00001	1	67.50
	1	63.00
	1	52.50
	1	35.00
90.00000	1	67.50
	1	66.75
	1	51.75
	1	31.50
13.48148	3	56.00
	1	50.00
263.00000	1	64.00
	2	51.75
	1	50.00
54.59259	1	67.50
	1	66.00
	1	52.50
	1	31.50
197.00000	1	81.00
	1	70.00
	1	66.00
47.00001	1	67.50
	1	60.00
	1	52.00
	1	38.50

TABLE 1. SUPPLEMENT

Supplemental Table of Percentage Waste for Incomplete Runs

Data	B	C	D
A4–1	0.3918	0.3921	0.9355
–2		0.4353	1.0887
–3			0.9747
A5–1	0.5459	0.4927	

Figure 4-1 shows the size of the knapsack problem actually solved versus iteration count for a typical problem. Using identical prices, the average problem had 18.2 variables versus an original number of 30. This results in a substantial time saving in solving the subproblems.

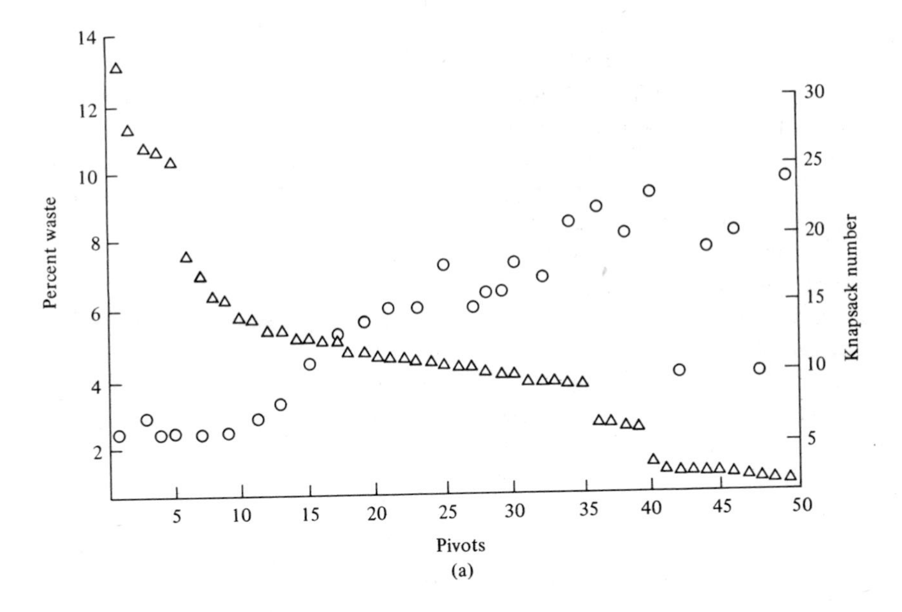

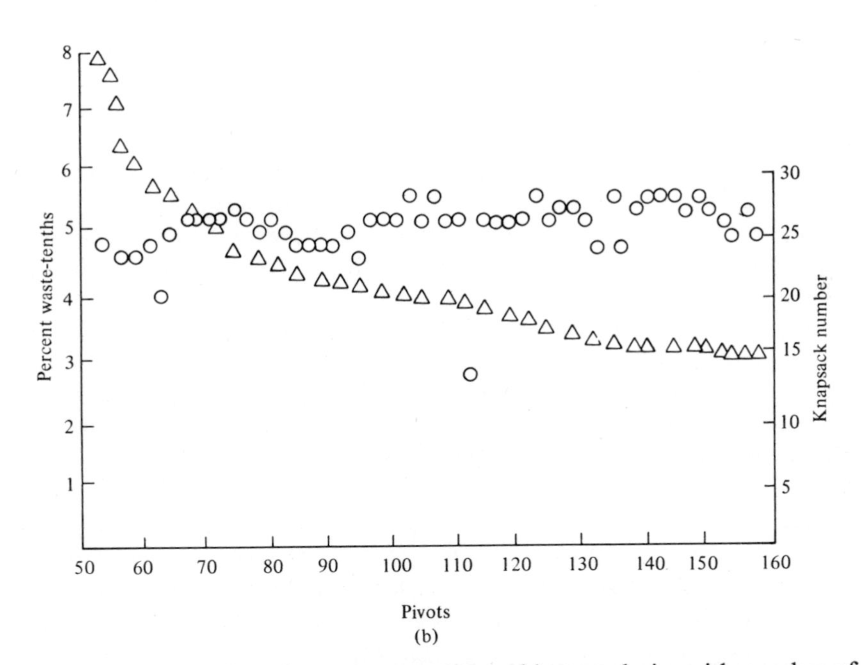

FIGURE 4-1 Variation of percentage waste and knapsack size with number of pivots executed. Crosses and the left-hand scale give the waste; note the change of scale in percentage waste in the two parts. Circles and the right-hand scale give the number of variables in the knapsack problem.

Another device used by Gilmore and Gomory to reduce computations is termed the "median method." Its purpose is to ease the following problem: In bringing the column with min $\bar{c}_j$ into the basis, one obtains maximum decrease of z per unit increase in x_j but does not necessarily obtain the largest possible decrease in z. This is because x_j may be small, and the decrease in z is $\bar{c}_j x_j$. Often less negative $\bar{c}_j$ may lead to larger x_j. This requires too much computation to correct in general, but in this problem something could be done. From Table 1 it is seen that the demands vary widely. Consider a cutting pattern containing both high- and low-demand lengths. The number of times this pattern may be used tends to be limited by the lowest demand, since otherwise this cutting pattern alone would exceed this demand. To permit large x_j and large changes in z, lengths were divided into a high- and a low-demand group (greater and less than the median demand). Every other pivot step, a knapsack problem was solved with the variables a_i for the low-demand group set to zero. This tended to yield patterns with large x_j, and proved of much value.

Computational results are shown in Table 2. All problems used the improved knapsack method developed in [2] and took advantage of identical prices. Solution was on an IBM 7094. Comparison of columns A and B shows the value of the median method, especially for series A4. Column C did not complete the knapsack maximization, taking instead the first knapsack solution yielding $\bar{c}_j < 0$. This was an attempt to see if the effort expended to minimize $\bar{c}_j$ was worthwhile. Although this question has been discussed elsewhere [4] with varying conclusions, in this case finding min $\bar{c}_j$ is best. The number of pivots is, of course, reduced but, more important, the saving in running time is significant, especially for longer-run problems.

Column D was run to examine the effect of choosing any column promising some improvement at each iteration, i.e., of making a blind choice from among the millions of possible improving columns. This is the dual of the current situation in integer programming, where a row which cuts off part of the constraint set must be chosen blindly from among many possible such rows. The result in integer programming is erratic behavior; some problems terminate quickly, others run for thousands of pivots without optimality. The same phenomenon is evidenced here. The series averaged over $20m$ pivots for solution (m is the number of rows) versus $2.5m$ for the usual simplex estimate. The A2 series averaged $4.4m$, while the A4 series required $27m$, with only one of the five problems complete at that point. These results suggest that such behavior is not particular to integer programming but will arise in any situation in which a random choice must be made from many possible rows or columns.

Extension to a Rational Objective Function. As pointed out in [2], customers in the paper industry will often accept a range of order quantities. That is, instead of requiring n_i pieces of length l_i, any quantity between n_i' and n_i'' is allowed, where $n_i'' > n_i'$. According to Gilmore and Gomory [2,

TABLE 2. DATA ON RUNS TO TEST METHODS OF COMPUTATION

Data	No. of ordered lengths	% Waste	A Max. knapsack; median method		B Max. knapsack; no median method		C No max. on knapsack; median method		D No max. on knapsack; no median method	
			Pivots	Time (min)	Pivots	Time (min)	Pivots	Time (min)	Pivots	Time (min)
A1-1	40	3.0474	148	1.34	194	2.64	333	1.47	541	2.07
-2	35	0.0184	233	4.50	347	9.24	560	6.30	1104	11.03
-3	30	0.0616	161	6.32	171	6.75	418	7.19	586	7.84
-4	25	0.6227	101	0.47	135	0.93	177	0.62	545	1.51
-5	20	0.0539	90	1.32	111	2.23	214	2.25	370	3.08
A2-1	40	4.7232	110	0.52	167	0.62	122	0.50	316	1.06
-2	35	7.8219	90	0.31	76	0.28	93	0.30	126	0.40
-3	30	8.6921	87	0.25	71	0.22	87	0.24	111	0.25
-4	25	9.6407	47	0.14	73	0.15	64	0.14	76	0.17
-5	20	5.1748	21	0.09	19	0.06	28	0.07	28	0.08
A3-1	40	5.0845	57	0.33	73	0.32	66	0.33	80	0.52
-2	35	0.4825	141	3.22	157	2.34	348	3.35	450	3.83
-3	30	0.2114	83	0.59	124	0.78	371	1.97	518	1.69
-4	25	0.1484	90	0.54	95	0.58	259	0.81	380	0.92
-5	20	2.7297	24	0.12	27	0.07	173	0.30	476	0.93
A4-1	40	0.3869	138	5.53	210	41.58[a]	637	10.65[a]	727	15.87[a]
-2	35	0.4326	125	4.20	208	7.79	648	4.58[a]	878	17.75[a]
-3	30	0.4494	92	1.24	140	4.52	489	3.62	817	19.23[a]
-4	25	0.1415	98	2.51	141	6.30	415	3.66	878	9.19
-5	20	0.5490	67	1.74	77	1.57	536	5.13	786	18.10[a]
A5-1	50	0.4862	291	14.51	540	16.97[a]	980	12.51[a]	2123	19.87

[a] Indicates a problem that was not run to completion. The percentage waste for these runs are given in the supplemental table.

p. 881], $n_i'' \doteq 1.1n_i'$ is common. If this is the case, then the number of rolls used is no longer a suitable objective, since this is generally smallest when the quantities n_i' are produced. Minimum percentage waste, which measures the true efficiency of the operation, may require greater production. Thus, to take advantage of the ranges of demand, percentage waste is introduced as the objective function. The waste incurred in cutting pattern j is

$$w_j = l - \sum_i l_i a_{ij} \tag{9}$$

If pattern j is used x_j times, then the total percentage waste is

$$\frac{z_1}{z_2} = 100\left(\frac{\sum_j w_j x_j}{l \sum_j x_j}\right) \tag{10}$$

Thus the problem has a rational objective:

$$\text{minimize } \frac{\sum_j w_j x_j}{\sum_j x_j} \tag{11}$$

subject to

$$\sum_j p_j x_j \geq n' \tag{12}$$

$$\sum_j p_j x_j \leq n'' \tag{13}$$

$$x_j \geq 0 \tag{14}$$

Introducing a vector of slack variables, s, in (12) and subtracting (12) from (13) yields a system with upper bounded slack variables,

$$\sum_j p_j x_j - s = n' \tag{15}$$

$$0 \leq s \leq (n'' - n') \tag{16}$$

Consider the application of the Gilmore–Gomory algorithm, described in Section 3.11.1, to this fractional program. Since this procedure is so similar to the simplex method, the upper bounds on s are easily dealt with without increasing basis size by the upper bounding method of Section 6.3. The only difficulty occurs in choosing among the many columns p_j. If one follows the usual simplex criterion, the variable with most negative directional derivative is chosen. This derivative is, by (3.11.1-39),

$$\bar{c}_j = \frac{\bar{z}_2 \bar{c}_{1j} - \bar{z}_1 \bar{c}_{2j}}{\bar{z}_2^2} \tag{17}$$

where $\bar{z}_1/\bar{z}_2$ is the current value of the ratio and, from (3.11.1-43) and (3.11.1-44),

$$\bar{c}_{1j} = c_{1j} - \pi_1 p_j = w_j - \pi_1 p_j \tag{18}$$

$$\bar{c}_{2j} = c_{2j} - \pi_2 p_j = 1 - \pi_2 p_j \tag{19}$$

Using the above in (17) yields

$$\bar{c}_j = \frac{\bar{z}_2(w_j - \pi_1 p_j) - \bar{z}_1(1 - \pi_2 p_j)}{\bar{z}_2^2}$$

or, using (9),

$$\bar{c}_j = \frac{l\bar{z}_2 - \bar{z}_1 + \sum_i (\bar{z}_1\pi_{2i} - \bar{z}_2(\pi_{1i} + l_i))a_{ij}}{\bar{z}_2^2} \tag{20}$$

Let

$$k = l\bar{z}_2 - \bar{z}_1 \tag{21}$$

$$\bar{\pi}_i = \bar{z}_1\pi_{2i} - \bar{z}_2(\pi_{1i} + l_i) \tag{22}$$

Then the problem is to compute

$$\min \bar{c}_j = \frac{\min(k - \sum_i \bar{\pi}_i a_{ij})}{\bar{z}_2^2} \tag{23}$$

To find the cutting pattern which minimizes the above, we solve the subproblem

$$\text{maximize} \sum_i \bar{\pi}_i a_i$$

subject to $\sum_i l_i a_i \leq l, \quad a_i = \text{nonnegative integer}$

This is again a knapsack problem. The only change from the original case is that the objective coefficients $\bar{\pi}_i$ are now functions of the simplex multipliers π_1 and π_2. Once the subproblem is solved, computations proceed precisely as indicated in Section 3.11.1.

A further extension of the ideas in this section to cutting-stock problems in two or more dimensions is described in [5].

4.2 Column-Generation and Multi-item Scheduling

In Section 3.10, approximate solutions to a class of resource allocation problems, including multi-item production scheduling, were obtained through formulation as large linear programs. The Dantzig–Wolfe decomposition principle was then applied to these programs. Although this is a useful approach, it creates some difficulties. Perhaps the most serious of these stems from the fact that the linear program being decomposed is only an approximation to an integer program whose solution is actually desired, and only a good approximation when the number of items is much greater than the number of time periods. Under these conditions, basic solutions of the linear program have the property that most of the variables will have integral values. This property is not shared by the master program of the decomposition principle. Thus solution by the decomposition principle cannot guarantee a majority of integer-valued variables until optimality. Intermediate solutions may then not be meaningful, and any lower bound on the minimal cost which may be developed has limited value. Since, as mentioned in [7], programs with very many columns often have a tremendous number of near-optimal solutions prior to optimality, these limitations can be serious.

In this section an alternative approach to the linear program formulated by Manne, proposed by Lasdon and Mackey [8], is discussed. This program is attacked directly, in that no equivalent "master program" is formulated. Instead, the large number of columns is handled by column generation via subproblems. The multiplicity of constraints is dealt with by using the generalized upper bounding procedure described in Section 6.4. The subproblems are the same as are solved using the decomposition principle, and the size of the basis matrix that must be manipulated is one less than that of the master program (although the operations performed using this matrix are somewhat more complex). Since only basic solutions of the original problem occur, a majority of integer values is always present. A lower bound may be derived to aid in terminating computations, and a number of additional degrees of freedom inherent in this new approach permits further computational refinement. Computational results suggest that these refinements are among the most attractive features of the method. Comparison with the decomposition principle shows an improvement by at least a factor of two in the number of iterations required for all problems studied, with work per iteration approximately the same.

Solution Procedure. The linear program to be solved is that in (3.10.1-12)–(3.10.1-15), rewritten here as

$$\text{minimize} \sum_{i,j} c_{ij}\theta_{ij} \tag{1}$$

subject to

$$\sum_{j} \theta_{ij} = 1, \qquad i = 1, \ldots, I \tag{2}$$

$$\sum_{i,j} Y_{ij}\theta_{ij} \leq b \tag{3}$$

$$\theta_{ij} \geq 0 \tag{4}$$

where

$$c_{ij} = c_i(x_i^j), \qquad Y_{ij} = Y_i(x_i^j) \tag{5}$$

and the x_i^j are elements of a finite set, V_i.

This linear program has, for realistic problems, a tremendous number of columns, and may have many rows if I is large, e.g., in the hundreds or thousands. The columns are easily dealt with. Let B be a basis matrix for the problem and let $\pi = (\pi_1, \pi_2)$ be the set of simplex multipliers corresponding to this basis, with π_1 associated with the constraints (2) and π_2 with (3). Pricing out a nonbasic column, the relative cost coefficient is

$$\bar{c}_{ij} = c_{ij} - \pi_2 Y_{ij} - \pi_{1i} \tag{6}$$

To choose a column to enter the basis, we must find

$$\min_{i,j} \bar{c}_{ij} = \min_i \min_j \bar{c}_{ij} \tag{7}$$

The inner minimization in (7) may be accomplished by solving the subproblem

$$\underset{j}{\text{minimize}} \; [c_{ij} - \pi_2 Y_{ij}] \tag{8}$$

or, using (5) and the fact that all x_i^j are elements of V_i:

$$\text{minimize } c_i(x_i) - \pi_2 Y_i(x_i) \tag{9}$$

subject to

$$x_i \in V_i \tag{10}$$

These subproblems are single-activity problems with a penalty term, $\pi_2 Y_i$, for use of the resource and are precisely as occur with the decomposition principle, (3.10.1-42)–(3.10.1-43). Since V_i is assumed finite and the functions c_i and Y_i are assumed to be bounded below on V_i, the subproblems have finite optimal solutions for all values of π_2. To decide on which column is to enter the basis, subtract π_{1i} from the optimal objective value in (9)–(10) and find the minimum of these quantities over i.

The problem of many constraints in (1)–(4) still remains. The solution lies in the special form of the constraints (2). In their work on generalized upper bounding techniques (see reference [9]), Dantzig and Van Slyke show that any linear program which includes constraints like those in (2) can be solved by utilizing a working basis whose dimension is the number of remaining constraints, (3), in our case ($T \times T$). For any basis, B, of (1)–(4), the simplex multipliers, values for the basic variables, and the vector $\bar{P}_s = B^{-1}P_s$ can all be easily derived from the corresponding quantities associated with the working basis.

Although the updating of the working basis is more complex than in the standard simplex method, the procedure is much more efficient (see [9]) than applying the simplex method directly, especially if I is much greater than T. The generalized upper bounding procedure is discussed in Section 6.4. It suffices to say here that its use greatly reduces computational problems caused by many rows in (1)–(4).

Lower Bounds. A lower bound on the optimal cost, similar to that obtained with the decomposition principle, may be developed here. Let $\pi = (\pi_1, \pi_2)$ again be a vector of simplex multipliers for some basis, B, of (1)–(4), rewritten here as

$$\text{minimize} \sum_{i,j} c_{ij}\theta_{ij} = z \tag{11}$$

subject to

$$\sum_{j} \theta_{ij} = 1, \qquad i = 1, \ldots, I \tag{12}$$

$$\sum_{i,j} Y_{ij}\theta_{ij} + s = b \tag{13}$$

$$\theta_{ij} \geq 0, \qquad \text{all } i, j, \qquad s_t \geq 0, \qquad t = 1, \ldots, T \tag{14}$$

where the s_t are slack variables. Multiply (12) by π_1 and (13) by π_2 and subtract from (11), yielding

$$\sum_{i,j} (c_{ij} - \pi_2 Y_{ij} - \pi_{1i})\theta_{ij} - \pi_2 s = z - \pi_2 b - \sum_{i} \pi_{1i} \tag{15}$$

The quantity in parentheses in (15) is, by (6), the reduced cost coefficient $\bar{c}_{ij}$. Replacing $\bar{c}_{ij}$ by $\min_j \bar{c}_{ij}$ yields

$$\sum_{i} \left(\min_{j} \bar{c}_{ij}\right) \sum_{j} \theta_{ij} - \pi_2 s \leq z - \pi_2 b - \sum_{i} \pi_{1i} \tag{16}$$

Using (12) in (16),

$$z \geq \sum_{i} \left(\min_{j} \bar{c}_{ij}\right) - \pi_2 s + \left(\pi_2 b + \sum_{i} \pi_{1i}\right) \tag{17}$$

Since

$$(\pi_1, \pi_2) = c_B B^{-1} \tag{18}$$

and, if e_I is the I-dimensional sum vector,

$$z_B = c_B x_B = c_B B^{-1}\left[\frac{e_I}{b}\right] = (\pi_1, \pi_2)\left[\frac{e_I}{b}\right] \tag{19}$$

then

$$z_B = \pi_2 b + \sum_i \pi_{1i} \tag{20}$$

and, since (17) holds for all values of z obtainable from (11)–(14),

$$\min z \geq z_B + \sum_i \left(\min_j \bar{c}_{ij}\right) - \pi_2 s \tag{21}$$

which is the desired lower bound. The behavior of this bound in computations is illustrated at the end of this section.

Multiple Pricing. In applying the Dantzig–Wolfe decomposition principle to (1)–(4), master programs having anywhere from one to I convexity constraints (summing subsets of the variables to unity) may be formed. Each corresponds to a partitioning of the constraints (2) into a corresponding number of subsets, then writing any solution of each subset as a convex combination of its extreme points. The approach of Section 3.10 uses one convexity constraint, while the direct approach described here has I of them, thus reproducing the original program. The same subproblems are solved in each case (although with different multipliers, π), yielding solutions $x_i(\pi)$, $i = 1, \ldots, I$. If a master program with p convexity constraints is formed, $1 \leq p \leq I$, then, at each iteration, p new columns are formed from these subproblem solutions, with the one with the lowest value of $(\min_j \bar{c}_{ij})$ entering the basis. A pivot operation is then performed, yielding new multipliers, $\hat{\pi}$. Often π and $\hat{\pi}$ will not differ radically, so that some of the columns corresponding to the $x_i(\pi)$ may again price out negative, with the most negative entering the basis. This cycle may be repeated until all reduced costs for these columns are nonnegative. Such a procedure corresponds to a partial solution of Dantzig's restricted master program (see Section 3.4), in which columns leaving the basis are dropped, and makes fullest use of the candidates proposed by the subproblems.

Since pricing out a single column, i.e., forming the scalar product of two vectors, may be expected to involve orders of magnitude less computation than solving a subproblem of the form (9)–(10), this option is quite attractive. As more convexity constraints are used, more new columns are formed at each cycle, and large decreases in the objective become more likely. For the problems considered later, the cycle described above is used an average of three times for each set of subproblem solutions, leading to very large reductions in computing time. Of course, the greater number of convexity constraints is a computational disadvantage if a revised simplex code is used.

This difficulty is greatly reduced, however, if the generalized upper bounding method is employed.

It should be noted that this multiple pricing option is available even if a master program with only one convexity constraint is used. One simply generates those columns with the k most negative reduced costs, $k \geq 1$, and forms a restricted master program with these. The experience cited here may suggest that doing this would improve the rate of convergence in this case, although this remains to be demonstrated. Of course, only a direct application of the simplex method, as proposed here, works with basic solutions of the original program. As noted earlier, this is of special importance when solving the program (1)–(4).

The excellent computational experience with this option has some implications for general use of the Dantzig–Wolfe decomposition principle. For a constraint matrix of the form

$$\begin{bmatrix} A_1 & A_2 & \cdots & A_n \\ B_1 & & & \\ & B_2 & & \\ & & \ddots & \\ & & & B_n \end{bmatrix}$$

one may form master programs having $1 \leq p \leq n$ convexity rows (see Section 3.4). The computational experience cited later indicates that choosing $p = n$ and solving a restricted master program at each cycle with n nonbasic columns may lead to the most rapid convergence.

Lot-Size Problems with "Continuous" Setups. To test the efficiency of the proposed approach, a number of test problems were solved. These involve the production of I items over T time periods, with a specified number of machines available to produce these items. Demands for each item in each time period are assumed known. Costs are (1) inventory holding and shortage costs and (2) setup costs. Setups occur when a given machine is changed from the production of one item to the production of another. We assume that any machine can, if suitably set up, produce any of the items, and that the setup cost is independent of which two items are involved in the changeover. A machine is set up to produce an item by installing in the machine a piece of equipment particular to the item. These pieces of equipment will be called dies. The problem is to allocate the machines to the items so that the sum of inventory and setup costs over all items is minimized.

This problem can be placed in the form of the resource allocation problem (3.10.1-5)–(3.10.1-7). Define

$$m_{it} = \text{number of machines used to produce item } i \text{ in time period } t$$
$$m_i = (m_{i1}, m_{i2}, \ldots, m_{iT})$$

b_t = number of machines available in time period t
$b = (b_1, \ldots, b_T)$
d_{it} = demand for item i in time period t, assumed known
k_i = production rate of a machine producing item i
y_{it} = inventory of item i at the end of time period t
$= y_{i,t-1} + k_i m_{it} - d_{it},\ t = 1, \ldots, T,\ y_{i0}$ given
p_s = cost of one setup
$(y_{it})_{\max}, (y_{it})_{\min}$ = given upper and lower bounds on y_{it}
n_i = number of dies available for item i

Since the setup cost is assumed independent of the items involved, the total setup cost may be written

$$c_s = p_s \sum_{i=1}^{I} s_i(m_i) \tag{22}$$

where

$$s_i(m_i) = \sum_{t=1}^{T} (m_{it} - m_{i,t-1})_+, \qquad m_{i0} \text{ given} \tag{23}$$

and where

$$(x)_+ = \begin{cases} x, & x \geq 0 \\ 0, & x < 0 \end{cases} \tag{24}$$

The inventory cost for item i is written

$$\gamma_i(m_i) = \sum_{t=1}^{T} \gamma_{it}(y_{it}) \tag{25}$$

where the γ_{it} represent holding costs for $y_{it} > 0$ and shortage costs for $y_{it} < 0$, and may have any convenient functional form. Let

$$c_i(m_i) = p_s s_i(m_i) + \gamma_i(m_i) \tag{26}$$

Then the problem is to choose $m_1 \cdots m_I$ to minimize

$$c = \sum_{i=1}^{I} c_i(m_i) \tag{27}$$

subject to

$$(y_{it})_{\min} \leq y_{it} \leq (y_{it})_{\max}, \qquad \text{all } i, t \tag{28}$$

$$m_{it} \leq n_i, \qquad \text{all } i, t \tag{29}$$

$$m_{it} = \text{nonnegative integer}, \qquad \text{all } i, t \tag{30}$$

$$\sum_{i=1}^{I} m_{it} \leq b_t, \qquad t = 1, \ldots, T \tag{31}$$

Let

$$V_i = \{m_i \mid m_i \text{ satisfies (28)–(30)}\} \tag{32}$$

Since the n_i are finite, V_i is finite and is assumed nonempty. Let the elements of V_i be indexed by j, with m_i^j the jth element. Defining

$$c_i(m_i^j) \equiv c_{ij} \tag{33}$$

the linear program (1)–(4) assumes the form

$$\text{minimize} \sum_{i,j} c_{ij}\theta_{ij} \tag{34}$$

subject to

$$\sum_j \theta_{ij} = 1, \qquad i = 1, \ldots, I \tag{35}$$

$$\sum_{i,j} m_i^j \theta_{ij} \leq b \tag{36}$$

$$\theta_{ij} \geq 0, \qquad \text{all } i, j \tag{37}$$

The ith subproblem is

$$\text{minimize } p_s s_i(m_i) + \gamma_i(m_i) - \pi_2 m_i \tag{38}$$

subject to

$$(y_{it})_{\min} \leq y_{it} \leq (y_{it})_{\max}, \qquad t = 1, \ldots, T \tag{39}$$

$$m_{it} \leq n_i \qquad \text{all } t \tag{40}$$

$$m_{it} = \text{nonnegative integer}, \qquad \text{all } t \tag{41}$$

Each subproblem may be solved by dynamic programming with two state variables, m_{it} and y_{it}. Since the initial states m_{i0} and y_{i0} are given, a forward recursion is used.

Computational Results. To test the efficiency of the proposed approach, a number of test problems of the form just discussed were solved. It was found early in the research that use of a full set of artificial variables in phase 1 led to very poor starting points for phase 2. To avoid this, the single-item subproblems (38)–(41) were first solved with $\pi_2 = 0$. If these solutions satisfy (31), they are optimal. If not, an initial basic feasible solution is constructed using a single θ_{ij} for each of the subproblem solutions, plus T slack and artificial variables. The inverse of a basis so formed is easily found without numerical inversion, and the procedure has led to much better initial points for phase 2.

Four test problems were solved by each of three methods: (1) the decomposition principle, (2) the approach proposed here, and (3) the new approach with subproblem solutions priced out many times until none price out negative. These problems had 6 time periods and 6, 10, 10, and 15 items, respectively. Problems 2 and 3 have 22 and 24 machines available in each time period, respectively. Other parameters, e.g., demands for each item, machine availabilities, etc., were chosen to approximate reality in the production environment under consideration and to yield nontrivial subproblems. Note that, even though the number of items is small, the number of columns in the program (34)–(37) can be very large. If the upper and lower bounds in (39) are absent, then each item generates $(n_i + 1)^T$ different columns, which is 262,124 if $T = 6$ and $n_i = 7$. Each problem had at least one such item, with problem 4 having two items with $n_i = 7$, one with $n_i = 5$, and two with $n_i = 4$. Upper and lower bounds were included to reduce the time required to solve the subproblems, but these were "loose," i.e., never binding in any cycle. Thus the problems solved here are of substantial size.

Figures 4-2 through 4-5 show the behavior of phase 2 cost versus number of phase 2 iterations. Although the initial cost values differ somewhat, owing to the different solutions found in phase 1, this difference is approximately equalized after the first phase 2 iteration. The decomposition principle exhibits the expected "tail," a characteristic shown to a lesser degree by

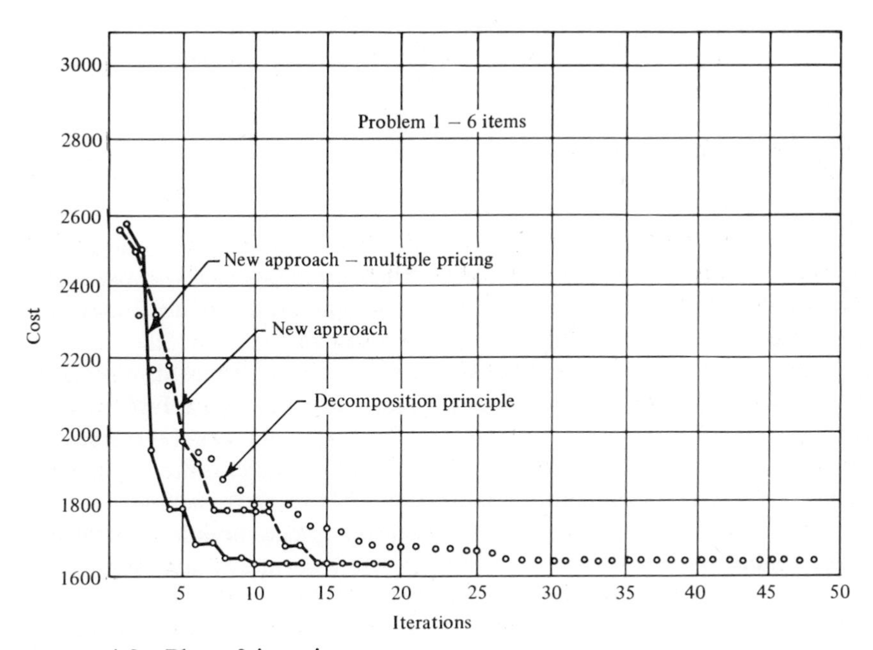

FIGURE 4-2 Phase 2 iterations.

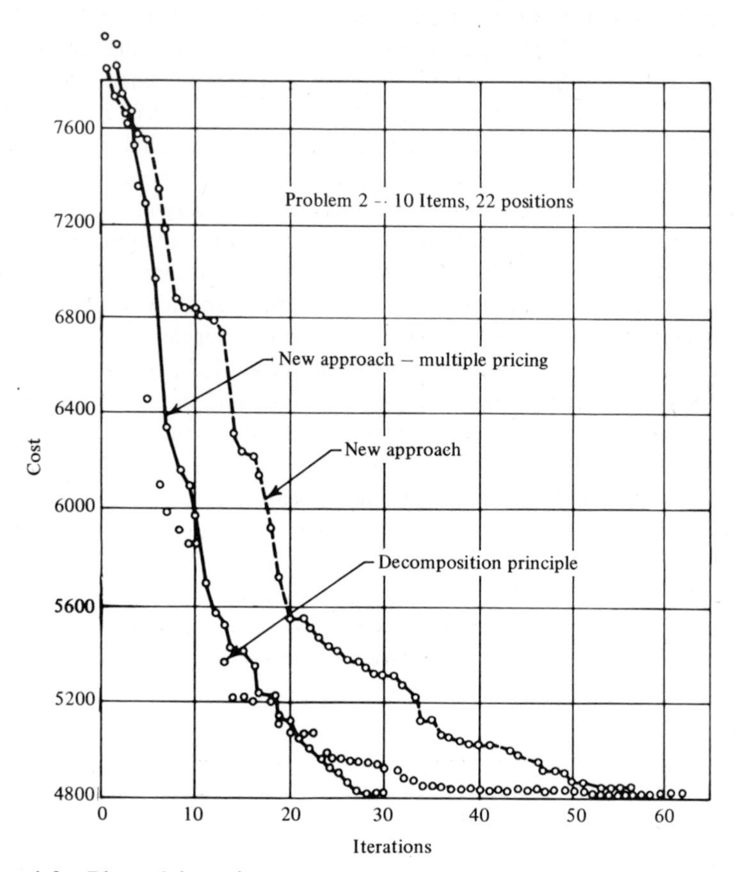

FIGURE 4-3 Phase 2 iterations.

the proposed method, probably due to the smaller number of columns involved. Continued pricing out of subproblem proposals yields a very significant decrease over the number of iterations required by the decomposition principle, the reduction being by factors of from about 2 to 9. Since, for the problems solved, the time required to price out a given subproblem solution is at least two orders of magnitude less than to solve a subproblem, only when method 3 re-solves the subproblems is the iteration count increased by one. If this is done, then, for the problems solved here, computer time is approximately proportional to iteration count for all three methods.

Tables 1 and 2 show that, with the multiple pricing option, subproblem solutions are used many times after being generated by the dynamic programs. Figures 4-6 and 4-7 show the behavior of the lower bound using method 2. If one is willing to terminate computations when the predicted improvement in cost is less than 5 per cent of current costs, problems 1 through 4, using

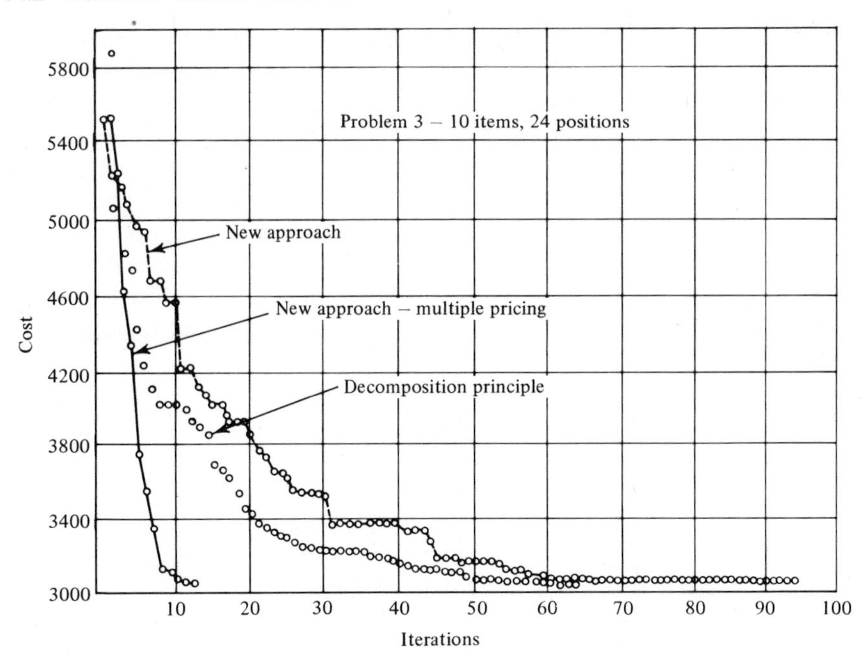

FIGURE 4-4 Phase 2 iterations.

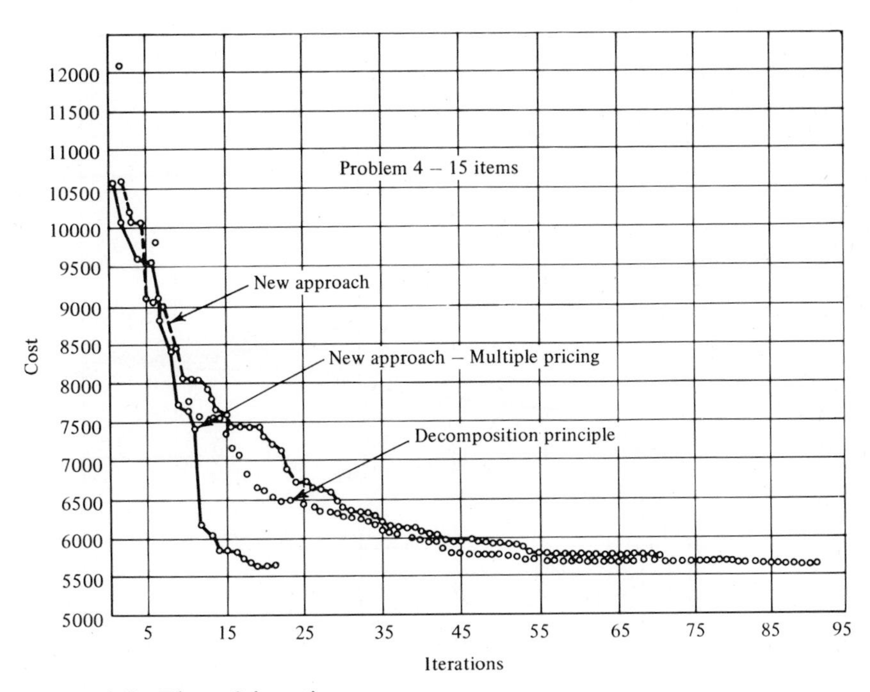

FIGURE 4-5 Phase 2 iterations.

method 3, could have been terminated in 16, 16, 36, and 26 iterations, respectively, rather than the 20, 18, 43, and 31 iterations required to meet the optimality test.

A number of other ideas to shorten computations were also tested. One of these involves solving the subproblems by a dynamic programming algorithm with only one state variable, the other (the variable m_{it}) being fixed at its initial value, m_{i0}. This is faster by factors of from 2 up, and often yields either optimal solutions or good approximations to these. As long as this, or any other technique, yields columns with negative reduced cost, it is useful. In three test problems, using this option until it met the optimality

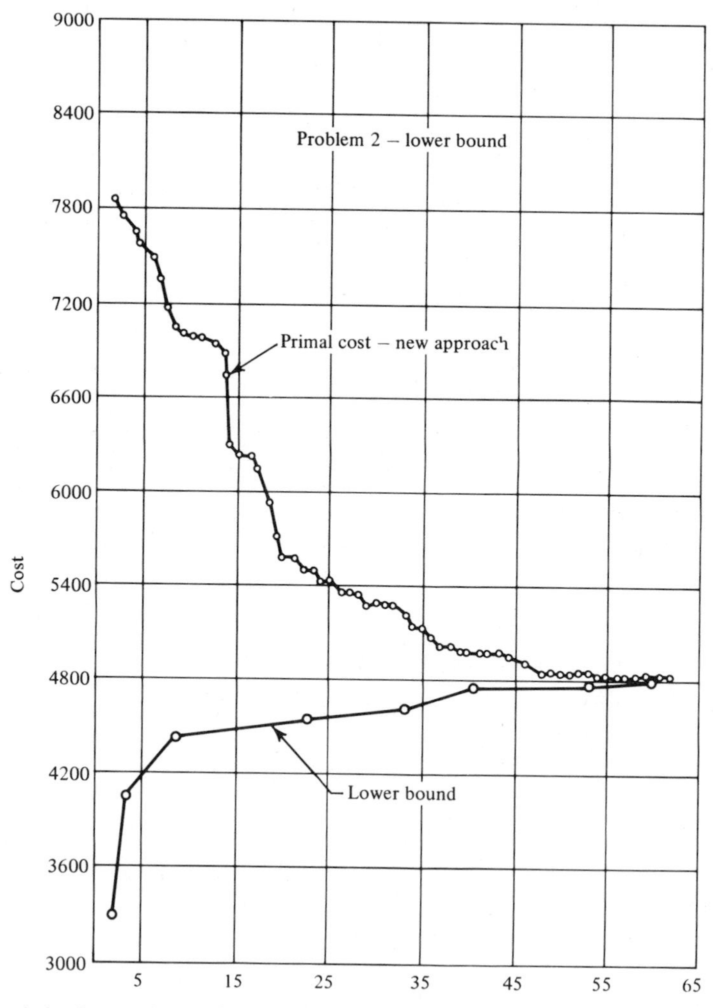

FIGURE 4-6 Lower bound.

test, then switching to the correct, two-state variable algorithm, has led to reductions in running time of about one half.

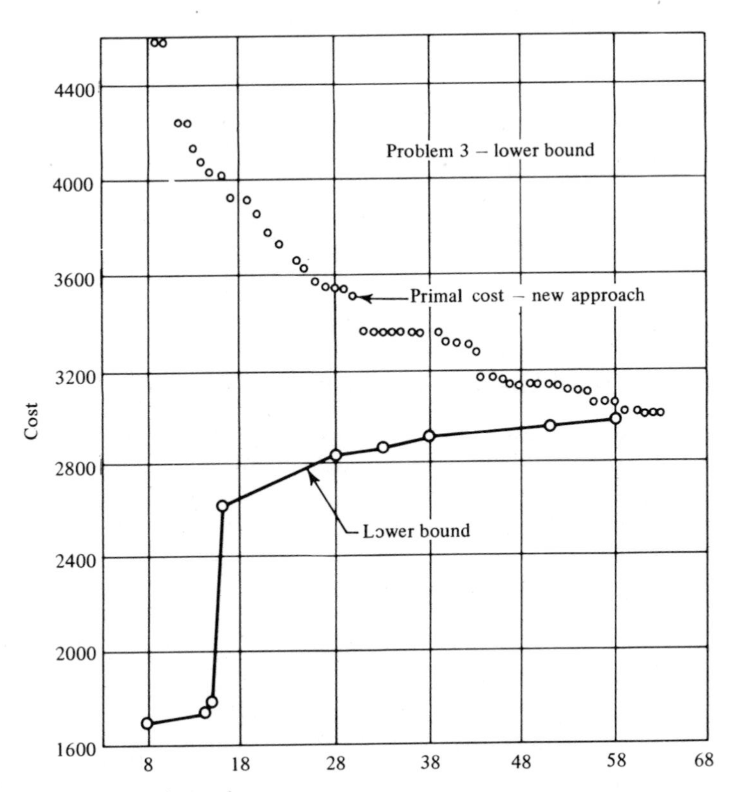

FIGURE 4-7 Lower bound.

TABLE 1. MULTIPLE PRICING, PROBLEM 3

Subproblem solutions phase 2	Cumulative number of multiple pricings
1	7
2	10
3	12
4	15
5	19
6	24
7	25
8	26
9	28
10	30
11	30

TABLE 2. MULTIPLE PRICING, PROBLEM 4

Subproblem solutions phase 2	Cumulative number of multiple pricings
2	5
4	11
6	18
8	21
10	33
12	40
14	44
16	47
18	47

4.3 Generalized Linear Programming

A generalized linear program, as defined by Wolfe [10] allows not only the activity levels, x_i, but also their associated columns, a_i, and cost coefficients, c_i, to vary. The problem is: Choose nonnegative x_i and vectors (a_i, c_i) for each i to minimize

$$z = \sum_i c_i x_i \tag{1}$$

subject to

$$\sum_i a_i x_i = b, \qquad x_i \geq 0 \tag{2}$$

$$(a_i, c_i) \in S_i \tag{3}$$

where S_i is assumed here to be a bounded convex polyhedron. This problem is not, as it stands, linear, since it involves products of unknown quantities. It may, however, be transformed into a linear program by the same device used in the decomposition principle. Since the S_i are bounded, any $(a_i, c_i) \in S_i$ may be written

$$(a_i, c_i) = \sum_k \lambda_{ik}(a_i^k, c_{ik}), \qquad \sum_k \lambda_{ik} = 1, \qquad \lambda_{ik} \geq 0 \tag{4}$$

where the (a_i^k, c_{ik}) are extreme points of S_i. Substituting (4) into (1) and (2), the problem becomes

$$\text{minimize } z = \sum_{i,k} c_{ik}\lambda_{ik}x_i \tag{5}$$

subject to

$$\sum_{i,k} a_i^k \lambda_{ik} x_i = b \tag{6}$$

Let

$$u_{ik} = \lambda_{ik} x_i \geq 0 \tag{7}$$

Then (5)–(6) becomes a linear program in the variables u_{ik}:

$$\text{minimize} \sum_{i,k} c_{ik} u_{ik} \tag{8}$$

subject to

$$\sum_{i,k} a_i^k u_{ik} = b, \qquad u_{ik} \geq 0 \tag{9}$$

Let $\{u_{ik}^0\}$ be an optimal solution of (8)–(9). To obtain the associated optimal solution of (1)–(3), if, for given i, all $u_{ik}^0 = 0$ then, by (7), $x_i = 0$ and (a_i, c_i) may assume any value. If for given i, at least one u_{ik}^0 is positive, then

$$\lambda_{ik}^0 = \frac{u_{ik}^0}{x_i^0} \tag{10}$$

$$\sum_k \lambda_{ik}^0 = \frac{\sum_k u_{ik}^0}{x_i^0} = 1 \tag{11}$$

so

$$x_i^0 = \sum_k u_{ik}^0 \tag{12}$$

$$(a_i^0, c_i^0) = \sum_k \lambda_{ik}^0 (a_i^k, c_{ik}) \tag{13}$$

Of course, in solving (8)–(9), one would not wish to tabulate the extreme points (a_i^k, c_{ik}) in advance. This is avoided by using column generation. Pricing out a column of (8)–(9) yields a relative cost coefficient

$$\bar{c}_{ik} = c_{ik} - \pi a_i^k$$

which is to be minimized to find a column to enter the basis. Since the vectors (a_i^k, c_{ik}) are extreme points of S_i, this is equivalent to the subproblem

$$\text{minimize } c_i - \pi a_i$$

subject to

$$(a_i, c_i) \in S_i$$

which is itself a linear program. Having solved this, further computations proceed exactly as in the revised simplex method. Barring cycling, an optimal solution is found in a finite number of iterations.

These ideas may be extended to the case where the S_i are general convex sets, not necessarily polyhedral. As pointed out by Dantzig [10], this extension permits the formulation of nonlinear convex programs as generalized programs. Consider the problem

$$\text{minimize } f(x)$$

$$\text{subject to} \qquad g_i(x) \leq 0, \qquad i = 1, \ldots, m$$

where the g_i are convex functions. An equivalent problem is obtained by introducing new variables $(y_0, y_1, \ldots, y_m)$:

$$\text{minimize } y_0$$

$$\begin{aligned} \text{subject to} \qquad & f(x) \leq y_0 \\ & g_i(x) \leq y_i \leq 0, \qquad i = 1, \ldots, m \end{aligned}$$

This may in turn be transformed into a generalized linear program with a single variable column $(y_0, y) = (y_0, y_1, \ldots, y_m)$ and variables $\lambda, \lambda_1, \ldots, \lambda_m$:

$$\text{minimize } z$$

$$\text{subject to} \qquad \begin{array}{llllll} \lambda & & & & & = 1 \\ y_1\lambda + \lambda_1 & & & & & = 0 \\ y_2\lambda & + \lambda_2 & & & & = 0 \\ \vdots & & \ddots & & & \vdots \\ y_m\lambda & & & & \lambda_m & = 0 \\ y_0\lambda & & & & & = z \\ \lambda_i \geq 0, & i = 1, \ldots, m & & & & \end{array} \tag{14}$$

where (y_0, y) is contained in the set

$$R = \{(y_0, y) \mid y_0 \geq f(x), y \geq g(x) \text{ for some } x\} \tag{15}$$

This set may also be expressed as the set of all points in E^{m+1} on and above the graph of the function

$$w(y) = \min\{f(x) \mid g(x) \leq y\}$$

It is not difficult to show (Problem 12) that if f and the g_i are convex functions, R is a convex set and w is a convex function. Let points $(f(x_t), g(x_t)) \in R$ be chosen.[1] Since R is convex, the convex hull of these points is in R. Generalized

[1] The problem of how to choose such points is discussed in Section 4.4.

linear programming approximates R by its "inner linearization" IL(R), the set of all points in, above, and to the right of this convex hull

$$\text{IL}(R) = \left((y_0, y) \mid (y_0, y) \geq \sum_t \lambda_t(f(x_t), g(x_t)), \sum_t \lambda_t = 1, \lambda_t \geq 0\right) \quad (16)$$

This is illustrated in Figure 4-8.

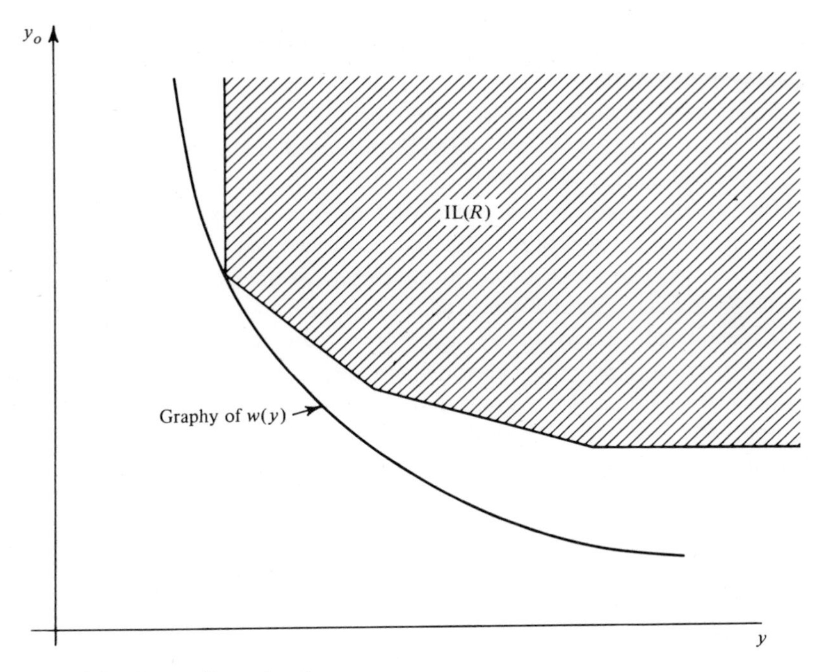

FIGURE 4-8 Inner linearization.

In solving (14) over $(y_0, y) \in \text{IL}(R)$, since y_0 is the objective, optimal solutions have y_0 equal to $\sum_t \lambda_t f(x_t)$. Then the objective involves only the variables λ_t, so (14) is equivalent to the linear program

$$\text{minimize} \sum_t \lambda_t f(x_t)$$

subject to

$$\sum_t \lambda_t g_i(x_t) \leq 0, \qquad i = 1, \ldots, m$$

$$\sum_t \lambda_t = 1$$

$$\lambda_i \geq 0$$

A difficulty arising here is that, no matter what the points $\{x_t\}$, IL(R) and R will, in general, not be equal unless R is polyhedral. Because of this, solving a convex nonlinear program by generalized linear programming is usually an infinite but convergent procedure. This is discussed in detail in Section 4.4 under the title of grid linearization. The set R and its various linearizations are used extensively in Chapters 8 and 9.

Application to Decision Problems in a Markov Chain. Sequential decision problems often arise in which the state transitions of the system being controlled are probabilistic. Such a problem could arise in a dynamic single item inventory context, as discussed by Manne [11]. Assume that demand for the item, d, is a random variable, independent from time period to time period and identically distributed within each period, having a finite set of values, $(0, 1, 2, \ldots, D)$. The quantity ordered or produced in a period is also discrete, with values $(0, 1, \ldots, M)$. Further, assume that unsatisfied demand is not backlogged but results in lost sales, and there is a positive integral upper bound, n, on inventory. It follows that if initial inventory is a nonnegative integer, less than or equal to n, the inventory at the beginning of any time period can assume only the values $(0, 1, 2, \ldots, n)$.

A decision rule or stationary inventory policy is a function $m(i)$ which gives the amount to order in a period when the inventory level at the beginning of a period is i units. We assume that quantities ordered are delivered with essentially no time lag, so the behavior of the inventory variable in a period is as shown in Figure 4-9. The decision rule is selected so that, for any i,

$$i + m(i) \leq n \tag{17}$$

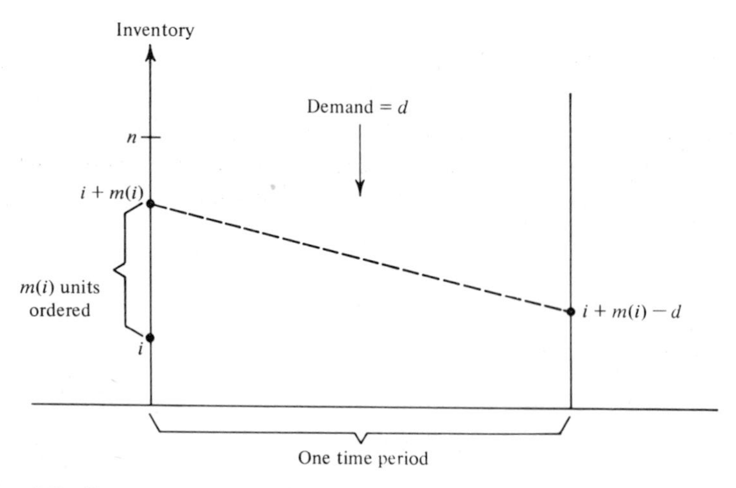

FIGURE 4-9 Inventory versus time.

As seen from Figure 4-9, inventory at the end of a period is

$$j = \text{final inventory}$$
$$= \begin{cases} i + m(i) - d, & 0 < i + m(i) - d \leq n \\ 0, & i + m(i) - d \leq 0 \end{cases} \tag{18}$$

It is evident that the conditional probability that final inventory is j units given the entire past history of the process is independent of all this history except the value of initial inventory, i. The inventory process is thus a finite Markov chain, whose states are the inventory levels $(0, 1, 2, \ldots, n)$. The transition probabilities for this process are defined as

$$p_{ij} = \Pr[\text{final inventory} = j \mid \text{initial inventory} = i] \tag{19}$$

For a given decision rule, $m(i)$, these may be determined from the probability distribution of demand

$$p_{ij} = \begin{cases} \Pr[d = i + m(i) - j], & 0 < j \leq n \\ \Pr[d \geq i + m(i)], & j = 0 \end{cases} \tag{20}$$

We assume that this process eventually attains statistical equilibrium. That is, if the process is allowed to run for a sufficiently long time, the probability that the current state is i is independent of time and of the initial conditions. Let x_i denote this probability:

$$\lim_{t \to \infty} \Pr[\text{state} = i] = x_i \tag{21}$$

The costs per period incurred in operating the process are assumed to be of three types. There is a holding cost for positive inventory, $c_h(i)$, an ordering cost, $c_0(m(i))$, and a penalty cost for unsatisfied demand, $c_p(d - i - m(i))$, positive if its argument is positive and zero otherwise. The equilibrium probabilities, x_i, can be used to form the long-run expectation of the sum of these costs

$$\text{Exp}[C] = \sum_{i=1}^{n} x_i \left\{ c_h(i) + c_0(m(i)) + \sum_{r} \Pr[d = r] c_p(r - i - m(i)) \right\} \tag{22}$$

A decision problem can now be posed, requiring that the decision rule $m(i)$ be chosen to minimize (22).

This problem is easily rephrased as one of choosing transition probabilities for the Markov chain from a finite set of possibilities. From (20) we see that if the probability distribution of demand is known, each choice of decision rule, $m(\cdot)$, determines the transition probabilities p_{ij}. As an example, consider a process in which the demand probabilities are

$$\Pr[d = r] = \tfrac{1}{3}, \qquad r = 0, 1, 2 \tag{23}$$

The quantity ordered is either 0 or 1,

$$m(i) \in [0, 1] \qquad \text{for all } i \tag{24}$$

and the upper bound on inventory is 2, so

$$i \in [0, 1, 2] \tag{25}$$

CASE 1. $i = 0$. If $m(0) = 0$ and $i = 0$, then no matter what the demand, final inventory is zero. The transition probabilities from $i = 0$ under the decision $m(0) = 0$ are

$$p_{00}^0 = 1, \qquad p_{01}^0 = 0, \qquad p_{02}^0 = 0 \tag{26}$$

These may be incorporated in a vector

$$p_0^0 = (1, 0, 0) \tag{27}$$

If $m(0) = 1$, then final inventory is zero if demand is either 1 or 2 units, and is 1 if demand is zero. Thus

$$p_0^1 = (\tfrac{2}{3}, \tfrac{1}{3}, 0) \tag{28}$$

Thus, for the state $i = 0$, there is a finite set of transition probability vectors, one for each value of $m(0)$. This set is given by

$$S_0 = \{\overbrace{(1, 0, 0)}^{m(0)=0}, \overbrace{(\tfrac{2}{3}, \tfrac{1}{3}, 0)}^{m(0)=1}\} \tag{29}$$

CASES 2, 3. $i = 1, 2$. We leave the reader to verify that the sets of possible transition probability vectors for $i = 1$ and 2 are given by

$$S_1 = \{\overbrace{(\tfrac{2}{3}, \tfrac{1}{3}, 0)}^{m(1)=0}, \overbrace{(\tfrac{1}{3}, \tfrac{1}{3}, \tfrac{1}{3})}^{m(1)=1}\} \tag{30}$$

$$S_2 = \{\overbrace{(\tfrac{1}{3}, \tfrac{1}{3}, \tfrac{1}{3})}^{m(2)=0}\} \tag{31}$$

Given a single vector from each of S_0, S_1, and S_2, the steady-state probabilities are determined (as will be shown shortly), and the expected cost (22) can be evaluated. Once a choice of transition probability vectors is made which minimizes this cost, the optimal decision rule $m(\cdot)$ is found by choosing those values for $m(i)$ which generated these vectors.

Of course, problems of this type can arise in situations which do not involve inventories. Wolfe and Dantzig [12] formulate the problem in general terms as follows. Let the vector of transition probabilities from state i be

$$p_i = (p_{i1}, p_{i2}, \ldots, p_{in}), \qquad i = 1, \ldots, n \tag{32}$$

where

$$p_{ij} \geq 0, \qquad \sum_{j=1}^{n} p_{ij} = 1 \tag{33}$$

The vector p_i may be chosen from a set S_i:

$$p_i \in S_i$$

with a choice of p_i incurring a cost $c_i(p_i)$. Either of two assumptions may hold regarding S_i and c_i.

(a) S_i is a finite set and c_i an arbitrary function defined on S_i.

(b) S_i is a bounded convex polyhedron, defined by a finite set of linear equalities and inequalities, every element of the polyhedron being a distribution. The cost of any element $p \in S_i$, $c_i(p)$, is a linear function defined on S_i.

The relation between cases (a) and (b) may be seen by introducing the notion of mixed strategies. A mixed strategy is a set of probability distributions, one for each state, where the kth element in distribution i is the probability that decision k is made when the current state is state i. Thus in a mixed strategy we select from among pure strategies with given probabilities. Let these probabilities be λ_{ik}:

$$\lambda_{ik} = \Pr[\text{decision} = k \mid \text{state} = i] \tag{34}$$

where

$$\sum_{k} \lambda_{ik} = 1, \qquad \lambda_{ik} \geq 0 \tag{35}$$

Now consider the set S_i in (a). Let p_{ij}^k be the probability of the transition $i \rightarrow j$ when the decision made is k, and define p_i^k as the vector of elements

p_{ij}^k as j goes from one to n. If mixed strategies are allowed, vectors of transition probabilities other than the vectors $p_i^k \in S_i$ may be generated. Suppose the initial state is state i. The transition probability p_{ij} is

$$p_{ij} = \sum_k \lambda_{ik} p_{ij}^k \tag{36}$$

or, in vector form,

$$p_i = \sum_k \lambda_{ik} p_i^k, \qquad p_i^k \in S_i \tag{37}$$

The element p_i in (37) is simply any vector in the convex hall of the set S_i of case (a), $C(S_i)$, and $C(S_i)$ is a bounded convex polyhedron, as in case (b). Thus, by admitting mixed strategies, any distribution in $C(S_i)$ can be generated.

Although cases (a) and (b) are closely related, the techniques of generalized linear programming apply directly only to (b). We will show shortly, however, that both cases can be treated, since when (b) is formulated as a generalized program, there is an optimal solution to this program in which only a single element of S_i appears for each i, that being an extreme point. This means that, even when mixed strategies are admitted, only pure strategies (decision rules) ever need be used.

Proceeding with our general formulation, let the equilibrium distribution for the Markov chain be

$$x = (x_1, \ldots, x_n) \tag{38}$$

where x_i is the (unconditional) steady-state probability of being in state i. To obtain relations defining this distribution, note that one can attain state i by being initially in state j, and then making the transition $(j \rightarrow i)$, a compound event with probability $x_j p_{ji}$. Since j could be any state, x_i is the sum of these probabilities:

$$x_i = \sum_j x_j p_{ji} \tag{39}$$

or, in vector form,

$$x = \sum_j x_j p_j \tag{40}$$

The expected cost per stage under steady-state conditions is

$$\sum_i c_i(p_i) x_i \tag{41}$$

and the problem is to choose $p_i \in S_i$, $i = 1, \ldots, n$ to minimize this cost subject to (40) and

$$\sum_i x_i = 1, \qquad x_i \geq 0 \tag{42}$$

Equations (40) have unknown quantities on both right- and left-hand sides. Bringing all these to one side, this problem may be written

$$\text{minimize } z = \sum_i \hat{c}_i(v_i)x_i \tag{43}$$

subject to

$$\sum_i x_i = 1 \tag{44}$$

$$\sum_i v_i x_i = 0 \tag{45}$$

$$x_i \geq 0, \qquad v_i \in T_i \tag{46}$$

where T_i is the set of all vectors

$$v_i = (p_{i1}, p_{i2}, \ldots, p_{ii} - 1, \ldots, p_{in}) \tag{47}$$

such that $p_i = (p_{i1}, \ldots, p_{in}) \in S_i$, and

$$\hat{c}_i(v_i) = c_i(p_i) \tag{48}$$

The problem (43)–(46) is a generalized linear program if the assumptions of case (b) hold. Let us proceed under these assumptions, pausing to consider (a) later. The set T_i is a bounded polyhedron if S_i is, and extreme points of S_i correspond, through (47), to extreme points of T_i. We write any $v_i \in T_i$ as a convex combination of extreme points, v_i^k:

$$v_i = \sum_k \lambda_{ik} v_i^k, \qquad \lambda_{ik} \geq 0, \qquad \sum_k \lambda_{ik} = 1 \tag{49}$$

Substituting (49) into (43) yields

$$z = \sum_i \hat{c}_i\left(\sum_k \lambda_{ik} v_i^k\right) x_i \tag{50}$$

Since, in case (b), c_i is linear,

$$\hat{c}_i\left(\sum_k \lambda_{ik} v_i^k\right) = \sum_k \lambda_{ik} \hat{c}_i(v_i^k) \tag{51}$$

so (50) becomes

$$z = \sum_{i,k} \hat{c}_i(v_i^k)\lambda_{ik}x_i \tag{52}$$

Multiplying (44) by $\sum_k \lambda_{ik}$ yields

$$\sum_{i,k} \lambda_{ik}x_i = 1 \tag{53}$$

Substituting (49) into (45) and writing the result together with (52) and (53) yields the linear program

$$\text{minimize } z = \sum_{i,k} c_{ik}u_{ik} \tag{54}$$

subject to

$$\sum_{i,k} u_{ik} = 1 \tag{55}$$

$$\sum_{i,k} v_i^k u_{ik} = 0 \tag{56}$$

$$u_{ik} \geq 0 \tag{57}$$

where

$$u_{ik} = \lambda_{ik}x_i, \qquad c_{ik} = \hat{c}_i(v_i^k) \tag{58}$$

This linear program also has an interpretation in terms of mixed strategies, as shown by Manne [11]. If λ_{ik} is given the intepretation in (34), then

$$u_{ik} = \Pr[\text{decision} = k \mid \text{state} = i] \Pr[\text{state} = i] \tag{59}$$

$$u_{ik} = \Pr[\text{state} = i, \text{decision} = k] \tag{60}$$

i.e., u_{ik} is the joint probability that the state is i and the decision made is k. These probabilities must satisfy (55) and (57) by definition. Equation (56) is the requirement of statistical equilibrium in terms of the u_{ik}. Equilibrium requires that the (unconditional) probabilities of being in any state, j, be equal at the beginning and end of a period. At the beginning of a period this probability is $\sum_k u_{jk}$, and at the end it is

$$\sum_{i,k} u_{ik}p_{ij}^k$$

with p_{ij}^k the jth component of p_i^k. Equality requires

$$\sum_{i,k} u_{ik}p_{ij}^k = \sum_k u_{jk}, \qquad \text{all } j \tag{61}$$

which is (56) in component form. Thus (54)–(57) could have been written directly by admitting mixed strategies as alternatives.

To apply the linear program (54)–(57) to case (a), where S_i (and hence T_i) are finite, view the vectors v_i^k as elements of T_i. Then, for (54)–(57) and (43)–(46) to be equivalent under case (a), we must show that there is an optimal solution of (54)–(57) in which at most one u_{ik} is positive for each i [12]. If this is true, the optimal solution of (43)–(46) is obtained from that of (54)–(57) by choosing

$$x_i = u_{ik},\ v_i = v_i^k, \text{ if } u_{ik} > 0 \text{ for some } k$$
$$x_i = 0,\ v_i \text{ arbitrary in } S_i, \text{ if } u_{ik} = 0 \text{ for all } k$$

In case (b), the fact that at most one v_i^k is selected for each i means that there is an optimal solution to (54)–(57) in which only pure strategies appear.

THEOREM. *There is an optimal solution of* (54)–(57) *with at most one* u_{ik} *positive for each* i.

PROOF. Any linear program with a finite optimal solution has a basic feasible solution which is optimal. Let such a solution for (54)–(57) have $r \leq n + 1$ positive components, and let $\bar{B}$ be the $(n + 1)$ by r submatrix of the basis associated with these. Since the basis matrix itself is nonsingular, $\bar{B}$ has rank r. In tableau form, (55)–(56) are as shown in Table 1. Let s be

TABLE 1

$u_{11}, \ldots, u_{1m_1}$	$u_{21}, \ldots, u_{2m_2}$	$\cdots$	$u_{n1}, \ldots, u_{nm_n}$	Constants
$1 \cdots 1$	$1 \cdots 1$		$1 \cdots 1$	1
$p_{11}^1 - 1 \cdots p_{11}^{m_1} - 1$	$p_{21}^1 \cdots p_{21}^{m_2}$	$\cdots$	$p_{n1}^1 \cdots p_{n1}^{m_n}$	0
$p_{12}^1 \quad p_{12}^{m_1}$	$p_{22}^1 - 1 \quad p_{22}^{m_2} - 1$		$\cdot \quad \cdot$	0
$\vdots \quad \vdots$	$\vdots \quad \vdots$		$\vdots \quad \vdots$	$\vdots$
$p_{1n}^1 \quad p_{1n}^{m_1}$	$p_{2n}^1 \quad p_{2n}^{m_2}$	$\cdots$	$p_{nn}^1 - 1 \quad p_{nn}^{m_n} - 1$	0

the number of rows of $\bar{B}$ containing at least one entry $p_{ii}^k - 1$. Excluding the first row, the remaining $n - s$ rows have nonnegative entries and zero right-hand sides; hence these rows may contain only zeros, and at most $s + 1$ of the $n + 1$ rows of $\bar{B}$ do not vanish. At most s of these are linearly independent (since the sum of all rows but the first is zero), so

$$\text{rank}(\bar{B}) = r \leq s \tag{62}$$

Now note the structure of the columns in the tableau. Each has exactly one entry $p_{ii}^k - 1$, so it is impossible to place such an entry in $s > r$ rows with r columns, hence $s = r$. Then each column must have the entry $p_{ii}^k - 1$ in a different row to fill all r rows. This means that each column of $\bar{B}$ corresponds to a variable u_{ik} with different index i, proving the theorem.

To solve (54)–(57) using column generation, assume a feasible basis for this program is available, and let (π_0, π) be the simplex multipliers of this basis, with π_0 associated with (55) and π with (56). Pricing out a generic column leads to the subproblem:

$$\text{minimize } c_i(p_i) - \pi p_i \tag{63}$$

$$\text{subject to} \qquad p_i \in S_i \tag{64}$$

If S_i is finite and c_i arbitrary then this is a discrete programming problem which might be solved by a branch and bound method when S_i has many elements. If the domain of definition of c_i can be extended to the convex hull of S_i, $C(S_i)$, and if this extended c_i is concave over $C(S_i)$, then (63) is minimized at an extreme point of $C(S_i)$, and all other points in S_i may be ignored. If S_i is a convex polyhedron and c_i linear, the problem may be solved by linear programming. In this case, according to Theorem 1, only a single extreme point of each S_i need appear in an optimal solution.

4.4 Grid Linearization and Nonlinear Programming

4.4.1 General Development. Since linear programming is a well-developed, efficient tool, it is natural to attempt to use it in nonlinear problems through linearization. One means of accomplishing this is through what Wolfe [13] has termed grid linearization. Let the nonlinear program to be solved be

$$\text{minimize } f(x) \tag{1}$$

$$\text{subject to} \qquad g_i(x) \leq 0, \qquad i = 1, \ldots, m \tag{2}$$

Choose a set of grid points, $x_1, \ldots, x_T$. The convex hull of this set has elements

$$x = \sum_t \lambda_t x_t \tag{3}$$

$$\lambda_t \geq 0, \qquad \sum_t \lambda_t = 1 \tag{4}$$

Assuming, for the moment, that x is a scalar, a function f may be linearized on this grid by drawing its piecewise linear approximation between grid

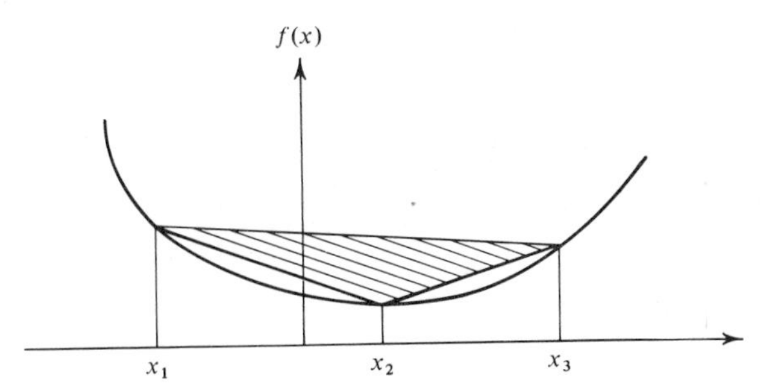

FIGURE 4-10 Piecewise linear approximation.

points, as shown in Figure 4-10. Any point, f, on the darkened lines may be written

$$f = \sum_t \lambda_t f(x_t) \tag{5}$$

where the λ_t satisfy (4). Unfortunately, other points are also represented by (4) and (5), namely, all points in the convex hull of the set $\{f(x_t), x_t\}$, shown shaded in Figure 4-10. Only points on the darkened lines constitute useful interpolations between grid points. To obtain only these, the λ_t must be restricted so that at most two are positive and these have adjacent indices. (The statement of this criterion for many dimensions is considered later.) It will be shown that, if f and the g_i are convex, the method we propose always yields a valid interpolation.

Replacing f and the g_i by their grid linearizations yields a linear program in the variables λ_t:

$$\text{minimize } z = \sum_{t=1}^{T} \lambda_t f(x_t) \tag{6}$$

subject to

$$\sum_{t=1}^{T} \lambda_t g_i(x_t) \leq 0, \qquad i = 1, \ldots, m \tag{7}$$

$$\sum_{t=1}^{T} \lambda_t = 1 \tag{8}$$

$$\lambda_t \geq 0 \tag{9}$$

In general, many grid points will be required to obtain an accurate approximation to the nonlinear functions f and g_i. Each grid point yields a column in (6)–(9), and it would be wasteful to choose these in advance and compute the coefficients of this column for each. What is desired is an accurate approximation only in the vicinity of the optimum. By letting the simplex method itself choose new grid points via subproblems, this may be accomplished.

Assume that, at the beginning of iteration k, a set of grid points $x_1, \ldots, x_T$ is available. The linear program (6)–(9) is then solved, yielding an optimal solution $\{\lambda_t^k\}$ and simplex multipliers (π^k, π_0^k). The current approximation to the optimal solution of (1)–(2) is

$$x^k = \sum_t \lambda_t^k x_t$$

Of all new grid points which might be introduced to improve this approximation, the simplex method would choose one whose associated column priced out most negative, i.e., that point, x_t, which minimizes

$$\bar{c}_t = f(x_t) - \sum_{i=1}^{m} \pi_i^k g_i(x_t) - \pi_0^k$$

The task of finding such a point is that of solving the unconstrained, nonlinear subproblem

$$\text{minimize } f(x) - \sum_i \pi_i^k g_i(x) \tag{10}$$

This subproblem may be solved by any appropriate method[2]. If

$$\min\left\{ f(x) - \sum_i \pi_i^k g_i(x) \right\} - \pi_0^k < 0 \tag{11}$$

then, barring degeneracy, the objective of (6)–(9) may be reduced by introducing a new column into the basis. If the above minimum is attained at $\hat{x}^k$, the new column has cost coefficient $f(\hat{x}^k)$ and remaining coefficients

$$p^k = (g_1(\hat{x}^k), \ldots, g_m(\hat{x}^k), 1)$$

This column is added to the linear program (6)–(9), which is then re-solved, and a new cycle begins. Note that (6)–(9) plays the role of the restricted master program in the Dantzig–Wolfe decomposition principle (see Section 3.4). If the objective (1) and constraints (2) are linear functions defined over a bounded convex polyhedral domain, any point in this domain can be represented as a convex combination of extreme points (which constitute a finite grid), and approximations such as (5) are exact. The grid linearization process

[2] See Section 1.1 for some possible approaches.

then becomes the Dantzig–Wolfe decomposition algorithm for linear programs. In that case, retaining all previously generated columns in the restricted master program is optional. In the nonlinear case, the convergence proofs which presently exist [16] require that all columns be retained.

A geometric interpretation of this procedure in terms of the set R [equation (4.3-15)] is enlightening. Let λ^k and

$$u^k = -(\pi_0^k, \pi^k)$$

be optimal for (6)–(9) and its dual respectively at cycle k. The vector (λ^k, u^k) is then a saddle point for the Lagrangian function associated with (6)–(9) (see Section 1.3.3 and Chapter 8). As shown in Theorem 9.3-3, π^k is a subgradient[3] of the function

$$\tilde{w}(y) = \min\left\{\sum_t \lambda_t f(x_t) \,\middle|\, \sum_t \lambda_t g(x_t) \le y, \sum_t \lambda_t = 1, \lambda_t \ge 0\right\}$$

at

$$\bar{y} = \sum_t \lambda_t^k g(x_t)$$

That is, π^k is the vector of slopes of a supporting hyperplane to the graph of $\tilde{w}$ at $\bar{y}$. Since this graph is the boundary of the set IL(R) in (4.3-16), the situation is as illustrated in Figure 4-11. As shown in Section 8.4, if $\hat{x}^k$ solves (10), then $(f(\hat{x}^k), g(\hat{x}^k))$ is a point of contact of R with a supporting hyperplane whose slopes are π^k. The vector $(f(\hat{x}^k), g(\hat{x}^k))$ is thus on the boundary of R. Adding the new column is equivalent to forming an improved inner linearization, as shown by the dashed lines in Figure 4-11. If the current solution is near optimal, then, by using π^k as the slopes of the hyperplane, the approximation of R is improved where it is most needed, in the vicinity of the optimum.

The procedure has a number of desirable properties if f and the g_i are convex. First, the points x^k are feasible, since

$$g_i(x^k) \le \sum_t \lambda_t^k g_i(x_t) \le 0$$

Second, if $\pi_i \le 0$, $i = 1, \ldots, m$, then (10) is convex and any local minimum of the subproblem is global. The condition $\pi_i \le 0$ may be guaranteed by minimizing z over only the slack variables in (7) before a new grid point is sought, since their relative costs are $-\pi_i$, and optimality implies that these be nonnegative. Finally, upper and lower bounds on $\min f$ are available.

[3] See Appendix 2.

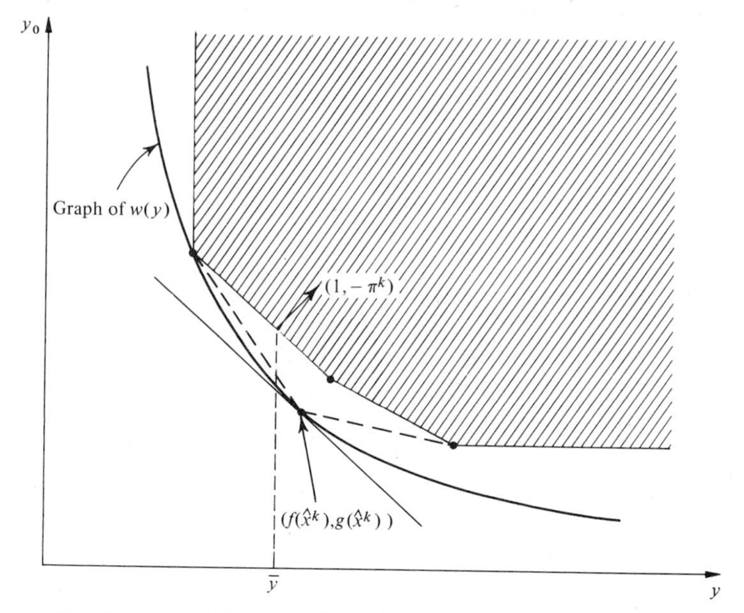

FIGURE 4-11 Improved linearization of R.

Let

$$S = \{x \mid g_i(x) \leq 0, i = 1, \ldots, m\}$$

Since f is convex and x^k is feasible

$$\min_{x \in S} f(x) \leq f(x^k) \leq \sum_t \lambda_t^k f(x_t) = z^k \tag{12}$$

which provides an upper bound. If $\pi_i \leq 0$, $i = 1, \ldots, m$, the minimal value in (11) is always a lower bound on $\min f$. To see this, let

$$u_i = -\pi_i, \qquad i = 1, \ldots, m$$

and define

$$L(x, u) = f(x) + \sum_i u_i g_i(x), \qquad u_i \geq 0 \tag{13}$$

By definition

$$\min_{x \in E^n} L(x, u) \leq L(x, u) \tag{14}$$

the above holding for all x. For those x satisfying $g_i \leq 0$, the sum in (13) is nonpositive and (14) becomes

$$\min_{x \in E^n} L(x, u) \leq f(x) \qquad \text{for all } x \in S \tag{15}$$

Thus

$$\min_{x \in S} f(x) \geq \min_{x \in E^n} L(x, u) \tag{16}$$

which is the lower bound. Note that (16) holds even if the functions f and g_i are not convex.

The Lagrangian function $L(x, u)$ plays a central role in mathematical programming. Its use in nonlinear decomposition procedures is discussed in Chapter 8. There it is shown that the minimal value in (15), when viewed as a function of u, is the objective function of the dual of the nonlinear primal (1)–(2). In fact, the optimality test given below is simply the condition that primal and dual objective values be equal.

If condition (11) is not met and all $\pi_i \leq 0$, then there is no grid point or slack variable which will reduce the cost, z. The current approximation, x^k, is then optimal for (1)–(2), as is shown by the following theorem.

THEOREM 1 (DANTZIG [16]). *Let* (π^k, π_0^k) *be simplex multipliers corresponding to an optimal solution of the restricted master program* (6)–(9) *at iteration* k. *If*

$$\min\left\{ f(x) - \sum_{i=1}^{m} \pi_i^k g_i(x) \right\} - \pi_0^k \geq 0$$

then the point

$$x^k = \sum_t \lambda_t^k x_t$$

is optimal; i.e.,

$$\min_{x \in S} f(x) = f(x^k)$$

PROOF. By definition, the relative cost factors for basic variables λ_t^k are zero:

$$\bar{c}_t^k = f(x_t) + \sum_i u_i^k g_i(x_t) + u_0^k = 0, \ t \text{ basic}$$

where $u_i^k = -\pi_i^k \geq 0$. Multiplying the above by λ_t^k and summing,

$$\sum_t \lambda_t^k f(x_t) + \sum_i u_i^k \sum_t \lambda_t^k g_i(x_t) + u_0^k = 0$$

By complementary slackness,

$$\sum_i u_i^k \sum_t \lambda_t^k g_i(x_t) = 0$$

Thus

$$u_0^k = -\sum_t \lambda_t^k f(x_t) = -z^k$$

Now add $u_0^k + z^k$, which is zero, to the right-hand side of (16) and combine the result with (12) yielding

$$\min_{x \in E^n} \{L(x, u^k) + u_0^k\} + z^k \leq \min_{x \in S} f(x) \leq f(x^k) \leq z^k \qquad (17)$$

By the hypothesis of the theorem

$$\min_{x \in E^n} \{L(x, u^k) + u_0^k\} \geq 0$$

so the above quantity may be removed from the left-hand side of (17), yielding

$$z^k \leq \min_{x \in S} f(x) \leq f(x^k) \leq z^k$$

This implies

$$\min_{x \in S} f(x) = f(x^k) = z^k$$

and the theorem is proved.

Dantzig [16] proves that, for a convex program, this algorithm converges to an optimal solution as the number of iterations tends to infinity. The proof again requires that all columns generated in past iterations be retained in each new restricted master program.

Separable Programs. A significant advantage arises if f and the g_i are additively separable, i.e., if

$$f = \sum_j f_j(x_j)$$

$$g_i = \sum_j g_{ij}(x_j)$$

Then a set of grid points, $\{x_{jt}\}$, may be defined for each variable x_j, and each of the functions f_j and g_{ij} linearized separately, yielding the program

$$\text{minimize} \sum_{j,t} \lambda_{jt} f_j(x_{jt}) \tag{18}$$

subject to

$$\sum_{j,t} \lambda_{jt} g_{ij}(x_{jt}) \leq 0, \qquad i = 1, \ldots, m \tag{19}$$

$$\sum_t \lambda_{jt} = 1, \qquad j = 1, \ldots, n \tag{20}$$

$$\lambda_{jt} \geq 0 \tag{21}$$

The subproblem (11) separates, in this case, into j independent, single-variable problems:

$$\text{minimize } f_j(x_j) - \sum_i \pi_i g_{ij}(x_j) \tag{22}$$

which, being unconstrained, may often be solved in closed form by setting the derivative to zero.

Admissible Interpolations. Returning to the question of insuring that only valid interpolations occur, consideration of the three-dimensional case will convince the reader that what is desired is this: The optimal solution of the restricted master program (6)–(9) must be such that the convex hull of those grid points x_t for which $\lambda_t > 0$ contains no other grid points. Under a few additional assumptions, this happens automatically when the functions f and g_i are convex, as is shown in the following theorem, due to Wolfe [14].

THEOREM 2. *If f and all g_i are convex and either* (a) *f is strictly convex or* (b) *the restricted master program* (6)–(9) *has a unique solution, then any optimal solution of* (6)–(9) *has the property that the convex hull of those grid points, x_t, for which $\lambda_t > 0$ contains no other grid point.*

PROOF. Let the x_t for which $\lambda_t > 0$ be $x_1, \ldots, x_p$, and assume there is another grid point, x_{t^*}, which is contained in the convex hull of $x_1, \ldots, x_p$. The point x_{t^*} is then in the relative interior of the convex hull, H, of some subset of these points, say $x_1, \ldots, x_q$, $q \leq p$ (see Figure 4-12), i.e.,

$$x_{t^*} = \sum_{i=1}^{q} s_i x_i, \qquad s_i > 0, \qquad \sum_i s_i = 1 \tag{23}$$

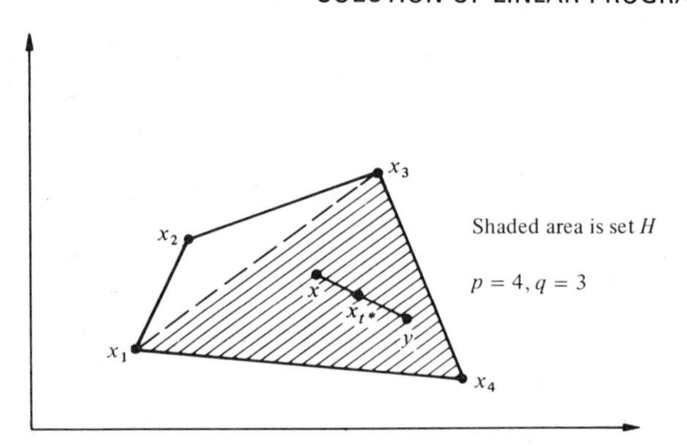

FIGURE 4-12 Inadmissible interpolation.

Let (π, π_0) be the simplex multipliers associated with an optimal solution of (6)–(9). Then the optimality conditions for (6)–(9) are

$$\bar{c}_t = F(x_t) = f(x_t) - \sum_i \pi_i g_i(x_t) - \pi_0 \geq 0, \qquad \text{all } t \tag{24}$$

$$\bar{c}_t = 0 \qquad \text{if } \lambda_t > 0 \tag{25}$$

$$\pi_i \leq 0 \tag{26}$$

From (24) and (26), $F(x)$ is convex so, by (25),

$$F\left(\sum_{i=1}^{q} \beta_i x_i\right) \leq \sum_{i=1}^{q} \beta_i F(x_i) = 0 \tag{27}$$

Thus $F(x)$ is nonpositive for all $x \in H$. In fact, we can show $F(x)$ is zero on H. For assume not; then $(F(x) < 0,\ x \in H)$ and $(x_{t^*} \in r \text{ int } H)$ imply there exists $y \in H$ such that

$$x_{t^*} = \alpha x + (1 - \alpha) y, \qquad 0 < \alpha < 1 \tag{28}$$

as shown in Figure 4-12. Then

$$F(x_{t^*}) \leq \alpha F(x) + (1 - \alpha) F(y) < 0 \tag{29}$$

contradicting (24). Thus, setting $F(x)$ to zero for all $x \in H$,

$$f(x) = \sum_i \pi_i g_i(x) + \pi_0 \tag{30}$$

But the left-hand side of (30) is convex, and the right-hand side concave (since $\pi_i \leq 0$). The only way they may be equal is if both sides are linear on H. But, since H consists of more than one point, strict convexity of f means this is impossible so (a) has been proved.

To prove (b) we show that the existence of x_{t^*} implies that another solution to (6)–(9), $\{\bar{\lambda}_t\}$, may be constructed, a contradiction. Let

$$\bar{\lambda}_t = \lambda_t - \theta s_t, \qquad t = 1, \ldots, q \tag{31}$$

$$\bar{\lambda}_{t^*} = \lambda_{t^*} + \theta \tag{32}$$

$$\bar{\lambda}_t = \lambda_t, \qquad \text{all other } t \tag{33}$$

for $\theta > 0$ and sufficiently small, all $\bar{\lambda}_t \geq 0$, and

$$\sum_{\text{all } t} \bar{\lambda}_t = \sum_{\text{all } t} \lambda_t - \theta \sum_{t=1}^{q} s_t + \theta = 1 \tag{34}$$

Substituting the $\bar{\lambda}_t$ into (6) and (7),

$$\sum_t \bar{\lambda}_t f(x_t) = \sum_t \lambda_t f(x_t) + \theta\left(f(x_{t^*}) - \sum_{t=1}^{q} s_t f(x_t)\right) \tag{35}$$

$$\sum_t \bar{\lambda}_t g_i(x_t) = \sum_t \lambda_t g_i(x_t) + \theta\left(g_i(x_{t^*}) - \sum_{t=1}^{q} s_t g_i(x_t)\right) \tag{36}$$

Since $f(x)$ is linear on H and $x_{t^*} \in H$,

$$f(x_{t^*}) = \sum_{t=1}^{q} s_t f(x_t) \tag{37}$$

so the z-value of $\{\lambda_t\}$ and $\{\bar{\lambda}_t\}$ is the same. To show that the $\bar{\lambda}_t$ satisfy (7), i.e., make the expression in (36) nonpositive, note that the first term on the right of (36) is nonpositive [since the λ_t are an optimal solution to (6)–(9)] as is the term multiplied by θ (by convexity of the g_i), so (36) is nonpositive for all $\theta \geq 0$. Thus, for sufficiently small $\theta > 0$, the $\bar{\lambda}_t$ constitute another optimal solution to (6)–(9), contradicting (b).

Miller[15] has generalized these ideas to the case of possibly nonconvex but separable functions. Here the requirement that all interpolations be valid is met by allowing a variable, λ_{jt}, in (18)–(21) to enter the basis only if either $\lambda_{j,t-1}$, $\lambda_{j,t+1}$, or no other λ_{jt} are already basic. This restriction means that not even a global solution to the linear program (18)–(21) may be ensured, but the method has, in practice, produced local solutions which are quite valuable.

4.4.2 Nonlinear Version of the Dantzig–Wolfe Decomposition Principle. By using grid linearization, Sekine [17] has obtained an extension of the Dantzig–Wolfe decomposition principle to the case of convex nonlinear subsystems. The problem to be solved is

$$\text{minimize } f(x) = \sum_{i=1}^{p} f_i(x_i) \tag{1}$$

where each f_i is a convex function, subject to a set of linear coupling constraints:

$$\sum_{i=1}^{p} A_i x_i = b \tag{2}$$

and to a set of possibly nonlinear subsystem constraints involving each x_i independently:

$$x_i \in S_i, \qquad i = 1, \ldots, p \tag{3}$$

where the S_i are convex sets, not necessarily polyhedral. This formulation includes linear programs of angular structure as a special case, when $f_i(x_i) = c_i x_i$ and S_i is determined by linear constraints.

Since the constraints on each x_i are independent, we may choose a separate set of grid points for each. Let the tth grid point for x_i be

$$x_i^t \in S_i \tag{4}$$

Each function $f_i(x_i)$ is replaced by its linearization on the grid $\{x_i^t\}$

$$f_i(x_i) \doteq \sum_t \lambda_i^t f_i(x_i^t) \tag{5}$$

where

$$\sum_t \lambda_i^t = 1, \qquad \lambda_i^t \geq 0 \tag{6}$$

The associated value of x_i is

$$x_i^a = \sum_t \lambda_i^t x_i^t \tag{7}$$

To obtain an approximating linear program for (1)–(3), each f_i in (1) is replaced by its linearization and (7) is substituted into (2), yielding

$$\text{minimize } z = \sum_{i,t} \lambda_i^t f_i(x_i^t) \tag{8}$$

subject to

$$\sum_{i,t} (A^i x_i^t)\lambda_i^t = b \tag{9}$$

$$\sum_t \lambda_i^t = 1, \qquad i = 1, \ldots, p \tag{10}$$

$$\lambda_i^t \geq 0 \tag{11}$$

Since $x^t \in S_i$ and S_i is convex, the point x_i^a in (7) is also an element of S_i. If the λ_i^t satisfy (9), then x_i^a also satisfies (2) and thus is feasible for the original problem.

As before, one would not wish to choose many grid points in advance. This is again avoided by having the simplex method generate them when needed, by using column generation. The column associated with λ_i^t is

$$p_i^t = \{f_i(x_i^t) \mid A_i x_i^t \mid u_i\}' \tag{12}$$

with u_i the ith unit vector. Assume that a feasible basis for (8)–(11) is available, and let (π_1, π_2) be the simplex multipliers of this basis associated with (9) and (10), respectively. Pricing out the column p_i^t yields a relative cost factor

$$\bar{c}_i^t = f_i(x_i^t) - \pi_1 A_i x_i^t - \pi_{2i} \tag{13}$$

The problem of finding a point $x_i^t \in S_i$ minimizing this becomes the ith subproblem:

$$\left.\begin{aligned} &\text{minimize } f_i(x_i) - \pi_1 A_i x_i && (14)\\ &\text{subject to } \quad x_i \in S_i && (15)\end{aligned}\right\} \text{subproblem } i$$

This is a convex program, which we assume has a solution for any vector π_1. If f_i is linear and S_i is polyhedral, the above becomes the subproblem obtained in the decomposition principle of Chapter 3.

Let $\hat{x}_i$ solve subproblem i, and assume that at least one of the relative cost factors is negative:

$$f_i(\hat{x}_i) - \pi_1 A_i \hat{x}_i - \pi_{2i} < 0 \qquad \text{for some } i \tag{16}$$

Form the columns

$$\hat{p}_i = \{f_i(\hat{x}_i) \mid A_i \hat{x}_i \mid u_i\}, \qquad i = 1, \ldots, p \tag{17}$$

These may be used to form a restricted master program, analogous to that in Section 3.4. This program has as variables the current basic set $\{\lambda_i^t\}$ plus p new variables $\hat{\lambda}_i$, one for each of the columns $\hat{p}_i$:

$$\left.\begin{aligned} &\text{minimize } z = \sum_i \sum_{t \text{ basic}} \lambda_i^t f_i(x_i^t) + \sum_i \hat{\lambda}_i f_i(\hat{x}_i) \\ &\text{subject to} \quad \sum_i \sum_{t \text{ basic}} (A_i x_i^t)\lambda_i^t + \sum_i (A_i \hat{x}_i)\hat{\lambda}_i = b \\ &\qquad \sum_{t \text{ basic}} \lambda_i^t + \sum_i \hat{\lambda}_i = 1, \qquad i = 1, \ldots, p \\ &\qquad \lambda_i^t \geq 0, \qquad \hat{\lambda}_i \geq 0 \end{aligned}\right\} \text{restricted master program}$$

This linear program has the basis associated with the variables λ_i^t as an initial feasible basis, and has the p columns $\hat{p}_i$ in (17) as nonbasic columns. Solving it yields a new basis and new simplex multipliers with which to begin the next cycle.

All the results regarding grid linearization quoted earlier in this section are applicable here. The convergence results of Dantzig [16] also apply, as does a similar proof by Sekine, found in [17].

4.5 Design of Multiterminal Flow Networks

A directed network (N, A) is a set of elements $\{i, j, \ldots\}$ called nodes together with a subset, A, of the set of ordered pairs (i, j) of elements from N, called arcs. Associated with each arc (i, j) is a capacity v_{ij}, the maximal amount of some homogeneous quantity which may flow from i to j per unit time. Ford and Fulkerson [20] define a flow in a network as follows. Let $A(i)$ ("after i)" denote the set of all nodes $j \in N$ such that $(i, j) \in A$;

$$A(i) = \{j \in N \mid (i, j) \in A\} \tag{1}$$

Similarly, let $B(i)$ ("before i)" denote the set of all $j \in N$ such that $(j, i) \in A$;

$$B(i) = \{j \in N \mid (j, i) \in A\} \tag{2}$$

Let us single out two nodes s and t as a source node and sink node, respectively. Then a flow of value f_{st} from s to t is a function $\bar{f}_{ij}$ from A to the nonnegative real numbers satisfying

$$\sum_{j \in A(i)} \bar{f}_{ij} - \sum_{j \in B(i)} \bar{f}_{ji} = \begin{cases} 0, & i \neq s, t \\ f_{st}, & i = s \\ -f_{st}, & i = t \end{cases} \tag{3}$$

$$\bar{f}_{ij} \leq v_{ij}, \qquad \text{all } (i, j) \in A \tag{4}$$

Equations (3) are conservation equations, specifying that, for nodes which are neither source nor sink, the flow in equals the flow out, while the flow out of the source is f_{st} and the flow out of the sink is $-f_{st}$. Relation (4) is the capacity constraint. The maximal flow problem is, given the capacities v_{ij}, find a flow from s to t whose value, f_{st}, is maximal among all such flows.

This problem may be viewed as one of *analysis*: Given the capacities, find a maximal flow. A problem of *synthesis*, defined by Gomory and Hu [18, 19] is: Given a set of flow requirements r_{ij} and per unit costs of capacity c_{ij}, specify the (nonnegative) arc capacities v_{ij} such that the *maximal* flow values satisfy

$$f_{ij} \geq r_{ij}, \qquad i \neq j \tag{5}$$

and such that the cost of capacity

$$z = \sum_{(i,j)\in A} c_{ij}v_{ij} \tag{6}$$

is minimized. Such a problem might arise, for example, in designing a communications network. It is assumed in what follows that if $(i, j) \in A$, then $(j, i) \in A$, and that $v_{ij} = v_{ji}$, $r_{ij} = r_{ji}$, $c_{ij} = c_{ji}$. This implies that $f_{ij} = f_{ji}$.

Given two nodes i and j, let X be a subset of N which contains i but not j and $\bar{X}$ its complement relative to N, so that $j \in \bar{X}$. A cut separating i and j is that subject of arcs (x, y) of the network directed from some node in X to some node in $\bar{X}$. The capacity of a cut is the sum of its arc capacities. Ford and Fulkerson prove in [20] that the maximal flow value f_{ij} from i to j is equal to the minimum of the capacities of all cuts separating i and j.

Let K_{ij}^{α} be the capacity of the αth cut separating i and j. Then K_{ij}^{α} is a partial sum of the arc capacities v_{ij}, and inequalities (5) may be written

$$K_{ij}^{\alpha} \geq r_{ij}, \qquad \begin{matrix} i \neq j \\ \text{all } \alpha \end{matrix} \tag{7}$$

These are a very large set of constraints, one for each cut separating each node pair (i, j).

As a first step toward easing this difficulty, Gomory and Hu show that if (7) is satisfied for a certain subset of arcs in the network, it is satisfied for all arcs. The development proceeds as follows. First, necessary and sufficient conditions are given that a given symmetric matrix (f_{ij}) be realizable, i.e., be the matrix of maximal flow values for some network. The result is as follows.

THEOREM 1. *A necessary and sufficient condition that a symmetric, nonnegative matrix (f_{ij}) be realizable is*

$$f_{ik} \geq \min(f_{ij}, f_{jk}), \qquad \textit{all } i, j, k \tag{8}$$

PROOF. Necessity—let (f_{ij}) be realizable and assume

$$f_{ik} < \min(f_{ij}, f_{jk}) \tag{9}$$

The flow value f_{ik} is equal to the minimal capacity of all cuts separating i and k. Let X be a set of nodes containing i but not k such that a minimal cut connects X and $\bar{X}$. If node j is in X, then this cut separates j and k and has a capacity less than f_{jk}, a contradiction. If j is in $\bar{X}$, the cut separates i and j, and a similar contradiction is obtained. Thus (8) must hold.

Note that (8) generalizes by induction to the relation

$$f_{ip} \geq \min(f_{ij}, f_{jk}, \ldots, f_{op}) \tag{10}$$

where $i, j, k, \ldots, o, p$ is any path in the network between i and p. It is (10) that will be used in the following development.

To prove sufficiency, we assume that (10) is true and give a realization for (f_{ij}). The notion of maximal spanning tree is needed. A spanning tree in a graph is a tree (subset of the arcs containing no loops) which connects all nodes. If each arc has a "weight," a number associated with it, then the spanning tree for which the sum of the weights on all tree links is largest is called a maximal spanning tree.

Let the weights be the numbers f_{ij}, and let $i, j, k, \ldots, o, p$ be a path between i and p in a maximal spanning tree. Then

$$f_{ip} \leq \min(f_{ij}, f_{jk}, \ldots, f_{op}) \tag{11}$$

for, if not, and say $f_{jk} < f_{ip}$, arc (j, k) could be deleted from the tree and (i, p) inserted, yielding another spanning tree with an increase in the sum of the arc weights, a contradiction. (For an example, see Figure 4-13.) Relations (10) and (11) imply that, for any path in a maximal spanning tree,

$$f_{ip} = \min(f_{ij}, f_{jk}, \ldots, f_{op}) \tag{12}$$

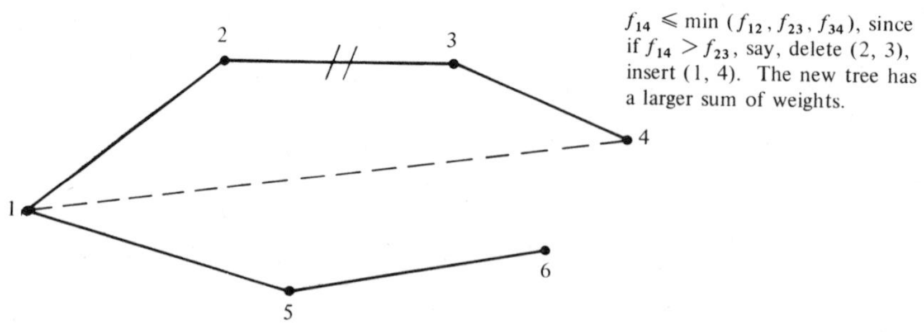

FIGURE 4-13 Maximal spanning tree.

Now consider a network with arc capacities f_{ij} for arcs in the tree and zero capacities for all other arcs. The maximal flow value between any nodes i and p in this network is given by (12), since the only way flow can move from i to p is along a path in the tree, and there it is limited by the arc with minimal capacity. Thus (f_{ij}) has been realized.

Given the preceding theorem and the notion of maximal spanning tree introduced in its proof, define a dominant requirement tree, T, as a maximal spanning tree where the arc weights are the flow requirements r_{ij}. Then the following is true.

THEOREM 2. *Let the network under consideration have n nodes. If the flow constraints* (5) *hold for all* $(n - 1)$ *arcs* (i, j) *in the dominant requirement tree, T, they hold for all node pairs in the network.*

PROOF. Consider any node pair (i, p) not in T and a path $(i, j, k, \ldots, o, p)$ in T from i to p. Since the f_{ij} are maximal flow values, (10) holds:

$$f_{ip} \geq \min(f_{ij}, f_{jk}, \ldots, f_{op}) \tag{13}$$

and since the path is in T (11) holds, with the f_{ij} replaced by r_{ij}:

$$r_{ip} \geq \min(r_{ij}, r_{jk}, \ldots, r_{op}) \tag{14}$$

Then if all the requirements in T are satisfied,

$$f_{ij} \geq r_{ij}, f_{jk} \geq r_{jk}, \ldots, f_{op} \geq r_{op} \tag{15}$$

Then (13) and (14) imply

$$f_{ip} \geq r_{ip} \tag{16}$$

so all flow requirements not in T are satisfied as well.

We have now shown that all node pairs not in the dominant requirement tree may be disregarded, and the flow constraints (7) may be written

$$K_{ij}^{\alpha} \geq r_{ij}, \qquad (i, j) \in T, \qquad \text{all } \alpha \tag{17}$$

Since there are only $n - 1$ node pairs in T, this can be a substantial saving. Still, (17) involves capacities of all cuts between these node pairs, so many inequalities are still present. Of course, for given $(i, j) \in T$, if the minimal cut satisfies (17), then all do, so for any set of arc capacities only $(n - 1)$ constraints need be checked. However, which $(n - 1)$ these are change as the capacities vary. To see how various subsets of constraints may be chosen efficiently, we write the dual problem, in which the choice of constraints or rows to add to the primal program becomes a choice of columns to enter the dual basis.

Let $p'_{ij\alpha}$ be the row coefficient vector for the inequality in (17) with indices (i, j, α). Recalling that K^{α}_{ij} is the sum of arc capacities in the αth cut separating nodes i and j, $p'_{ij\alpha}$ has ones in positions where an arc is present in this cut and zeros elsewhere. The dual of (6), (17) is

$$\text{maximize} \sum_{\alpha} \sum_{(i,j)\in T} r_{ij}\pi_{ij\alpha} \tag{18}$$

subject to

$$\sum_{\alpha} \sum_{(i,j)\in T} p_{ij\alpha}\pi_{ij\alpha} \leq c \tag{19}$$

$$\pi_{ij\alpha} \geq 0 \tag{20}$$

where c is a vector with components c_{ij}. This problem has as many rows as there are arcs in the network and as many columns as there are cuts separating node pairs $(i, j) \in T$. An initial feasible solution is $\pi_{ij\alpha} = 0$, and slack variables equal to c_{ij}. The simplex multipliers for (19) are, of course, just the primal variables, the arc capacities v_{ij}. Letting v be a vector of these capacities and pricing out a column of (17) yields

$$\underset{\substack{(i,j)\in T \\ \text{all } \alpha}}{\text{maximize}}\ \bar{c}_{ij\alpha} = r_{ij} - vp_{ij\alpha} \tag{21}$$

This maximization is equivalent to minimizing $vp_{ij\alpha}$, which, by definition of $p_{ij\alpha}$, is the capacity of the αth cut separating i and j. But minimizing this value over α yields the maximal flow between i and j when the capacities are the multipliers, v_{ij}. Thus, for each $(i, j) \in T$, we must solve a maximal flow problem, $n - 1$ problems in all.

The optimality criterion for the dual is

$$\bar{c}_{ij\alpha} \leq 0 \Rightarrow f_{ij} \geq r_{ij}, \qquad (i, j) \in T \tag{22}$$

If this is not satisfied, columns are formed for each violation, and these are used, together with the current basic variables, to form a restricted master program over which (18) is maximized using the simplex method. This yields new multipliers, which are used as capacities in forming new network flow problems. At optimality the simplex multipliers of the dual are the desired optimal capacities.

Note that, since the dual is being solved, none of the sequence of network capacities generated have associated flows satisfying all flow requirements until optimality is achieved. A primal approach, which maintains the flow requirements satisfied throughout, is discussed in [18].

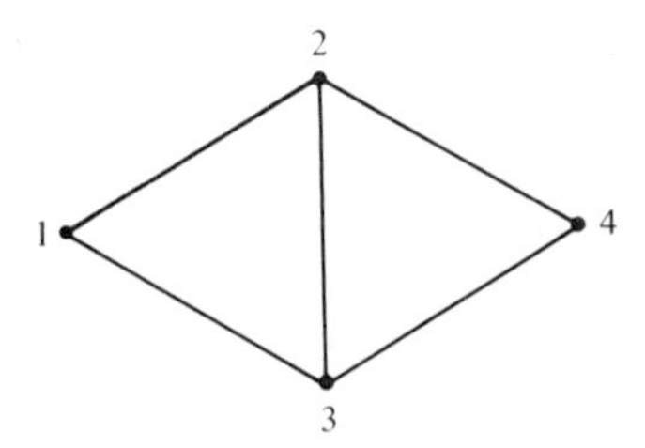

FIGURE 4-14

Example. Consider the network shown in Figure 4-14 with flow requirement and cost matrices

$$(r_{ij}) = \begin{bmatrix} — & 3 & 1 & 2 \\ & — & 4 & 1 \\ & & — & 2 \\ & & & — \end{bmatrix}, \qquad (c_{ij}) = \begin{bmatrix} — & 3 & 1 & — \\ & — & 4 & 1 \\ & & — & 2 \\ & & & — \end{bmatrix}$$

First, a dominant requirement tree must be found. As pointed out in [18, p. 261], this may be done as follows. Select the link with the largest value. Then, of those remaining, select the link with next largest value which does not form a loop when added to those already chosen. After $n - 1$ repetitions of this process in an n-node network, a dominant requirement tree is produced. For this example, the weighted network and dominant requirement tree are shown in Figure 4-15. Thus, only the flow constraints

$$f_{12} \geq 3, \qquad f_{23} \geq 4, \qquad f_{34} \geq 2$$

need be enforced.

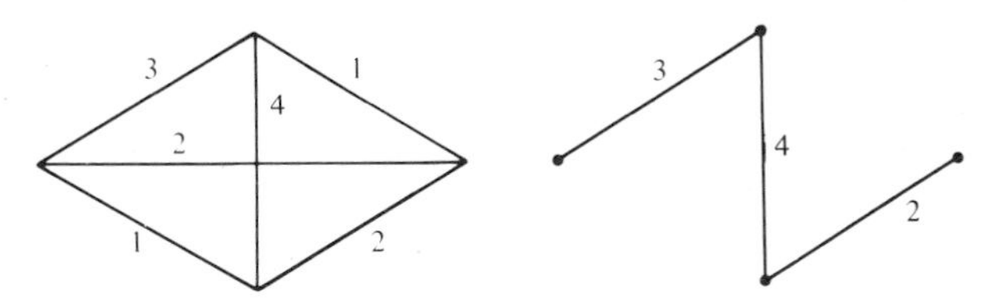

FIGURE 4-15 Weighted network and dominant requirement tree.

The dual program (18)–(20) takes the form

$$\text{maximize } 3 \sum_{\alpha} \pi_{12\alpha} + 4 \sum_{\alpha} \pi_{23\alpha} + 2 \sum_{\alpha} \pi_{34\alpha}$$

subject to

$$\begin{pmatrix} v_{12} \\ v_{13} \\ v_{23} \\ v_{24} \\ v_{34} \end{pmatrix} \begin{pmatrix} (1,2) \\ (1,3) \\ (2,3) \\ (2,4) \\ (3,4) \end{pmatrix} \left| \quad \sum_{\alpha} p_{12\alpha}\pi_{12\alpha} + \sum_{\alpha} p_{23\alpha}\pi_{23\alpha} + \sum_{\alpha} p_{34\alpha}\pi_{34\alpha} + s = \begin{pmatrix} 3 \\ 1 \\ 1 \\ 2 \\ 2 \end{pmatrix} \right.$$

with all variables nonnegative. The correspondence between the rows of this program and the arcs of the network is indicated above; e.g., $p_{12\alpha}$ has a one in position 3 if arc (2, 3) is included in the αth cut separating nodes 1 and 2, etc. The simplex multipliers v_{ij} will, at optimality, be the desired arc capacities.

Iteration 1. The initial basic variables are the slacks, s, and the initial vector of simplex multipliers is $v = 0$. We generate an initial set of columns for the dual by using v as arc capacities of the network, and solving three network flow problems to find f_{12}, f_{23}, and f_{34}. Since the capacities are currently zero, these flows are zero and any cut is minimal. We arbitrarily select the following:

$$\begin{aligned}
&\text{cut separating 1 and 2} = \{(1, 2) \quad (1, 3)\} \\
&\text{cut separating 2 and 3} = \{(2, 1) \quad (2, 3) \quad (2, 4)\} \\
&\text{cut separating 3 and 4} = \{(3, 4) \quad (2, 4)\}
\end{aligned}$$

The associated columns are

$$\begin{aligned}
p'_{121} &= (1 \quad 1 \quad 0 \quad 0 \quad 0 \quad 3) \\
p'_{231} &= (1 \quad 0 \quad 1 \quad 1 \quad 0 \quad 4) \\
p'_{341} &= (0 \quad 0 \quad 0 \quad 1 \quad 1 \quad 2)
\end{aligned}$$

Restricted Master 1

s_1	s_2	s_3	s_4	s_5	$-z$	π_{121}	π_{231}	π_{341}	$\bar{c}$
					Cycle 0				
1						1	1		3
	1					1			1
		1					(1)		1
			1				1	1	2
				1				1	2
					1	3	4	2	0
							*		

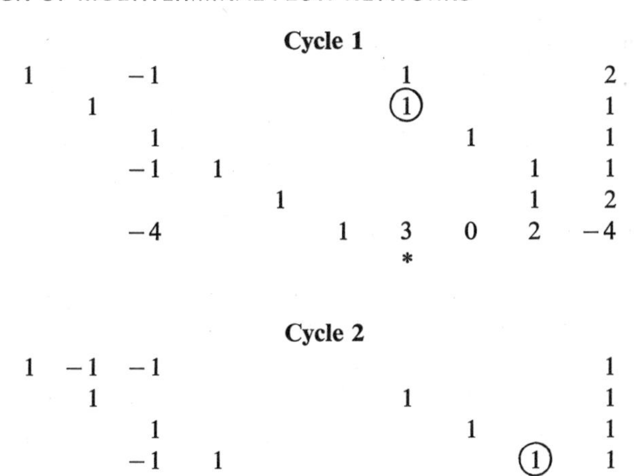

Cycle 1

1		−1				1			2
	1					(1)			1
		1					1		1
		−1	1					1	1
				1				1	2
		−4			1	3	0	2	−4
						*			

Cycle 2

1	−1	−1							1
	1					1			1
		1					1		1
		−1	1					(1)	1
				1				1	2
	−3	−4			1			2	−7
								*	

Cycle 3: optimal

1	−1	−1							1
	1					1			1
		1					1		1
		−1	1					1	1
		1	−1	1					1
	−3	−2	−2		1				−9

Simplex Multipliers. $v = (0 \quad 3 \quad 2 \quad 2 \quad 0)$.

Iteration 2. Using these multipliers as capacities, we obtain the network shown in Figure 4-16. Since this network is a tree when arcs of zero capacity are deleted, solving the required flows problems is easy. The solutions are

$$f_{12} = 2, \ \bar{c}_{122} = 3 - 2 = 1, \qquad \text{minimal cut} = \{(1, 2) \quad (2, 3) \quad (3, 4)\}$$

$$f_{23} = 2, \ \bar{c}_{232} = 4 - 2 = 2, \qquad \text{minimal cut} = \{(1, 2) \quad (2, 3) \quad (3, 4)\}$$

$$f_{34} = 2, \ \bar{c}_{342} = 2 - 2 = 0, \qquad \text{minimal cut} = \{(2, 4) \quad (3, 4)\}$$

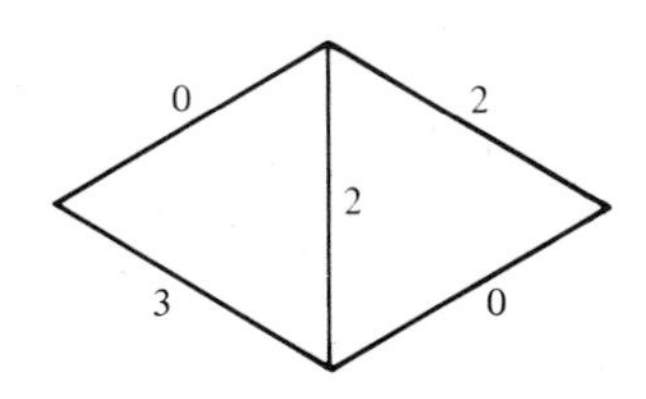

FIGURE 4-16

The columns generated are

$$p_{122} = (1 \quad 0 \quad 1 \quad 0 \quad 1 \quad 3)$$
$$p_{232} = (1 \quad 0 \quad 1 \quad 0 \quad 1 \quad 4)$$
$$p_{342} = (0 \quad 0 \quad 0 \quad 1 \quad 1 \quad 2)$$

These are multiplied by the current basis inverse (found in columns 1 through 6 of the final simplex tableau) and are added to the restricted master program.

Restricted Master 2

s_1	π_{121}	π_{231}	π_{341}	s_5	$-z$	π_{122}	π_{232}	π_{342}	s_2	s_3	s_4	$\bar{c}$
						Cycle 0						
1									−1	−1		1
	1								1			1
		1				1	1			1		1
			1			−1	−1	1		−1	1	1
				1		2	(2)			1	−1	1
					1	1	2		−3	−2	−2	−9
							*					
						Cycle 1: optimal						
1									−1	−1		1
	1								1			1
		1		$-\frac{1}{2}$						$\frac{1}{2}$	$\frac{1}{2}$	$\frac{1}{2}$
			1	$\frac{1}{2}$				1		$-\frac{1}{2}$	$\frac{1}{2}$	$\frac{3}{2}$
				$\frac{1}{2}$		1	1			$\frac{1}{2}$	$-\frac{1}{2}$	$\frac{1}{2}$
				−1	1	−1			−3	−3	−1	−10

Simplex Multipliers. $v = (0 \quad 3 \quad 3 \quad 1 \quad 1)$.

Iteration 3. Using the above as capacities, the new network is as shown in Figure 4-17. Solving the new flow problems in this network yields

$$f_{12} = 3, \qquad \min \bar{c}_{12\alpha} = 3 - 3 = 0$$
$$f_{23} = 4, \qquad \min \bar{c}_{23\alpha} = 4 - 4 = 0$$
$$f_{34} = 2, \qquad \min \bar{c}_{34\alpha} = 2 - 2 = 0$$

The current capacity vector, v, is therefore optimal.

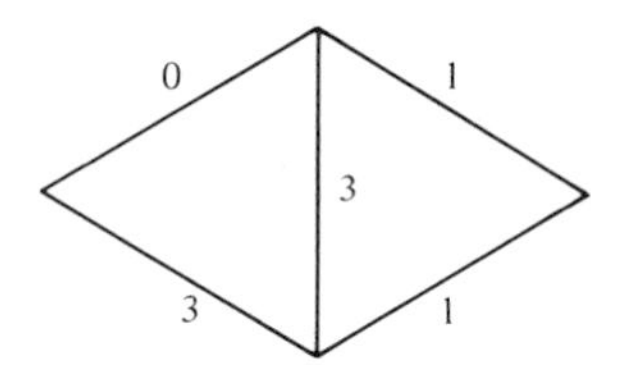

FIGURE 4-17

Problems

1. Write the dynamic programming recursion for solving the knapsack problem and outline the computational procedure. See [2, p. 866] for an estimate of the computer time required to perform the computations, and for a more efficient algorithm.
2. Extend the cutting-stock formulation to the following cases:
   (a) There are rolls of different lengths $L_1, L_2, \ldots, L_k$ available for cutting. A roll of length L_r costs c_r dollars.
   (b) We again have lengths $L_1, \ldots, L_k$, but now assume there are k machines to do the cutting, the rth of which can cut only u_r rolls of length L_r in a given period of time. The problem is now to determine patterns which minimize cost, meet or exceed demands, and overload no machine.

   Note that, in (b), a restricted master program can be formed with one or more patterns for each length L_r in it. Why can this be done in (b) but not in (a)? Also consider the application of the generalized upper bounding method of Section 6.4 to (b) and outline its advantages.
3. Investigate the extension of this formulation to a two-dimensional situation, where the stock to be cut has both a length and width (see [5]).
4. Construct the initial basic feasible solution for phase 1 suggested in the portion of Section 4.2 dealing with computational results. Write explicitly the basis inverse associated with this solution.
5. Write the forward dynamic programming recursion relations for the subproblems given in relations (4.2-38)–(4.2-41).
6. For those interested in computations, complete data for a number of scheduling problems is given in Section 8.8.3. The procedures of sections 4.2 and 3.10 may be coded and applied to these problems and comparisons made. Other sample problems are not difficult to construct.
7. Show how the optimal decision rule $m(i)$ in the inventory problem may be recovered from an optimal solution of (4.3-54)–(4.3-57) when that solution satisfies Theorem 1 of Section 4.3.
8. For the inventory process specified in relations (4.3-23)–(4.3-25), let the costs be

$$c_h(i) = i$$
$$c_0(m(i)) = 2m(i)$$
$$c_p(r - i - m(i)) = \max[0, 4(r - i - m(i))]$$

Write the linear program (4.3-54)–(4.3-57) in detached coefficient form, solve it, and verify that the optimal solution has the property given in Theorem 1 of Section 4.3. Recover the optimal decision rule and the steady-state probabilities x_i under this rule from this solution. Compare the optimal cost with that of the "do-nothing" policy $m(i) = 0$ for all i. For a similar problem, see [11, p. 263].

9. (Manne [11].) Consider an inventory problem in which the costs include a "setup" cost, incurred when the level of production is changed from one period to the next. Here the state has two components, the initial inventory and the production level in the previous period. Extend the formulations of this section to include such a problem. Verify that, if there are r possible production levels, the resulting linear program has $r(n + 1)$ rows.

10. Give a pricing interpretation for the optimal values of the simplex multipliers of the linear program (4.3-54)–(4.3-57).

11. Extend the formulation of generalized linear programming to the case where the sets S_i are polyhedral but may not be bounded.

12. Show that, if f and the g_i are convex functions, the set R in relation (4.3-15) is convex.

13. Let x^k solve the subproblem

$$\text{minimize } L(x, u^k) = f(x) + \sum_{i=1}^{m} u_i^k g_i(x)$$

where $u_i^k = -\pi_i^k$, and the π_i^k are the simplex multipliers for an optimal solution of the restricted master program at iteration k. Show that if x^k satisfies

$$g_i(x^k) \leq 0, \qquad i = 1, \ldots, m$$
$$u_i^k g_i(x^k) = 0, \qquad i = 1, \ldots, m$$

then x^k solves the original problem (4.4-1)–(4.4-2).

14. Given the problem

$$\text{minimize } f = (x - 2)^2 + 2y$$

subject to

$$g_1 = x^2 - y \leq 0$$
$$g_2 = x + y \leq 2$$
$$y \geq 0$$

(a) Solve the problem.

(b) Choose an initial grid $x_1 = 0$, $x_2 = 1$. Using an initial basis of the two slack variables and λ_1 (the weight associated with x_1), solve the initial restricted master program. Note that y appears linearly and need not be linearized.

(c) Formulate and solve the first subproblem. Note that the conditions of Problem (4.4-13) are met and the new grid point generated is the optimal x-value.

(d) Introduce a new column and solve the new restricted master program, using the optimal basis from (b) as a starting basis. Verify that the resulting solution is optimal.

15. Show that if there is an x such that $g_i(x) < 0$, $i = 1, \ldots, m$, then there is a choice of grid points x_t for which the restricted master program (4.4-6)–(4.4-9) has a nondegenerate basic feasible solution. Construct such a solution.

16. The following is a case where the grid linearization procedure is finite. Consider the problem

$$\text{minimize} \sum_{j=1}^{n} f_j(x_j)$$

$$\text{subject to} \qquad Ax = b, \qquad 0 \leq x \leq u$$

where u is a positive vector and each function f_j is piecewise linear on the interval $[0, u_j]$ with breakpoints $\{x_{jk}\}$. Linearize each f_j on the grid $\{x_{jk}\}$ by writing

$$f_j(x_j) \doteq \sum_k \lambda_{jk} f_j(x_{jk}), \qquad \sum_k \lambda_{jk} = 1, \qquad \lambda_{jk} \geq 0$$

and let $\{\lambda_{jk}^0\}$ solve the resulting linear program. Show that

$$x_j^0 = \sum_k \lambda_{jk}^0 x_{jk}, \qquad j = 1, \ldots, n$$

solves the original problem.

17. Assuming relation (4.5-8) is true, establish relation (4.5-10) by induction.

18. In order that the maximization in (4.5-21) be solvable as a set of maximal flow problems, the multipliers v must be nonnegative. How is this guaranteed at the beginning of iterations 2, 3, ...? How can it be guaranteed at the beginning of iteration 1?

References

1. P. C. Gilmore and R. E. Gomory, "A Linear Programming Approach to the Cutting Stock Problem," *Operations Res.*, **9**, 1961, pp. 849–859.
2. P. C. Gilmore and R. E. Gomory, "A Linear Programming Approach to the Cutting Stock Problem—Part II," *Operations Res.*, Nov.–Dec. 1963, pp. 863–887.
3. M. L. Balinski, "Integer Programming: Methods, Uses, Computation," *Management Sci.*, **12**, No. 3, 1965, pp. 253–313.
4. E. M. L. Beale, P. A. B. Hughes, and R. E. Small, "Experiences in Using a Decomposition Program," *Computer J.*, **8**, No. 1, 1965, pp. 13–18.
5. P. C. Gilmore and R. E. Gomory, "Multistage Cutting Stock Problems of Two or More Dimensions," *Operations Res.*, **13**, 1965, pp. 94–120.

6. A. S. Manne, "Programming of Economic Lot Sizes," *Management Sci.*, **4**, 1958, pp. 115–135.
7. B. P. Dzielinski and R. E. Gomory, "Optimal Programming of Lot Sizes, Inventory and Labor Allocations," *Management Sci.*, **11**, No. 9, 1965, pp. 874–890.
8. L. S. Lasdon and J. E. Mackey, "An Efficient Algorithm for Multi-item Scheduling," Systems Research Center Report SRC 68–9, Case Western Reserve University, 1968.
9. G. B. Dantzig and R. M. Van Slyke, "Generalized Upper Bounding Techniques," *J. Computer System Sci.*, **1**, 1967, pp. 213–226.
10. G. B. Dantzig, *Linear Programming and Extensions*, Princeton University Press, Princeton, N.J., 1963, chap. 22.
11. A. S. Manne, "Linear Programming and Sequential Decisions," *Management Sci.*, **6**, No. 3, 1960, pp. 259–269.
12. P. Wolfe and G. B. Dantzig, "Linear Programming in a Markov Chain," *Operations Res.*, **10**, Sept.–Oct. 1962, pp. 702–710.
13. P. Wolfe, "Methods of Nonlinear Programming," in *Nonlinear Programming* J. Abadie, ed., John Wiley & Sons, Inc., New York, 1967, pp. 100–142.
14. P. Wolfe, "Foundations of Nonlinear Programming: Notes on Linear Programming and Extensions—Part 65," *DDC Rept. AD–619–968*, also *RAND Memorandum RM–4669–PR*, 1965.
15. C. E. Miller, "The Simplex Method for Local Separable Programming," in *Recent Advances in Mathematical Programming*, R. L. Graves and P. Wolfe, eds., McGraw-Hill, Inc., New York, 1963, pp. 89–100.
16. G. B. Dantzig, *Linear Programming and Extensions*, Princeton University Press, Princeton, N.J., 1963, chaps. 22 and 24.
17. Y. Sekine, "Decentralized Optimization of an Interconnected System," *IEEE Trans. Circuit Theory*, June 1963, pp. 161–168.
18. R. E. Gomory and T. C. Hu, "An Application of Generalized Linear Programming to Network Flows," *J. Soc. Ind. Appl. Math.*, **10**, No. 2, 1962, pp. 260–283.
19. R. E. Gomory and T. C. Hu, "Multi-terminal Network Flows," *J. Soc. Ind. Appl. Math.*, **9**, No. 4, 1961, pp. 551–570.
20. L. R. Ford and D. R. Fulkerson, *Flows in Networks*, Princeton University Press, Princeton, N.J., 1962.

5

Partitioning and Relaxation Procedures in Linear Programming

5.1 Introduction

In this chapter, linear programs whose constraint matrices have a block-diagonal structure linked by coupling variables or by coupling variables and constraints will be studied. A program coupled in both ways is shown below:

$$\text{maximize } z = c_1x_1 + \cdots + c_px_p + c_0y \tag{1}$$

subject to

$$A_1x_1 + \cdots + A_px_p + D_0y = b_0 \tag{2}$$

$$\begin{aligned} B_1x_1 \quad\quad\quad\quad\quad\quad\quad\quad &+ D_1y = b_1 \\ B_2x_2 \quad\quad\quad\quad &+ D_2y = b_2 \\ \ddots \quad\quad &\quad\;\; \vdots \quad\quad \vdots \\ B_px_p &+ D_py = b_p \end{aligned} \tag{3}$$

$$x_i \geq 0, \quad i = 1, \ldots, p, \quad y \geq 0 \tag{4}$$

The Dantzig–Wolfe decomposition principle of Chapter 3 can be used to solve this problem. If the problem has only coupling variables, the Dantzig–Wolfe procedure may be applied to the dual (which has only coupling constraints). If both coupling variables and constraints are present, the constraint matrix is partitioned between (2) and (3), as shown. Relations (3) become the constraints of the subproblem, and (2) becomes the master program. Since the subproblem has dual-angular form, it may be dealt with by the methods of this chapter or by dualizing and again applying the Dantzig–Wolfe method. In either case, the overall result is a "three-level" algorithm, involving an iteration (solving the subproblem) within an outer iterative loop (solving the master program). Since the inner iteration is quite complex, this seems computationally undesirable, especially since the Dantzig–Wolfe method is often slowly convergent (see Sections 4.2, 3.9, and 3.10).

The algorithms of this section provide useful alternative means for solving (1)–(4). They are called partitioning methods because the problem variables are first partitioned into two subsets, coupling variables, y, and block variables, x_i. The x_i are further partitioned into dependent and independent sets by specifying basis matrices within each matrix B_i. This partitioning permits the dependent variables and block constraints to be eliminated from the program, yielding a smaller reduced problem. The adjective relaxation is employed because, in eliminating the dependent variables, control is lost over their nonnegativity constraints. These are thus relaxed, and the iterative step defines new partitions so that they are eventually satisfied.

Since, as pointed out by Geoffrion [1], the concept of constraint relaxation provides a unifying framework in which the workings of these (and other) methods may be understood, we now pause to discuss this concept.

5.2 Relaxation

Consider the general concave program

$$\left.\begin{aligned} &\text{maximize } f(x) && (1)\\ \text{subject to}\quad &g_i(x) \geq 0, \quad i = 1, \ldots, m && (2)\\ &x \in S && (3)\end{aligned}\right\}\text{problem } P$$

where f and the g_i are concave functions of an n vector x and S is a convex subset of E^n. We consider here the case where the constraints $g_i \geq 0$ are numerous enough to be troublesome. Then a reasonable solution strategy is to relax (i.e., temporarily drop) some of these constraints, and solve a program involving the remainder. If this problem is infeasible, so is the original. If it is feasible, and if the resulting solution satisfies the relaxed constraints, the solution is optimal for P. If not, we enforce one or more of the relaxed constraints and repeat the procedure. Relaxation would appear to be an effective strategy if relatively few of the constraints are known to be binding at optimality. This is true in nondegenerate linear programs (exactly n are binding) and appears, in general, to hold even more strongly in nonlinear programs [1].

Geoffrion [1] has formalized this idea as follows. Let M be the set of integers $\{1, \ldots, m\}$, let R denote a subset of M, and let P^R denote the program

$$\text{maximize } f(x) \tag{4}$$

$$\text{subject to} \quad g_i(x) \geq 0, \quad i \in R \tag{5}$$

$$x \in S \tag{6}$$

We assume that an initial set R can be found such that P^R has a finite supremum, and that all finite suprema are attained. Then a relaxation strategy which permits constraints to be deleted as well as added proceeds as follows:

0. Set $\bar{f} = +\infty$ and choose an initial set R such that f is bounded above by the constraints of P^R.
1. Solve P^R. If P^R is infeasible, P is infeasible. Otherwise, obtain an optimal solution x^R.
2. If $g_i(x^R) \geq 0$, $i \in M - R$, x^R is optimal for the original problem, P.
3. Otherwise, let V be a subset of $M - R$ containing the indices of at least one violated constraint, and let
$$D = \{i \mid g_i(x^R) > 0,\ i \in R\}$$
4. If $f(x^R) = \bar{f}$, replace R by $R' = RUV$, and return to step 1.
5. If $f(x^R) < \bar{f}$, replace R by $R' = RUV - D$, replace $\bar{f}$ by $f(x^R)$, and return to step 1.

This procedure adds one or more violated constraints and, if $f(x^R)$ has decreased, deletes those constraints in R which are not binding at x^R. Insisting that no constraints be deleted if $f(x^R) = \bar{f}$ guarantees that only a finite number of problems P^R need be solved, as we shall show. The proof requires two preliminary results.

THEOREM 1. *If x^R solves P^R and the constraints in D are dropped, then x^R also solves P^{R-D}. Further, if x^R is the unique solution to P^R, then it is the unique solution to P^{R-D}.*

PROOF. Assume there is a vector, $\hat{x}$, satisfying the constraints in $R - D$ such that

$$f(\hat{x}) > f(x^R) \tag{7}$$

Consider points on the line segment joining x^R and $\hat{x}$:

$$x_\lambda = \lambda\hat{x} + (1 - \lambda)x^R, \qquad 0 < \lambda < 1 \tag{8}$$

Since S is convex and all g_i are concave, $x_\lambda \in S$ and x_λ satisfies the constraints in $R - D$ for all $0 < \lambda < 1$. Consider the constraints in D. Since the g_i are concave,

$$g_i(x_\lambda) \geq \lambda g_i(\hat{x}) + (1 - \lambda)g_i(x^R)$$

Since $g_i(x^R) > 0$, $i \in D$, then for $\lambda > 0$ and sufficiently small, $g_i(x_\lambda) > 0$, $i \in D$. The point x_λ is therefore feasible in P^R for such λ. But, by concavity of f,

$$f(x_\lambda) \geq \lambda f(\hat{x}) + (1 - \lambda)f(x^R) > f(x^R) \tag{9}$$

which contradicts the optimality of x^R.

To prove uniqueness, if $f(\hat{x}) = f(x^R)$, then (9) yields

$$f(x_\lambda) = f(x^R)$$

for all $\lambda > 0$ and sufficiently small, which contradicts the uniqueness of x^R.

The following result shows that the sequence of subproblem maxima obtained by relaxation is monotone nonincreasing.

THEOREM 2

$$f(x^{R'}) \leq f(x^R) \tag{10}$$

PROOF. If $R' = RUV$, the theorem is obviously true. By Theorem 1,

$$f(x^{R-D}) = f(x^R) \tag{11}$$

Since

$$f(x^{(R-D)UV}) \leq f(x^{R-D}) \tag{12}$$

the theorem is proved.

Note that concavity of f and the g_i and convexity of S permits us to drop nonbinding constraints. Theorem 2 is, of course, true in the absence of concavity if constraints are added while none are deleted. This is the case in Benders' partitioning algorithm of Section 7.3. However, deleting constraints is essential if the linear programming methods of Ritter and Rosen, to be considered later, are to be efficient.

THEOREM 3. *The relaxation procedure terminates after a finite number of problems P^R have been solved, either with an optimal solution to P or with a set of constraints such that the current P^R is infeasible.*

PROOF. Since M is finite, it has only a finite number of different subsets. As long as $f(x^R)$ decreases from iteration to iteration, no subset can be repeated. Since no constraints are deleted if $f(x^{R'}) = f(x^R)$, and at least one constraint is added, f can remain constant for only a finite number of iterations. Thus, in a finite number of steps, the procedure must terminate either in step 1 or 2 (possibly with $R = M$).

We note that relaxation has already been utilized in Section 4.5, where unsatisfied primal constraints were generated and added (as columns in the dual) to yield a new P^R.

Relation to the Dual Simplex Method. It is not difficult to show that, if the program P is linear, then, with proper choice of V, relaxation is equivalent to Lemke's dual simplex method (see Section 1.2.4). To show this, it is convenient to state P in the form

$$\text{maximize } cx \tag{13}$$

subject to

$$Ax = b \tag{14}$$

$$x_i \geq 0, \qquad i = 1, \ldots, n \tag{15}$$

where A is $m \times n$, $n > m$, and $\text{rank}(A) = m$. The constraints to be relaxed are the nonnegativities, (15). Let these be indexed by the indices of their associated variables, and let

$$N = \{1, 2, \ldots, n\} \tag{16}$$

The equalities (14) will always be satisfied; i.e., the set S in (3) is

$$S = \{x \mid Ax = b\} \tag{17}$$

In what follows, we assume the following.

Assumption 1. All feasible bases of the dual of (13)–(15) are nondegenerate; i.e., exactly m of the dual constraints are binding.

THEOREM 4. *Let the indices of the nonbasic variables at the initial cycle of the dual method comprise the initial set, R, of relaxation. Further, let the set V in step 3 of relaxation be the index of the most violated nonnegativity constraint. Then, at cycle k of the dual method and of relaxation, the index set of nonbasic variables in the dual method is equal to the set R in relaxation, and $x^R = x^k$, where x^k is the vector of values for x in the dual method.*

The proof requires three lemmas.

LEMMA 1. *If R is the index set of nonbasic variables at cycle k of the dual method, then x^k is the unique optimal solution to P^R.*

PROOF. Let B be the basis at cycle k of the dual method. It suffices to show that $\pi_B = c_B B^{-1}$ is feasible for the dual of P^R, with dual objective value equal to $c_B x_B$. P^R is shown below, along with its dual:

P^R	Dual of P^R
maximize $z = cx$	minimize $v = \pi b$
subject to	subject to
$Ax = b$	$\pi B = c_B$
$x_i \geq 0, \quad i \in R$	$\pi A^i \geq c_i, \quad i \in R$

where A^i is the ith column of A. In the dual simplex method, the basis B is dual feasible, so π_B satisfies all the dual constraints and

$$v(\pi_B) = c_B B^{-1} b = c_B x_B \tag{18}$$

Thus x^k solves P^R. By our assumption 1 of dual nondegeneracy, all dual constraints in R are satisfied strictly, so this optimal solution is unique.

LEMMA 2. *Let R be the index set of nonbasic variables at cycle k of the dual simplex method, and assume that the algorithm terminates at this cycle with the information that the primal is infeasible (i.e., with* $\min \bar{b}_i = \bar{b}_r < 0$ *and all* $\bar{a}_{rj} \geq 0$*). Then the subproblem of relaxation formed by adding the index j_r (the index of the rth basic variable) to R is also infeasible.*

PROOF. It suffices to produce an unbounded solution to the dual of P^{R+j_r}. The primal–dual pair under consideration is

P^{R+j_r}	Dual of P^{R+j_r}
maximize $z = cx$	minimize $v = \pi b$
subject to	subject to
$Ax = b$	$\pi A^i = c_i, \quad i \in N - R - j_r$
$x_i \geq 0, \quad i \in R + j_r$	$\pi A^i \geq c_i, \quad i \in R + j_r$

Consider the dual solution

$$\pi = c_B B^{-1} + \theta B_r^{-1}, \qquad \theta \geq 0 \tag{19}$$

where B_r^{-1} is row r of B^{-1}. Since

$$\begin{aligned} \bar{a}_{rj} = (B^{-1}A^j)_r = B_r^{-1}A^j &\geq 0, & j &\in R \\ &= 0, & j &\in N - R,\ j \neq j_r \\ &= 1, & j &= j_r \end{aligned}$$

then B_r^{-1} is a solution to the homogeneous dual constraints, i.e., with all c_i replaced by zero. Thus π in (19) is dual feasible for all $\theta \geq 0$, with objective value

$$v(\pi) = c_B x_B + \theta \bar{b}_r \tag{20}$$

Since $\bar{b}_r < 0$, $v(\pi) \to -\infty$ as $\theta \to \infty$.

LEMMA 3. *If x_s enters the basis in cycle k of the dual method, then $x_s > 0$ in the resulting basic solution.*

PROOF. By the pivot operation

$$x_s = \frac{\bar{b}_r}{\bar{a}_{rs}}$$

By construction, $\bar{b}_r < 0$ and $\bar{a}_{rs} < 0$.

PROOF OF THEOREM 4. By choice of the initial set R, Lemma 1 implies that the theorem is true for $k = 1$. We assume that the result is true for k and prove it for $k + 1$.

Cycle k of the dual method either (1) terminates with an optimal solution, (2) terminates with the information that the problem is infeasible, or (3) continues to the next cycle. In case (1), according to Lemma 1, relaxation also

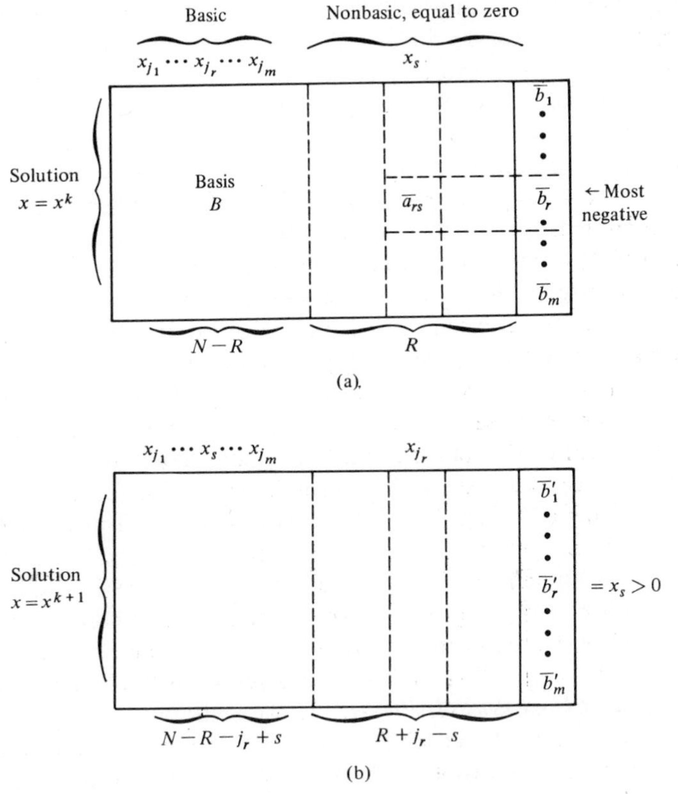

FIGURE 5-1 Dual method at (a) cycle k and (b) cycle $k + 1$.

terminates. By Lemma 2, in case (2) relaxation also terminates at cycle $k + 1$ with an infeasible subproblem. In case (3), the situations at cycle k and $k + 1$ of the dual method are shown in Figure 5-1. By the induction hypothesis, R is as shown and $x^R = x^k$. Since all variables in R are zero, relaxation deletes none. It selects x_{j_r} as the most negative variable in $N - R$ and adds the index j_r to R; i.e., $R \leftarrow R + j_r$. It then solves P^{R+j_r}. By Lemma 1, the solution x^{k+1} is the unique solution to P^{R+j_r-s}. By Lemma 3, x_s is positive in this solution, so x^{k+1} is the unique solution to P^{R+j_r}. Since the dual is nondegenerate, $cx^{R+j_r} < cx^R$, so relaxation deletes all positive variables from $R + j_r$. The only such variable is x_s, so $R + j_r \leftarrow R + j_r - s$, and the theorem is proved.

Computational Aspects. In order for relaxation to be efficient, effective procedures for identifying violated constraints (i.e., choosing the set V) and for re-solving successive problems P^R must be available. There is wide latitude allowed in the choice of V. A common choice is the index of the most violated constraint, as in the dual simplex method. This choice is often used when the number of constraints is truly large, and when the set of constraint functions has some special structure. Then the most violated constraint may often be generated without evaluating all constraint functions, by solving a suitable subproblem (see Sections 4.5 and 7.3). This "row generation" is dual to the column-generation procedures discussed earlier. Thus computational experience with various tactics for pricing out columns in linear programs has bearing here. As illustrated by the cutting-stock computations of Section 4.1, choosing a single violated constraint at random is likely to be much less effective than choosing the most violated constraint. Further, as V is taken to contain more indices, e.g., the indices of the k most violated constraints, the change in $f(x^R)$ from iteration to iteration will increase. Of course, the labor required to re-solve the successive problems P^R should increase also. An option which is useful when there are many constraints is to identify the k most violated constraints, and to apply relaxation to a restricted problem consisting of these k plus those currently in $R - D$. Within this restricted problem, V can be chosen as, say, the index of the most violated constraint. In its primal form, this tactic is that of finding those columns with the k most negative reduced costs, and performing standard simplex iterations on a problem which has these as its initial nonbasic columns. This is called "multiple pricing" or "suboptimization" (see Section 6.2) and yields a distinct reduction in computer time when the product form of the inverse is used [2].

The option of enforcing more than a single violated constraint is easily implemented when relaxation is applied to linear problems with coupling constraints and/or coupling variables, as is shown in the following sections.

Generation of Dual Feasible Points. Under some mild additional assumptions (see Sections 1.3.2 and 1.3.3), a vector of optimal Lagrange multipliers is associated with each problem P^R. It is easy to show that these vectors are feasible for the minimax dual of P and generate a monotone nonincreasing sequence of dual objective values. This dual is (see Section 8.4 for further details)

$$\underset{u \geq 0}{\text{minimize}} \sup_{x \in S} \left(f(x) + \sum_{i=1}^{m} u_i g_i(x) \right)$$

By the saddle-point theorem, Theorem 1.3.3-1, any Lagrange multiplier vector $u^R \geq 0$ associated with P^R satisfies

$$f(x^R) = \max_{x \in S} \left(f(x) + \sum_{i \in R} u_i^R g_i(x) \right)$$

By choosing $u_i^R = 0$, $i \notin R$, we obtain a feasible point for the dual of P. Since the associated dual objective value is equal to $f(x^R)$, the sequence of these values is nonincreasing.

Dual feasibility will be important in the relaxation algorithms which follow. Not only will each cycle generate a dual-feasible point for P, but the optimal dual solution for any P^R will be feasible for the dual of the next P^R. This permits effective solution of successive problems P^R by a dual linear programming method.

Restriction. It is interesting to note that the tactic of relaxing a primal constraint corresponds to restricting a dual constraint $u_i \geq 0$ to be $u_i = 0$. As pointed out in [1], restriction is a useful tactic when applied in the primal, and may be formalized in much the same way as relaxation. To do this, define the subproblem P^R as

$$\text{maximize } f(x)$$

subject to

$$\begin{aligned} g_i(x) &= 0, \quad & i &\in R \\ g_i(x) &\geq 0, \quad & i &\in M - R \\ & x \in S \end{aligned}$$

where R is again a subset of M. To ensure that any P^R is concave, all g_i must be assumed linear. Any nonlinear constraints are incorporated into the set S. Assume that an initial nonempty set R can be found such that P^R is feasible, that all finite suprema are attained, and that if P^R has an optimal solution it has an optimal vector of Lagrange multipliers. Restriction then proceeds as follows:

0. Set $\bar{f} = -\infty$ and choose an initial set R such that P^R is feasible.
1. Solve P^R. If the objective is unbounded, the original problem P is unbounded. Otherwise, obtain an optimal solution x^R.
2. Let u_i^0 be the optimal Lagrange multiplier associated with the ith constraint. If $u_i^0 \geq 0$, $i \in R$, x^R solves P.

3. Otherwise, let V be a subset of R containing the indices of at least one negative u_i^0, and let

$$D = \{i \mid g_i(x^R) = 0, i \in M - R\}$$

4. If $f(x^R) = \bar{f}$, replace R with $R' = R - V$ and return to step 1.
5. If $f(x^R) > \bar{f}$, replace R with $R' = R \cup D - V$, replace $\bar{f}$ by $f(x^R)$ and return to step 1.

It is not difficult to show that the sequence of objective values $f(x^R)$ produced by this procedure is nondecreasing and that the procedure terminates in a finite number of cycles.

Restriction is obviously dually related to relaxation. Thus it is not surprising that it is useful in problems with many variables rather than many constraints. To see this, note that the constraints $g_i \geq 0$ will often contain the conditions $x_i \geq 0$, and may always be made to do so by appropriate transformations. Then restricting $x_i = 0$ simply eliminates the variable from the problem. If P is linear, $g_i(x) = x_i$, and if R is chosen as the index set of the current nonbasic variables, restriction becomes identical with the primal simplex method (at least if all bases of the primal are nondegenerate). In linear programs with many columns, restriction thus becomes a column-generation scheme (see Chapter 4). The set V in step 3 is then a subset of the nonbasic variables having negative relative cost factors, and the efficiency of the algorithm depends on how easily such a subset may be found. Of course, $g_i(x)$ need not equal x_i in order to reduce the number of variables. Restricting any subset of p independent linear constraints permits p variables to be eliminated.

Although restriction, like relaxation, has found widest application in linear programs, its usefulness is not limited to such problems. An excellent example of the application of restriction to nonlinear programs is found in Rosen's partitioning algorithm, Section 7.2. There the constraints $g_i(x)$ are not linear in all variables, leading to a nonconvex problem P^R. Thus additional steps beyond those outlined here are taken, including a linearization of P^R.

Combined use of restriction and relaxation is discussed in [11], and some computational experience with this is found in [10].

5.3 Problems with Coupling Constraints and Coupling Variables

Ritter [3] has developed a partitioning procedure for problems with a block-diagonal constraint matrix linked by both coupling constraints and coupling variables. At each iteration, the program variables are partitioned into two subsets, a set S_1 in which the nonnegativity constraints are relaxed, and a set S_2 in which they are enforced. When the variables in S_1 become nonnegative, the current solution is optimal. The iterative step takes an initial partition and refines it until an optimal partition is found. The procedure is basically a dual method, in that a sequence of dual-feasible extreme points with nonincreasing objective values is generated.

The problem considered and its dual are shown below.

Primal:

$$\text{maximize } z = c_1'x_1 + \cdots + c_p'x_p + c_0'y \tag{1}$$

$$\text{subject to} \quad A_1x_1 + \cdots + A_px_p + D_0y = b_0 \tag{2}$$

$$\begin{aligned} B_1x_1 \qquad\qquad\qquad\qquad + D_1y &= b_1 \\ B_2x_2 \qquad\qquad + D_2y &= b_2 \\ \ddots \qquad\qquad \vdots \quad & \quad \vdots \\ B_px_p + D_py &= b_p \end{aligned} \tag{3}$$

$$x_i \geq 0, \qquad i = 1, \ldots, p, \qquad y \geq 0 \tag{4}$$

Dual:

$$\text{minimize } v = b_1'u_1 + \cdots + b_p'u_p + b_0'u_0 \tag{5}$$

$$\begin{aligned} \text{subject to} \quad D_1'u_1 + \cdots + D_p'u_p + D_0'u_0 &\geq c_0 \\ B_1'u_1 \qquad\qquad\qquad + A_1'u_0 &\geq c_1 \\ B_2'u_2 \qquad\qquad + A_2'u_0 &\geq c_2 \\ \ddots \qquad\qquad + \quad \vdots \quad & \quad \vdots \\ B_p'u_p + A_p'u_0 &\geq c_p \end{aligned} \tag{6}$$

In the above, A_i is $m_0 \times n_i$, B_i is $m_i \times n_i$, and D_i is $m_i \times n_0$.

We assume that each matrix B_i is of rank m_i, and hence contains a nonsingular submatrix B_{i1}. Note that this need not be the case, even if the system (2)–(3) is of full rank. If rank $(B_i) < m_i$, artificial variables (with unit vector columns) may be added in the rows of B_i until the augmented matrix has rank m_i. If these artificial variables are assigned costs which are sufficiently large and negative, they will all appear at zero level in an optimal solution. If rank $(B_i) = m_i$ for each i, and if an initial vector $y_0 \geq 0$ is known such that the constraints

$$B_ix_i = b_i - D_iy_0, \qquad x_i \geq 0, \quad i = 1, \ldots, p \tag{7}$$

are feasible, then a "good" set of initial bases may be found by solving the subproblems

$$\begin{aligned} &\text{maximize } c_i'x_i \\ \text{subject to} \quad &B_ix_i = b_i - D_iy_0, \qquad x_i \geq 0 \end{aligned} \tag{8}$$

and using the optimal basis matrices as the B_{i1}.

Once the B_{i1} are available, the constraints (3) may be written

$$B_{i1}x_{i1} + B_{i2}x_{i2} + D_i y = b_i \tag{9}$$

Since each B_{i1} is nonsingular, (9) may be solved for x_{i1}:

$$x_{i1} = B_{i1}^{-1}b_i - B_{i1}^{-1}B_{i2}x_{i2} - B_{i1}^{-1}D_i y \tag{10}$$

The primal objective (1) and coupling constraints (2) are similarly partitioned

$$z = \sum_{i=1}^{p} (c'_{i1}x_{i1} + c'_{i2}x_{i2}) + c'_0 y \tag{11}$$

$$\sum_{i=1}^{p} (A_{i1}x_{i1} + A_{i2}x_{i2}) + D_0 y = b_0 \tag{12}$$

Then (10) is substituted into (11) and (12), thereby eliminating the constraints (3) and the variables x_{i1}. Let

$$d'_i = c'_{i2} - c'_{i1}B_{i1}^{-1}B_{i2} \tag{13}$$

$$d'_0 = c'_0 - \sum_{i=1}^{p} c'_{i1}B_{i1}^{-1}D_i \tag{14}$$

$$\alpha = \sum_{i=1}^{p} c'_{i1}B_{i1}^{-1}b_i \tag{15}$$

$$M_i = A_{i2} - A_{i1}B_{i1}^{-1}B_{i2} \tag{16}$$

$$M_0 = D_0 - \sum_{i=1}^{p} A_{i1}B_{i1}^{-1}D_i \tag{17}$$

$$b = b_0 - \sum_{i=1}^{p} A_{i1}B_{i1}^{-1}b_i \tag{18}$$

Then (11) and (12) become a new smaller problem, the reduced problem

$$\left.\begin{aligned} &\text{maximize } z - \alpha = \sum_{i=1}^{p} d'_i x_{i2} + d'_0 y \\ &\text{subject to} \quad \sum_{i=1}^{p} M_i x_{i2} + M_0 y = b \\ &\qquad x_{i2} \geq 0, \quad y \geq 0 \end{aligned}\right\} \text{reduced problem} \tag{19}$$

This problem has only as many equality constraints as there are coupling constraints in the primal. The equalities of (19), along with (10), have the same solution set as the coupling and block constraints, (2)–(3). However, since the reduced problem does not explicitly contain the variables x_{i1}, there is no way to enforce the constraints $x_{i1} \geq 0$. Thus these constraints are relaxed. The solution set of (19) and (10) then contains the solution set of the primal, so if the reduced problem is infeasible, then so is the primal. However, the reduced problem may have an unbounded solution in cases where the primal has a finite optimal solution. To eliminate this possibility, a constraint of the form

$$\sum_{j=1}^{p} e_j' x_{j2} + e_0' y \leq t, \qquad t \text{ large, positive} \tag{20}$$

with the e_j vectors with all components unity, may be added to the reduced problem. If t is sufficiently large and this constraint is binding when the algorithm terminates, then the primal optimum is unbounded.

Let the solution of the reduced problem be $(\{x_{i2}^0\}, y^0)$. Substituting this solution into (10) yields new values for x_{i1}:

$$x_{i1}^0 = B_{i1}^{-1} b_i - B_{i1}^{-1} B_{i2} x_{i2}^0 - B_{i1}^{-1} D_i y^0 \tag{21}$$

The solution $(\{x_{i1}^0, x_{i2}^0\}, y^0)$ solves a modified primal problem in which the conditions $x_{i1} \geq 0$ are dropped. If these constraints happen to be satisfied, this solution solves the primal.

THEOREM 1 (OPTIMALITY TEST). *The solution* $(\{x_{i1}^0, x_{i2}^0\}, y^0)$ *solves the primal problem if and only if*

$$x_{i1}^0 \geq 0, \qquad i = 1, \ldots, p \tag{22}$$

Modifying the Reduced Problem. If the optimality test is not passed, then, following the relaxation strategy of Section 5.2, a new reduced problem is formed in which some violated nonnegativities are enforced, while some nonnegativities on positive x_{i2} are relaxed.

Consider any vector x_{j1} with the property that at least one component of x_{j1}^0 is negative. Let the first l components of x_{j1}^0 be negative,

$$(x_{j1}^0)_1 < 0, (x_{j1}^0)_2 < 0, \ldots, (x_{j1}^0)_l < 0, \qquad l \geq 1 \tag{23}$$

Let x_{j2}^0 have its first q components positive,

$$(x_{j2}^0)_1 > 0, (x_{j2}^0)_2 > 0, \ldots, (x_{j2}^0)_q > 0, \qquad q \geq 0 \tag{24}$$

For convenience, the relation between x_{j1} and x_{j2} is rewritten here:

$$x_{j1} + B_{j1}^{-1}B_{j2}x_{j2} = B_{j1}^{-1}b_j - B_{j1}^{-1}D_j y \tag{25}$$

There are two cases.

CASE 1. At least one element of the submatrix formed by the first l rows and first q columns of $B_{j1}^{-1}B_{j2}$ is not zero.

In this case, a pivot operation may be performed on any nonzero element of this submatrix. Such an operation transforms (25) by interchanging a component of x_{j2}, positive in x_{j2}^0, with a component of x_{j1}, negative in x_{j1}^0. This interchange yields new basis matrices, $\hat{B}_{j1}$, which are used to form a new reduced problem.

CASE 2. The submatrix formed by the first l rows and first q columns of $B_{j1}^{-1}B_{j2}$ contains all zero elements.

In this case components of x_{j1} and x_{j2} cannot be interchanged by pivoting. Instead, choose some negative component of x_{j1}^0, say $(x_{j1}^0)_k$, $1 \leq k \leq l$ (a good choice may be the most negative), and use (25) to write the constraint that this variable be nonnegative explicitly in terms of x_{j2} and y:

$$(x_{j1})_k = \alpha_{jk} - \bar{b}_{jk}x_{j2} - \bar{d}_{jk}y \geq 0 \tag{26}$$

with α_{jk}, $\bar{b}_{jk}$, and $\bar{d}_{jk}$ the kth rows of $B_{j1}^{-1}b_j$, $B_{j1}^{-1}B_{j2}$, and $B_{j1}^{-1}D_j$, respectively. This constraint is added to the reduced problem, (19).

Repeating either case 1 or 2 for each vector x_{i1}^0 with negative components yields a new reduced problem, in which the matrices B_{i1} may have been changed and constraints of the form (26) may have been added. If the current iteration is not the first, the reduced problem may contain constraints like (26) added in previous iterations. Before the procedures of cases 1 and 2 are applied, these constraints are treated as outlined below.

CASE 3. If a constraint (26) is not binding at the optimal solution, i.e., if

$$\alpha_{jk} - \bar{b}_{jk}x_{j2}^0 - \bar{d}_{jk}y^0 > 0 \tag{27}$$

then the constraint is dropped from the reduced problem. By Theorem 5.2-1, the optimal solution is unchanged by this procedure.

CASE 4. If a constraint (26) is binding at the new optimal solution, i.e., if

$$\alpha_{jk} - \bar{b}_{jk}x_{j2}^0 - \bar{d}_{jk}y^0 = 0 \tag{28}$$

then the constraint is dropped if there is a change of basis permitting interchange of $(x_{j1})_k$ and some component of x_{j2} which is positive in x_{j2}^0. The elements of $\bar{b}_{jk}$ associated with positive components of x_{j2}^0 are scanned. If any such element is nonzero, a pivot operation is performed on that element and the corresponding constraint (26) is dropped. Since this constraint has merely been rewritten in more convenient form, the optimal solution is again unchanged.

CASE 5. If (28) holds, and if all potential pivot elements above are zero, the constraint (26) is retained. In this case $\bar{b}_{jk}x_{j2}^0 = 0$ and (28) becomes

$$\bar{d}_{jk}y^0 = \alpha_{jk} \tag{29}$$

We wish to show that, if the solution $\{x_{i2}^0\}$, y^0 of the reduced problem is nondegenerate, there can be at most n_0 (the number of coupling variables) of these equations. To see this, note that cases 3 and 4 yield a modified reduced problem with some rows deleted, no slacks in the optimal basis, and the same optimal solution as the original. The structure of this modified problem for a two-block primal is shown in Figure 5-2. The q bottom rows are those

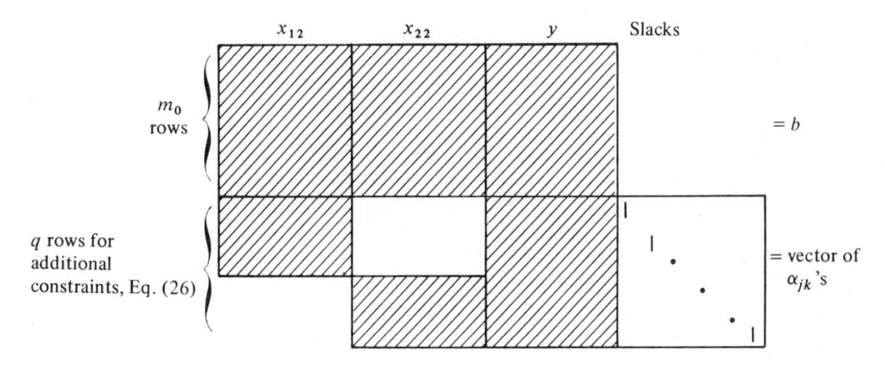

FIGURE 5-2

remaining after cases 3 and 4 are applied. If the original system (2)–(3) has full rank, then the top m_0 rows of the system in Figure 5-2 are independent. With the additional rows, the system has rank $m_0 + q$. Assume now that the solution $\{x_{i2}^0\}$, y^0 has $m_0 + q$ positive components, and consider the basis matrix, B^0, associated with this solution. It must have the form shown in Figure 5-3. Since all basic variables are positive and $\bar{b}_{jk}x_{j2}^0 = 0$, the coefficients of x_{12} and x_{22} in the last q rows are zero. If $q > n_0$, then, since the number of y columns is at most n_0, the last q rows of B^0 are dependent, which is impossible. Hence $q \leq n_0$.

Note that, if some basic components of x_{j2}^0 are zero, then the submatrix of B^0 which is currently zero may contain some nonzero components. In

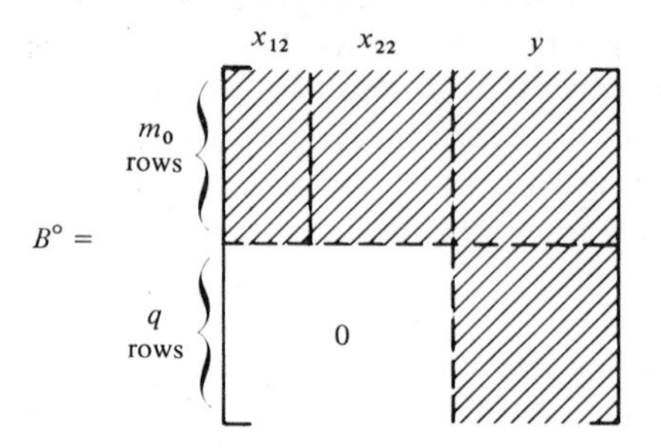

FIGURE 5-3

this case, if there are d columns corresponding to zero components of x_{12}^0 and x_{22}^0, we group these with the y columns and obtain $q \leq n_0 + d$. Thus, in the general case, there need never be more than $p + n_0 + d$ [p is the number of blocks in (2), n_0 the number of coupling variables] additional constraints like (26) in the reduced problem. This is because at most $n_0 + d$ need be retained and at most p are added.

Summary. A summary of the algorithm follows:

1. Choose initial basis matrices B_{i1}, $i = 1, \ldots, p$. This may be done arbitrarily or by solving the subproblems (8) and using their optimal bases as the B_{i1}. Using these matrices, form a reduced problem (19).
2. Solve the reduced problem, yielding solutions $\{x_{i2}^0\}$, y^0. Compute x_{i1}^0 using (21). If $x_{i1}^0 \geq 0$, $i = 1, \ldots, p$ the solution $(\{x_{i1}^0, x_{i2}^0\}, y^0)$ is optimal.
3. If $x_{i1}^0 \ngeq 0$ for $i \in I$ and if there are additional constraints of the form (26) in the reduced problem, deal with these as in cases 3–5.
4. For each $i \in I$, perform either the pivot operations of case 1 or add constraints as in case 2.
5. Using the new basis matrices and constraints from steps 3–4, form a new reduced problem and return to step 2.

Convergence. Since this is a relaxation procedure, the theorems of Section 5.2 apply. By Theorem 5.2-2, the sequence of optimal reduced problem objective values is monotone nonincreasing, while Theorem 5.2-3 yields conditions for finite convergence:

THEOREM 2. *If nonnegativities on positive x_{i2} are dropped only when the reduced problem objective decreases, this algorithm terminates in a finite number of iterations, either with an optimal solution to* (1)–(4) *or with a reduced problem that is infeasible.*

Since the constraints added to the reduced problem are not satisfied by some optimal solution to this problem, the objective decreases at each iteration except when there is an alternative optimal solution which satisfies all the new constraints. If, in this case, constraints are deleted, the possibility of cycling exists, unless removed by a perturbation scheme.

Dual Feasibility. The following theorem, combined with the facts that Ritter's algorithm generates a sequence of infeasible primal solutions converging to the maximal value from above, shows that the algorithm is a dual method.

THEOREM 3. *To each solution* $(\{x_{i1}^0, x_{i2}^0\}, y^0)$ *obtained at iteration k, there corresponds an extreme point of the dual program* (5)–(6), $\{u_i^0\}$, *with equal objective value:*

$$\sum_{i=1}^{p} b_i' u_i^0 = c_0' y^0 + \sum_{i=1}^{p} (c_{i1}' x_{i1}^0 + c_{i2}' x_{i2}^0) \tag{30}$$

PROOF. The solution $(\{x_{i1}^0, x_{i2}^0\}, y^0)$ solves a modified primal problem given by (1)–(3) along with the nonnegativities

$$x_{i2} \geq 0, \qquad i = 1, \ldots, p, \quad y \geq 0 \tag{31}$$

$$(x_{i_k 1})_{j_k} \geq 0 \tag{32}$$

where (32) corresponds to the kth constraint of the form (26). The dual of this modified primal has each of the inequalities of (6) corresponding to components of $\{x_{i1}\}$ not constrained in sign replaced by equalities. By the duality theorem, if the modified primal has a finite optimal solution, this (modified) dual has one also. The modified dual thus has an optimal extreme point solution, $\{u_i^0\}$, satisfying (30). Since any extreme point of the modified dual is an extreme point of the original dual, (5)–(6), the theorem is proved.

Computational Questions. As a relaxation procedure, this algorithm permits many options in selecting the unsatisfied constraints to be enforced at each cycle. By Theorem 5.2-4, if only the single most violated nonnegativity (over all blocks) is enforced, the procedure is identical to the dual simplex method. However, the block structure of the primal makes it easy to enforce more, at least if the required pivot elements are nonzero. Otherwise, additional constraints must be added to the reduced problem, making its solution more difficult. Thus there is a tradeoff between enforcing many nonnegativities at each cycle and decreasing the size of the reduced problem. If one is willing to pay the price, positive x_{i2}, which might be used to pivot out constraints of the form (26), may instead be used to pivot in additional negative x_{i1}. Alternatively, more constraints of the form (26) may be added.

An extension of this algorithm to a nonlinear concave objective, along with some computional experience in the linear case, is found in [9].

5.4 Rosen's Partitioning Procedure for Angular and Dual-Angular Problems

5.4.1 Development of the Algorithm. The partitioning method of Rosen [4] solves a primal–dual pair of problems of the form shown below.

Dual–Angular Problem

$$\text{maximize } v = \sum_{i=1}^{p} b_i' u_i + b_0' y \quad (1)$$

subject to

$$B_i' u_i + A_i' y \leq c_i, \quad i = 1, \ldots, p \quad (2)$$

Angular Problem

$$\text{minimize } z = \sum_{i=1}^{p} c_i' x_i \quad (3)$$

subject to

$$\sum_{i=1}^{p} A_i x_i = b_0 \quad (4)$$

$$B_i x_i = b_i, \quad i = 1, \ldots, p \quad (5)$$

$$x_i \geq 0, \quad i = 1, \ldots, p \quad (6)$$

The matrices A_i are $m_0 \times n_i$, while the B_i are $m_i \times n_i$. This technique may be viewed as a specialization of Ritter's method, discussed in the previous section, to problems with only coupling constraints or coupling variables. Thus it is also a relaxation procedure. It works in the angular problem, generating a sequence of infeasible basic solutions there and a corresponding sequence of feasible basic solutions in the dual. We will show that, by specializing to this primal–dual pair of problems, a number of simplifications occur. Additional constraints are not required in the reduced problem, so this problem need never have more than m_0 rows. Further, successive reduced problems may be solved efficiently by the dual simplex method, because the optimal basis at cycle i is dual feasible for the problem at $i + 1$.

Throughout this section, we make the following assumption.

Assumption 1. The constraint matrix of the angular problem has full rank. This implies the following result.

THEOREM 1. *Under assumption* 1, *each matrix B_i contains a nonsingular submatrix, B_{il}, of dimension m_i.*

PROOF. If B_i had rank less than m_i, then the corresponding rows of the angular problem would be dependent (since they are formed by adding zeros to the rows of B_i), contradicting assumption 1. Thus $n_i \geq m_i$, and B_i contains the desired submatrix.

We may as well assume that $n_i > m_i$, since if $n_i = m_i$, the variables x_i are uniquely determined and may be eliminated.

An initial set of basis matrices, B_{i1}, may always be found by employing a phase 1 procedure in each angular block. A better starting point can usually be determined by solving either of the following pairs of subproblems.

Dual-Angular Subproblem

$$\text{maximize } b_i'u_i$$

subject to (7)

$$B_i'u_i \leq c_i - A_i'y^*$$

Angular Subproblem

$$\text{minimize } c_i'x_i$$

subject to (8)

$$B_ix_i = b_i, \qquad x_i \geq 0$$

In the above, y^* is a vector for which the constraints of (7) are feasible. In all cases, we will insist that the initial basis matrices be chosen so that their associated basic solutions are nonnegative, and thus are feasible for the ith angular block. This condition is maintained thereafter by the algorithm. As we shall see in Theorem 2, this choice guarantees the existence of a sequence of pivot operations in each nonoptimal angular block leading to the desired interchange of positive and negative variables.

Assumption 2. The initial basis matrices B_{i1} satisfy

$$B_{i1}^{-1}b_i \geq 0, \qquad i = 1, \ldots, p$$

The basis matrices B_{i1} are used to partition the angular and dual-angular problems as follows:

Dual-Angular

$$\text{maximize } \sum_i b_i'u_i + b_0'y$$

subject to

$$B_{i1}'u_i + A_{i1}'y \leq c_{i1} \tag{9}$$

$$B_{i2}'u_i + A_{i2}'y \leq c_{i2} \tag{10}$$

Angular

$$\text{minimize } \sum_i (c_{i1}'x_{i1} + c_{i2}'x_{i2})$$

subject to

$$\sum_i (A_{i1}x_{i1} + A_{i2}x_{i2}) = b_0 \tag{11}$$

$$B_{i1}x_{i1} + B_{i2}x_{i2} = b_i \tag{12}$$

$$x_{i1} \geq 0, \qquad x_{i2} \geq 0, \qquad i = 1, \ldots, p$$

As in Section 5.3, relaxing the constraints $x_{i1} \geq 0$ and using (12) to eliminate x_{i1} from the coupling constraints (11) leads to the reduced problem. Solving (12) for x_{i1} yields

$$x_{i1} = B_{i1}^{-1}b_i - B_{i1}^{-1}B_{i2}x_{i2} \tag{13}$$

Then the reduced problem is

$$\left.\begin{aligned} &\text{minimize} \sum_i d_i' x_{i2} \\ \text{subject to} \quad &\sum_i M_i x_{i2} = \hat{b}_0 \\ &x_{i2} \geq 0 \end{aligned}\right\} \text{reduced problem} \tag{14}$$

where

$$d_i' = c_{i2}' - c_{i1}' B_{i1}^{-1} B_{i2} \tag{15}$$

$$M_i = A_{i2} - A_{i1} B_{i1}^{-1} B_{i2} \tag{16}$$

$$\hat{b}_0 = b_0 - \sum_{i=1}^{p} A_{i1} B_{i1}^{-1} b_i \tag{17}$$

This problem has only m_0 rows. As in Section 5.3, the solution set of (14) and (13) contains that of the angular problem, so if the reduced problem is infeasible, the other is also. The reduced problem may be unbounded when the primal is not, which may again be dealt with by adding a constraint

$$\sum_j e_j' x_{j2} \leq t, \qquad t \text{ large, positive}$$

Let $\{x_{i2}^0, i = 1, \ldots, p\}$ solve the reduced problem. Substituting these vectors into (13) yields values for the x_{i1}:

$$x_{i1}^0 = B_{i1}^{-1} b_i - B_{i1}^{-1} B_{i2} x_{i2}^0, \qquad i = 1, \ldots, p \tag{18}$$

According to our previous results regarding relaxation, the solutions above are optimal if and only if they are feasible.

THEOREM 2 (OPTIMALITY TEST). *The vectors* $\{x_{i1}^0, x_{i2}^0\}$ *solve the angular problem if and only if*

$$x_{i1}^0 \geq 0, \qquad i = 1, \ldots, p \tag{19}$$

If the optimality test is not passed, then, again following the relaxation strategy of Section 5.2, at least one violated nonnegativity is enforced while a corresponding nonnegativity on a positive variable is relaxed. In the problem with both coupling variables and constraints, pivot operations could not always accomplish this interchange, and additional constraints were necessary. Here, since there are no coupling variables in the angular problem, a sequence of pivot steps always exists which leads to the desired solution.

THEOREM 3 (ROSEN [4]). *If, for some i, x_{i1}^0 has negative components, a sequence of pivot operations may be performed in the ith angular block, interchanging a component of x_{i2} which is positive in x_{i2}^0 and a component of x_{i1} which is negative in x_{i1}^0.*

The proof of this theorem requires the following lemma.

LEMMA 1 [5]. *Consider the linear system*

$$Ax = b$$

where A is $m \times n$, $n \geq m$, and rank$(A) = m$. *If this system has a nonnegative solution, it has a basic nonnegative solution.*

PROOF OF THEOREM 3. For simplicity we drop the block subscript i, and concentrate on a particular nonoptimal block, using subscripts to denote particular components of x_1, x_2, etc.

We have available two solutions to the system

$$B_1 x_1 + B_2 x_2 = b \qquad (20)$$
$$x_1 \geq 0, \qquad x_2 \geq 0$$

These are

$$\underset{\text{feasible}}{\begin{bmatrix} x_{B_1} \\ \hdashline 0 \end{bmatrix}} \quad \text{and} \quad \underset{\text{infeasible}}{\begin{bmatrix} x_1^0 \\ \hdashline x_2^0 \end{bmatrix}} \qquad (21)$$

where

$$x_{B_1} = B_1^{-1} b \qquad (22)$$

The first is nonnegative, the second has some negative components in x_1^0. All vectors of the form

$$(1 - \theta)\begin{bmatrix} x_{B_1} \\ \hdashline 0 \end{bmatrix} + \theta \begin{bmatrix} x_1^0 \\ \hdashline x_2^0 \end{bmatrix} \qquad (23)$$

also satisfy $B_1 x_1 + B_2 x_2 = b$, $x_2 \geq 0$. When $\theta = 0$, this solution is feasible for (20), while for $\theta = 1$ it is infeasible. Let

$$J = \{j \mid x_{1j}^0 < 0\} \qquad (24)$$

The largest θ which maintains the solution in (23) nonnegative is limited by those components of (23) with indices $j \in J$. Setting these components equal to zero yields

$$x_{B_1 j} + \theta(x_{1j}^0 - x_{B_1 j}) = 0, \qquad j \in J \tag{25}$$

or

$$\theta_j = \frac{x_{B_1 j}}{x_{B_1 j} - x_{1j}^0} \tag{26}$$

Note that $0 \leq \theta_j \leq 1$. The largest θ maintaining (23) nonnegative is

$$\theta_s = \min_{j \in J} \theta_j \tag{27}$$

Let

$$K = \{k \mid x_{2k}^0 > 0\}$$

Note that K must be nonempty, since if all x_{2k}^0 were zero, then $x_1^0 = B_1^{-1}b \geq 0$, contrary to our hypothesis that x_1^0 has negative components. When $\theta = \theta_s$, (23) satisfies

$$B_1(x_{B_1} + \theta_s(x_1^0 - x_{B_1})) + \sum_{k \in K} B_2^k(\theta_s x_{2k}^0) = b \tag{28}$$

with B_2^k the kth column of B_2. By definition of θ_s, the coefficient of the sth column of B_1, B_1^s, in (28) is zero.

Since B_1 was assumed $m \times m$ and nonsingular, the matrix $[B_1 \mid B_2]$ has rank m. Let us delete the column B_1^s from (28) and examine the rank of the resulting system. Multiplying the deleted system by B_1^{-1} yields

$$\sum_{\substack{i=1 \\ i \neq s}}^{m} v_i^0 e_i + \sum_{k \in K} \bar{B}_2^k(\theta_s x_{2k}^0) = B_1^{-1}b \tag{29}$$

with v_i^0 the ith component of $x_{B_1} + \theta_s(x_1^0 - x_{B_1})$, e_i the ith unit vector, and $\bar{B}_2^k$ the kth column of $B_1^{-1}B_2$. The above system has the same rank as the system (28) without B_1^s. Moreover, at least one of the vectors $\bar{B}_2^k$, $k \in K$, say $\bar{B}_2^{k*}$, must have its sth component nonzero. This is because for $j \in J$ the vector

$$x_1^0 = x_{B_1} - B_1^{-1}B_2 x_2^0 \tag{30}$$

has negative components. Since $s \in J$, if all components of $B_1^{-1}B_2$ in row s and columns $k \in K$ were zero, then

$$x_{1s}^0 = x_{B_{1s}} \geq 0 \tag{31}$$

which is a contradiction. The vectors e_i, $i = 1, \ldots, m$, $i \neq s$, and $\bar{B}_2^{k*}$ are thus independent, so the system (28) with B_1^s deleted has rank m. Since this system has a nonnegative solution then, by Lemma 1, it has a basic nonnegative solution. Any such solution does not involve B_1^s and can include only those columns B_2^k with $k \in K$, so the theorem is proved.

Note that the preceding theorem does not assert the existence of a single pivot operation leading to a new basis, $\hat{B}_{i1}$. However, these bases may be found by solving the linear program

$$\text{minimize } x_{1s}$$

$$\text{subject to} \qquad \sum_{k=1}^{m} B_1^k x_{1k} + \sum_{k \in K} B_2^k x_{2k} = b \tag{32}$$

$$x_{1k} \geq 0, \qquad x_{2k} \geq 0$$

where s is the index selected in (27). The initial basis for this program is B_1, and there is an optimal solution with x_{1s} nonbasic. Only a few pivot steps should be required to solve this program. Of course, one such program is solved for each nonoptimal block in which it is desired to enforce violated constraints.

Solving the programs (32) yields new basis matrices, $\hat{B}_{i1}$, for each nonoptimal angular block. These are used to form a new reduced problem, and a new cycle begins. Since the algorithm is again a particular realization of the relaxation procedure of Section 5.2, finite convergence follows under conditions similar to those given in Theorem 5.3-2. Here, however, no provision has been made for enforcing violated nonnegativities by including additional constraints like (5.3-26) in the reduced problem. This could be done if desired, and then Theorem 5.3-2 would hold here. If no additional constraints are added, the assumption that each reduced problem has no alternative optimal solutions satisfying the newly enforced nonnegativities guarantees finite convergence.

Summary

1. Choose initial feasible bases for each angular block.
2. Using the current set of bases, form and solve the reduced problem (14), obtaining optimal solutions $\{x_{i2}^0\}$.
3. Compute x_{i1}^0 from (18).

4. If $x_{i1}^0 \geq 0$, $i = 1, \ldots, p$, $\{x_{i1}^0, x_{i2}^0\}$ solves the angular problem, (3)–(6).
5. Otherwise, for each i such that $x_{i1}^0 \ngeq 0$, compute the index s given by (27), and solve the linear program (32).
6. Using the new basis matrices $\hat{B}_{i1}$ obtained from solving (32), return to step 2.

Generation of Dual Feasible Points. The reduced problem, taken together with the constraints (13), is equivalent to an angular problem in which the constraints $x_{i1} \geq 0$ are relaxed. The dual of this problem has the corresponding dual angular constraints (9) holding as equalities; i.e.,

$$B'_{i1}u_i + A'_{i1}y = c_{i1}, \qquad i = 1, \ldots, p \tag{33}$$

These may be used to solve for u_i in terms of y:

$$u_i = (B'_{i1})^{-1}c_{i1} - (B'_{i1})^{-1}A'_{i1}y \tag{34}$$

It is easily verified that substituting the above into the bottom partition of the dual angular constraints, (10), yields the dual of the reduced problem:

$$\left.\begin{array}{lll} & \text{maximize } b'_0 y & \\ \text{subject to} \quad M'_i y \leq d_i, & i = 1, \ldots, p & \end{array}\right\} \text{dual of reduced problem} \tag{35}$$

Let y^0 be the vector of simplex multipliers corresponding to an optimal solution, $\{x_{i2}^0\}$, of the reduced problem. Substituting y^0 into (34) yields values for the remaining dual-angular variables, u_i:

$$u_i^0 = (B'_{i1})^{-1}(c_{i1} - A'_{i1}y^0), \qquad i = 1, \ldots, p \tag{36}$$

THEOREM 4. *The vectors $\{u_i^0\}$, y^0 are a feasible extreme-point solution to the dual-angular problem. The sequence of dual-angular objective values generated by these solutions in successive iterations is nondecreasing.*

PROOF. The solution $\{u_i^0\}$, y^0 is feasible by construction. To show that it is an extreme point note that, as a vector of simplex multipliers, y^0 satisfies m_0 of the constraints (35) as equalities, and the $m_0 \times m_0$ submatrix associated with these equalities is nonsingular. Letting M be this submatrix and d be the corresponding right-hand side, we have

$$My^0 = d \tag{37}$$

Thus $\{u_i^0\}$, y^0 satisfies as equalities a subset of the dual-angular constraints having the form

$$
\begin{array}{llll}
u_1 & & + (B'_{11})^{-1}A'_{11}y & = (B'_{11})^{-1}c_{11} \\
& u_2 & \vdots & \vdots \\
& \ddots & \vdots & \vdots \\
& & u_p + (B'_{p1})^{-1}A'_{p1}y & = (B'_{p1})^{-1}c_{p1} \\
& & My & = d
\end{array}
\tag{38}
$$

This system has as many equations as unknowns and its coefficient matrix is upper block triangular with nonsingular blocks. Thus this matrix is nonsingular. According to Simonnard [6], this proves that $\{u_i^0\}$, y^0 is an extreme point.

To demonstrate monotonicity, we reason as follows. The vector $\{u_i^0\}$, y^0 solves the dual of a modified angular problem, one with the constraints $x_{i1} \geq 0$ relaxed. The objective value $v^0 = \sum_i b_i' u_i^0 + b_0' y^0$ is thus equal to the optimal objective value of this modified problem, z^0. By Theorem 5.2-2, the sequence of z^0 values obtained is nondecreasing, so the theorem is proved.

We note that the previous theorem is a specialization of the general result on dual feasibility given in Section 5.2. A similar comment applies to Theorem 5.3-3.

5.4.2 Computational Considerations.

Selection of Violated Constraints to be Enforced. As in Ritter's method of Section 5.3, there are many ways to choose which violated constraints to enforce. The same comments made there apply here, except for the fact that the size of the reduced problem is constant, independent of the number of nonnegativities enforced. This makes enforcing as many of these as possible at each cycle more attractive. If this option is chosen, it may be implemented by solving the linear program (32) with objective function

$$z = \sum_{j \in J} x_{1j} \tag{39}$$

which J is given in (24). Again, there is one such program for each nonoptimal angular block.

Solving Successive Reduced Problems. In applying this partitioning algorithm, most of the computation time will probably be spent solving the reduced problems. It is therefore important that information from the solution at cycle k be used effectively in solving the problem at cycle $k + 1$. The following results show how this may be done.

For simplicity, we first consider a one-block angular problem

$$\text{minimize } z = c_1 x_1 + \hat{c}_2 x_2$$

subject to

$$\begin{aligned} &\begin{matrix} m_0 \\ \text{rows} \end{matrix}\left\{ A_1 x_1 + \hat{A}_2 x_2 = \hat{b}_0 \right. \\ &\begin{matrix} m_1 \\ \text{rows} \end{matrix}\left\{ B_1 x_1 + \hat{B}_2 x_2 = \hat{b}_1 \right. \\ &x_1 \geq 0, \qquad x_2 \geq 0 \end{aligned} \tag{40}$$

where B_1 is a basis matrix of dimension m_1. Note that any linear program may be written in this way, so Rosen's partitioning algorithm is completely general. Using the matrix B_1 to eliminate x_1 from the first m_0 rows and from the objective yields the following tableau:

$$\text{minimize } c_2 x_2$$

subject to

$$\begin{aligned} A_2 x_2 &= b_0 \\ x_1 + B_2 x_2 &= b_1 \\ x_1 \geq 0, \qquad x_2 &\geq 0 \end{aligned} \tag{41}$$

The reduced problem is obtained by relaxing the restrictions $x_1 \geq 0$:

$$\text{minimize } c_2 x_2$$

subject to

$$A_2 x_2 = b_0, \qquad x_2 \geq 0 \tag{42}$$

Let x_2^0 solve this problem and let B_0 be the basis matrix associated with x_2^0. Again for simplicity, we assume that B_0 is composed of the first m_0 columns of A_2:

$$B_0 = [A_2^1, \ldots, A_2^{m_0}] \tag{43}$$

Let

$$x_2 = (y \mid w) \tag{44}$$

with y containing the basic components of x_2^0 and w the nonbasic. Placing the reduced problem in canonical form with respect to y yields the following tableau (with $x \equiv x_1$, $b \equiv b_1$):

$$\begin{aligned} y \qquad + Ew &= y^0 \\ -z + \bar{c}w &= -z^0 \\ x + Cy \qquad + Dw &= b \end{aligned} \tag{45}$$

Assume now that component r of the vector

$$x^0 = b - Cy^0 \tag{46}$$

is negative and that a single pivot operation suffices to interchange x_r with some component of y which is positive in y^0, say y_s. If this pivot is extended into the first m_0 rows of (45), y_s is eliminated from the reduced problem, x_r is introduced, and the matrix elements of the reduced problem tableau change. Let x_r now replace y_s as the sth basic variable of the reduced problem, and let the corresponding basis matrix be $\bar{B}$. We will show that $\bar{B}$ has the same simplex multipliers as B_0, and is an optimal but infeasible basis for the new reduced problem. This result holds when a number of components of y and x are interchanged. It implies that the new reduced problem may be solved efficiently by using $\bar{B}$ as an initial basis for the dual simplex method.

The important relationships between successive reduced problems are given in the following theorem, which is an extension of an earlier result due to Gass [7].

THEOREM 5. *In the tableau* (45), *let a number of pivot operations be performed, which interchange one or more components of* x *(at least one of which is negative in* x^0*) with some components of* y*, all positive in* y^0*. If* $\bar{B}$ *denotes the pivot transformation of the previous reduced problem optimal basis,* B_0*, then*

(a) $\bar{B}$ *has the same simplex multipliers as* B_0 *and these cause all nonbasic columns of the new reduced problem to price out nonnegative. Thus* $\bar{B}$ *is an optimal basis.*

(b) *The basic solution associated with* $\bar{B}$ *has* $y_i = y_i^0$ *if* y_i *is basic and* $x_j = x_j^0$ *if* x_j *is basic. Thus this solution is infeasible only in components where* $x_j^0 < 0$.

PROOF. Consider first a single pivot operation on c_{rs}, which interchanges x_r and y_s in the tableau below, which is a detailed version of (45):

x_1	$\cdots$	x_r	$\cdots$	x_{m_1}	y_1	$\cdots$	y_s	$\cdots$	y_{m_0}	$-z$	w_1	$\cdots$	w_t	rhs
					1						e_{11}	$\cdots$	e_{1t}	y_1^0
						$\ddots$								
							1				$\vdots$			$\vdots$
								$\ddots$						
									1		$e_{m_0 1}$	$\cdots$	$e_{m_0 t}$	$y_{m_0}^0$
										1	$\bar{c}_1$	$\cdots$	$\bar{c}_t$	$-z^0$
1					c_{11}	$\cdots$	c_{1s}	$\cdots$	c_{1m_0}		d_{11}	$\cdots$	d_{1t}	b_1
	$\ddots$				$\vdots$		$\vdots$		$\vdots$		$\cdot$		$\cdot$	$\vdots$
		1			c_{r1}	$\cdots$	c_{rs}	$\cdots$	c_{rm_0}		$\cdot$		$\cdot$	b_r
			$\ddots$		$\vdots$		$\vdots$		$\vdots$		$\cdot$		$\cdot$	$\vdots$
				1	$c_{m_1 1}$	$\cdots$	$c_{m_1 s}$	$\cdots$	$c_{m_1 m_0}$		$d_{m_1 1}$	$\cdots$	$d_{m_1 t}$	b_{m_1}

Note that $\bar{c}_i \geq 0$, all i, since y^0 is an optimal solution to the reduced problem. After the pivot, the upper $m_0 + 1$ rows (the reduced problem tableau) appear as follows:

$$
\begin{array}{ccccccccccccc|c}
x_1 & \cdots & x_r & \cdots & x_{m1} & y_1 & \cdots & y_s & \cdots & y_{m_0} & -z & w_1 \cdots & w_t & \text{rhs} \\
\hline
 & & & & & 1 & & & & & & & & y_1^0 \\
 & & & & & & \ddots & & & & & & & \vdots \\
 & & -\dfrac{1}{c_{rs}} & \cdots & & -\dfrac{c_{r1}}{c_{rs}} & \cdots & 0 & \cdots & -\dfrac{c_{rm_0}}{c_{rs}} & & \cdots & \cdot & y_s^0 - b_r/c_{rs} \\
 & & & & & & & & \ddots & & & & & \vdots \\
 & & & & & & & & & 1 & & & & y_{m_0}^0 \\
 & & & & & & & & & & 1 & \bar{c}_1 \cdots & \bar{c}_t & -z^0 \\
\hline
\end{array}
$$

As indicated, x_r and y_s are interchanged, yielding the new reduced problem tableau. To place this tableau in canonical form relative to basic variables $y_1 \cdots y_{s-1}, x_r, y_{s+1} \cdots y_{m_0}, -z$ we multiply all columns by the $m_0 + 1$ dimensional elementary matrix

$$
E_{rs} = \begin{bmatrix}
1 & & & & & & \\
 & \ddots & & & & & \\
 & & 1 & & & & \\
-c_{r1} & -c_{r2} & \cdots & -c_{rs} & \cdots & -c_{rm_0} & 0 \\
 & & & 1 & & & \\
 & & & & & \ddots & \\
 & & & & & & 1
\end{bmatrix} \tag{47}
$$

Since this operation leaves the z row unchanged, the new tableau still satisfies the optimality conditions $\bar{c}_i \geq 0$. The inverse of the new basis is

$$
\bar{B}^{-1} = E_{rs} B_0^{-1} \tag{48}
$$

The matrix E_{rs} has the unit vector u_{m_0+1} as its last row, so $\bar{B}^{-1}$ and B_0^{-1} have the same last row. Since this row contains the simplex multipliers, part (a) of the theorem is proved for this single pivot case.

The extension of the above argument to the case of multiple pivots is straightforward. The new canonical form is obtained by multiplying the original by a product of elementary matrices, each differing from the identity in only one row. The new basis inverse is given by

$$
\bar{B}^{-1} = E_{r_k s_k} E_{r_{k-1} s_{k-1}} \cdots E_{r_1 s_1} B_0^{-1} \tag{49}
$$

Again, since all variables y_i have zero coefficients in the z row, this row is unchanged by the pivot operations and each elementary matrix has the unit vector u_{m_0+1} as its last row. Thus $\bar{B}^{-1}$ has the same simplex multipliers as B_0^{-1} and is an optimal basis, so part (a) of the theorem is proved.

To prove part (b) we reason as follows. With $w = 0$, equations (45) have the unique solution $y = y^0$, $z = z^0$, $x = x^0 = b - Cy^0$. After any number of pivot operations is performed this is still the unique solution (again with $w = 0$), whatever the form of the transformed system. Thus any variable x_i introduced into the top m_0 rows of (45) has the value x_i^0 in the new canonical form, while variables y_i have the values y_i^0.

Extension to More than One Angular Block. The situation for the multi-block problem can be understood by considering a two-block primal:

$$\text{minimize } z$$

subject to

$$\begin{array}{l} \text{reduced} \\ \text{problem} \\ \text{tableau} \end{array}\left\{\begin{bmatrix} y_1 \\ \hdashline 0 \end{bmatrix} + \begin{bmatrix} E_{11} \\ \hdashline E_{21} \end{bmatrix} w_1 + \begin{bmatrix} 0 \\ \hdashline y_2 \end{bmatrix} + \begin{bmatrix} E_{12} \\ \hdashline E_{22} \end{bmatrix} w_2 = \begin{bmatrix} y_1^0 \\ \hdashline y_2^0 \end{bmatrix}\right. \tag{50}$$

$$\begin{aligned} \bar{c}_1 w_1 + \bar{c}_2 w_2 - z &= -z^0 \\ x_1 + C_1 y_1 + D_1 w_1 &= b_1 \\ x_2 + C_2 y_2 + D_2 w_2 &= b_2 \end{aligned}$$

The reduced problem is in canonical form with respect to the variables $y = (y_1, y_2)$. It is evident that, as before, pivot operations which interchange components of any y_i and x_i do not affect the z row, since y_1 and y_2 have zero coefficients there. Thus the new canonical form will remain dual feasible, i.e., optimal. Owing to the block structure, pivot operations in one block do not affect the submatrices and vectors associated with the other. Because of this, the new basis inverse for the reduced problem, $\bar{B}^{-1}$, is related to the old by

$$\bar{B}^{-1} = \left[\begin{array}{cc:c} E_1 & & \\ & E_2 & 0 \\ \hdashline & 0 & 1 \end{array}\right] B_0^{-1} \tag{51}$$

where E_1 and E_2 are products of elementary matrices such as (47), each differing from the identity in only one row. Since rows $m_0 + 1$ of $\bar{B}^{-1}$ and B_0^{-1} are identical, these bases have the same simplex multipliers and part (a) of Theorem 5 is true for the multiblock case. The reasoning in the proof of part (b) is unaltered here. Thus Theorem 5 holds in the case of multiple angular blocks.

It is not difficult to show that Theorem 5 applies essentially unchanged to Ritter's algorithm of Section 5.3. Minor modifications must be made to account for the fact that constraints are added to and deleted from the reduced problem.

Because of the preceding results, the dual simplex method is the logical candidate for solving the reduced problem. If only a single pivot operation is performed among all angular blocks, the initial basis will be dual feasible and primal infeasible in only one component. Thus only a few dual simplex cycles should be required to solve the reduced problem. If more pivots are made, there are more infeasibilities, more cycles required, but a greater change in the objective after each re-solution of the reduced problem. In choosing between these alternatives it should be noted that, for large problems, all data not currently being processed will probably be maintained outside of core memory in auxiliary storage. Transfers between solving the reduced problem and operations in the various blocks are thus likely to be relatively slow, and setting up the data for solution can also be time consuming. Because of this, to minimize overall computation time, it seems best to perform as many interchanges of positive and negative variables per cycle as possible. This should minimize the number of cycles, hence the number of transfer and setup operations.

Interpretation as a Restriction Procedure. Although both this algorithm and that of the previous section have been presented as relaxation procedures, they have dual interpretations in terms of restriction. As pointed out earlier, relaxation of a nonnegativity $x_{ij} \geq 0$ implies that the corresponding dual constraint holds as an equality, i.e., is restricted. We have already shown that the sequence of feasible dual solutions obtained by solving these equalities is improving. In fact, it is not difficult to show that, if relaxation is applied to the primal (3)–(6), the primal and dual solutions generated are equal to those obtained if restriction is applied to the dual, (1)–(2). We must only ensure that the rule for choosing the set V in the two procedures is the same: If $x_{ij}^0 < 0$ and $x_{ij} \geq 0$ is enforced, then the corresponding dual constraint, which was restricted, is dropped from the restricted set, and conversely.

While the restriction viewpoint leads to the same algorithm as relaxation in the linear case, it will prove more useful in the nonlinear problems of Section 7.2. There, all the primal–dual correspondences of linear programming are not available, and restriction, properly applied, takes best advantage of problem structure.

5.4.3 Computational Experience. There has been extensive computational experience with this algorithm, as described by Rosen [4]. Six problems were solved on an IBM 7090/94 computer, with problem size, matrix density, and solution data given in Table 1. The 15-block problem had submatrices B_i which were 50 per cent dense with nonzero elements in a checkerboard pattern. The coupling matrices A_i were 100 per cent dense. All nonzero matrix elements and the cost vectors c_i were generated as pseudo-random

TABLE 1. SUMMARY OF TEST-PROBLEM RESULTS

Name	Size			No. of nonzero matrix elements	Matrix density (%)	No. of cycles for solution	Time (min) for solution (IBM 7094)
	Blocks	Constraints	Variables				
PP problem 1	1	4	5	19	95	2	1.3
Beale 2-block	2	9	15	39	29	4	2.0
Rosen 2-block	2	60	100	1,660	27.6	7	3.6
Benders 6-block	6	71	164	2,490	19.3	14	6.0
Rosen 10-block	10	220	420	8,220	8.9	39	75
Random-element 15-block	15	930	1,200	72,000	6.4	16	179

numbers uniformly distributed over the range (0, 100). The right-hand-side vectors b_i were chosen to ensure that the angular problem was feasible. Additional data on the solution of this problem are given in Table 2.

An application of this algorithm to transportation problems, which includes further computational experience, is described in [8].

TABLE 2. SOLUTION HISTORY FOR RANDOM-MATRIX 15-BLOCK PROBLEM

Cycle	No. of optimal blocks	No. of $x_{i1}^0 < 0$	Function value
1			39,365.24
2	2	121	39,968.80
3	[a]	[a]	40,109.06
4	4	70	40,132.92
5	3	72	40,167.84
6	1	151	40,258.11
7	2	57	40,335.48
8	4	92	40,367.09
9	3	56	40,395.18
10	7	46	40,440.69
11	8	17	40,470.01
12	7	18	40,478.85
13	10	7	40,486.65
14	11	6	40,488.97
15	12	3	40,489.39
16	15	0	40,489.55

[a] Restart.

5.4.4 Example of Rosen's Partitioning Method. Consider the problem

$$\text{minimize } -x_1 - x_2 - 2x_5 - x_6$$

subject to

$$\begin{aligned}
x_1 + 2x_2 \qquad\qquad + 2x_5 + x_6 \qquad\qquad\qquad + x_{10} \qquad &= 40\\
x_1 + x_2 \qquad\qquad + 4x_5 + 2x_6 \qquad\qquad\qquad\qquad + x_{11} &= 50\\
x_1 + 3x_2 + x_3 \qquad\qquad\qquad\qquad\qquad\qquad\qquad &= 30\\
2x_1 + x_2 \qquad + x_4 \qquad\qquad\qquad\qquad\qquad\qquad &= 20\\
x_5 \qquad + x_7 \qquad\qquad\qquad &= 10\\
x_6 \qquad + x_8 \qquad\qquad &= 10\\
x_5 + x_6 \qquad\qquad + x_9 \qquad &= 15\\
x_i \geq 0, \quad \text{all } i
\end{aligned}$$

This is derived from the example of Section 3.5 by adding the constraint

$$x_1 + x_2 + 4x_5 + 2x_6 \leq 50$$

CYCLE 1. To obtain initial basis matrices we solve the following subproblems:

Block 1

$$\text{minimize} \; - x_1 - x_2$$

subject to

$$\begin{aligned} x_1 + 3x_2 + x_3 \qquad &= 30 \\ 2x_1 + x_2 \qquad + x_4 &= 40 \\ x_i \geq 0 \end{aligned}$$

SOLUTION.

$$x_1 = 6, \qquad x_2 = 8,$$

all other $x_i = 0$

Block 2

$$\text{minimize} \; - 2x_5 - x_6$$

subject to

$$\begin{aligned} x_5 \qquad + x_7 \qquad\qquad &= 10 \\ x_6 \qquad + x_8 \qquad &= 10 \\ x_5 + x_6 \qquad\qquad + x_9 &= 15 \\ x_i \geq 0 \end{aligned}$$

SOLUTION.

$$x_5 = 10, \qquad x_6 = 5, \qquad x_8 = 5,$$

all other $x_i = 0$

The initial bases are

$$B_{11} = \begin{bmatrix} 1 & 3 \\ 2 & 1 \end{bmatrix}, \qquad B_{21} = \begin{bmatrix} 1 & 0 & 0 \\ 0 & 1 & 1 \\ 1 & 1 & 0 \end{bmatrix}$$

$$B_{11}^{-1} = \frac{1}{5}\begin{bmatrix} -1 & 3 \\ 2 & -1 \end{bmatrix}, \qquad B_{21}^{-1} = \begin{bmatrix} 1 & 0 & 0 \\ -1 & 0 & 1 \\ 1 & 1 & -1 \end{bmatrix}$$

The angular problem can now be reduced to canonical form relative to the basic variables in each block by a sequence of pivot operations in these bases. The resulting system is

First Angular System

x_1	x_2	x_3	x_4	x_5	x_6	x_7	x_8	x_9	x_{10}	x_{11}	
		$\frac{1}{5}$	$\frac{2}{5}$			1		1			$= z + 39$
		$-\frac{3}{5}$	$-\frac{1}{5}$			-1		-1	1		$= -7$
		$-\frac{1}{5}$	$-\frac{2}{5}$			-2		-2		1	$= -14$
1		$-\frac{1}{5}$	$\frac{3}{5}$								$= 6$
	1	$\frac{2}{5}$	$-\frac{1}{5}$								$= 8$
				1		1					$= 10$
						$\textcircled{1}$	1	-1			$= 5$
					1	-1		1			$= 5$

The initial reduced problem is obtained from the top three rows.

Reduced Problem 1

$$\text{minimize } \hat{z} = \tfrac{1}{5}x_3 + \tfrac{2}{5}x_4 + x_7 + x_9$$

subject to

$$\begin{aligned} -\tfrac{3}{5}x_3 - \tfrac{1}{5}x_4 - x_7 - x_9 + x_{10} &= -7 \\ -\tfrac{1}{5}x_3 - \tfrac{2}{5}x_4 - 2x_7 - 2x_9 + x_{11} &= -14 \\ x_i &\geq 0 \end{aligned}$$

The optimal canonical form is as follows:

x_3	x_4	x_7	x_9	x_{10}	x_{11}	$-\hat{z}$	Constants
1	0	0	0	-2	0	0	0
0	$\frac{1}{5}$	1	1	$\frac{1}{5}$	$-\frac{3}{5}$	0	7
0	$\frac{1}{5}$	0	0	$\frac{1}{5}$	$\frac{2}{5}$	1	-7

Thus

$$x_3^0 = 0, \qquad x_7^0 = 7, \qquad \text{all other } x_i^0 = 0$$

Optimality Test. Substituting the x_i^0 into each transformed angular block yields

$$\begin{gathered} x_1^0 = 6, \qquad x_2^0 = 8 \\ x_5^0 = 3, \qquad x_6^0 = 12, \qquad x_8^0 = -2 \end{gathered}$$

with objective value $z^0 = -32$. Since $x_8^0 < 0$, the solution is not optimal.

Pivoting in Block 2. Since x_7 is the only positive nonbasic variable in block 2, x_8 must leave the basis and x_7 enters. Pivoting on the element circled in the first angular system yields the updated angular system.

Second Angular System

x_1	x_2	x_3	x_4	x_5	x_6	x_7	x_8	x_9	x_{10}	x_{11}	
		$\frac{1}{5}$	$\frac{2}{5}$				-1	2			$= z + 34$
		$-\frac{3}{5}$	$-\frac{1}{5}$				1	-2	1		$= -2$
		$-\frac{1}{5}$	$-\frac{2}{5}$				2	-4		1	$= -4$
1		$-\frac{1}{5}$	$\frac{3}{5}$								$= 6$
	1	$\frac{2}{5}$	$-\frac{1}{5}$								$= 8$
				1			-1	1			$= 5$
						1	1	-1			$= 5$
					1		1				$= 10$

Reduced Problem 2

$$\text{minimize } \tfrac{1}{5}x_3 + \tfrac{2}{5}x_4 - x_8 + 2x_9$$

subject to

$$\begin{aligned} -\tfrac{3}{5}x_3 - \tfrac{1}{5}x_4 + x_8 - 2x_9 + x_{10} &= -2 \\ -\tfrac{1}{5}x_3 - \tfrac{2}{5}x_4 + 2x_8 - 4x_9 + x_{11} &= -4 \\ x_i &\geq 0 \end{aligned}$$

The transform of the previous optimal basis is

$$\bar{B} = \begin{array}{c} \begin{array}{ccc} x_3 & x_8 & -\hat{z} \end{array} \\ \begin{bmatrix} -\frac{3}{5} & 1 & 0 \\ -\frac{1}{5} & 2 & 0 \\ \frac{1}{5} & -1 & 1 \end{bmatrix} \end{array}$$

Placing the reduced problem into canonical form with respect to x_3, x_8, $-\hat{z}$ yields

x_3	x_4	x_8	x_9	x_{10}	x_{11}	$-\hat{z}$	Constants
1				-2	1		0
	$-\frac{1}{5}$	1	$\boxed{-2}$	$-\frac{1}{5}$	$\frac{3}{5}$		-2
	$\frac{1}{5}$			$\frac{1}{5}$	$\frac{2}{5}$	1	-2

Since all $\bar{c}_j \geq 0$, $\bar{B}$ is dual feasible, as shown earlier, with a single infeasibility. Performing a single dual simplex iteration on the above yields the following optimal tableau:

x_3	x_4	x_8	x_9	x_{10}	x_{11}	$-\hat{z}$	Constants
1				-2	1		0
	$\frac{1}{10}$	$-\frac{1}{2}$	1	$\frac{1}{10}$	$\frac{3}{10}$		1
	$\frac{1}{5}$			$\frac{1}{5}$	$\frac{1}{5}$	1	-2

The optimal solution is $x_9^0 = 1$, all other $x_i = 0$. Substituting these values into the blocks of the second angular system yields

$$x_1^0 = 6, \qquad x_2^0 = 8$$
$$x_5^0 = 4, \qquad x_6^0 = 10, \qquad x_7^0 = 6$$

with objective value

$$z^0 = -32$$

Since all $x_i^0 \geq 0$, this solution is optimal.

Problems

1. Apply relaxation to the problem given in (1) and (2) of Section 1.2.4 and verify the correspondences between the resulting operations and those of the dual simplex method.
2. For the restriction procedure of Section 5.2, prove that
   (a) the optimality test is valid.
   (b) the sequence of objective values is nondecreasing.
   (c) the procedure requires solution of only a finite number of problems P^R.
3. Show that, with the choices of g_i and R given at the end of Section 5.2, restriction can be made identical with the primal simplex method.
4. Show how the dual solution in Theorem 5.3-3 may be constructed from the optimal simplex multipliers of the reduced problem and the matrices B_{i1}^{-1}.
5. Show that if the solution to the reduced problem is unique at each cycle, Ritter's algorithm converges to an optimal solution in a finite number of cycles.
6. Discuss the conversion of the algorithm of Section 5.3 into one generating a sequence of primal feasible solutions by applying it to the dual of the problem whose solution is desired. Show how basis matrices in the dual blocks may be obtained immediately if such matrices are known in the primal blocks. What are the dimensions of these bases, and what is the size of the reduced problem? Outline the relative advantages and disadvantages between applying the algorithm in the primal and in the dual.
7. Consider the extension of Ritter's algorithm to problems with constraints of the form (5.3-2)–(5.3-4) but with a general differentiable concave objective.
8. Solve the dual of the example of Section 3.5 using Rosen's method.
9. Extend the results of Theorem 5.4-5 to Ritter's algorithm of Section 5.3.
10. Consider a linear program with upper bounded variables. Convert this to a one-block primal by adding slacks in the bound inequalities and apply Rosen's partitioning algorithm. Outline the steps of the resulting procedure, indicating any simplifications which occur. What is the simplest choice for the initial set of variables whose nonnegativities shall be relaxed?
11. As an extension of an upper bounded program, consider the problem

$$\text{minimize } c_1x_1 + c_2x_2 + dy$$

subject to

$$A_1x_1 + A_2x_2 + ay = b$$
$$x_2 \leq ey$$
$$x_1 \geq 0, \qquad x_2 \geq 0, \qquad y \geq 0$$

where e is a vector with all components unity and y is a scalar variable. Repeat the procedure of Problem 10 for this case.

12. Show that, if the choices of the set V are related as indicated at the end of Section 5.4, then applying relaxation to the primal (5.4-3)–(5.4-6) yields the same results as applying restriction to the dual (5.4-1)–(5.4-2).

References

1. A. M. Geoffrion, "Relaxation and the Dual Method in Mathematical Programming," *Working Paper 135*, Western Management Science Institute, University of California at Los Angeles, 1968.
2. R. Graves and P. Wolfe, eds., *Recent Advances in Mathematical Programming*, McGraw-Hill, Inc., New York, 1963, pp. 177–200.
3. K. Ritter, "A Decomposition Method for Linear Programming Problems with Coupling Constraints and Variables," Mathematics Research Center, University of Wisconsin, *Rept. 739*, 1967.
4. J. B. Rosen, "Primal Partition Programming for Block Diagonal Matrices," *Numerische Mathematik*, **6**, 1964, pp. 250–260.
5. G. Hadley, *Linear Programming*, Addison-Wesley Publishing Company, Inc., Reading, Mass., 1962, pp. 80–83.
6. M. Simonnard, *Linear Programming*, Prentice-Hall, Inc., Englewood Cliffs, N.J., 1966, p. 397.
7. S. I. Gass, "The Dualplex Method for Large-Scale Linear Programs," Ph.D. Thesis, University of California at Berkeley, 1966, *ORC Rept. 66–15*.
8. M. D. Grigoriadis and W. F. Walker, "A Treatment of Transportation Problems by Primal Partition Programming," *Management Sci.*, **14**, No. 9, 1968, pp. 565–599.
9. M. D. Grigoriadis and K. Ritter, "A Decomposition Method for Structured Linear and Nonlinear Programs," *Computer Sciences Tech. Rept. 10*, Computer Sciences Department, University of Wisconsin, 1968.
10. A. M. Geoffrion, "Constrained Maximum Likelihood Estimation of Several Stochastically Ordered Distributions," The RAND Corporation, forthcoming.
11. A. M. Geoffrion, "Reducing Concave Programs with Some Linear Constraints," *SIAM J. Appl. Math.*, **15**, 1967, pp. 653–664.

6

Compact Inverse Methods

6.1 Introduction

Most existing techniques for solving large mathematical programs involve some measure of decentralization. This is certainly true in all other chapters of this book. There the methods studied are either of the decomposition or partitioning type. All require iterative solution of a number of smaller subproblems, which may be optimization problems or may simply require, say, a pivot operation. These exchange information with some coordinating or master program.

Effective solution procedures for large-scale systems need not be of this type. A number of efficient "centralized" methods exist for solving large linear programs of special structure. These are all specializations of the revised simplex method, which reduce computer time and storage requirements. They do so by taking advantage of special properties of basis matrices in the class of problems being solved, using these to store the inverse or manipulate the associated equations in compact form. The revised simplex method itself, with the basis inverse in product form, is a compact inverse technique, which takes advantage of the low density of nonzero matrix elements in large problems.

This chapter begins with a discussion of product-form codes. Upper bounding methods [1] and their extensions follow. These include Dantzig and Van Slyke's generalized upper bounding procedure [2] and its extension to angular problems [3].

6.2 Revised Simplex Method with Inverse in Product Form

Before considering the product form, the steps of the revised simplex method with inverse in explicit form are briefly reviewed. The problem being solved is

$$\text{minimize } z = cx$$

subject to
$$Ax = b, \qquad x \geq 0 \tag{1}$$

where

$$A = [P_1, P_2, \ldots, P_n] \tag{2}$$

is an $m \times n$ matrix of rank m, b an m vector of constants, and c and n vector of objective coefficients. The equation

$$cx - z = 0 \tag{3}$$

is added to the system, with $-z$ taken as an additional basic variable, here the $(m + 1)$st. Let B be a feasible basis for the resulting $m + 1$ row $\times$ $n + 1$ column augmented system. At the beginning of cycle k, assume that B^{-1}, the associated basic solution $x_B = B^{-1}b$, and the data of the original problem (A, b, and c) are available. Cycle k proceeds as follows:

1. Row $m + 1$ of B^{-1} has the form

$$B_{m+1}^{-1} = (-\pi_1, -\pi_2, \ldots, -\pi_m, 1) \tag{4}$$

where the π_i are simplex multipliers. Form, for each nonbasic variable, the relative cost factor $\bar{c}_j$, using the equation

$$\bar{c}_j = c_j - \sum_i \pi_i a_{ij} = c_j - \pi P_j \tag{5}$$

2. Assuming the standard column-selection rule is employed, find

$$\min_j \bar{c}_j = \bar{c}_s \tag{6}$$

3. If $\bar{c}_s \geq 0$, stop. The current basic solution is optimal.
4. If $\bar{c}_s < 0$, compute the transformed column

$$\begin{pmatrix} \bar{P}_s \\ \bar{c}_s \end{pmatrix} = B^{-1} \begin{pmatrix} P_s \\ c_s \end{pmatrix} \tag{7}$$

5. Let

$$\bar{P}_s = (\bar{a}_{1s}, \bar{a}_{2s}, \ldots, \bar{a}_{ms})' \tag{8}$$

If all $\bar{a}_{is} \leq 0$, stop. The optimum is unbounded. Let

$$\hat{x} = (x_B \mid 0) + x_s(-\bar{P}_s \mid u_s) \tag{9}$$

where u_s is the sth unit vector. Then $\hat{x}$ is a solution, feasible for all $x_s \geq 0$, whose objective value

$$\hat{z} = c_B x_B + \bar{c}_s x_s \tag{10}$$

approaches $-\infty$ as $x_s \to +\infty$.

6. Otherwise, compute

$$\frac{\bar{b}_r}{\bar{a}_{rs}} = \min_{\bar{a}_{is} > 0} \frac{\bar{b}_i}{\bar{a}_{is}} = \theta \tag{11}$$

7. Construct the augmented matrix

$$\left[\begin{array}{c|c} B^{-1} & \begin{matrix} \bar{P}_s \\ \\ \bar{c}_s \end{matrix} \end{array}\right]$$

and transform it by pivoting on $\bar{a}_{rs}$. The first $m + 1$ columns of the result are the inverse of the new basis. Update the basic solution by

$$\begin{aligned} (x_B)_i &\leftarrow (x_B)_i - \theta\bar{a}_{is}, \qquad i \neq r \\ (x_B)_r &\leftarrow \theta \end{aligned} \tag{12}$$

and return to step 1.

Product Form of the Inverse. In what follows, it will be convenient to let the cost coefficient c_i be the $(m + 1)$st element of P_i, and $\bar{c}_i$ the $(m + 1)$st element of $\bar{P}_i$; i.e.,

$$a_{m+1,i} = c_i, \qquad \bar{a}_{m+1,i} = \bar{c}_i \tag{13}$$

An elementary matrix is defined here as a square matrix differing from the identity in only one row or column. In product-form codes, B^{-1} is not available explicitly, but is stored as a product of elementary matrices. To see how these are developed, let B_c^{-1} be the current inverse and assume that the new inverse is to be computed by a pivot on $\bar{a}_{rs}$. The following operations are performed on B_c^{-1}:

1. Replace row r by $1/\bar{a}_{rs}$ (row r).
2. For $i = 1, \ldots, m + 1$, $i \neq r$, replace row i by (row i) $-$ $\bar{a}_{is}/\bar{a}_{rs}$ (row r).

It is easily verified by direct matrix multiplication that multiplying B_c^{-1} on the left by the following elementary matrix performs these operations:

$$E = \begin{bmatrix} 1 & & & \eta_1 & & & \\ & 1 & & \eta_2 & & & \\ & & \ddots & \vdots & & & \\ & & & \eta_r & & & \\ & & & \vdots & 1 & & \\ & & & & & \ddots & \\ & & & \eta_{m+1} & & & 1 \end{bmatrix} \tag{14}$$

$$\uparrow$$
$$\text{column } r$$

where

$$\eta_i = \begin{cases} \dfrac{-\bar{a}_{is}}{\bar{a}_{rs}}, & i = 1, \ldots, m+1, \quad i \neq r \\[2ex] \dfrac{1}{\bar{a}_{rs}}, & i = r \end{cases} \tag{15}$$

That is,

$$B_n^{-1} = EB_c^{-1} \tag{16}$$

where B_n^{-1} is the new inverse. If the initial basis is the identity matrix, and if k pivot operations have been performed, the inverse at cycle k, B_k^{-1}, is given by

$$B_k^{-1} = E_k E_{k-1} \cdots E_1 \tag{17}$$

with each E_i an elementary matrix like (14). The above representation of B_k^{-1} as a product of elementary matrices is called the product form of the inverse.

An important property of elementary matrices is that they can be stored in a computer memory by recording only the elements of the nonunit vector column and its position in the matrix (in practice, only the nonzero elements and their row positions need be stored). These columns are often called "eta vectors." Since the inverse is now not available explicitly, all computations involving it require a sequence of matrix multiplications. We consider these computations in their order of occurrence in the simplex method.

Evaluating the Simplex Multipliers. The multipliers are given by

$$\pi = c_B B^{-1} \tag{18}$$

so, if B^{-1} is given in product form, we must evaluate the matrix product

$$\pi = (\cdots((c_B E_k)E_{k-1})\cdots)E_1 \tag{19}$$

The operations are carried out as indicated by the parentheses, i.e., first evaluate the row vector $c_B E_k$, then $(c_B E_k)E_{k-1}$, etc. Each of these requires the multiplication of an elementary column matrix on the left by a row vector. Let

$$E = [u_1, u_2, \ldots, u_{r-1}, \eta, u_{r+1}, \ldots, u_{m+1}] \tag{20}$$

with u_i the ith unit vector, and let

$$v = (v_1, \ldots, v_{m+1}) \tag{21}$$

be an arbitrary row vector. Then

$$vE = (v_1, \ldots, v_{r-1}, \delta, v_{r+1}, \ldots, v_{m+1}) \tag{22}$$

where

$$\delta = v\eta = \sum_{i=1}^{m+1} v_i \eta_i \tag{23}$$

That is, vE is a row vector differing from v in only one element, δ, which is the scalar product of v with the nonunit column η. Thus computation of π when the product form contains k elementary matrices requires the evaluation of k scalar products.

Computing the Transformed Column, $\bar{P}_s$. The vector $\bar{P}_s$ is given by

$$\bar{P}_s = B^{-1}P_s \tag{24}$$

$$\bar{P}_s = E_k(\cdots(E_2(E_1P_s))\cdots) \tag{25}$$

Here the multiplications proceed in forward sequence, first E_1P_s, then $E_2(E_1P_s)$, etc. Each operation requires the evaluation of a product of the form Ev. If E and v are as in (20)–(21) with v now a column vector

$$Ev = (\alpha_1, \ldots, \alpha_{m+1})' \tag{26}$$

where

$$\alpha_i = \begin{cases} v_i + \eta_i v_r, & i = 1, \ldots, m+1, \quad i \neq r \\ \eta_r v_r, & i = r \end{cases} \tag{27}$$

Thus evaluating Ev requires $m + 1$ multiplications and m additions. Since, on digital computers, the add time is significantly less than the multiply time, the time required is essentially that for $m + 1$ multiplications, equal to that in evaluating vE.

Updating the Inverse. This updating requires only adding a new eta vector of the form (14)–(15) to the list currently in storage, the elements of this vector being derived from the current $\bar{P}_s$.

Relative Efficiencies and Advantages. The revised simplex method, especially with product form, has a number of advantages over the standard form. These relate to accuracy, speed, and storage requirements. Most of them are based on the fact that large linear programs (e.g., with $n > m > 100$) are almost always sparse, i.e., contain a small percentage of nonzero matrix elements. Densities of 5 per cent or less are common. The revised form of the simplex method is better able to take advantage of this fact. It computes the relative costs $\bar{c}_j$ and transformed vector $\bar{P}_s$ directly from the original problem data. Since this is sparse, and multiplications only need be performed when both operands are nonzero, computation time is significantly reduced. In addition, use of the original data tends to reduce cumulative roundoff error. In contrast, the standard simplex tableau, even if originally sparse, tends to fill up with nonzero elements rapidly as the iterations progress. Computational speed is thus reduced and, since each tableau is computed from the previous one, cumulative error can become more serious.

A tabulation of the number of multiplications and divisions per iteration for the three forms of the simplex method is revealing (operations common to all three are omitted). The standard form does its arithmetic while pivoting in the entire $(m + 1) \times n$ tableau. The operation count, assuming only nonbasic columns are transformed, is as follows:

1. Divide row r by $\bar{a}_{rs}$: $n - m$ divisions.
2. For $i = 1, \ldots, m + 1$, $i \neq r$, multiply row r by $-\bar{a}_{is}$ and add to row i: $m(n - m)$ multiplications:

$$\text{total} = n_s = (m + 1)(n - m) \text{ operations} \tag{28}$$

No reduction factor for zero elements has been included, the assumption being that enough iterations have been performed so that the tableau is essentially full.

For the revised method with inverse in explicit form, assume that the original problem data has nonzero density d, but that B^{-1} is full. Pricing out the $n - m$ nonbasic columns requires one scalar product, πP_j, for each column, $dm(n - m)$ operations. Having selected P_s to enter, computing

$\bar{P}_s = B^{-1}P_s$ requires dm^2 operations. Pivoting on B^{-1} requires $(m + 1)^2$ operations, so the total is

$$n_{re} = dm(n - m) + dm^2 + (m + 1)^2 \tag{29}$$

If the inverse is stored in product form, computation time depends on the number of eta vectors present and on their sparseness. As we will show, nonzero densities for these vectors significantly less than one may be achieved by efficient inversion routines. Thus let us assume an average density (over both vectors and time) of d_1 for the eta vectors, and let the average number of eta vectors present[1] be m. Then, in computing the simplex multipliers, m scalar products are formed, $d_1m(m + 1)$ operations. Pricing requires $dm(n - m)$. Computing $B^{-1}P_s$ requires m scalar products, $d_1m(m + 1)$ operations, while computing the new eta vector involves $m + 1$. The total is

$$n_{rp} = d_1m(m + 1) + dm(n - m) + d_1m(m + 1) + m + 1 \tag{30}$$

Note that, if d_1 and d are unity, the standard method requires less work than the other two. However, for problems with low density, the revised forms become more efficient. Considering the revised method with explicit inverse, the relation $n_{re} < n_s$ is

$$dm(n - m) + dm^2 + (m + 1)^2 < (m + 1)(n - m) \tag{31}$$

or, neglecting quantities of order less than nm or m^2,

$$dmn < nm - 2m^2 \tag{32}$$

or

$$d < 1 - 2\left(\frac{m}{n}\right) \tag{33}$$

For a ratio $n/m > 4$, $d < 50$ per cent, which is usually achieved for large m. In comparing the explicit and product forms, if terms of order less than m^2 are dropped we have

$$n_{re} \doteq dm^2 + dm(n - m) + m^2 = m^2(1 + d) + dm(n - m) \tag{34}$$

$$n_{rp} \doteq d_1m^2 + dm(n - m) + d_1m^2 = m^2(2d_1) + dm(n - m)\,. \tag{35}$$

If

$$1 + d > 2d_1 \tag{36}$$

or

$$d_1 < \frac{1 + d}{2} \tag{37}$$

[1] When the number gets much above m, the basis is reinverted, as is discussed shortly.

Then the product form would be expected to require fewer operations per cycle. For d small this says essentially $d_1 < \frac{1}{2}$. In sparse problems this can usually be attained by periodic reinversions.

The storage advantages of the revised forms are fairly evident. Only B^{-1} need be stored, rather than the entire transformed tableau. Further, in large problems, B^{-1} must be kept in some form of backup storage, disk, tape, etc. Transferral of information from these devices to core memory is currently a "bottleneck" operation, significantly slower than arithmetic speeds. Thus transferring B^{-1} rather than the entire tableau saves time as well as storage. In product-form codes where the eta vectors are kept sparse by periodic reinversions, storage demands are further reduced. Wolfe and Cutler [4] report that, using a product-form code, a problem with 245 rows could be solved all in core on a 7090 with 32,768 words of core. A similar code with explicit inverse would require 75,000 core words, and the standard form would use 118,000 words.

Multiple Pricing. The fact that the relative costs $\bar{c}_j$ must be computed separately in the revised simplex method yields a great deal of flexibility. In product-form codes, computation time depends largely on the number of times the complete problem data (A, b, c) must be consulted. This is certainly true when the data are kept in auxiliary storage, since it must then be transferred to core each time. Even in all in core routines, the bulk of the labor occurs in computing the multipliers, pricing, and finding $\bar{P}_s$, all of which require the original data. Let each referral to original problem data be termed a "pass" [4]. The number of passes can be reduced by the following tactic, called multiple pricing or suboptimization. Compute the multipliers π, and find the nonbasic columns with the k most negative reduced costs, $k > 1$. Transform these columns, and form a restricted problem (in core) involving only the current basic variables and these k. Apply the standard simplex method to this problem, again in core. When the restricted problem is solved, begin another pass.

Solving the restricted problem requires no referrals to the original data and can yield large decreases in objective value. The number of passes, hence overall solution time, can be significantly reduced [5]. In the current IBM MPS-360 linear programming code [6], k is an integer between 1 and 10. Too high a value for k can slow convergence by wasting time on columns which are relatively unprofitable.

Multiple pricing was considered earlier, in Section 4.2, where it produced excellent results.

Reinversion. When using the product form, a new eta vector is added to the current set at each step. The more eta vectors we have in this set, the greater will be the storage requirements and the longer the computing time per cycle. These problems are compounded by the fact that the eta vectors generated in later iterations tend to contain more nonzero elements. Thus, after a certain number of cycles, it is desirable to compact the representation of the current inverse by reinversion, using the original problem data. This has the additional advantage of eliminating cumulative error. Efficient routines exist which can significantly reduce both the number and density of the eta vectors. Such a routine is a key element in an efficient product-form code.

Efficient reinversion algorithms are based on the fact that, although the basis inverse is unique, its representation as a product of elementary matrices is not. Thus a representation can be sought which has certain desirable features, e.g., is sparse. A simple but effective procedure has been suggested by Larsen [7], modifications of which are contained in a number of current operating codes, e.g., LP/90 for IBM 7090 computers [5].

Let the basis matrix to be inverted be

$$B = [P_1, P_2, \ldots, P_{m+1}] \tag{38}$$

Larsens procedure for computing B^{-1} is as follows:

1. Initialize $i = 1$.
2. Choose P_i as the vector to enter the basis. Multiply P_i by the current product of elementary matrices generated thus far, yielding $\overline{P}_i$.
3. Select a pivot row in $\overline{P}_i$. First determine the admissible pivot rows, those which have not previously been selected, and in which the elements of $\overline{P}_i$ have absolute value greater than some tolerance T (a small positive number; Larsen chooses $T = 10^{-3}$). Then, from this admissible set, choose the pivot row as the one in which the matrix $[P_i \cdots P_{m+1}]$ has a minimal number of nonzero elements.
4. Knowing the pivot row and the elements of $\overline{P}_i$, form a new eta vector using (15), and add it to the current set.
5. Replace i by $i + 1$ and return to step 2.

In this routine, vectors are chosen to enter the basis in the order that they appear in storage, first P_1, then P_2, etc. At the end of cycle i, the current eta vectors represent the inverse of a matrix containing $P_1, \ldots, P_i$ and unit vectors. The vector P_{i+1} is introduced into this matrix by multiplying it by the current inverse (in product form) and pivoting on an admissible pivot element. P_{i+1} must replace some unit vector, and the pivot element must not be so near zero as to seriously impair digital accuracy, hence the tolerance T. Within these restrictions, the pivot choice rule can be rationalized as follows. The eta vector computed from $\overline{P}_s$ with pivot element $\bar{a}_{rs}$ has components

$$\eta_i = \frac{-\bar{a}_{is}}{\bar{a}_{rs}}, \qquad i = 1, \ldots, m + 1, \quad i \neq r \tag{39}$$

$$\eta_r = \frac{1}{\bar{a}_{rs}} \tag{40}$$

The only way to create zero components η_i is to create zero elements $\bar{a}_{is}$. The elements $\bar{a}_{is}$ at cycle i are formed as follows. Let B_c^{-1} be the product of the current elementary matrices, and consider the updated matrix

$$B_c^{-1}[P_i, P_{i+1}, \ldots, P_{m+1}] \tag{41}$$

The $\bar{a}_{is}$ are formed by multiplying some row of this matrix, the pivot row, by a nonzero constant and adding multiples of it to other rows. If the pivot row is the one with minimal number of nonzeros, it is likely that few new nonzero elements will be created. Unfortunately, the columns $B_c^{-1}P_j$, $j = i, \ldots, m + 1$ are not all available (only $B_c^{-1}P_i$ is). Instead, in step 3, the row of

$$[P_i, P_{i+1}, \ldots, P_{m+1}]$$

with minimal nonzeros is chosen, using the heuristic that this row of (41) should also have few nonzero elements.

Larsen reports excellent results with this routine. In the two problems shown in Table 1, the inversion method described here is compared with the standard method, which pivots on the elements with maximum absolute value. Reductions in density and computation time of over 40 per cent are achieved. Larsen states that these data are representative of the overall experience obtained over several months of operation.

TABLE 1

Problem 1	
Rows	354
Vectors	529
Elements	6170
Density	3.29%

	Standard	Modified	% Reduction
Inversion 1 (2333 nonzeros in basis)			
Minutes	3.79	2.05	46
Transformation vector nonzeros	17,950	7,496	53
60 simplex iterations (min)	14.80	9.58	35
Inversion 2 (2084 nonzeros in basis)			
Minutes	3.64	2.06	43
Transformation vector nonzeros	17,382	7,626	56
Total minutes	22.23	11.69	48

Problem 2	
Rows	360
Vectors	1239
Elements	7168
Density	1.61%

	Standard	Modified	% Reduction
Inversion 1 (1060 nonzeros in basis)			
Minutes	1.71	1.43	16
Transformation vector nonzeros	4,368	1,694	61
50 simplex iterations (min)	4.41	3.50	21
Inversion 2 (1117 nonzeros in basis)			
Minutes	1.70	1.53	10
Transformation vector nonzeros	4,868	1,958	60
Total minutes	7.82	6.46	17

Newer Reinversion Procedures[2]. Improved reinversion routines have been developed since the publication of Larsen's procedure in 1962. These use more sophisticated criteria for choosing a pivot element. All, in effect, reorder the basis matrix to be inverted so that it has the form shown in Figure 6-1, where only the shaded portions contain nonzero elements. The $UT \times UT$ and $LT \times LT$ partitions are nonsingular lower triangular matrices, while the $K \times K$ partition is called the kernel. Since the upper right partition is zero, pivots in the upper triangle affect no other rows and columns. Thus, eta vectors representing these pivots are formed and are transferred to auxiliary memory, e.g., disk, tape, etc. Similarly, pivots in the LT rows and columns affect no rows above them. These pivots are deferred until later, for two reasons. They would tend to increase the density of the $LT \times K$ submatrix, causing eta vectors formed from columns intersecting it to have more nonzero elements. Also, the kernel will be triangularized, so the lower triangle will eventually extend to include the kernel.

The procedure described here for pivoting in the kernel is currently used in the "Allegro" linear programming code for Control Data 3600 and 3800 machines. In contrast to Larsen's method, after each elementary matrix is formed it updates (i.e., multiplies by the current elementary matrix) all kernel columns which have not yet been pivoted in. Selection of a pivot element is based on a result due to Markowitz. Let

r_i = number of nonzero elements in row i

k_j = number of nonzero elements in column j

If a pivot is performed using a_{ij} as a pivot element, then multiples of row i are added to $k_j - 1$ other rows and create at most $r_i - 1$ new nonzero

[2] This section was communicated to the author by David Sommer.

elements in each.[3] Thus an upper bound on the number of new nonzero elements in the updated matrix is

$$u_{ij} = (r_i - 1)(k_j - 1) \tag{42}$$

One could then choose as pivot term the nonzero element with smallest u_{ij}. In practice, the search for minimum u_{ij} may be done only approximately.

In implementing these ideas, it is not necessary to actually reorder the matrix to be inverted into the form of Figure 6-1. Define all rows and columns

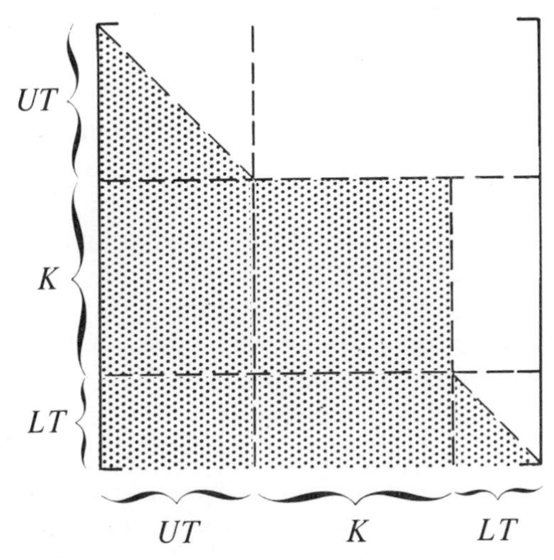

FIGURE 6-1 Upper and lower triangles and kernel.

initially as admissible, let the row count be the number of nonzero elements a row has in admissible columns, and similarly for the column count. To recognize a row belonging to the forward triangle, find one with row count of one, and make this row inadmissible. Let the nonzero element be in column r_1 and make that column inadmissible. Modify the row and column counts to reflect the removal of this row and column from the admissible set. Then any other row with row count of one is also in the forward triangle. Repeating this procedure until all admissible rows have counts of two or more locates all forward triangle rows. The same procedure with the roles of rows and columns reversed locates the columns of the backward triangle. In the example of Figure 6-2, where x represents a nonzero element, row 3 has one nonzero element. A pivot on a_{32} affects no other columns except column 2, so an eta vector is formed from column 2 and is transferred to

[3] Fewer may be created due to the fact that nonzero elements may already exist in these $k_j - 1$ other rows.

	1	2	3	4	5	6	
1			x	x			i
2	x		x		ⓧ		i
3		ⓧ					f
4	x		x				
5	x	x		x		x	i
6	x	x	x		x		
		f		i	i	i	

FIGURE 6-2 Basis matrix.

auxiliary storage. Row 3 and column 2 are labeled inadmissible, (with an f) and column 2 becomes a unit vector. All other rows have row counts of 2 or more, so the forward triangle operations are complete. Turning to the backward triangle, column 6 has a count of one, with the nonzero element appearing in row 5. Labelling row 5 and column 6 inadmissible (with an "i") column 4 has a count of one, with the nonzero in row 1. Making column 4 and row 1 inadmissible, all other columns have counts of 2 or more, so we proceed to triangularize the kernel.

A procedure for selecting a pivot element in the kernel proceeds as follows:

1. Find an admissible row with smallest row count. From the remaining rows, again find a smallest one and repeat the procedure k times (typical value $k = 3$).
2. For each of the k rows found in step one, find an admissible column with minimal column count intersecting (i.e. having a nonzero element in) that row.
3. Repeat steps one and two with the roles of rows and columns reversed.
4. From the $2k$ pairs of row and column indices found in steps one through three, select a pair with smallest u_{ij} value. This determines the pivot element.

Choosing $k = 2$ for our small example yields a number of elements with smallest u_{ij} of 2. Choosing a_{25} arbitrarily, row 2 and column 5 are labeled inadmissible. Since the kernel is being triangularized, the pivot on a_{25} need only normalize a_{25} to unity and drive elements of column 5 which are in admissible rows (here the (6, 5) element) to zero. We form an eta vector representing this transformation, use it to update columns one and three, and store it.

After the pivots on a_{32} and a_{25}, the basis appears as in Figure 6-3. Now all of the four admissible pivot elements remaining have $u_{ij} = 1$, so a_{41} is chosen arbitrarily, and row 4 and column one are marked inadmissible. We form an

$$
\begin{array}{c|cccccc|c}
 & 1 & 2 & 3 & 4 & 5 & 6 & \\
\hline
1 & & & x & x & & & i \\
2 & x & & x & & 1 & & i \\
3 & & 1 & & & & & f \\
4 & \textcircled{x} & & x & & & & i \\
5 & x & & & x & & x & i \\
6 & x & & \textcircled{x} & & & & \\
\hline
 & i & f & & i & i & i &
\end{array}
$$

FIGURE 6-3 Updated basis matrix.

eta vector to make the pivot term unity and the (6, 1) element zero, use it to update column 3, and store it. The (6, 3) entry is the only possible pivot element remaining. Since the only nonzero element of column 3 which is in an admissible row is the pivot element, we need only divide row 6 by this element. Forming and storing an eta vector which does this completes the kernel operations.

The resulting basis matrix appears in Figure 6-4. The submatrix composed of

$$
\begin{array}{c|cccccc|c}
 & 1 & 2 & 3 & 4 & 5 & 6 & \\
\hline
1 & & & x & x & & & i \\
2 & x & & x & & 1 & & i \\
3 & & 1 & & & & & f \\
4 & 1 & & x & & & & i \\
5 & x & & & x & & x & i \\
6 & & & 1 & & & & i \\
\hline
 & i & f & i & i & i & i &
\end{array}
$$

FIGURE 6-4 Triangularized basis matrix.

the columns and rows labeled "i" is lower triangular. The final stage of the inversion process applies the forward triangle operations outlined earlier to this submatrix. Generating and storing the appropriate eta vectors completes the procedure.

In the actual "Allegro" code, the elements of the basis matrix are stored in a doubly threaded list, with a pointer going from each nonzero element a_{ij} to this core addresses of the next nonzero elements in row i and in column j. An element is defined to be nonzero if its absolute value is greater than some positive ϵ, a typical value being 10^{-11}. More stringent nonzero criteria must be adopted when selecting a pivot element. Assume a pivot is to be made in column s, whose components are $\bar{a}_{is}$. Compute

$$p = \max_{i} |\bar{a}_{is}| \tag{43}$$

where i ranges over the set of currently admissible rows. Then an element $\bar{a}_{is}$ is deemed large enough to pivot on only if

$$|\bar{a}_{is}| \geq \delta \tag{44}$$

and

$$|\bar{a}_{is}| \geq \gamma p \tag{45}$$

with typical values for δ and γ being 10^{-6} and 10^{-2} respectively. Relation (45) protects against the generation of very large elements in the eta vectors. If a pivot element chosen in the forward triangle fails to meet this criterion, the matrix being inverted is judged to be singular. The column being pivoted in is replaced by an artificial unit vector column, and a Phase I is initiated. If the test is failed elsewhere, the next best pivot element in the column (in terms of smallest row count) is selected. If all potential pivot elements are less than δ, a Phase I is again initiated.

In operating on the kernel, once a column has been pivoted in, its nonzero elements in inadmissible rows may be moved to auxiliary storage to await the lower triangle operations. Despite this, the list of matrix elements in admissible rows and columns may be initially too large for core memory or may later come to be too large due to the creation of new nonzero elements. At any stage of operations on the kernel, if the list of matrix elements exceeds available core memory, the columns selected to reside outside of core are those with largest column counts. This is based on the heuristic that the pivot selection rule will select them last. These columns are not updated until there is room for them in core. Thus, at least in some iterations, they play no role in selecting a pivot element. This can lead to bad pivot selections; the procedure often yields poor results when more than one core load of columns must reside in auxiliary storage.

It is worth noting that the number of new nonzero elements formed by a given pivot (disregarding cancellation of existing elements) can be computed without excessive difficulty if the updated matrix is available. An option which selects the pivot element for which this number is smallest is included in the Allegro code. However, results using it have generally not been significantly better than those obtained using the upper bound u_{ij}.

6.3 Upper Bounding Methods

Linear programs in which some of the variables have upper bounds occur often in practice. We assume here, for simplicity, that all the variables have such bounds, so the program has the form

$$\text{minimize } z = cx \tag{1}$$

subject to

$$Ax = b \tag{2}$$

$$x \geq 0 \tag{3}$$

$$x \leq u \tag{4}$$

where A is $m \times n$ of rank m, and u is a positive n vector. The results which follow are easily modified when only some of the variables are bounded.

The most straightforward way of dealing with such a program is to convert the bounds (4) into equalities through slack variables s:

$$x + s = u, \qquad s \geq 0 \tag{5}$$

These are then taken together with the equalities (2), and the program (1)–(3) and (5) is solved by the simplex method. Doing this increases both the size of the basis matrix and the number of program variables by n. Since $n > m$, and often $n \gg m$, this represents a serious increase in computational requirements.

A more attractive alternative lies in special techniques developed for such problems. By taking advantage of the special structure of the constraints (4), these upper bounding procedures solve (1)–(4) while maintaining a basis inverse of dimension only $m \times m$. The computations proceed as in the simplex method for problems without bounds, but the optimality test and the criteria for entering and leaving the basis are modified. In addition, the values of the independent or nonbasic variables are not zero, but are either zero or u_j. As we shall see, the operations of the simplex method are not seriously affected by permitting positive values for nonbasic variables.

Theoretical Basis of the Method. The algorithm is based on two main theorems. For convenience, we define the constraint set of (1)–(4) as

$$S = \{x \mid Ax = b, 0 \leq x \leq u\} \tag{6}$$

THEOREM 1. *A vector $x \in S$ is an extreme point of S if and only if $n - m$ components of x are either zero or u_j while the remaining m are between zero and u_j and are associated with a basis matrix for* (2).

PROOF. $\Rightarrow$: Rewriting (2), (3), and (5) in component form,

$$
\begin{array}{llllll}
P_1x_1 + P_2x_2 + \cdots + P_nx_n & & & & = b \\
x_1 & & + s_1 & & = u_1 \\
& x_2 & & + s_2 & = u_2 \\
& \ddots & & \ddots & \vdots \\
& & x_n & & + s_n = u_n \\
\end{array}
\tag{7}
$$

$$x_i \geq 0, \qquad s_i \geq 0$$

If (x, s) is an extreme point solution of the above, then x is an extreme point of S. A vector (x, s) is an extreme point solution for (7) if and only if it is a basic feasible solution, i.e., is nonnegative, has $n - m$ variables equal to zero, and the remainder are associated with a nonsingular submatrix. Since each u_i is positive, each of the last n constraints must contain at least one positive variable. Thus only m of these constraints can contain two basic variables, leaving $n - m$ with exactly one. The value of x_j in each of these is either zero or u_j.

Consider now those of the last n constraints containing two basic variables. For simplicity, let these be the constraints $m + 1, \ldots, 2m$. Then the columns of the basis include the vectors

$$
\left\{\begin{bmatrix} P_1 \\ \hline e_1 \end{bmatrix}, \ldots, \begin{bmatrix} P_m \\ \hline e_m \end{bmatrix}\right\} \quad \text{and} \quad \left\{\begin{bmatrix} 0 \\ \hline e_1 \end{bmatrix}, \ldots, \begin{bmatrix} 0 \\ \hline e_m \end{bmatrix}\right\} \tag{8}
$$

with e_i the ith unit vector. For these vectors to be independent, $\{P_1, \ldots, P_m\}$ must be independent, so the forward implication is proved.

$\Leftarrow$: Let the variables associated with a basis matrix for (2) be $x_1, \ldots, x_m$ and consider the $n + m$ square submatrix of (7):

$$
\begin{array}{ccc}
s_1 \cdots s_m & x_{m+1} \cdots x_n & x_1 \cdots x_m \\
\end{array}
$$
$$
\left[\begin{array}{c|c|c}
0 \cdots 0 & P_{m+1} \cdots P_n & P_1 \cdots P_m \\
\hline
e_1 \cdots e_m & e_{m+1} \cdots e_n & e_1 \cdots e_m
\end{array}\right] \tag{9}
$$

By subtracting the s_i columns from the x_i columns for $i = 1, \ldots, m$, this matrix is transformed into

$$
\left[\begin{array}{cc|c}
0 \cdots 0 & P_{m+1} \cdots P_n & P_1 \cdots P_m \\
\hline
e_1 \cdots & \cdots e_n & 0
\end{array}\right] \tag{10}
$$

These two matrices have the same rank. Since $P_1, \ldots, P_m$ are independent, the above matrix is nonsingular. Hence the variables $(x_1, \ldots, x_n, s_1, \ldots, s_m)$ are a set of basic variables for (7), and $(x_1, \ldots, x_n)$ is an extreme point of S.

Assume now that we have an extreme point of S. The following theorem shows how this solution may be tested for optimality.

THEOREM 2. *Let the basic variables of Theorem* 1 *be* $x_1, \ldots, x_m$ *and let these variables be eliminated from the z equation, so that*

$$z = z_0 + \sum_{j=m+1}^{n} \bar{c}_j x_j \tag{11}$$

with $\bar{c}_j$ *the relative cost factors. Then this basic solution is optimal for* (1)–(4) *if and only if*

$$\left.\begin{aligned} x_j = 0 &\Rightarrow \bar{c}_j \geq 0 \\ x_j = u_j &\Rightarrow \bar{c}_j \leq 0 \end{aligned}\right\} j = m + 1, \ldots, n \tag{12}$$

PROOF. The above expression for z is independent of $x_1, \ldots, x_m$. For $j = m + 1, \ldots, n$, if $x_j = 0$, increasing x_j can only increase z, while if $x_j = u_j$, decreasing x_j can only increase z. Thus the current point is a local, hence global, minimum.

Computational Procedure. Assume that the system (1)–(3) is in canonical form relative to basic variables $x_1, \ldots, x_m$:

$$\begin{array}{llllll} x_1 & & + \bar{a}_{1,m+1}x_{m+1} & + \cdots + & \bar{a}_{1n}x_n & = \bar{b}_1 \\ & \ddots & \vdots & & \vdots & \quad \vdots \\ & x_m & + \bar{a}_{m,m+1}x_{m+1} & + \cdots + & \bar{a}_{mn}x_n & = \bar{b}_m \\ & & -z + \bar{c}_{m,m+1}x_{m+1} & + \cdots + & \bar{c}_n x_n & = -z_0 \end{array} \tag{13}$$

In the above, $x_{m+1}, \cdots, x_n$ are either zero or u_j. The values of $x_1, \ldots, x_m$ are obtained by substituting the values of $x_{m+1}, \ldots, x_n$ into (13). Thus the $\bar{b}_i$ are not the current values of the basic variables, and may be negative or greater than u_i.

STEP 1. *Choosing a Variable to Enter the Basis.* If the optimality test is not passed, there is a nonbasic variable x_k such that either

$$\text{(a) } x_k = 0 \text{ and } \bar{c}_k < 0 \tag{14}$$

or

$$\text{(b) } x_k = u_k \text{ and } \bar{c}_k > 0 \tag{15}$$

If $\bar{c}_k$ is replaced by $-\bar{c}_k$ in case (b), then choose x_s to enter the basis, where

$$\min_j \bar{c}_j = \bar{c}_s \tag{16}$$

If such a variable can enter at a positive level, z must decrease.

STEP 2. *Choosing a Variable to Leave the Basis.* As x_s increases from zero (decreases from u_s) the values of the basic variables also change. There are two possibilities.

CASE 1. x_s reaches its upper (lower) bound without any basic variable violating its upper or lower bound.

CASE 2. Some basic variable reaches one of its bounds before x_s does.

In case 1, set x_s to its upper (lower) bound, record the new values of the basic variables, and return to step 1 without changing the basis. After perhaps several repetitions of this case, we must reach either optimality or case 2.

In case 2, let x_r be one of the basic variables to reach its upper or lower bound first. Perform a basis change, with x_s entering and x_r leaving, and return to step 1.

Let us assume that x_s is decreasing from its upper bound, u_s. From (13) the basic and nonbasic variables are related by

$$x_i = \bar{b}_i - \sum_{j=m+1}^{n} \bar{a}_{ij} x_j, \qquad i = 1, \ldots, m \tag{17}$$

Let

$$R = \{j \mid x_j = u_j, j \in \{m + 1, \ldots, n\}\} \tag{18}$$

Then the current values of the basic variables are

$$x_i = \bar{b}_i - \sum_{j \in R} \bar{a}_{ij} u_j = x_B^i \tag{19}$$

Consider now decreasing some x_s with $s \in R$ from its upper bound. By (17), the resulting values of x_i are

$$x_i = \bar{b}_i - \sum_{\substack{j \in R \\ j \neq s}} \bar{a}_{ij} u_j - \bar{a}_{is} x_s \tag{20}$$

or, adding and subtracting $\bar{a}_{is} u_s$,

$$x_i = x_B^i + \bar{a}_{is}(u_s - x_s) \tag{21}$$

If $\bar{a}_{is} > 0$, when x_s decreases x_i increases, so x_i is in danger of violating its upper bound. Since we wish

$$x_B^i + \bar{a}_{is}(u_s - x_s) \leq u_i \tag{22}$$

then

$$x_s \geq u_s + \frac{x_B^i - u_i}{\bar{a}_{is}}, \qquad \bar{a}_{is} > 0 \tag{23}$$

so the smallest allowed value of x_s is

$$\theta_1 = \max_{\bar{a}_{is} > 0} \left\{ u_s + \frac{x_B^i - u_i}{\bar{a}_{is}} \right\}, \qquad \text{attained for } i = r_1 \tag{24}$$

If $\bar{a}_{is} < 0$, when x_s decreases x_i decreases, so x_i is in danger of violating its lower bound. Since we wish

$$x_B^i + \bar{a}_{is}(u_s - x_s) \geq 0, \qquad \bar{a}_{is} < 0 \tag{25}$$

then

$$x_s \geq u_s + \frac{x_B^i}{\bar{a}_{is}} \tag{26}$$

The smallest allowed value of x_s here is

$$\theta_2 = \max_{\bar{a}_{is} < 0} \left\{ u_s + \frac{x_B^i}{\bar{a}_{is}} \right\}, \qquad \text{attained for } i = r_2 \tag{27}$$

Since x_s must also be nonnegative, its new value is

$$x_s = \theta = \max\{0, \theta_1, \theta_2\} \tag{28}$$

(a) If $\theta = 0$, there is no basis change. The new values of the basic variables are, by (21),

$$x_i = x_B^i + \bar{a}_{is} u_s, \qquad i = 1, \ldots, m \tag{29}$$

(b) If $\theta = \theta_1$, x_{r_1} leaves the basis.
(c) If $\theta = \theta_2$, x_{r_2} leaves the basis.

In both cases (b) and (c), new values for the basic variables are computed from (21). Ties are resolved arbitrarily, but always in favor of (a).

It is easy to show that the analogous relations for the case where x_s is currently zero and increasing are

$$x_s = \theta = \min\{u_s, \theta_1, \theta_2\} \tag{30}$$

where

$$\theta_1 = \min_{\bar{a}_{is} > 0} \frac{x_B^i}{\bar{a}_{is}}, \qquad \text{achieved for } i = r_1 \tag{31}$$

$$\theta_2 = \min_{\bar{a}_{is} < 0} \frac{x_B^i - u_i}{\bar{a}_{is}}, \qquad \text{achieved for } i = r_2 \tag{32}$$

(a) If $\theta = u_s$, there is no change of basis.
(b) If $\theta = \theta_1$, x_{r_1} leaves the basis.
(c) If $\theta = \theta_2$, x_{r_2} leaves the basis.

If a basis change is in order, the vector $\bar{P}_s = B^{-1}P_s$ is computed, and the current basis inverse is transformed by a pivot in either row r_1 or r_2 of $\bar{P}_s$. Assuming that the values of the basic variables have been updated, a new cycle now begins.

6.4 Generalized Upper Bounding

6.4.1 Development of the Algorithm. This section deals with linear programs of the form

$$\text{minimize } x_0$$

subject to

$$\begin{array}{l} m \text{ rows}\left\{ \overbrace{A^0x_0 + \cdots + A^{n_0}x_{n_0}}^{S_0} + \overbrace{A^{n_0+1}x_{n_0+1} + \cdots + A^{n_1}x_{n_1}}^{S_1} + \cdots + \overbrace{A^{n_{p-1}+1}x_{n_{p-1}+1} + \cdots + A^{n_p}x_{n_p}}^{S_p} = b \right. \quad (1) \\ p \text{ rows}\left\{ \begin{array}{lr} x_{n_0+1} + \cdots + x_{n_1} & = 1 \\ \qquad x_{n_1+1} + \cdots + x_{n_2} & = 1 \\ \qquad\qquad \ddots & \\ \qquad\qquad\qquad x_{n_{p-1}+1} + \cdots + x_{n_p} & = 1 \end{array} \right. \quad (2) \end{array}$$

$$x_j \geq 0, \qquad \text{all } j$$

where the columns A^j and the vector b have m components. Transportation and generalized transportation problems have this structure, as do Dantzig–Wolfe master programs (see Chapter 3). Any linear program in which each variable has at most one nonzero coefficient in the last p equations, with this coefficient and the right-hand side of the equation both positive, may be placed in this form by dividing each equation by its right-hand side and scaling the variables. Modifications for negative coefficients are considered later.

The procedure studied here, described by Dantzig and Van Slyke in [2], is a specialization of the simplex method which solves such programs while maintaining a "working basis" of dimension $m \times m$. This is a substantial saving when p is large compared to m. All the quantities needed to carry out an iteration of the simplex method, e.g., the values of the basic variables, simplex multipliers, and the representation of the incoming column in terms of the basic columns, are derived from the corresponding quantities associated with the working basis. This generalized upper bounding algorithm has been referred to earlier, in Section 4.2, where it was used to reduce the size of the basis matrix in a multi-item scheduling problem.

Let the set of columns with unity as their $(m + i)$th coefficient be S_i, $i = 1, \ldots, p$ [see (1)–(2)]. The set of columns with zeros as their $(m + 1)$st to $(m + p)$th coefficients is S_0. These sets will also be used for the variables associated with these columns and for the indices of these variables. Vectors with $m + p$ components are denoted by bold face type, so the jth column of the system (1)–(2) is $\mathbf{A}^j$, while the corresponding vector with the last p components deleted is A^j.

To develop the technique, two theorems are needed.

THEOREM 1. *Any feasible basis for* (1)–(2) *must include at least one column from each set* S_i, $i = 0, \ldots, p$.

PROOF. The variable $x_0 \in S_0$ is always basic. Since, in each of the other sets, the variables are summed to unity and are nonnegative, at least one must be positive in any basic solution.

THEOREM 2. *The number of sets containing two or more basic variables is at most* $m - 1$.

PROOF. For any set of $m + p$ basic variables, each set S_i, $i = 0, \ldots, p$, must contain at least one. Allocating these to their respective sets leaves $m - 1$ variables. If each is in a different set, then $m - 1$ sets have two basic variables; otherwise fewer than $m - 1$ sets have two or more.

Assume that an initial feasible basis, $\mathbf{B}$, for the program (1)–(2) is available:

$$\mathbf{B} = [\mathbf{A}^{j_1}, \mathbf{A}^{j_2}, \ldots, \mathbf{A}^{j_{m+p}}] \tag{3}$$

Such a basis may be found by applying the technique to be described in phase 1 of the simplex method. Since each set must have at least one column in $\mathbf{B}$, we can choose one basic column from each set $S_1, \ldots, S_p$ and call it the *key column*, the associated variable being the key variable. S_0 has no key column. If the key column in S_i is denoted by $\mathbf{A}^{k_i}$, then the basis in (3) may be partitioned as shown in Figure 6-5, with $I_{p \times p}$ the identity matrix of dimension p. The m vectors in the partition $C_{p \times m}$ are either zero vectors

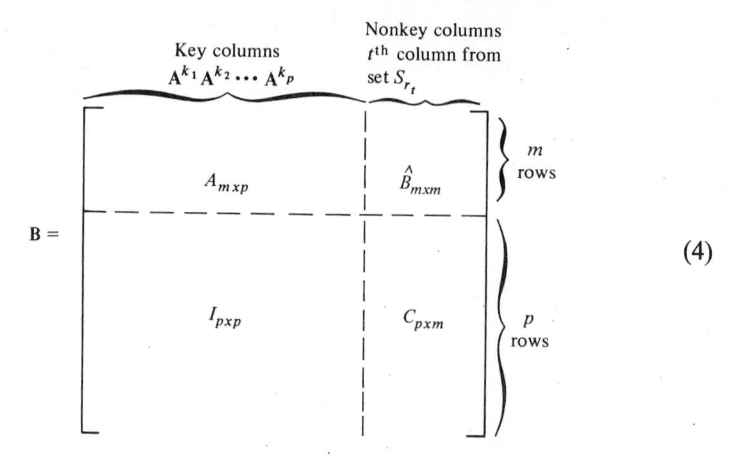

(4)

FIGURE 6-5

(for columns in S_0) or unit vectors. The $(p + t)$th column of **B** is from the set S_{r_t}, and the column $\mathbf{A}^0$ is assumed to be the $(m + p)$th column of **B**. Thus, if each of the key columns $\mathbf{A}^{k_i}$ is subtracted from each of the other basic columns in its set, $C_{p \times m}$ is reduced to a matrix of zeros, and **B** becomes upper block triangular. Let **T** be the matrix which, when multiplying **B** on the right, performs this subtraction. Since

$$\mathbf{B}\mathbf{x}_B = \mathbf{b} \tag{5}$$

if

$$\mathbf{x}_B = \mathbf{T}\mathbf{y}_B \tag{6}$$

then

$$[\mathbf{B}\mathbf{T}]\mathbf{y}_B = \mathbf{b} \tag{7}$$

If **BT** is to have the form

$$\mathbf{BT} = \begin{bmatrix} A_{m \times p} & B_{m \times m} \\ I_{p \times p} & O_{p \times m} \end{bmatrix} \tag{8}$$

then **T** must have the form

$$\mathbf{T} = \begin{bmatrix} I_{p \times p} & -C_{p \times m} \\ O_{m \times p} & I_{m \times m} \end{bmatrix} \tag{9}$$

Since $\det(\mathbf{T}) = 1$, **T** is nonsingular.

If, in the transformed system (7), all key columns are brought to the right-hand side, the remaining system is

$$B_{m \times m} \begin{bmatrix} y_B^{p+1} \\ \vdots \\ y_B^{p+m} \end{bmatrix} = d \tag{10}$$

where

$$d = b - \sum_i A^{k_i} \tag{11}$$

Defining

$$D^j = A^j - A^{k_i}, \qquad \text{if } A^j \in S_i \tag{12}$$

then, since $B_{m \times m} \equiv B$ was formed from $\hat{B}_{m \times m}$ by subtracting key columns, the columns of B are given by

$$B = \{D^j \mid A^j \text{ is basic and not key}\} \tag{13}$$

Let us use each of the equations (2), which sum subsets of variables to unity, to solve for the key variable in each set in terms of the nonkey

$$x_{k_i} = 1 - \sum_{j \neq k_i} x_{n_{i-1}+j} \tag{14}$$

Replacing x_{k_i} by the above in the first m equations, (1), yields

$$\sum_j D^j y_j = d \tag{15}$$

This is called the reduced system. Note that the operations leading to this system were also used to obtain the reduced problems in Sections 5.3 and 5.4. Each equation summing variables to unity plays the role of an angular block, and x_{k_i} is the basic or dependent variable in block i.

Clearly, the working basis, B, is composed of columns from the reduced system. We can now prove the following theorem.

THEOREM 3. *The working basis, B, is a basis for the reduced system.*

PROOF. B must be proved nonsingular. If the columns of B are dependent, then so are the last m columns of $\mathbf{BT}$, since they are formed by adding zeros to the columns of B. Thus $\mathbf{BT}$ is singular and, since $\det(\mathbf{T}) = 1$, $\det(\mathbf{BT}) = \det(\mathbf{B}) \det(\mathbf{T}) = 0$, implying $\det(\mathbf{B}) = 0$. But $\mathbf{B}$ is a basis matrix, hence nonsingular, so this is a contradiction.

At this point, we have associated with each feasible basis, **B**, of the original system, a set of key columns, $\{A^{k_i}\}$, and a working basis, B, of reduced dimension. We now show that each operation required in applying the simplex method to the original problem (1)–(2) may be performed by using quantities associated with B.

Computing the Simplex Multipliers. In pricing out the nonbasic columns in (1)–(2), the simplex multipliers $(\pi \mid \mu)$ for the basis **B** are required, with π associated with the first m equations and μ with the last p. These satisfy

$$(\pi \mid \mu)\mathbf{B} = \mathbf{c}_B = (0, 0, \ldots, 0, 1) \tag{16}$$

Multiplying on the right by **T**,

$$(\pi \mid \mu)(\mathbf{BT}) = \mathbf{c}_B\mathbf{T} = (0, 0, \ldots, 0, 1) \tag{17}$$

In partitioned form the above is

$$(\pi \mid \mu)\left[\begin{array}{c|c} A_{m \times p} & B \\ \hline I_{p \times p} & O_{p \times m} \end{array}\right] = (0, 0, \ldots, 0, 1) \tag{18}$$

Since **BT** is block triangular, this system is easily solved, yielding

$$\pi B = \overbrace{(0, \ldots, 0, 1)}^{m \text{ components}} \tag{19}$$

and

$$\pi A^{k_i} + \mu_i = 0, \qquad i = 1, \ldots, p \tag{20}$$

or

$$\mu_i = -\pi A^{k_i}, \qquad i = 1, \ldots, p \tag{21}$$

Since B is nonsingular, (19) implies that π is the mth row of B^{-1}, and μ_i is given in terms of π by (21). Thus if B^{-1} is available, a full set of prices for the original system is easily generated.

Determining the Column to Enter the Basis. This is done in the usual manner, by computing

$$\bar{c}_j = -(\pi \mid \mu)\mathbf{A}^j = -\pi A^j - \mu_i, \qquad \text{if } A^j \in S_i \tag{22}$$

for nonbasic columns $\mathbf{A}^j$. If

$$\min \bar{c}_j = \bar{c}_s \geq 0 \tag{23}$$

the current basic solution is optimal. If not, then column $\mathbf{A}^s$ enters the basis, $\mathbf{B}$. Let $\mathbf{A}^s \in S_\sigma$.

Finding the Representation of the Entering Column in Terms of the Current Basis. To see which column leaves the basis, the vector

$$\bar{\mathbf{A}}_s = \mathbf{B}^{-1}\mathbf{A}^s \tag{24}$$

must be computed. Multiplying both sides of the above relation by $\mathbf{B}$:

$$\mathbf{B}\bar{\mathbf{A}}^s = \mathbf{A}^s \tag{25}$$

We again make a transformation of variables to reduce $\mathbf{B}$ to block triangular form:

$$\bar{\mathbf{A}}^s = \mathbf{T}\mathbf{Z}^s \tag{26}$$

so

$$(\mathbf{BT})\mathbf{Z}^s = \mathbf{A}^s \tag{27}$$

In partitioned form, the above is

$$\left[\begin{array}{c|c} A_{m \times p} & B \\ \hline I_{p \times p} & O_{p \times m} \end{array}\right] \begin{bmatrix} Z_1^s \\ \vdots \\ Z_p^s \\ Z_{p+1}^s \\ \vdots \\ Z_{p+m}^s \end{bmatrix} = \begin{bmatrix} A_s \\ 0 \\ \vdots \\ 1 \\ \vdots \\ 0 \end{bmatrix} \begin{matrix} \\ \\ \\ \leftarrow \text{position } \sigma \\ \\ \\ \end{matrix} \tag{28}$$

The bottom p equations yield

$$Z_i^s = 0, \qquad 1 \leq i \leq p, \quad i \neq \sigma \tag{29}$$

$$Z_\sigma^s = 1 \tag{30}$$

Using these values in the top m equations,

$$A^{k_\sigma} + B \begin{bmatrix} Z_{p+1}^s \\ \vdots \\ Z_{p+m}^s \end{bmatrix} = A^s \tag{31}$$

Defining

$$\bar{D}^s = \begin{bmatrix} Z^s_{p+1} \\ \vdots \\ Z^s_{p+m} \end{bmatrix} = \begin{bmatrix} \bar{D}^s_1 \\ \vdots \\ \bar{D}^s_m \end{bmatrix} \tag{32}$$

(31) becomes

$$B\bar{D}^s = A^s - A^{k_\sigma} \tag{33}$$

which has the solution

$$\bar{D}^s = B^{-1}(A^s - A^{k_\sigma}) \tag{34}$$

The vector $\bar{D}^s$ is simply the representation, in terms of the working basis, of the vector $(A^s - A^{k_\sigma})$ entering the basis of the reduced system. This is easily computed if B^{-1} is available. Then, from (26),

$$\bar{\mathbf{A}}^s = \begin{bmatrix} \bar{A}^s_1 \\ \cdot \\ \cdot \\ \cdot \\ \bar{A}^s_p \\ \hline \bar{A}^s_{p+1} \\ \vdots \\ \bar{A}^s_{p+m} \end{bmatrix} = \left[\begin{array}{c|c} I_{p\times p} & -C_{p\times m} \\ \hline O_{m\times p} & I_{m\times m} \end{array}\right] \begin{bmatrix} 0 \\ \vdots \\ 1 \\ \vdots \\ 0 \\ \hline \bar{D}^s \end{bmatrix} \begin{matrix} \\ \\ \leftarrow \text{position } \sigma \\ \\ \\ \\ \end{matrix} \tag{35}$$

By (35), $\bar{A}^s_i$ is the scalar product of row i of $\mathbf{T}$ with the vector $\mathbf{Z}^s$. Recall that the tth column of $-C_{p\times m}$ has a -1 in position r_t (where column $p + t$ of $\mathbf{B}$ is an element of S_{r_t}) and zeros elsewhere, and is all zeros if $r_t = 0$. Thus, when taking the scalar product of row i of $-C_{p\times m}$ with $\bar{D}^s$, a nonzero product is formed whenever $r_t = i$. If we define the sets

$$R(i) = \{t \mid t \in \{1, 2, \ldots, m\}, r_t = i\}$$

then the $\bar{A}^s_i$ are given by the relation

$$\bar{A}^s_i = \begin{cases} -\sum\limits_{t\in R(i)} \bar{D}^s_t, & 1 \le i \le p, \quad i \ne \sigma \\ 1 - \sum\limits_{t\in R(i)} \bar{D}^s_t, & i = \sigma \end{cases} \tag{36}$$

$$\bar{A}^s_{p+t} = \bar{D}^s_t, \qquad t = 1, \ldots, m \tag{37}$$

If the vector $\bar{D}^s$ is available, the quantities $\bar{A}_i^s$ are easily computed from (36)–(37).

Choosing the Column to Leave the Basis. This is done in the usual way, by computing

$$\min_{\bar{A}_i^s > 0} \frac{\bar{b}_i}{\bar{A}_i^s} = \frac{\bar{b}_r}{\bar{A}_r^s} = \theta \tag{38}$$

with $\bar{b}_i$ the current value for the ith basic variable. If all $\bar{A}_i^s \leq 0$, the program (1)–(2) has an unbounded solution. Otherwise, column r of $\mathbf{B}$, $\mathbf{A}^{j_r}$, leaves the basis. Let $\mathbf{A}^{j_r} \in S_\rho$.

Updating the Values of the Basic Variables. This operation is performed according to the standard formulas

$$\begin{aligned} (\bar{b}_i)_{\text{new}} &= (\bar{b}_i)_{\text{old}} - \theta \bar{A}_i^s, \qquad i = 1, \ldots, m + p, \quad i \neq r \\ (\bar{b}_r)_{\text{new}} &= \theta \end{aligned} \tag{39}$$

Alternatively, the new basic variables may be computed from the values of the basic variables in the reduced system, $B^{-1}d$, via formulas similar to (36)–(37).

Updating the Inverse of the Working Basis. For convenience, (4) and (8) are rewritten here:

$$\mathbf{B} = \left[\begin{array}{c|c} A_{m \times p} & \hat{B}_{m \times m} \\ \hline I_{p \times p} & C_{p \times m} \end{array}\right], \qquad \mathbf{BT} = \left[\begin{array}{c|c} A_{m \times p} & B_{m \times m} \\ \hline I_{p \times p} & O_{p \times m} \end{array}\right] \tag{4), (8}$$

Recall that the column entering $\mathbf{B}$ is $\mathbf{A}^s \in S_\sigma$ and the column leaving is $\mathbf{A}^{j_r} \in S_\rho$. There are two cases to be considered.

CASE 1. $\mathbf{A}^{j_r}$ *is not a key column.* In this case, the column leaving is one of the last m columns of $\mathbf{B}$, say column $p + i_2$. If $\mathbf{A}^{j_r}$ is replaced in $\mathbf{B}$ by $\mathbf{A}^s$ and $\mathbf{B}$ is then transformed to the block triangular form in (8), the only change in the working basis B is that the column of B corresponding to $\mathbf{A}^{j_r}$, $A^{j_r} - A^{k_\rho}$, is replaced by $A^s - A^{k_\sigma}$. Thus, to update B^{-1}, adjoin to it the column

$$\bar{D}^s = B^{-1}(A^s - A^{k_\sigma}) \tag{40}$$

and perform a pivot operation, using component i_2 of $\bar{D}^s$ as pivot element. Note that $\bar{D}^s$ has already been computed in (34). Of course, all other quantities used in the next cycle must be updated as well. These are the columns A^{k_i}, which appear in equations (21) for the simplex multipliers u_i and the indices r_t, which are used to define the sets $R(i)$ in the equation preceding

(36). The only one of these which changes here is the index r_{i_2}, which becomes σ, so column i_2 of C becomes the σth unit vector. Recording this change completes Case 1.

CASE 2. $\mathbf{A}^{j_r}$ *is a key column. Subcase* (a): $\mathbf{A}^{j_r}$ *and* $\mathbf{A}^s$ *are from different sets, i.e.,* $\rho \neq \sigma$. In this subcase, we are replacing the key column in S_ρ with a column, $\mathbf{A}^s$, not in S_ρ. A new key column for S_ρ must be found. By theorem 1, S_ρ must contain at least one basic variable after $\mathbf{A}^{j_r}$ leaves. Thus one of the nonkey columns of $\mathbf{B}$ is in S_ρ. Choose any one such column, say column $p+i_2$, and call it $\mathbf{A}^k$, where $k=p+i_2$. It will become key by interchanging it with $\mathbf{A}^{j_r} \equiv \mathbf{A}^{k_\rho}$. If this interchange is made, and the resulting $\mathbf{B}$ is transformed to the block-triangular form in (8), the columns of B having the form $A^{j_i} - A^{k_\rho}$ with $j_i \in S_\rho$ are changed as follows:

$$A^{j_i} - A^{k_\rho} \leftarrow A^{j_i} - A^k, \qquad j_i \in S_\rho,\ j_i \neq k \tag{41}$$

$$A^k - A^{k_\rho} \leftarrow A^{k_\rho} - A^k \tag{42}$$

These replacements can be accomplished by multiplying $A^k - A^{k_\rho}$ by -1 and adding the results to each column $A^{j_i} - A^{k_\rho}$, $j_i \in S_\rho$. In matrix form

$$B \leftarrow BT_1 \tag{43}$$

where

$$T_1 = \begin{bmatrix} 1 & & & & & & & \\ & 1 & & & & & & \\ & & \ddots & & & & & \\ & & & 1 & & & & \\ 0\cdots -1 \cdots 0 & & & 0 \cdots -1 \cdots 0 \cdots -1 \cdots 0 & & & & \\ & & & & 1 & & & \\ & & & & & \ddots & & \\ & & & & & & & 1 \end{bmatrix} \leftarrow \text{row } i_2 \tag{44}$$

The -1's occur in the columns corresponding to $\mathbf{A}^{j_i} \in S_\rho$. Since the ones in the ρth row of C also appear in these positions, row i_2 of T_1 is the negative of row ρ of C. The matrix B^{-1} is thus replaced by

$$B^{-1} \leftarrow T_1^{-1} B^{-1} \tag{45}$$

Since performing the column transformations represented by T_1 twice leads back to the original matrix B

$$T_1^{-1} = T_1 \tag{46}$$

so

$$B^{-1} \leftarrow T_1 B^{-1} \tag{47}$$

The interchange of $\mathbf{A}^{j_r}$ and $\mathbf{A}^k$ also causes the vector A^{j_r} to be replaced by A^k. Once this change is made, the situation becomes that of Case 1. Applying

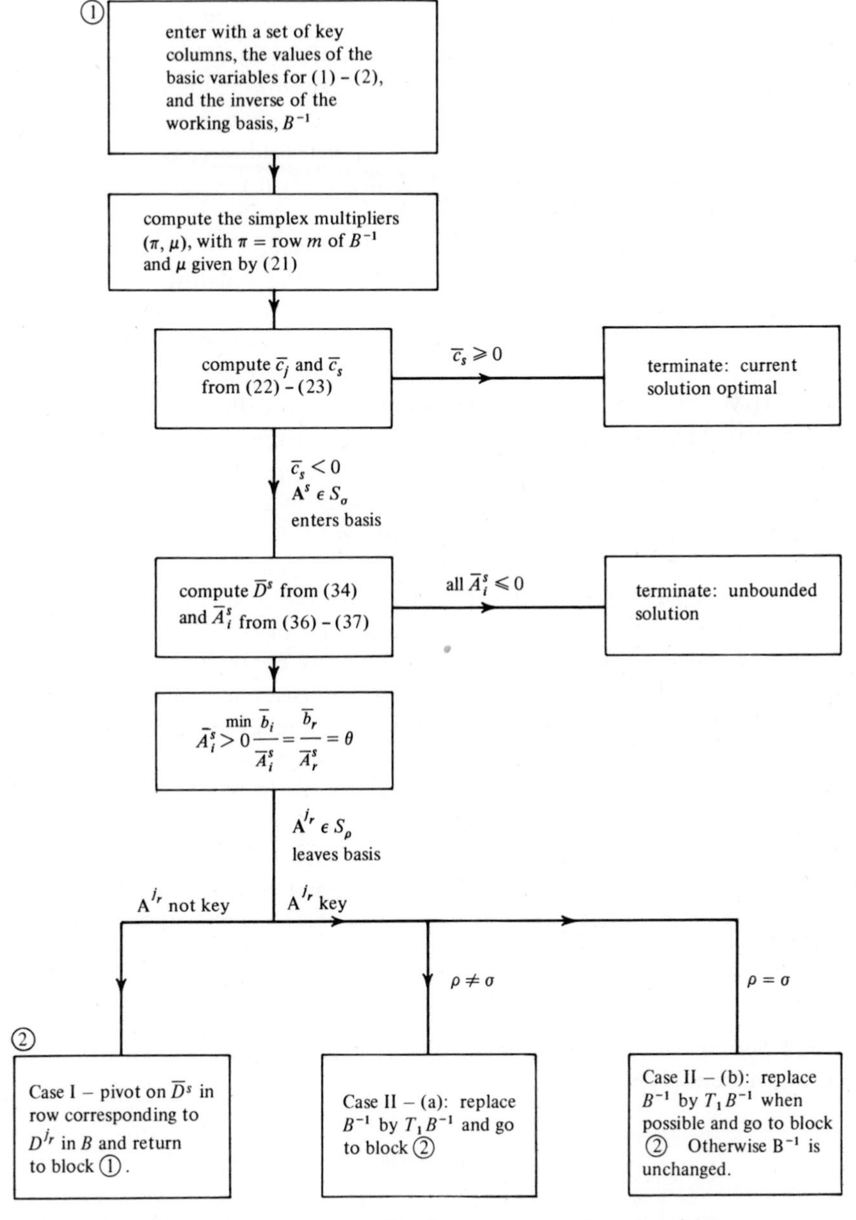

FIGURE 6-6 Flow chart for generalized upper bounding algorithm.

the procedures indicated there to the output of this subcase completes the updating.

Subcase (b): $\rho = \sigma$. Here the key column in S_ρ is being replaced by another column in S_ρ. If there are nonkey columns from S_ρ interchange one of these with $\mathbf{A}^{j_r}$ as described in subcase (a). Then apply the operations of case I. If no nonkey column is from S_ρ, then $\mathbf{A}^s$ replaces $\mathbf{A}^{j_r}$ in $\mathbf{B}$. Let the resulting $\mathbf{B}$ be transformed to the block triangular form of (8). Since the last m columns of $\mathbf{B}$ contain no column from S_ρ, the submatrix B in (8) is unchanged, i.e. the working basis remains the same. The only other change is that A^s replaces A^{j_r} in the upper left partition of $\mathbf{B}$. This completes subcase (b).

Performing the interchange described above when possible simplifies this subcase considerably. When the interchange is possible we go to previous cases and, when not, the working basis is unchanged.

Summary of the Algorithm. After completing the updating of B^{-1}, a new iteration can begin. A flow chart of the procedure is shown in Figure 6-6.

Computational Experience. Dantzig and Van Slyke report [2] that in a problem with 2,813 variables, $p = 780$, $m = 39$, this procedure, implemented on an IBM 7094 computer, required 15 minutes to solve the problem. Iterations were carried out at the rate of 50 per minute. The estimated running time for a general linear programming code on this problem is 150 minutes, so an estimated decrease in computer time by a factor of 10 was achieved.

Modifications for Negative Coefficients in the Last p Equations. By dividing each of the last p equations by its right-hand side and scaling the variables, the right-hand-side coefficients in the last p equations become $+1$, and the coefficients for the variables become $+1$ or -1. Theorems 1 and 2 still hold. Each of the key columns in the basis $\mathbf{B}$ can be made to have a coefficient of $+1$, since each set S_i must have at least one such column in any basis. When forming the working basis, if a nonkey basic column has a coefficient of -1, the key column is added to it rather than subtracted. Theorem 3 still holds for the basis so formed, as do relations (19)–(21) for the simplex multipliers. When pricing out a column with a -1 coefficient, μ_i is added to $-\pi A^j$ rather than subtracted, as in (22). Similar minor modifications must also be made in the formulas for computing $\overline{A}_i^s$ and updating the inverse of the working basis (see the problems at the end of this chapter). Other than these changes, the algorithm operates as before.

6.4.2 Example of the Generalized Upper Bounding Method. This problem is taken from [2]:

$$\text{maximize } x_0$$

subject to (in detached coefficient form)

S_0		S_1		S_2	S_3	S_4		S_5		
$\mathbf{A}^0$	$\mathbf{A}^1$	$\mathbf{A}^2$	$\mathbf{A}^3$	$\mathbf{A}^4$	$\mathbf{A}^5$	$\mathbf{A}^6$	$\mathbf{A}^7$	$\mathbf{A}^8$	$\mathbf{A}^9$	b
1	0	2	0	3	4	5	1	−1	−12	15
1	1	−1	0	2	1	4	2	−3	6	7
0	0	0	1	0	0	0	0	0	0	0
	1	1	1							1
				1						1
					1					1
						1	1			1
								1	1	1
*	*	*	*	*	*	*		*		
	√			√	√	√		√		

with all variables nonnegative. The initial basis is composed of the columns with an $*$. Choose as key columns those with a √. The initial basis in partitioned form is

$$\mathbf{B} = \begin{array}{c} \text{Key columns} \\ \overbrace{\qquad\qquad\qquad\qquad\qquad\qquad}^{} \end{array}$$

$$\mathbf{B} = \left[\begin{array}{ccccc:ccc} \mathbf{A}^1 & \mathbf{A}^4 & \mathbf{A}^5 & \mathbf{A}^6 & \mathbf{A}^8 & \mathbf{A}^2 & \mathbf{A}^3 & \mathbf{A}^0 \\ 0 & 3 & 4 & 5 & -1 & 2 & 0 & 1 \\ 1 & 2 & 1 & 4 & -3 & -1 & 0 & 1 \\ 0 & 0 & 0 & 0 & 0 & 0 & 1 & 0 \\ \hdashline 1 & & & & & 1 & 1 & \\ & 1 & & & & & & \\ & & 1 & & & & & \\ & & & 1 & & & & \\ & & & & 1 & & & \\ \mathbf{A}^{k_1} & \mathbf{A}^{k_2} & \mathbf{A}^{k_3} & \mathbf{A}^{k_4} & \mathbf{A}^{k_5} & & & \end{array}\right]$$

with basic variables

$$\bar{\mathbf{b}} = \mathbf{x}_B = [\tfrac{1}{2} \quad 1 \quad 1 \quad 1 \quad 1 \quad \tfrac{1}{2} \quad 0 \quad 3]$$

Note that $\mathbf{A}^2$ and $\mathbf{A}^3$ are in S_1 and $\mathbf{A}^0$ is in S_0, so

$$r_1 = 1, \qquad r_2 = 1, \qquad r_3 = 0$$

To transform $\mathbf{B}$ to block-triangular form, subtract $\mathbf{A}^1$ from $\mathbf{A}^2$ and $\mathbf{A}^3$. The matrix $\mathbf{T}$ which does this is

$$\mathbf{T} = \left[\begin{array}{ccccc|ccc} 1 & & & & & -1 & -1 & \\ & 1 & & & & & & \\ & & 1 & & & & & \\ & & & 1 & & & & \\ & & & & 1 & & & \\ \hline & & & & & 1 & & \\ & & 0 & & & & 1 & \\ & & & & & & & 1 \end{array}\right]$$

The matrix **BT** is

$$\left[\begin{array}{ccccc|ccc} 0 & 3 & 4 & 5 & -1 & 2 & 0 & 1 \\ 1 & 2 & 1 & 4 & -3 & -2 & -1 & 1 \\ 0 & 0 & 0 & 0 & 0 & 0 & 1 & 0 \\ \hline 1 & & & & & & & \\ & 1 & & & & & & \\ & & 1 & & & & 0 & \\ & & & 1 & & & & \\ & & & & 1 & & & \end{array}\right]$$

The initial working basis is the matrix in the upper right partition of **BT**, i.e.,

$$B = [A^2 - A^1, A^3 - A^1, A^0] = \begin{bmatrix} 2 & 0 & 1 \\ -2 & -1 & 1 \\ 0 & 1 & 0 \end{bmatrix}$$

whose inverse is

$$B^{-1} = \begin{bmatrix} \frac{1}{4} & -\frac{1}{4} & -\frac{1}{4} \\ 0 & 0 & 1 \\ \frac{1}{2} & \frac{1}{2} & \frac{1}{2} \end{bmatrix}$$

Iteration 1

Computing the Simplex Multipliers. The vector π is the bottom row of B^{-1}:

$$\pi = (\tfrac{1}{2}, \tfrac{1}{2}, \tfrac{1}{2})$$

The multipliers μ_i are given by

$$\mu_i = -\pi A^{k_i}, \qquad i = 1, \ldots, 5$$

yielding

$$\mu_1 = -\tfrac{1}{2}, \quad \mu_2 = -\tfrac{5}{2}, \quad \mu_3 = -\tfrac{5}{2}, \quad \mu_4 = -\tfrac{9}{2}, \quad \mu_5 = 2$$

***Pricing Out the Columns* $\mathbf{A}^7$, $\mathbf{A}^9$.** By (22),

$$\bar{c}_j = -\pi A_j - \mu_i, \qquad \text{if } A^j \in S_i$$

so

$$\bar{c}_7 = -\pi A^7 - \mu_4 = -\tfrac{3}{2} - (-\tfrac{9}{2}) = 3$$
$$\bar{c}_9 = -\pi A^9 - \mu_5 = -3 - 2 = -5$$

Since $\bar{c}_7 = \max \bar{c}_j > 0$, $\mathbf{A}^7 \in S_4$ enters the basis.

***Computing* $\bar{\mathbf{A}}^7$.** By (34),

$$\bar{D}^7 = B^{-1}(A^7 - A^6)$$
$$= \begin{bmatrix} \frac{1}{4} & -\frac{1}{4} & -\frac{1}{4} \\ 0 & 0 & 1 \\ \frac{1}{2} & \frac{1}{2} & \frac{1}{2} \end{bmatrix} \begin{bmatrix} -4 \\ -2 \\ 0 \end{bmatrix} = \begin{bmatrix} -\frac{1}{2} \\ 0 \\ -3 \end{bmatrix} = \begin{bmatrix} \bar{D}_1^7 \\ \bar{D}_2^7 \\ \bar{D}_3^7 \end{bmatrix}$$

Using (36)–(37) with $p = 5$, $\sigma = 4$, $r_1 = r_2 = 1$, $r_3 = 0$:

$$\bar{A}_1^7 = -(\bar{D}_1^7 + \bar{D}_2^7) = \tfrac{1}{2}$$
$$\bar{A}_2^7 = \bar{A}_3^7 = \bar{A}_5^7 = 0$$
$$\bar{A}_4^7 = 1$$
$$\bar{A}_6^7 = \bar{D}_1^7 = -\tfrac{1}{2}$$
$$\bar{A}_7^7 = \bar{D}_3^7 = 0$$
$$\bar{A}_8^7 = \bar{D}_3^7 = -3$$

Finding the Variable to Leave the Basis

$$\theta = \min_{\bar{A}_i^7 > 0} \frac{\bar{b}_i}{\bar{A}_i^7} = \min\left(\frac{\bar{b}_1}{\bar{A}_1^7}, \frac{\bar{b}_4}{\bar{A}_4^7}\right) = \min(1, 1) = 1$$

Thus either the first or fourth column of $\mathbf{B}$, $\mathbf{A}^1$ or $\mathbf{A}^6$, leaves the basis. We choose $\mathbf{A}^6 \in S_4$ to leave, since this leaves B^{-1} unchanged for the next iteration.

New Values of Basic Variables. Since $\mathbf{A}^6$ is the fourth column of $\mathbf{B}$,

$$\bar{b}_i \leftarrow \bar{b}_i - \theta \bar{A}_i^7, \qquad i \neq 4$$
$$\bar{b}_4 = \theta$$

so the new basis and basic variables are

$$\mathbf{B} = [\mathbf{A}^1 \quad \mathbf{A}^4 \quad \mathbf{A}^5 \quad \mathbf{A}^7 \quad \mathbf{A}^8 \mid \mathbf{A}^2 \quad \mathbf{A}^3 \quad \mathbf{A}^0]$$
$$\mathbf{x}_B = (0 \quad 1 \quad 1 \quad 1 \quad 1 \quad 1 \quad 0 \quad 6)$$

Updating the Inverse of the Working Basis. Since $\mathbf{A}^6$, the column leaving, is key, and the columns entering and leaving are from the same sets, this is case 2(a) of this section. The working basis is unchanged because no non-key column is from S_4.

Iteration 2

Simplex Multipliers. Since B^{-1} is unchanged,

$$\pi = (\tfrac{1}{2}, \tfrac{1}{2}, \tfrac{1}{2})$$

$$\mu_1 = -\tfrac{1}{2}, \quad \mu_2 = -\tfrac{5}{2}, \quad \mu_3 = -\tfrac{5}{2}, \quad \mu_4 = -\tfrac{3}{2}, \quad \mu_5 = 2.$$

Pricing out $\mathbf{A}^6$, $\mathbf{A}^9$

$$\begin{aligned} \bar{c}_6 &= -\pi A^6 - \mu_4 = -3 \\ \bar{c}_9 &= -\pi A^9 - \mu_5 = 1 \end{aligned}$$

Since $\bar{c}_9 > 0$, $\mathbf{A}^9 \in S_5$ enters the basis.

Computing $\bar{\mathbf{A}}^9$. By (34),

$$\begin{aligned} \bar{D}^9 &= B^{-1}(A^9 - A^8) \\ &= \begin{bmatrix} \frac{1}{4} & -\frac{1}{4} & -\frac{1}{4} \\ 0 & 0 & 1 \\ \frac{1}{2} & \frac{1}{2} & \frac{1}{2} \end{bmatrix} \begin{bmatrix} -11 \\ 9 \\ 0 \end{bmatrix} = \begin{bmatrix} -5 \\ 0 \\ -1 \end{bmatrix} = \begin{bmatrix} \bar{D}_1^9 \\ \bar{D}_2^9 \\ \bar{D}_3^9 \end{bmatrix} \end{aligned}$$

Using (36)–(37) with $p = 5$, $\sigma = 5$, $r_1 = r_2 = 1$, $r_3 = 0$:

$$\begin{aligned} \bar{A}_1^9 &= -(\bar{D}_1^9 + \bar{D}_2^9) = 5 \\ \bar{A}_2^9 &= \bar{A}_3^9 = \bar{A}_4^9 = 0 \\ \bar{A}_5^9 &= 1 \\ \bar{A}_6^9 &= \bar{D}_1^9 = -5 \\ \bar{A}_7^9 &= \bar{D}_2^9 = 0 \\ \bar{A}_8^9 &= \bar{D}_3^9 = -1 \end{aligned}$$

Finding the Variable to Leave the Basis.

$$\theta = \min\left(\frac{\bar{b}_1}{\bar{A}_1^9}, \frac{\bar{b}_5}{\bar{A}_5^9}\right) = \min(0, 1) = 0$$

so $\mathbf{A}^1 \in S_1$ leaves the basis.

Updating B^{-1}. Since the column leaving is key and is from a different set than the column entering we must interchange one of the last three columns of **B** which are in S_1 with $\mathbf{A}^1$. In this case, $\mathbf{A}^1$ may be interchanged with either $\mathbf{A}^2$ or $\mathbf{A}^3$. Let $\mathbf{A}^2$ be arbitrarily chosen. The current working basis

$$B_{\text{old}} = [A^2 - A^1, A^3 - A^1, A^0]$$

will change to

$$\tilde{B} = [A^1 - A^2, A^3 - A^2, A^0]$$

To perform this transformation, multiply $A^2 - A^1$ by -1 and add the result to $A^3 - A^1$; i.e., multiply B_{old} on the right by

$$T_1 = \begin{bmatrix} -1 & -1 & \\ & 1 & \\ & & 1 \end{bmatrix}$$

Then

$$\tilde{B} = B_{\text{old}} T_1$$

so

$$\tilde{B}^{-1} = T_1^{-1} B_{\text{old}}^{-1} = T_1 B_{\text{old}}^{-1}$$

$$= \begin{bmatrix} -\frac{1}{4} & \frac{1}{4} & -\frac{3}{4} \\ 0 & 0 & 1 \\ \frac{1}{2} & \frac{1}{2} & \frac{1}{2} \end{bmatrix}$$

After the interchange, the corresponding basis of the original system is

$$\mathbf{B} = [\mathbf{A}^2 \quad \mathbf{A}^4 \quad \mathbf{A}^5 \quad \mathbf{A}^7 \quad \mathbf{A}^8 \mid \mathbf{A}^1 \quad \mathbf{A}^3 \quad \mathbf{A}^0]$$

Now let $\mathbf{A}^1$ leave this basis and $\mathbf{A}^9$ enter. Since $\mathbf{A}^1$ is no longer key, this is case 1 of the updating procedure, where $\mathbf{A}^9 - \mathbf{A}^8$ replaces column 1 of $\tilde{B}$. To find the new inverse, multiply $A^9 - A^8$ by $\tilde{B}^{-1}$:

$$\tilde{B}^{-1}(A^9 - A^8) = \begin{bmatrix} -\frac{1}{4} & \frac{1}{4} & -\frac{3}{4} \\ 0 & 0 & 1 \\ \frac{1}{2} & \frac{1}{2} & \frac{1}{2} \end{bmatrix} \begin{bmatrix} -11 \\ 9 \\ 0 \end{bmatrix} = \begin{bmatrix} 5 \\ 0 \\ -1 \end{bmatrix}$$

and pivot on this vector in row 1, yielding

$$B_{\text{new}}^{-1} = \begin{bmatrix} -\frac{1}{20} & \frac{1}{20} & -\frac{3}{20} \\ 0 & 0 & 1 \\ \frac{9}{20} & \frac{11}{20} & \frac{7}{20} \end{bmatrix}$$

The new basis for the original problem is

$$\mathbf{B} = [\mathbf{A}^2 \quad \mathbf{A}^4 \quad \mathbf{A}^5 \quad \mathbf{A}^7 \quad \mathbf{A}^8 \mid \mathbf{A}^9 \quad \mathbf{A}^3 \quad \mathbf{A}^0]$$

with basic variables

$$\mathbf{x}_B = [1 \quad 1 \quad 1 \quad 1 \quad 1 \quad 0 \quad 0 \quad 6]$$

Iteration 3

Simplex Multipliers. The new π vector is row 3 of B_{new}^{-1}:

$$\pi = (\tfrac{9}{20}, \tfrac{11}{20}, \tfrac{7}{20})$$

$$\mu_1 = -\tfrac{7}{20}, \quad \mu_2 = -\tfrac{49}{20}, \quad \mu_3 = -\tfrac{47}{20}, \quad \mu_4 = -\tfrac{31}{20}, \quad \mu_5 = \tfrac{42}{20}$$

Pricing out $\mathbf{A}^1$, $\mathbf{A}^6$

$$\begin{aligned} \bar{c}_1 &= -\pi A^1 - \mu_1 = -\tfrac{11}{20} + \tfrac{7}{20} = -\tfrac{4}{20} \\ \bar{c}_6 &= -\pi A^6 - \mu_4 = -\tfrac{99}{20} - \tfrac{31}{20} = -\tfrac{130}{20} \end{aligned}$$

Since all $\bar{c}_i \leq 0$, the current solution is optimal.

6.5 Extension to Angular Structures

Generalized upper bounding techniques have a natural generalization to problems with angular structure. The following is an extension of the work done in [3]. Let the problem be:

$$\text{minimize } (x_0)_1$$

subject to

$$\begin{array}{llr} m_0 \text{ rows } \{ \overbrace{A_0x_0}^{S_0} + \overbrace{A_1x_1}^{S_1} + \overbrace{A_2x_2}^{S_2} + \cdots + \overbrace{A_px_p}^{S_p} = b & & (1) \\ m_1 \text{ rows } \{ \qquad B_1x_1 \qquad\qquad\qquad\qquad = b_1 & & \\ \qquad\qquad\qquad\qquad B_2x_2 \qquad\qquad\quad = b_2 & & \\ \qquad\qquad\qquad\qquad\qquad\qquad \ddots & & (2) \\ m_p \text{ rows } \{ \qquad\qquad\qquad\qquad\qquad B_px_p = b_p & & \\ \qquad x_i \geq 0, \qquad i = 0, 1, \ldots, p & & \end{array}$$

where each A_i is an $m_0 \times n_i$ matrix, each B_i is $m_i \times n_i$, and $(x_0)_1$ is the first component of x_0. The methods to be developed permit the simplex method to be applied to such problems while maintaining a working basis of dimension only $m_0 \times m_0$, plus basis matrices from each B_i. As in the previous section, the procedure may be derived by transforming an arbitrary basis matrix for the angular problem into block-triangular form.

Throughout this section we make the following assumption:

Assumption 1. The system (1)–(2) is of full rank.

Note that this implies that each B_i has rank m_i [otherwise some rows of the system (2) would be linearly dependent]. Hence $n_i \geq m_i$, and we may as well assume $n_i > m_i$, since otherwise the variables x_i could be eliminated. As before, it is convenient to define S_i as the set of columns or variables (depending on context) of the angular problem having components in common with B_i. The set S_0 contains those columns with zeros everywhere but in their first m_0 components.

The method is based on the following theorem.

THEOREM 1. *Any basis matrix,* **B**, *for the angular problem* (1)–(2) *can be partitioned to have the form*

Nonkey columns
tth column from
set S_{r_t}

Key columns

$$\mathbf{B} = \begin{array}{l} m_0 \text{ rows} \{ \\ \\ \\ \\ \end{array} \left[\begin{array}{cccc:c} A_{11} & A_{21} & \cdots & A_{p1} & \hat{B} \\ \hdashline B_{11} & & & & \\ & B_{21} & & & C \\ & & \ddots & & \\ & & & B_{p1} & \end{array}\right] \tag{3}$$

m_0
columns

where each B_{i1} *has dimension* $m_i \times m_i$ *and is nonsingular.*

PROOF. Consider the set of columns of **B** drawn from the set S_i. When taken together, these have the form

$$\begin{array}{r} m_0 \text{ rows} \{ \\ \\ \text{partition } i \{ \\ \\ \end{array} \left[\begin{array}{c} A \\ \hdashline 0 \\ \hdashline D \\ \hdashline 0 \end{array}\right]$$

All other basic columns have zero components in partition i. If D had rank less than m_i, then some rows of **B** would be dependent, which is impossible.

Thus D has rank m_i and contains a square nonsingular submatrix, which is B_{i1}.

We assume throughout that the column associated with $(x_0)_1$ is always basic and is the "last" column of **B**, i.e., the (m_0)th nonkey column.

Note that the identity matrix $I_{p\times p}$ in the basis **B** of the previous section [see equation (4) of that section] is here replaced by a block-diagonal matrix with nonsingular blocks. It will be necessary to store and update the inverses of these blocks. In the previous section, each of these blocks was the single element matrix {1}, so no such operations were necessary.

Our objective in what follows is to use the fact that the B_{i1} are nonsingular to develop a transformation matrix **T** such that **BT** is block triangular. In particular, the submatrix C in (3) is to be transformed into the zero matrix, whereupon $\hat{B}$ transforms into the working basis. Since the key columns are to be left unchanged, the first $m_1+m_2+\cdots+m_p$ columns of **T** are unit vectors. Thus **T** has the form

$$\mathbf{T} = \left[\begin{array}{c|c} I_1 & V \\ \hline 0 & E \end{array}\right] \tag{4}$$

where I_1 is an identity matrix, and V and E are to be chosen to satisfy the matrix equation

$$\left[\begin{array}{c|c} A_1 & \hat{B} \\ \hline \hat{B}_1 & C \end{array}\right]\left[\begin{array}{c|c} I_1 & V \\ \hline 0 & E \end{array}\right] = \left[\begin{array}{c|c} A_1 & B \\ \hline \hat{B}_1 & 0 \end{array}\right] \tag{5}$$

In the above, A_1 and B_1 represent the corresponding partitions of **B**:

$$[\quad A_1 \quad] = [A_{11}A_{21}\cdots A_{p1}] \tag{6}$$

$$\left[\quad \hat{B}_1 \quad\right] = \begin{bmatrix} B_{11} & & & \\ & B_{21} & & \\ & & \ddots & \\ & & & B_{p1} \end{bmatrix} \tag{7}$$

Setting corresponding partitions on the right and left hand sides of (5) equal yields

$$\hat{B}_1 V + CE = 0 \tag{8}$$
$$A_1 V + \hat{B}E = B \tag{9}$$

These are satisfied most simply by choosing

$$E = I \tag{10}$$
$$V = -\hat{B}_1^{-1} C \tag{11}$$

so the transformation matrix $\mathbf{T}$ has the form

$$\mathbf{T} = \left[\begin{array}{c|c} I_1 & V \\ \hline 0 & I_2 \end{array}\right] \tag{12}$$

Using (10) in (9) yields the matrix B as

$$B = \hat{B} - A_1 V \tag{13}$$

Since $\mathbf{T}$, and hence the submatrix V, play a key role in what follows, we examine the structure of V. The matrix $\hat{B}_1$ is block diagonal with nonsingular blocks, so (11) can be written

$$\begin{bmatrix} \\ V \\ \\ \end{bmatrix} = -\begin{bmatrix} B_{11}^{-1} & & & \\ & B_{21}^{-1} & & \\ & & \ddots & \\ & & & B_{p1}^{-1} \end{bmatrix} \begin{bmatrix} \\ C \\ \\ \end{bmatrix} \tag{14}$$

Since the tth nonkey column of $\mathbf{B}$ is from S_{r_t}, if $r_t = 0$ column t of C is a zero vector, as is column t of V. If $r_t \neq 0$, column t of C has the structure

$$\left[\begin{array}{c} 0 \\ \hline R^t \\ \hline 0 \end{array}\right] \}\text{partition } r_t$$

so column t of V has the form

$$\left[\begin{array}{c} 0 \\ \hline -\bar{R}^t \\ \hline 0 \end{array}\right] \}\text{ partition } r_t$$

where

$$\bar{R}^t = B_{r_t 1}^{-1} R^t \tag{15}$$

Thus the columns of V can be stored by recording only the indices r_t and the corresponding vectors $\bar{R}^t$.

The matrix B in (5) and (13) is the working basis of the procedure. To use it as such we must show:

THEOREM 2. *B is nonsingular.*

PROOF. If B is singular, then so is $\mathbf{BT}$, since the submatrix below B is all zeros. But $\det(\mathbf{BT}) = (\det \mathbf{B}) \det(\mathbf{T}) = \det(\mathbf{B})$, since $\det(\mathbf{T}) = 1$. Thus $\det(\mathbf{BT}) = 0 \Rightarrow \det(\mathbf{B}) = 0$, a contradiction.

To clarify the preceding material, consider the example problem of Section 5.4, rewritten here as

$$\text{minimize } x_0$$

subject to

$$\begin{array}{cccccccccccccc}
\mathbf{A}^0 & \mathbf{A}^1 & \mathbf{A}^2 & \mathbf{A}^3 & \mathbf{A}^4 & \mathbf{A}^5 & \mathbf{A}^6 & \mathbf{A}^7 & \mathbf{A}^8 & \mathbf{A}^9 & \mathbf{A}^{10} & \mathbf{A}^{11} & & \\
-x_0 & -x_1 & -x_2 & & & -2x_5 & -x_6 & & & & & & = & 0 \\
 & x_1 & +2x_2 & & & +2x_5 & +x_6 & & & & +x_{10} & & = & 40 \\
 & x_1 & +x_2 & & & +4x_5 & +2x_6 & & & & & +x_{11} & = & 50 \\
 & x_1 & +3x_2 & +x_3 & & & & & & & & & = & 30 \\
 & 2x_1 & +x_2 & & +x_4 & & & & & & & & = & 20 \\
 & & & & & x_5 & & +x_7 & & & & & = & 10 \\
 & & & & & & x_6 & & +x_8 & & & & = & 10 \\
 & & & & & x_5 & +x_6 & & & +x_9 & & & = & 15
\end{array}$$

$$x_i \geq 0, \qquad i = 1, \ldots, 11$$

Consider the following submatrix of the above:

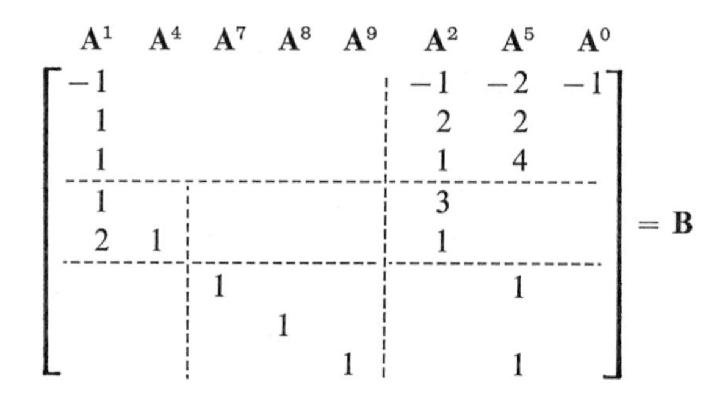

It is easily verified that **B** is nonsingular, so it is a (possibly infeasible) basis matrix for the problem. We compute

$$B_{11}^{-1} = \begin{bmatrix} 1 & 0 \\ 2 & 1 \end{bmatrix}^{-1} = \begin{bmatrix} 1 & 0 \\ -2 & 1 \end{bmatrix}$$

$$B_{21}^{-1} = \begin{bmatrix} 1 & & \\ & 1 & \\ & & 1 \end{bmatrix}$$

Then

$$\bar{R}^1 = B_{11}^{-1} R^1 = \begin{bmatrix} 1 & 0 \\ -2 & 1 \end{bmatrix} \begin{bmatrix} 3 \\ 1 \end{bmatrix} = \begin{bmatrix} 3 \\ -5 \end{bmatrix}$$

$$\bar{R}^2 = B_{21}^{-1} R^2 = \begin{bmatrix} 1 \\ 0 \\ 1 \end{bmatrix}$$

so

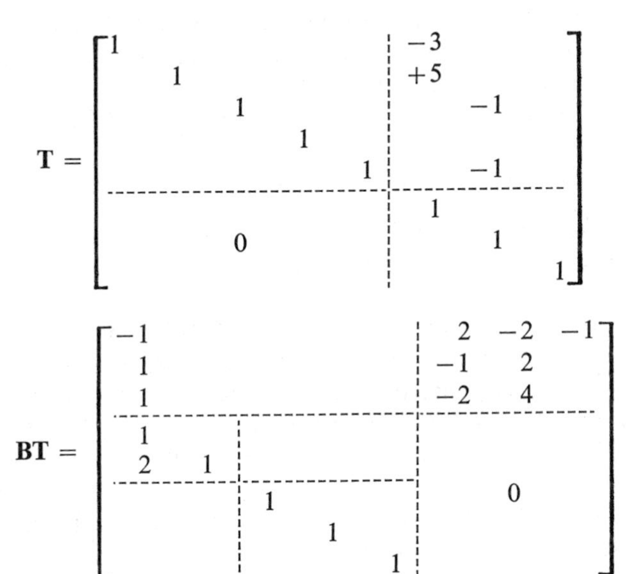

The working basis is

$$B = \begin{bmatrix} 2 & -2 & -1 \\ -1 & 2 & 0 \\ -2 & 4 & 0 \end{bmatrix}$$

As in the previous section, B has an interpretation as a basis matrix for a reduced system of m_0 equations. Let equations (2) be written

$$B_{i1}x_{i1} + B_{i2}x_{i2} = b_i, \qquad i = 1, \ldots, p \tag{16}$$

with (1) similarly partitioned:

$$A_0x_0 + \sum_{i=1}^{p} (A_{i1}x_{i1} + A_{i2}x_{i2}) = b \tag{17}$$

Since each B_{i1} is nonsingular, (16) may be used to eliminate the variables x_{i1}:

$$x_{i1} = B_{i1}^{-1}b_i - B_{i1}^{-1}B_{i2}x_{i2} \tag{18}$$

The reduced system is obtained by substituting (18) into (17):

$$A_0x_0 + \sum_{i=1}^{p} (A_{i2} - A_{i1}B_{i1}^{-1}B_{i2})x_{i2} = b - \sum_{i=1}^{p} A_{i1}B_{i1}^{-1}b_i \tag{19}$$

This is the same reduced system as is used in Rosen's partitioning method of Chapter 5. From (13), column t of B is equal to column t of $\hat{B}$ (call this Q^t) minus A_1 times column t of V. This column of V is zero except in partition r_t, where it contains the vector $-B_{r_t1}^{-1} R^t$, so column t of B has the form

$$Q^t - A_{r_t1} B_{r_t1}^{-1} R^t \tag{20}$$

Since Q^t is a column of A_{r_t2} and R^t a column of B_{r_t2}, this column is associated with some component of x_{r_t2} in (19). Thus B is a basis matrix for the reduced system.

We now examine how the operations of the revised simplex method may be carried out using quantities associated with the working basis. These operations require only that two sets of linear equations, with coefficient matrices $\mathbf{B}'$ and $\mathbf{B}$, be solved (one for the pricing vector, the other for the transform of the entering vector). Triangularizing $\mathbf{B}$ greatly simplifies their solution.

Determining the Simplex Multipliers. Here the vector of simplex multipliers $\boldsymbol{\pi} = (\pi_0, \pi_1, \ldots, \pi_p)$ is to be computed. These satisfy

$$\boldsymbol{\pi}\mathbf{B} = \mathbf{c}_B \tag{21}$$

or, since only $(x_0)_1$ has a nonzero objective coefficient, and its column is the rightmost column of $\mathbf{B}$,

$$\boldsymbol{\pi}\mathbf{B} = (0, 0, \ldots, 1) \tag{22}$$

Multiplying on the right by **T**,

$$\boldsymbol{\pi}(\mathbf{BT}) = (0, 0, \ldots, 1)\mathbf{T} = (0, 0, \ldots, 1) \tag{23}$$

In partitioned form (23) is

$$(\pi_0, \pi_1, \ldots, \pi_p)\left[\begin{array}{cccc|c} A_{11} & A_{21} & \cdots & A_{p1} & B \\ \hline B_{11} & & & & \\ & B_{21} & & & 0 \\ & & \ddots & & \\ & & & B_{p1} & \end{array}\right] = (0, 0, \ldots, 1) \tag{24}$$

Performing the matrix multiplication and equating terms:

$$\pi_0 A_{i1} + \pi_i B_{i1} = 0, \qquad i = 1, \ldots, p \tag{25}$$

$$\pi_0 B = (0, 0, \ldots, 1) = u_{m_0} \tag{26}$$

These sets of equations are easily solved:

$$\pi_0 = \text{row } m_0 \text{ of } B^{-1} \tag{27}$$

$$\pi_i = -\pi_0 A_{i1} B_{i1}^{-1}, \qquad i = 1, \ldots, p \tag{28}$$

Thus, if B^{-1} and the B_{i1}^{-1} are maintained, either in explicit or product form, the vectors π_0 and π_i are easily computed.

Determining the Column to Enter the Basis. As before, this is done by computing

$$\bar{c}_j = -\boldsymbol{\pi}\mathbf{A}^j, \qquad j \text{ nonbasic} \tag{29}$$

$$\bar{c}_j = \begin{cases} -\pi_0 A_i^{k_j} - \pi_i B_i^{k_j}, & \text{if } \mathbf{A}^j \in S_i, \quad i = 1, \ldots, p \\ -\pi_0 A_0^{k_j} & \text{if } \mathbf{A}^j \in S_0 \end{cases} \tag{30}$$

where $A_i^{k_j}$ and $B_i^{k_j}$ are columns k_j of A_i and B_i, respectively. If

$$\min \bar{c}_j \geq 0 \tag{31}$$

the current solution is optimal. If not, let

$$\min \bar{c}_j = \bar{c}_s < 0 \tag{32}$$

Then $\mathbf{A}^s$ enters the basis. Let $\mathbf{A}^s \in S_\sigma$.

Finding $\bar{\mathbf{A}}^s = \mathbf{B}^{-1}\mathbf{A}^s$**.** Here we must solve the linear system

$$\mathbf{B}\bar{\mathbf{A}}^s = \mathbf{A}^s \tag{33}$$

Let

$$\bar{\mathbf{A}}^s = \mathbf{T}\mathbf{Z}^s \tag{34}$$

Substituting (34) into (33) yields

$$(\mathbf{BT})\mathbf{Z}^s = \mathbf{A}^s \tag{35}$$

In partitioned form (35) is

$$\left[\begin{array}{cccc:c} A_{11} & A_{21} & \cdots & A_{p1} & B \\ \hdashline B_{11} & & & & \\ & B_{21} & & & 0 \\ & & \ddots & & \\ & & & B_{p1} & \end{array}\right] \begin{bmatrix} Z_1^s \\ \vdots \\ Z_p^s \\ \hdashline Z^s \end{bmatrix} = \begin{bmatrix} Q^s \\ \hdashline 0 \\ \hdashline R^s \\ \hdashline 0 \end{bmatrix} \}\text{ partition } \sigma$$

Writing these equations out,

$$\sum_{i=1}^{p} A_{i1}Z_i^s + BZ^s = Q^s \tag{36}$$

$$B_{i1}Z_i^s = 0, \qquad i = 1, \ldots, p, \quad i \neq \sigma \tag{37}$$

$$B_{\sigma 1}Z_\sigma^s = R^s \tag{38}$$

These have the solution

$$Z^s = 0, \qquad i = 1, \ldots, p, \quad i \neq \sigma \tag{39}$$

$$Z_\sigma^s = B_{\sigma 1}^{-1}R^s \tag{40}$$

$$Z^s = B^{-1}(Q^s - A_{\sigma 1}Z_\sigma^s) \tag{41}$$

Thus Z^s and Z_σ^s may be computed if $B_{\sigma 1}^{-1}$ and B^{-1} are known. Then the desired $\bar{\mathbf{A}}^s$ is computed from (34):

$$\begin{bmatrix} \bar{A}_1^s \\ \bar{A}_2^s \\ \vdots \\ \bar{A}_p^s \\ \hdashline \bar{A}^s \end{bmatrix} = \left[\begin{array}{c:c} & V_1 \\ I_1 & V_2 \\ & \vdots \\ & V_p \\ \hdashline 0 & I_2 \end{array}\right] \begin{bmatrix} 0 \\ \hdashline Z_\sigma^s \\ \hdashline 0 \\ \hdashline Z^s \end{bmatrix} \}\text{ partition } \sigma$$

(where the column block $\overbrace{V}^{\text{submatrix}}$ is labeled "submatrix V")

$$\bar{A}_i^s = V_i Z^s, \qquad i = 1, \ldots, p, \quad i \neq \sigma \tag{42}$$

$$\bar{A}_\sigma^s = Z_\sigma^s + V_\sigma Z^s \tag{43}$$

$$\bar{A}^s = Z^s \tag{44}$$

The fact that each column of V has at most one nonzero partition can be used to reduce (42)–(44) to more compact form. Recall that the nonzero partition of column t of V is partition r_t, and it contains the vector $-\bar{R}^t$, defined in (15). Thus column t of V_i is zero unless $r_t = i$, when it is $-\bar{R}^t$. As in equations (6.4–36), we define the sets

$$R(i) = \{t \mid t \in \{1, 2, \ldots, m_0\}, r_t = i\} \tag{45}$$

Then, letting $(Z^s)_t$ be component t of Z^s, (42)–(43) become

$$\bar{A}_i^s = -\sum_{t \in R(i)} (Z^s)_t \bar{R}^t, \qquad i = 1, \ldots, p, i \neq \sigma \tag{46}$$

$$\bar{A}_\sigma^s = Z_\sigma^s - \sum_{t \in R(\sigma)} (Z^s)_t \bar{R}^t \tag{47}$$

Note that, when the problem is that of section (6.4), the vectors $\bar{R}^t$ and Z_σ^s become scalars with value unity, and relations (44), (46), and (47) reduce to equations (6.4-36) and (6.4-37).

Choosing the Column to Leave the Basis. This is done according to the standard simplex formulas, as indicated in the previous section. If the solution is not unbounded, then column r of $\mathbf{B}$, $\mathbf{A}^{j_r}$, leaves the basis. We assume $\mathbf{A}^{j_r} \in S_\rho$. Since computing the new values of the basic variables also proceeds as in the previous section, we now consider updating the matrix B^{-1}, the matrices B_{i1}^{-1}, and any other quantities needed for the next iteration.

Updating the Inverse of the Working Basis.[4] An efficient procedure for updating B^{-1} may be derived by considering how the large basis matrix, $\mathbf{B}$, changes from cycle to cycle. Recall that the column $\mathbf{A}^{j_r} \in S_\rho$ is being replaced by $\mathbf{A}^s \in S_\sigma$. As in the previous section, $\mathbf{A}^s$ will not always replace $\mathbf{A}^{j_r}$ directly. In some cases, an interchange of $\mathbf{A}^{j_r}$ and some other basic column will be made first. Let $^*\mathbf{B}$ be the new basis matrix which results after either this interchange is made or after direct replacement of $\mathbf{A}^{j_r}$ by $\mathbf{A}^s$. The inverses of $\mathbf{B}$ and $^*\mathbf{B}$ are related by

$$^*\mathbf{B}^{-1} = \mathbf{E}\mathbf{B}^{-1} \tag{48}$$

where $\mathbf{E}$ is either an elementary matrix or a permutation matrix, whose precise structure will be developed shortly. Let $^*\mathbf{T}$ be the transformation matrix which brings $^*\mathbf{B}$ to block triangular form, and define

$$\mathbf{D} = \mathbf{B}\mathbf{T} \tag{49}$$

$$^*\mathbf{D} = {}^*\mathbf{B}^*\mathbf{T} \tag{50}$$

[4] This section is based on the work of James K. Hartman.

We write *$\mathbf{D}$ in partitioned form:

$$
{}^*\mathbf{D} = \left[\begin{array}{c|c} {}^*A_1 & {}^*B \\ \hline {}^*\hat{B}_1 & 0 \end{array}\right] \tag{51}
$$

Where *B is the new working basis. The inverse of *$\mathbf{D}$ is

$$
{}^*\mathbf{D}^{-1} = \left[\begin{array}{c|c} 0 & {}^*\hat{B}_1^{-1} \\ \hline {}^*B^{-1} & -{}^*B^{-1}{}^*A_1{}^*\hat{B}_1^{-1} \end{array}\right] \tag{52}
$$

An equation is now developed for *$\mathbf{D}^{-1}$, since it contains *B^{-1} in one partition. By (50) and (48)

$$
{}^*\mathbf{D}^{-1} = {}^*\mathbf{T}^{-1}{}^*\mathbf{B}^{-1} = {}^*\mathbf{T}^{-1}\mathbf{E}\mathbf{B}^{-1} \tag{53}
$$

Using (49)

$$
\mathbf{B} = \mathbf{D}\mathbf{T}^{-1} \tag{54}
$$

so

$$
\mathbf{B}^{-1} = \mathbf{T}\mathbf{D}^{-1} \tag{55}
$$

Using (55) in (53)

$$
{}^*\mathbf{D}^{-1} = {}^*\mathbf{T}^{-1}\mathbf{E}\mathbf{T}\mathbf{D}^{-1} \tag{56}
$$

The inverse of *$\mathbf{T}$ is

$$
{}^*\mathbf{T}^{-1} = \left[\begin{array}{c|c} I_1 & {}^*V \\ \hline 0 & I_2 \end{array}\right]^{-1} = \left[\begin{array}{c|c} I_1 & -{}^*V \\ \hline 0 & I_2 \end{array}\right] \tag{57}
$$

Thus, in partitioned form, (56) becomes

$$\left[\begin{array}{c:c} 0 & {}^*\hat{B}_1^{-1} \\ \hdashline {}^*B^{-1} & -{}^*B^{-1}{}^*A_1{}^*\hat{B}_1^{-1} \end{array}\right] = \left[\begin{array}{c:c} I_1 & -{}^*V \\ \hdashline 0 & I_2 \end{array}\right] \times \left[\begin{array}{c:c} E_1 & E_2 \\ \hdashline E_3 & E_4 \end{array}\right] \left[\begin{array}{c:c} I_1 & V \\ \hdashline 0 & I_2 \end{array}\right] \times \left[\begin{array}{c:c} 0 & \hat{B}_1^{-1} \\ \hdashline B^{-1} & -B^{-1}A_1\hat{B}_1^{-1} \end{array}\right] \tag{58}$$

To obtain the equation for ${}^*B^{-1}$ implied by the above, we retain only the first m_0 columns of the last matrix and the last m_0 rows of the first:

$${}^*B^{-1} = [\ 0 \ \vdots \ I_2] \left[\begin{array}{c:c} E_1 & E_2 \\ \hdashline E_3 & E_4 \end{array}\right] \left[\begin{array}{c:c} I_1 & V \\ \hdashline 0 & I_2 \end{array}\right] \left[\begin{array}{c} 0 \\ \hdashline B^{-1} \end{array}\right] \tag{59}$$

Multiplying the first two and last two matrices

$$*B^{-1} = [\ E_3 \ \vdots \ E_4] \begin{bmatrix} VB^{-1} \\ \hline B^{-1} \end{bmatrix} \tag{60}$$

or

$$*B^{-1} = (E_4 + E_3 V)B^{-1} \tag{61}$$

Equation (61) above is the basic transformation equation for $*B^{-1}$. We now develop the structure of $E_3V + E_4$ for the various types of basis changes that may occur. The subcases considered are as in Section (6.4), but the operations required are more complex.

CASE 1. $\mathbf{A}^{j_r}$ *is not a key column.* In this case, $\mathbf{A}^{j_r}$ is one of the last m_0 columns of $\mathbf{B}$, and the block diagonal structure of the submatrix $\hat{B}_1$ is unaltered if $\mathbf{A}^s$ replaces it. The matrix $\mathbf{E}$ is the usual elementary matrix developed in Section (6.2). Let $\mathbf{A}^{j_r}$ be the $m + l$th column of $\mathbf{B}$, where

$$m = \sum_{i=1}^{p} m_i \tag{62}$$

and let $\bar{a}_{is}$ be the ith component of $\bar{\mathbf{A}}^s = \mathbf{B}^{-1}\mathbf{A}^s$. Then $\mathbf{E}$ differs from an identity matrix only in column $m+l$, this column being an eta vector with components

$$\eta_i = \begin{cases} -\bar{a}_{is}/\bar{a}_{m+l,s}, & i = 1, \ldots, m + m_0,\ i \neq m + l \\ 1/\bar{a}_{m+l,s}, & i = m + l \end{cases} \tag{63}$$

Thus, referring to the partitioning of $\mathbf{E}$ in (59)

$$E_3 = [\ 0\ ] \tag{64}$$

$$E_4 = \begin{bmatrix} 1 & & & \eta_{m+1} & & \\ & \ddots & & \vdots & & \\ & & 1 & & & \\ & & & \eta_{m+l} & & \\ & & & \vdots & \ddots & \\ & & & \eta_{m+m_0} & & 1 \end{bmatrix} \tag{65}$$

Using (64) and (65) in our fundamental equation, (61)

$$*B^{-1} = E_4 B^{-1} \tag{66}$$

so in this case, the relation between the new and old inverses is very simple, involving only an elementary column matrix.

To complete this case, all other quantities required in the next cycle must be updated. The only one of these which changes is the lth column of V, which contains the vector $-\bar{R}^l$ in (46)–(47). This changes from a vector with nonzero elements in partition ρ to one whose nonzero elements are in partition σ, and are equal to the negative of $B_{\sigma 1}^{-1}$ times partition σ of $\mathbf{A}^s$. No additional work is needed, since this vector has already been computed as Z_σ^s in (40). Recording the elements of this vector and the index σ as the lth column of V completes this case.

CASE 2. $\mathbf{A}^{j_r}$ *is a key column. Subcase* (a): $\mathbf{A}^{j_r}$ *and* $\mathbf{A}^s$ *are from different sets, i.e.* $\rho \neq \sigma$. In this case, direct replacement of $\mathbf{A}^{j_r}$ by $\mathbf{A}^s$ would destroy the block diagonal structure of $\hat{B}_1$, which is crucial to the method. However, theorem 1 applies to $\mathbf{B}$ after $\mathbf{A}^{j_r}$ leaves, so one of the last m_0 columns of $\mathbf{B}$ must be such that when the vector in its ρth row partition replaces the corresponding subvector of $\mathbf{A}^{j_r}$ in $B_{\rho 1}$, the new $B_{\rho 1}$ which results is nonsingular.

As in the previous section, one such column is interchanged with $\mathbf{A}^{j_r}$, making $\mathbf{A}^{j_r}$ nonkey. Then the procedure of Case 1 is applied.

Let $\mathbf{A}^{j_r}$ be column i_1 of $\mathbf{B}$, and let it be interchanged with column $m + i_2$. The transformation for the column exchange is given by

$$
{}^*\mathbf{B} = \mathbf{B}\mathbf{E} \tag{67}
$$

where $\mathbf{E}$ is an identity matrix with columns i_1 and $m + i_2$ interchanged, i.e.

$$
\mathbf{E} = \left[\begin{array}{ccccccccc|ccc}
1 &&&&&&&&&&& \\
& 1 &&&&&&&&&& \\
&& \ddots &&&&&&&&& \\
&&& 1 &&&&&&&& \\
&&&& 0 & \cdot & \cdot & \cdot & & \cdot & 1 & \\
&&&& \cdot & 1 &&&&& \cdot & \\
&&&& \cdot && 1 &&&& \cdot & \\
&&&& \cdot &&& \ddots &&& \cdot & \\
&&&& \cdot &&&& 1 && \cdot & \\
\hline
&&&& \cdot &&&&& 1 & \cdot & \\
&&&& 1 & \cdot\odot & \cdot & \cdot & & \cdot & 0 & \\
&&&&&&&&&&& 1
\end{array}\right]
\begin{array}{l} \\ \\ \\ \\ \leftarrow \text{row } i_1 \\ \\ \\ \\ \\ \\ \leftarrow \text{row } m+i_2 \\ \\ \end{array}
\tag{68}
$$

↑ column i_1 ↑ column $m + i_2$

Since performing the interchange twice yields the identity matrix

$$\mathbf{E}^{-1} = \mathbf{E} \tag{69}$$

so, from (67)

$$^*\mathbf{B}^{-1} = \mathbf{E}^{-1}\mathbf{B}^{-1} \tag{70}$$

and, using (69)

$$^*\mathbf{B}^{-1} = \mathbf{E}\mathbf{B}^{-1} \tag{71}$$

From (68) we see that the bottom partitions of E are

$$E_3 = [\quad \underset{\substack{\uparrow \\ \text{column} \\ i_1}}{1} \quad \cdots \quad] \leftarrow \text{row } i_2 \tag{72}$$

$$E_4 = \begin{bmatrix} 1 & & & & & & \\ & \ddots & & & & & \\ & & 1 & & & & \\ & & & 0 & \cdots & \cdots & \cdots \\ & & & \vdots & 1 & & \\ & & & \vdots & & \ddots & \\ & & & \vdots & & & 1 \end{bmatrix} \begin{matrix} \\ \\ \\ \leftarrow \text{row } i_2 \\ \\ \\ \\ \end{matrix} \tag{73}$$

$$\qquad \uparrow \text{ column } i_2$$

If v is row i_1 of V then E_3V is an m_0-dimensional square matrix with row i_2 equal to v and all others zero. Hence

$$^*B^{-1} = (E_4 + E_3V)B^{-1} = \begin{bmatrix} 1 & & & & & & \\ & 1 & & & & & \\ & & \ddots & & & & \\ & & & 1 & & & \\ \hline & & & v & & & \\ \hline & & & & 1 & & \\ & & & & & \ddots & \\ & & & & & & 1 \end{bmatrix} \mathbf{B}^{-1} \quad \leftarrow \text{row } i_2 \tag{74}$$

so old and new inverses are related by an elementary row matrix.

In the above, $*B^{-1}$ is nonsingular if and only if $E_4 + E_3V$ is nonsingular, which in turn is true if and only if component i_2 of v is nonzero. That such a choice for i_2 always exists can be shown as follows. Row i_1 of V, v, is from the ρth row partition of V. The only nonzero columns in this partition are of the form $B_{\rho 1}^{-1}(\cdot)$, i.e. $B_{\rho 1}^{-1}$ times some other vector. For concreteness, let

$$i_1 = \sum_{j=1}^{\rho-1} m_j + 3 \tag{75}$$

so column 3 of $B_{\rho 1}$ is leaving $B_{\rho 1}$ and v is row 3 of the ρth row partition. If all vectors of the form $B_{\rho 1}^{-1}(\cdot)$ have zeros in position 3, then none could replace column 3 of $B_{\rho 1}$ and maintain $B_{\rho 1}$ nonsingular. This is impossible, by Theorem 1. Thus, some column, i_2, of V has its component in row i_1 nonzero, and can be interchanged with $\mathbf{A}^{j_r}$.

To complete this subcase, note that the interchange of columns causes $B_{\rho 1}$ and $A_{\rho 1}$ to change, so $B_{\rho 1}^{-1}$ must be updated by pivoting. Those columns of V containing vectors of the form $B_{\rho 1}^{-1}(\cdot)$ are also changed, and these are updated by the same operations used on $B_{\rho 1}^{-1}$.

Subcase (b): $\rho = \sigma$. Again, let $\mathbf{A}^{j_r}$ be column i_1 of $\mathbf{B}$. Search row i_1 of V, v, to see if it contains a nonzero element. If it does, and the element is in column i_2 of V, interchange columns i_1 and $m + i_2$ of $\mathbf{B}$ as described in subcase (a). Then perform the operations of Case 1. If $v = 0$, we replace $\mathbf{A}^{j_r}$ by $\mathbf{A}^s$ directly through a pivot operation. Here, $\mathbf{E}$ is an elementary matrix, with column i_1 an eta vector as in (63). Thus E_3 and E_4 are

$$E_3 = \begin{bmatrix} & \eta_{m+1} & \\ 0 & \vdots & 0 \\ & \eta_{m+m_0} & \end{bmatrix} \tag{76}$$
$$\uparrow$$
$$\text{column } i_1$$

$$E_4 = I \tag{77}$$

so

$$E_4 + E_3V = I + \begin{bmatrix} \eta_{m+1}v \\ \vdots \\ \eta_{m+m_0}v \end{bmatrix} \tag{78}$$

Since $v = 0$

$$E_4 + E_3V = I \tag{79}$$

and, by (61), there is no change in the working basis;

$$*B^{-1} = IB^{-1} = B^{-1} \tag{80}$$

Finally, we complete this last subcase, by updating $B_{\rho 1}^{-1}$, $A_{\rho 1}$, and vectors in V of the form $B_{\rho 1}^{-1}(\cdot)$ as described in subcase (a). Again, only one pivot operation is required.

Summarizing, each case replaces B^{-1} by itself times an elementary row or column matrix and either goes to Case 1 or is complete. Thus, at any iteration, k, the current B^{-1} is a product of elementary row and column matrices, each of dimension m_0. Updating the inverses B_{i1}^{-1} requires at most one pivot operation per cycle. Thus the total computational labor needed to compute all quantities needed for the next cycle is approximately that required in updating these quantities in two smaller linear programs, one of m_0 rows, the other having a number of rows equal to the size of some matrix B_{i1}.

Problems

1. Derive the expressions in (6.3-30)–(6.3-32).
2. Solve the problem given in relations (1.2.1-15), with the upper bounds $x_3 \leq 1$, $x_2 \leq \frac{1}{2}$, $x_1 \leq 4$.
3. Derive formulas similar to (6.4-36)–(6.4-37), for the updated values of the basic variables in terms of the vector $\bar{d} = B^{-1}d$.
4. Show in detail what charges are required in the algorithm of Section 6.4 when the last p equations contain some positive and some negative coefficients.
5. Show in detail how the algorithm of Section 6.4 would be applied in phase 1 of the simplex method.
6. Solve the example problem of Section 3.5 using the algorithm of Section 6.5 and compare the results at each iteration with those given in Section 3.5.
7. Consider the extension of the algorithm of Section 6.5 to block-diagonal problems with coupling rows and coupling columns. What important properties used in Section 6.5 are no longer present? Suggest an algorithm which circumvents these difficulties and uses only a working basis plus inverses from each block.
8. Show how the algorithm of Section 6.5 reduces to that of Section 6.4 and Section 6.3 when the problem being solved is of the form considered in those sections.

References

1. G. B. Dantzig, *Linear Programming and Extensions*, Princeton University Press, Princeton, N.J., 1963, chap. 18.
2. G. B. Dantzig and R. M. Van Slyke, "Generalized Upper Bounding Techniques," *J. Computer System Sci.*, **1**, 1967, pp. 213–226.

3. R. N. Kaul, "An Extension of Generalized Upper Bounding Techniques for Linear Programming," *Rept. ORC 65–27*, Operations Research Center, University of California at Berkeley, 1965.
4. P. Wolfe and L. Cutler, "Experiments in Linear Programming," in *Recent Advances in Mathematical Programming*, R. L. Graves and P. Wolfe, eds., McGraw-Hill, Inc., New York, 1963, pp. 177–200.
5. D. Smith and W. Orchard-Hays, "Computational Efficiency in Product Form LP Codes," in *Recent Advances in Mathematical Programming*, R. L. Graves and P. Wolfe, eds., McGraw-Hill, Inc., New York, 1963, pp. 211–218.
6. Mathematical Programming System/360 (360A-CO-14X) Linear and Separable Programming—Users Manual, *IBM Rept. H20–0476–0.*
7. L. J. Larsen, "A Modified Inversion Procedure for Product Form of the Inverse Linear Programming Codes," *Commun. ACM*, July 1962.

7

Partitioning Procedures in Nonlinear Programming

7.1 Introduction

As with linear programs, the term partitioning procedure is used here to mean an algorithm which partitions the variables of a problem into two subsets, changing first the values of those variables in one subset, then the other.. Of course, each of the subsets may be further partitioned, as when one contains variables which appear linearly and their constraint matrix is block diagonal. The partitioning to be used is, for the problems considered here, determined in an obvious way by the problems' structure.

Two related classes of mathematical programs are discussed. The first contains problems of the form

$$\text{minimize} \sum_{i=1}^{p} c_i'(y)x_i + c_0(y) \tag{1}$$

subject to

$$A_i(y)x_i \geq b_i(y), \qquad i = 1, \ldots, p \tag{2}$$

Here the matrices A_i, cost coefficients c_i, and right-hand-side vectors b_i are all continuous functions of a vector of coupling variables, y. If y is fixed, the problem is linear in the x_i, so it is natural to adjust the y and x_i variables separately. Problems of this type represent a nonlinear generalization of dual-angular linear programs. Rosen has extended his partitioning procedure for linear programs, described in Section 5.4, to deal with (1)–(2). This extension is discussed in the first part of this chapter.

The second class of problems considered here has elements of the form

$$\text{minimize } c'x + f(y) \tag{3}$$

subject to
$$Ax + F(y) \geq b \tag{4}$$
$$x \geq 0, \qquad y \in S \tag{5}$$

where f and F are scalar- and vector-valued functions of y, respectively, and S is an arbitrary subset of E^p. The freedom in specifying S admits some important new problems into this class. The most interesting case takes S as the set of all vectors in E^p with integral-valued components. If, in addition, F and f are linear, (3)–(5) is the well-known mixed-variable linear program. If A is block diagonal and $S = E^p$, then (3)–(5) is a special case of (1)–(2), where the A_i and c_i are not functions of y.

A partitioning algorithm presented by J. F. Benders, applicable to problems of the form (3)–(5), is discussed in the second part of this chapter. As with Rosen's method it varies first y, then solves a linear program for new x values, etc. However, the similarity ends here. Benders' algorithm adjusts the y variables by solving a mathematical program over S to which one or more constraints are added at each cycle. These constraints are constructed by using the dual solution of (3)–(5) for fixed y. When (3)–(5) is a mixed-variable linear program, the program which varies the y variables is an all-integer linear program. Thus, in this case, (3)–(5) is solved by separate variation of discrete and continuous variables. This is especially appealing when A has special structure, e.g., is block diagonal or of the transportation type.

As with linear problems, the dual concepts of constraint relaxation and restriction provide a unifying framework for these two algorithms. In Benders' procedure, the problem which determines the y variables has many constraints, so relaxation is applied to it. Since this program may not be convex, constraints can generally only be added but not deleted. Rosen's method uses restriction when it requires a subset of linear inequalities with nonsingular coefficient matrix to hold as equalities. These are then used to eliminate the variables x_i in terms of y, which permits construction of a coordinating problem in y.

7.2 Rosen's Partitioning Algorithm for Nonlinear Programs

The partitioning algorithm of Rosen for linear programs, discussed in Section 5.4, has been extended to a class of nonlinear programs [1]. The procedure is applicable to problems having the form

$$\text{minimize} \sum_{i=1}^{p} c_i'(y)x_i + c_0(y) \tag{1}$$

subject to
$$A_i(y)x_i \geq b_i(y), \qquad i = 1, \ldots, p, \quad p \geq 1 \tag{2}$$

Here the problem variables are partitioned into two subsets; coupling variables, y, and "linear" variables, x_i. The vectors of cost coefficients c_i, the constraint coefficient matrices A_i, and the right-hand-side vectors b_i may all be linear or nonlinear functions of the variables y. A wide variety of problems may be formulated to have this structure, as is discussed in Section 2.7.

The outstanding feature of such a structure is that, for fixed y, the problem decomposes into p independent linear subproblems in the variables x_i. Thus an iterative procedure which first varies y, then the x_i, is natural. As in the procedure of Section 5.4 for linear programs, basis matrices and shadow prices from the subproblem solutions are used to set up a coordinating program in the y variables. In contrast to the linear algorithms, the coordinating program here has a nonlinear objective.

Mathematical proofs of convergence to a global minimum exist only when, in (1)–(2), the coefficients c_i and matrices A_i are independent of y, so the problem becomes

$$\text{minimize} \sum_{i=1}^{p} c_i' x_i \tag{3}$$

$$\text{subject to} \qquad A_i x_i \geq b_i(y), \qquad i = 1, \ldots, p \tag{4}$$

This is the form for which the properties of the algorithm will be stated and proved, with the more general problem (1)–(2) considered briefly later. The following additional assumptions are made:

Assumption 1. For $i = 1, \ldots, p$, each component of $b_i(y)$ is a differentiable convex function of y. This implies that (3)–(4) is a convex program, for which any local minimum is global.

Assumption 2. There exist feasible vectors (x_i, y) interior to all nonlinear constraints, i.e., such that $A_i x_i \geq b_i(y)$ with strict inequality for nonlinear components of $b_i(y)$. This ensures satisfaction of the Kuhn–Tucker constraint qualification (see [2] and [3] and Section 1.3.2). With Assumption 1, it implies that the Kuhn–Tucker conditions are both necessary and sufficient for an optimal solution.

Assumption 3. For $i = 1, \ldots, p$, none of the rows of A_i are the zero vector. If the problem as originally formulated has zero rows, i.e., constraints $b_{ij}(y) \leq 0$ are present, then artificial variables x_{ij}^a with large positive costs are used to transform these into $b_{ij}(y) \leq x_{ij}^a$. If the costs are sufficiently large, $x_{ij}^a = 0$ in any optimal solution.

Assumption 4. A feasible vector, y_0, exists and is known, i.e., a vector for which there exist x_i satisfying (4) when $y = y_0$.

7.2.1 Development of the Algorithm. The problem (3)–(4) may be solved by minimizing over x_i for all (fixed) y, then minimizing the result over y. It may thus be written

$$\underset{y \in S}{\text{minimize}} \left\{ \sum_{i=1}^{p} \min\{c_i' x_i \mid A_i x_i \geq b_i(y)\} \right\} \tag{1}$$

where S is the set of y values for which each of the inner minimizations is feasible. If

$$S_i = \{y \mid \text{there exists } x_i \text{ such that } A_i x_i \geq b_i(y)\} \tag{2}$$

then

$$S = \bigcap_{i=1}^{p} S_i \tag{3}$$

Vectors $y \in S$ will be called feasible. Define

$$\psi_i(y) = \min_{x_i}\{c_i' x_i \mid A_i x_i \geq b_i(y)\} \tag{4}$$

Problem (1) then becomes

$$\text{minimize } \psi(y) = \sum_{i=1}^{p} \psi_i(y) \tag{5}$$

subject to

$$y \in S \tag{6}$$

The following theorem states that, under assumption 1, (5)–(6) is a convex program.

THEOREM 1. *If each $b_i(y)$ is convex, then the set S is convex and each function ψ_i is convex over S.*

The proof is left for Problem 1. By Theorem 1, the original problem has been reduced to solving a convex program in y. However, this program is made difficult by the fact that simply to evaluate one of the functions $\psi_i(y)$ requires solution of the linear subproblem

$$\left.\begin{array}{ll} \text{minimize } c_i' x_i & (7) \\ \text{subject to } A_i x_i \geq b_i(y) & (8) \end{array}\right\} \text{subproblem } i$$

Also, the constraints on y have not, as yet, been given explicitly in terms of constraint functions. To ease these difficulties, we try to extract as much information from the solution of the linear program (7)–(8) as possible. This information will determine the function ψ_i and the boundaries of S in the neighborhood of the current y vector.

Let y_0 be feasible, and assume that each of the linear programs (7)–(8) has been solved for $y = y_0$. This solution partitions A_i (which has more rows than columns) into a square nonsingular basis matrix A_{i1} and a nonbasic portion A_{i2}. The matrix A_{i1} determines the solution vector, x_i^0, through the relation

$$A_{i1}x_i^0 = b_{i1}(y_0), \qquad i = 1, \ldots, p \tag{9}$$

If we allow (9) to (temporarily) determine the relation between x_i and y, the subproblem (7)–(8) may be written

$$\text{minimize } c_i'x_i \tag{10}$$

subject to

$$A_{i1}x_i = b_{i1}(y) \tag{11}$$

$$A_{i2}x_i \geq b_{i2}(y) \tag{12}$$

Then (11) may be inverted to express x_i in terms of y,

$$x_i = A_{i1}^{-1}b_{i1}(y) \tag{13}$$

Equation (13) is used to eliminate y from (10) and (12), yielding

$$c_i'x_i = c_i'A_{i1}^{-1}b_{i1}(y) = (u_{i1}^0)'b_{i1}(y) \tag{14}$$

$$A_{i2}A_{i1}^{-1}b_{i1}(y) \geq b_{i2}(y), \qquad i = 1, \ldots, p \tag{15}$$

where

$$(u_{i1}^0)' = c_i'A_{i1}^{-1} \geq 0 \tag{16}$$

is the vector of dual variables or simplex multipliers associated with the optimal basis A_{i1}.

Given that (13) holds, $\psi_i(y)$ is given by (14) and the constraints (15) determine the set S_i in (2). Thus a coordinating problem in y space could be set up, with objective

$$f(y) = \sum_{i=1}^{p} (u_{i1}^0)'b_{i1}(y) \tag{17}$$

and constraints (15). This would be an application of the restriction procedure outlined in Section 5.2, where some constraints of the original problem [those in (11)] are required to hold as equalities. The objective (17) is convex, since the b_{i1} are convex and $u_{i1}^0 \geq 0$. However, the constraints (15) are nonlinear and may not determine a convex set, since the elements of $A_{i2}A_{i1}^{-1}$ may have any sign. The solution of programs with nonlinear constraints is considerably more difficult than if these are linear, and the absence of guaranteed convexity means that a global solution cannot be guaranteed. However, any coordinating program need only produce an improved feasible y vector, if one exists. Keeping this in mind, we will linearize the constraints (15) about the current point, y_0. The properties of the resulting coordinating problem are dealt with in Theorem 3.

Let

$$Q_i = A_{i2}A_{i1}^{-1} \tag{18}$$

and define the Jacobian matrix of $b_{ij}(y)$ $(j = 1, 2)$, $J_{ij}(y)$, as the matrix whose rows are gradients of the components of b_{ij} evaluated at the point y. Then linearizing (15) about y_0 yields

$$[Q_i J_{i1}(y_0) - J_{i2}(y_0)](y - y_0) \geq b_{i2}(y_0) - Q_i b_{i1}(y_0) \tag{19}$$

The coordinating problem has the convex objective (17) and the linear constraints (19):

$$\left.\begin{aligned} &\text{minimize } f(y) = \sum_{i=1}^{p} (u_{i1}^0)' b_{i1}(y) && (20)\\ &\text{subject to} \\ &[Q_i J_{i1}(y_0) - J_{i2}(y_0)](y - y_0) \geq b_{i2}(y_0) - Q_i b_{i1}(y_0), \quad i = 1, \ldots, p && (21)\end{aligned}\right\} \text{coordinating problem}$$

Let y^1 solve this problem. Since y_0 is feasible for the above,

$$f(y^1) \leq f(y_0) \tag{22}$$

Of course, it may be that equality holds in (22). In this case, the current solution $(\{x_i^0\}, y_0)$ may be tested for optimality.

THEOREM 2 (OPTIMALITY TEST). *Let x_i^0, $i = 1, \ldots, p$ solve the linear subproblems* (7)–(8) *for $y = y_0$. Then necessary and sufficient conditions that $(\{x_i^0\}, y_0)$ solve the original problem* [(7.2-3)–(7.2-4)] *are that y_0 solve the coordinating problem and that the conditions*

$$u_{i1}^* = u_{i1}^0 - Q_i u_{i2}^1 \geq 0, \qquad i = 1, \ldots, p \tag{23}$$

be satisfied. In (23), *u_{i2}^1 is the vector of dual variables corresponding to the ith set of constraints* (21) *at the coordinating problem optimum.*

PROOF. *Sufficiency*. Under assumptions 1–4, a given solution is optimal for the original problem if and only if the Kuhn–Tucker conditions are satisfied. Let this problem be written in partitioned form, as

$$\text{minimize} \sum_{i=1}^{p} c_i' x_i \tag{24}$$

subject to

$$\left.\begin{aligned} A_{i1}x_i &\geq b_{i1}(y) \\ A_{i2}x_i &\geq b_{i2}(y) \end{aligned}\right\} \quad i = 1, \ldots, p \qquad (25),\ (26)$$

Let u_{i1}, u_{i2} be multipliers for the constraints (25) and (26), respectively. Writing the Lagrangian function

$$L = \sum_{i=1}^{p} (c_i'x_i + u_{i1}'(b_{i1}(y) - A_{i1}x_i) + u_{i2}'(b_{i2}(y) - A_{i2}x_i)) \tag{27}$$

the Kuhn–Tucker conditions at $(\{x_i^0\}, y_0)$ are, for $i = 1, \ldots, p$:

$$A_{i1}'u_{i1} + A_{i2}'u_{i2} = c_i \tag{28}$$

$$J_{i1}'(y_0)u_{i1} + J_{i2}'(y_0)u_{i2} = 0 \tag{29}$$

$$u_{i1}'(b_{i1}(y_0) - A_{i1}x_i^0) = u_{i2}'(b_{i2}(y_0) - A_{i2}x_i^0) = 0 \tag{30}$$

$$A_{i1}x_i^0 \geq b_{i1}(y_0), \qquad A_{i2}x_i^0 \geq b_{i2}(y_0) \tag{31}$$

$$u_{i2} \geq 0 \tag{32}$$

$$u_{i1} \geq 0 \tag{33}$$

The optimality conditions for the linear subproblems are

$$u_{i1}^0 = (A_{i1}^{-1})'c_i \geq 0 \tag{34}$$

$$A_{i1}x_i^0 = b_{i1}(y_0), \qquad A_{i2}x_i^0 \geq b_{i2}(y_0) \tag{35}$$

and the Kuhn–Tucker conditions for the coordinating problem at $y = y_0$, obtained from the Lagrangian

$$\hat{L} = \sum_{i=1}^{p} \{(u_{i1}^0)'b_{i1}(y) + (u_{i2}^1)'[b_{i2}(y_0) - Q_i b_{i1}(y_0) - (Q_i J_{i1}(y_0) - J_{i2}(y_0))(y - y_0)]\} \tag{36}$$

are, for $i = 1, \ldots, p$,

$$J_{i1}'(y_0)[u_{i1}^0 - Q_i'u_{i2}^1] + J_{i2}'(y_0)u_{i2}^1 = 0 \tag{37}$$

$$(u_{i2}^1)'(b_{i2}(y_0) - Q_i b_{i1}(y_0)) = 0 \tag{38}$$

$$u_{i2}^1 \geq 0 \tag{39}$$

Comparing (28)–(33) and (34)–(35), (37)–(39), the only unsatisfied conditions are (28)–(29) and (33). Comparing (37) and (29) suggests that we let

$$u_{i1} = u_{i1}^* = u_{i1}^0 - Q_i' u_{i2}^1 \tag{40}$$

so that (29) is satisfied. Multiplying (40) by A_{i1}',

$$A_{i1}' u_{i1}^* + A_{i1}' Q_i' u_{i2}^1 = A_{i1}' u_{i1}^0 \tag{41}$$

Since, by (18), $Q_i' = (A_{i1}^{-1})' A_{i2}'$ and $u_{i1}^0 = (A_{i1}^{-1})'$, (41) becomes (28). Thus if

$$u_{i1}^* \geq 0, \qquad i = 1, \ldots, p \tag{42}$$

the complete set of Kuhn–Tucker conditions are satisfied at $(\{x_i^0\}, y_0)$, with u_{i1}^* and u_{i2}^1 the optimal values of the multipliers. This solution is therefore optimal.

Necessity is proved in a similar way, with the details left for Problem 2.

COROLLARY. *A sufficient condition that $(\{x_i^0\}, y_0)$ be optimal is that y_0 solve the coordinating problem with none of the linear constraints binding.*

PROOF. If y_0 is an interior minimum, then by (38), $u_{i2}^1 = 0$ and, from (40), $u_{i1}^* = u_{i1}^0$. Since $u_{i1}^0 \geq 0$, $(\{x_i^0\}, y_0)$ is optimal.

If the optimality test is failed, either $f(y^1) < f(y_0)$ or $f(y^1) = f(y^0)$ and some $u_{i1}^* \ngeq 0$. Consider first the case $f(y^1) < f(y_0)$. Since the coordinating problem was formed by assuming $A_{i1}x_i = b_{i1}(y)$ and linearizing the constraints (15), the coordinating problem objective $f(y)$ is equal to $\psi(y)$ only for y such that A_{i1} is an optimal basis for the linear subproblems (7)–(8). It is ψ, not f, which we wish to decrease. Further, owing to the linearization, y^1 may not satisfy (15) and so may not be in S. The following theorem deals with these difficulties. It shows that either y^1 is in S and satisfies $\psi(y^1) < \psi(y_0)$ or, if a nonzero move, θ_m can be made along the line from y_0 to y^1, then a new vector, $y_0 + \theta_m(y^1 - y_0)$ may be constructed which has this property.

THEOREM 3 (JUSTIFICATION OF CONSTRAINT LINEARIZATION). *Let the optimal solution of the coordinating program satisfy*

$$f(y^1) < f(y_0) \tag{43}$$

If y^1 satisfies

$$h_i(y^1) = Q_i b_{i1}(y^1) - b_{i2}(y^1) \geq 0, \qquad i = 1, \ldots, p \tag{44}$$

then

$$\psi(y^1) \leq f(y^1) < f(y_0) = \psi(y_0) \tag{45}$$

If y^1 does not satisfy (44), *define θ_m as the largest move from y_0 to y^1 which maintains feasibility:*

$$\theta_m = \max\{\theta \mid h_i(y_0 + \theta(y^1 - y_0)) \geq 0, i = 1, \ldots, p, 0 \leq \theta \leq 1\} \quad (46)$$

Then, if $\theta_m > 0$,

$$\psi(y_0 + \theta_m(y^1 - y_0)) < \psi(y_0) \quad (47)$$

PROOF. The inequalities (44) are, by (11)–(15), simply necessary and sufficient conditions that the basis A_{i1} be feasible for the ith linear subproblem when $y = y^1$. Since the A_{i1} may not be optimal bases when $y = y^1$,

$$\psi(y^1) \leq f(y^1) \quad (48)$$

Since the A_{i1} are optimal bases when $y = y_0$,

$$f(y_0) = \psi(y_0) \quad (49)$$

and combining (48)–(49) and (43), (45) is proved.

To prove (47), let

$$\hat{y} = y_0 + \theta_m(y^1 - y_0) \quad (50)$$

Since f is convex,

$$f(\hat{y}) \leq f(y_0) + \theta_m(f(y^1) - f(y_0)) \quad (51)$$

The basis matrices A_{i1} may not be optimal in the linear subproblems (7)–(8) when $y = \hat{y}$, so

$$\psi(\hat{y}) \leq f(\hat{y}) \quad (52)$$

Thus, using (49), we have

$$\psi(\hat{y}) \leq \psi(y_0) + \theta_m(f(y^1) - f(y_0)) \quad (53)$$

By the hypothesis of the theorem, $f(y^1) < f(y_0)$ and $\theta_m > 0$, so

$$\psi(\hat{y}) < \psi(y_0) \quad (54)$$

which is (47).

A situation not yet examined is $f(y^1) = f(y_0)$ and some $u_{i1}^* \ngeq 0$. Here a result very much like that for the linear case of Section 5.4 applies; there exists an alternative optimal basis for the ith linear subproblem in which at least one of the rows of A_{i1} corresponding to a negative component of u_{i1}^* is absent.

THEOREM 4 (ALTERNATIVE OPTIMAL BASES). *If $f(y^1) = f(y^0)$ and, for some i, u_{i1}^* has some negative components, an alternative optimal solution for the ith linear subproblem (with $y = y_0$) exists such that at least one row of the original basis A_{i1} corresponding to a negative component of u_{i1}^* is absent.*

PROOF. With $y = y_0$, a basis A_{i1} is optimal for the linear subproblem (7)–(8) if and only if it is optimal for the dual of (7)–(8), which is

$$\text{maximize } b_{i1}'(y_0)u_{i1} + b_{i2}'(y_0)u_{i2}$$

subject to

$$A_{i1}'u_{i1} + A_{i2}'u_{i2} = c_i \qquad (55)$$
$$u_{i1} \geq 0, \qquad u_{i2} \geq 0$$

The current basis A_{i1}' is feasible, i.e., $u_{i1}^0 = (A_{i1}^{-1})'c_i \geq 0$ and optimal, i.e., the relative cost factors for u_{i2} satisfy

$$b_{i2}(y_0) - A_{i2}x_i^0 \leq 0 \qquad (56)$$

The above relation is, of course, simply the requirement that (x_i^0, y_0) satisfy the bottom partition of the constraints of the ith linear block, (12). The vector u_{i1}^* is given by

$$u_{i1}^* = u_{i1}^0 - (A_{i1}^{-1})'A_{i2}'u_{i2}^1 \qquad (57)$$

Since u_{i1}^* has some negative components and $u_{i1}^0 \geq 0$, some components of u_{i2}^1 must be positive. By complementary slackness, the constraints of the coordinating problem corresponding to these components must be binding when $y = y_0$; i.e., some of the relative cost factors (56) must equal zero. If one or more nonbasic dual columns (primal rows) with zero relative costs can replace some current basic columns yielding a new feasible basis, the new basis is optimal. That such interchanges are possible has already been shown in Theorem 5.4-2, so the theorem is proved.

Summary. A summary of the algorithm follows:

1. Choose a vector $y_0 \in S$.
2. Solve the linear subproblems (7)–(8) with $y = y_0$, obtaining optimal solutions x_i^0, basis matrices A_{i1}, and dual variables u_{i1}^0.
3. Using the quantities obtained in step 2, form the coordinating problem (20)–(21) and solve it, obtaining an optimal solution y^1.
4. If $f(y^1) = f(y^0)$ and all u_{i1}^* in (23) are nonnegative, the solution $(\{x_i^0\}, y_0)$ is optimal.

5. If $f(y^1) = f(y^0)$ and $u_{i1}^* \ngeq 0$ for $i \in I$, then, by Theorem 4, there is an alternative optimal basis for the ith linear subproblem in which at least one row of A_{i1} corresponding to a negative component of u_{i1}^* is absent. Such a basis may be found by solving the linear program (32) of Section 5:4. Perform at least one change of basis and return to step 3 with new basis matrices, dual variables, etc.
6. If $f(y^1) < f(y_0)$ and all $h_i(y^1) \geq 0$, then, by Theorem 3, $\psi(y^1) < \psi(y_0)$. Go back to step 2 with y_0 replaced by y^1.
7. If $f(y^1) < f(y_0)$ and some $h_i(y^1) \ngeq 0$, compute
$$\theta_m = \max\{\theta \mid h_i(y_0 + \theta(y^1 - y_0)) \geq 0, i = 1, \ldots, p, 0 \leq \theta \leq 1\}$$
   (a) If $\theta_m > 0$, then, by Theorem 3, $\psi(y_0 + \theta_m(y^1 - y_0)) < \psi(y_0)$. Return to step 2 with y_0 replaced by $y_0 + \theta_m(y^1 - y_0)$.
   (b) If $\theta_m = 0$, then, for some i and j, $h_{ij}(y_0) = 0$ and $h_{ij}(y^1) < 0$. Then it can be shown that row j of A_{i2} may be interchanged with some row of A_{i1} to yield a new optimal basis for the ith linear subproblem. Using this new basis return to step 3.

To prove the statement made in step 7(b), note that $h_{ij}(y_0) = 0$ implies that the following relations hold:

$$A_{i1}x_i^0 = b_{i1}(y_0) \tag{58}$$
$$A_{i2}^j x_i^0 = b_{i2}^j(y_0) \tag{59}$$

where A_{i2}^j and b_{i2}^j are the jth rows of A_{i2} and b_{i2}, respectively. By assumption 3 $A_{i2}^j \neq 0$ so $A_{i2}^j A_{i1}^{-1} \neq 0$. Thus A_{i2}^j can be interchanged with some row of A_{i1} to yield a nonsingular matrix $\tilde{A}_{i1}$. The relation

$$\tilde{A}_{i1}x_i^0 = b_{i1}(y_0) \tag{60}$$

must hold and, since $\tilde{A}_{i1}$ is nonsingular, x_i^0 is the unique solution to (60). The dual objective $c_i x_i$ is thus unchanged, so $\tilde{A}_{i1}$ is a new optimal basis.

Since the linear subproblems are written with more inequalities than unknowns, a dual algorithm will be most efficient for their solution. A similar comment applies to the coordinating program, since its chief difficulty will ordinarily be in having a large number of constraints. The gradient projection method of Rosen [4] is thus appropriate for the coordinating problem, since it is a dual method whose efficiency depends principally upon the number of variables of a problem and only in a secondary way on the number of constraints. This method may also be used for the solution of the linear subproblems. It is shown in [4] that gradient projection will converge to the global solution of a program with convex objective and linear constraints as the number of iterations tends to infinity. A modification of this method which has exhibited more rapid convergence in experiments is described in [5].

An important computational aspect of the algorithm is that each iteration produces a feasible solution for the original problem with nonincreasing objective value. Thus computations may be terminated at any point with a solution which has an objective value no higher than all previous ones.

Finite convergence of this partitioning scheme is proved in the following theorem.

THEOREM 5 (FINITE CONVERGENCE). *If, in steps* 5 *and* 7, *no optimal basis is repeated, the procedure will find an optimal solution in a finite number of iterations.*

PROOF. Each cycle which does not satisfy the optimality test either reduces the coordinating problem objective $f(y)$ or examines alternative optimal bases in one or more subproblems. Each time $f(y)$ is reduced, we must have a different set of optimal bases $\{A_{i1}\}$ than at any previous cycle, since repetition of a set would imply repetition of a coordinating problem, hence of the minimal value of f. Consider a cycle when $f(y)$ remains the same. There are a finite number of optimal bases for a given value of y. If none of these optimal bases are repeated in steps 5 and 7, then, at any iteration, the current set of optimal bases cannot be the same as at any previous iteration. Since the total number of possible sets $\{A_{i1}\}$ is finite, an optimal set must be found in a finite number of iterations.

7.2.2 Use of Partition Programming in Refinery Optimization. Rosen, Ornea, and Eldredge [6, 7], with others, working at the Shell Development Company, Emeryville, California, have developed the basic partition programming scheme just described into a complete computing system for solving problems of optimal production and distribution in the petroleum industry. Inputs to the system are quantities such as quality specifications on products, availabilities and properties of input streams, customer demands, costs, etc., all in terms most convenient to the engineers and other analysts formulating the problem. These data are used to set up a mathematical program of the general form (7.2-1)–(7.2-2), with cost coefficients, right-hand-side vectors, and constraint matrices all allowed to be functions of y. This is then solved by a partition programming subroutine. Finally, the results are converted to a form most suitable to those formulating the problem, and are output. Because of these features, users need not be experts in mathematical programming.

The system, called SYMROS (System for Multi-Refinery Operations Scheduling), makes extensive use of process simulators. These are independent computer programs, defined within the system as subroutines. Their outputs are properties of the output streams of various refinery processing units, given the settings of the control variables associated with the unit (some of the y variables) and values of other parameters specific to the unit. This output is needed to form the matrices $A_i(y)$, right-hand-side vectors $b_i(y)$, and cost coefficients $c_i(y)$ in (7.2-1)–(7.2-2). Such simulators may be quite large, as a realistic description of processing units is usually quite complex. SYMROS is so structured that the simulators can be modified extensively without requiring any reprogramming in the rest of the system.

Since the functions comprising $A_i(y)$, $b_i(y)$, and $c_i(y)$ can be evaluated only by a large computer program, analytic determination of derivatives of these functions is not practical. The coordinating problem, (7.2.1-20)–(7.2.1-21), does require such derivatives, so these are evaluated by differentiating, entering the simulators once to evaluate the outputs at a base point y, and s more times to evaluate these at perturbed points $y + hu_i$, where h is positive and u_i is the ith unit vector. Care must be taken to ensure that the step size h is not too small for the limited accuracy of the simulators nor too large to yield a reasonable approximation to the derivative. Since SYMROS is too large to fit in core memory, repeated access to the simulators requires much transfer of programs from magnetic tape to core. Ornea and Eldredge [7] report that in large problems, more time is spent in these transfers than in actual computation.

As characterized in [6, 7] the partition programming subroutine uses Rosen's gradient projection method [4] to solve both the linear subproblems and the nonlinear coordinating program. A flow chart of the partition programming subroutine is shown in Figure 7-1 (for simplicity we assume that $\Delta y^j \neq 0$ if the optimality test is not passed). For a more detailed description of the system, the reader is referred to [6, 7].

7.3 Benders' Partitioning Algorithm for Mixed-Variable Programming Problems

7.3.1 Development of the Algorithm. In this section we discuss a method presented by Benders [8] for solving programming problems of the following form:

$$\left.\begin{array}{lr} \text{minimize } c'x + f(y) & \quad (1) \\ \text{subject to} \quad Ax + F(y) \geq b & \quad (2) \\ x \geq 0, \quad y \in S & \quad (3) \end{array}\right\} \text{problem P1}$$

The matrix A is $m \times n$, x and c n vectors, y a p vector, f a scalar-valued function of y, F an m vector whose components are functions of y, b an m vector, and S an arbitrary subset of E^p. The functions f and F need not be linear, and S may, for example, be the set of vectors in E^p with integral-valued components. In this case problem P1 is mixed, with some variables continuous, others discrete. An example of such a problem is given in Section 2.8.

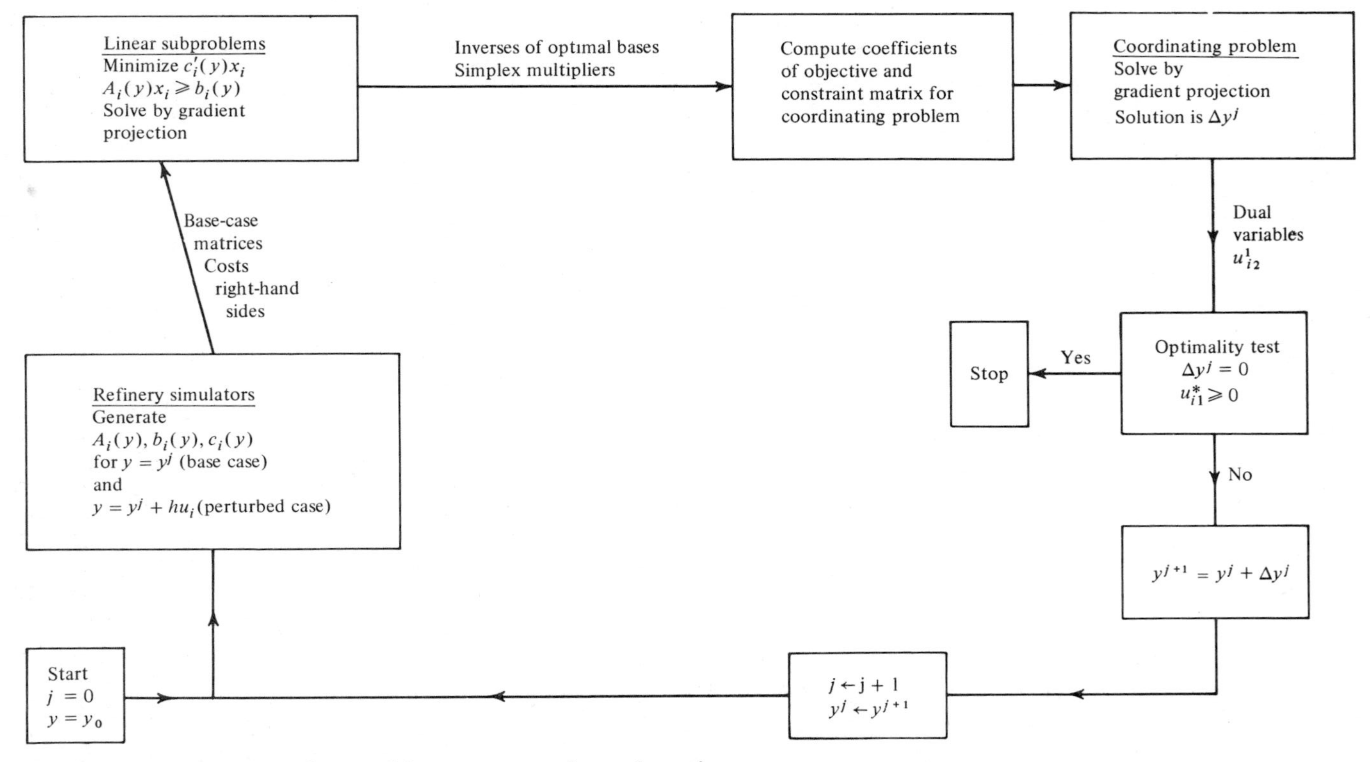

FIGURE 7-1 Flow chart for partition programming subroutine.

Since P1 is linear in x for fixed values of y, it is natural to attempt to solve it by fixing y, solving a linear program in x, obtaining a "better" y, etc. Of course, only values of y for which there exist x satisfying the resulting linear constraints may be considered. That is, y must lie in the set

$$R = \{y \mid \text{there exists } x \geq 0 \text{ such that } Ax \geq b - F(y),\ y \in S\} \tag{4}$$

Vectors $y \in R$ will be called feasible.

To formulate an explicit set of constraints determining R, Farkas' lemma [9] is used

FARKAS' LEMMA. *There exists a vector* $x \geq 0$ *satisfying* $Bx = a \Leftrightarrow a'u \geq 0$ *for all* u *satisfying* $B'u \geq 0$.

Applying this to the linear system (y fixed)

$$Ax - s = b - F(y) \tag{5}$$

$$x \geq 0, \qquad s \geq 0 \tag{6}$$

yields the conclusion that y is feasible if and only if

$$(b - F(y))'u \leq 0 \tag{7}$$

for all u satisfying

$$A'u \leq 0, \qquad u \geq 0 \tag{8}$$

Since the cone

$$C = \{u \mid A'u \leq 0,\ u \geq 0\} \tag{9}$$

is polyhedral, it has a finite number of generators. That is, there exist vectors u_i^r, $i = 1, \ldots, n_r$, such that any element $u \in C$ can be written

$$u = \sum_{i=1}^{n_r} \lambda_i u_i^r, \qquad \lambda_i \geq 0 \tag{10}$$

Substituting (10) into (7),

$$\sum_{i=1}^{n_r} \lambda_i (b - F(y))'u_i^r \leq 0 \tag{11}$$

which holds for all $\lambda_i \geq 0$ if and only if

$$(b - F(y))'u_i^r \leq 0, \qquad i = 1, \ldots, n_r \tag{12}$$

Thus a vector $y \in S$ is feasible if and only if it satisfies the finite system of constraints (12). The set R in (4) may therefore be written

$$R = \{y \mid (b - F(y))'u_i^r \leq 0, i = 1, \ldots, n_r, y \in S\} \tag{13}$$

If R is empty the original problem P1 is infeasible. Assuming R nonempty, we rewrite P1 as

$$\underset{y \in R}{\text{minimize}} \{f(y) + \min\{c'x \mid Ax \geq b - F(y), x \geq 0\}\} \tag{14}$$

This corresponds to our original notion of fixing y, solving a linear program in x, choosing better values for y, etc. For fixed y, the inner minimization above is a linear program, rewritten below along with its dual:

Primal

$$\text{minimize } c'x$$

subject to (15)

$$Ax \geq b - F(y)$$
$$x \geq 0$$

Dual

$$\text{maximize } (b - F(y))'u$$

subject to (16)

$$A'u \leq c$$
$$u \geq 0$$

If $y \in R$ the primal is feasible and, by the duality theorem of linear programming

$$\min\{c'x \mid Ax \geq b - F(y), x \geq 0\} = \max\{(b - F(y))'u \mid A'u \leq c, u \geq 0\} \tag{17}$$

where the max is taken as $-\infty$ if the dual is infeasible. Relation (17) then summarizes the facts that if the primal is unbounded, the dual is infeasible, while if both are feasible they have finite optimal solutions with equal objective values.

Substituting (17) into (14) yields a new form for P1:

$$\underset{y \in R}{\text{minimize}} \{f(y) + \max\{(b - F(y))'u \mid A'u \leq c, u \geq 0\}\} \tag{18}$$

Consider the constraint set of the dual, the polyhedron

$$P = \{u \mid A'u \leq c, u \geq 0\} \tag{19}$$

It is independent of y and the u_i^r in (9) are its extreme rays. If P is empty the primal, (15), is unbounded, as is the original problem P1. If P is nonempty, the inner maximum in (18) is taken on at an extreme point of P or approaches $+\infty$ along an extreme ray of P. The latter implies that the primal (15) is infeasible, contrary to our prior assumption, so we need only consider extreme points of P. There are a finite number of these, written u_i^p, $i = 1, \ldots, n_p$. The problem (18) may now be rewritten

$$\underset{y \in R}{\text{minimize}} \left\{ f(y) + \max_{1 \leq i \leq n_p} (b - F(y))' u_i^p \right\} \tag{20}$$

But this problem is equivalent to the following:

$$\text{minimize } z \tag{21}$$

subject to

$$z \geq f(y) + (b - F(y))' u_i^p, \qquad i = 1, \ldots, n_p \tag{22}$$

$$y \in R \tag{23}$$

Using the definition of R in (13) we may rewrite (21)–(23) and obtain a new problem, P2:

$$\text{minimize } z \tag{24}$$

subject to

$$z \geq f(y) + (b - F(y))' u_i^p, \qquad i = 1, \ldots, n_p \tag{25}$$

$$(b - F(y))' u_i^r \leq 0, \qquad i = 1, \ldots, n_r \tag{26}$$

$$y \in S \tag{27}$$

(24)–(27): problem P2

Previous statements regarding the relationship between problems P1 and P2 are summarized in the following theorem.

THEOREM 1 (EQUIVALENCE OF P2 AND P1)

(a) *P2 has a feasible solution* $\Leftrightarrow$ *P1 has a feasible solution.*
(b) *P2 is feasible without having an optimal solution* $\Leftrightarrow$ *P1 is feasible without having an optimal solution.*
(c) *If* (z^0, y^0) *solves P2 and* x^0 *solves the linear program*

$$\textit{minimize } c'x \tag{28}$$

subject to

$$Ax \geq b - F(y^0), \qquad x \geq 0 \tag{29}$$

then (x^0, y^0) *solves P1 and*

$$z^0 = c'x^0 + f(y^0) \tag{30}$$

(d) *If* (x^0, y^0) *solves P1 and* $z^0 = c'x^0 + f(y^0)$, *then* (z^0, y^0) *solves P2.*

The formal proof is left as an exercise.

For ease of discussion we let G be the constraint set of P2; i.e.,

$$G = \{(z, y)\} \mid (z, y) \text{ satisfies (25)–(27)}\} \tag{31}$$

so P2 may be written

$$\left.\begin{aligned} &\text{minimize } z \\ \text{subject to } \quad &(z, y) \in G \end{aligned}\right\} \text{problem P2}$$

Iteration by Successively Adding Constraints. Solution of the original problem has been reduced to solving P2 for (z^0, y^0) and then solving the linear program (28)–(29) to obtain the optimal values of x. Unfortunately, solution of P2 is hindered by the fact that it has one constraint for each extreme point and extreme ray of the polyhedron P, and this may be an enormous number in a problem of even moderate dimension. However, only a small fraction of the constraints will be binding at an optimal solution. This is a natural setting for applying the relaxation strategy of Section 5.2. We begin with only a few (or no) constraints, and solve the resulting modified P2, MP2. A relatively simple test is made to see if this solution satisfies the remaining constraints. If so, the solution is optimal, since the objective has been minimized over a set containing G. If not, a constraint is added which is not satisfied by the current solution, and the problem re-solved. The resulting algorithm involves iteration between two problems. The first is the problem MP2 in the variables (z, y) to which constraints are successively added. The second is the dual linear program (16) [or the primal (15)] which tests the optimality of a solution to MP2 and, if necessary, provides a new constraint. Note that, since we do not assume concavity of $-f(y)$ or $F(y)$, those steps of relaxation dealing with deletion of nonbinding constraints are no longer valid. This is because Theorem 5.2-1 no longer holds.

To develop the procedure in detail, the following assumptions are made.

Assumption 1. S is closed and bounded.

Assumption 2. $f(y)$ and the components of $F(y)$ are continuous on S.

These are satisfied in most applications and rule out the difficulties of feasible programs which have no optimal solutions. If S is not bounded, upper and lower bounds on the components of y may be added which are so large that either they are known to include the optimal solution or any solution exceeding these bounds has no realistic interpretation.

The following lemma is needed for future developments.

LEMMA 1. *Under assumptions* 1 *and* 2, *if P2 is feasible, then* z *has no finite lower bound on G if and only if P in* (19) *is empty.*

PROOF. $\Leftarrow$: By assumption, there is a point $(z^*, y^*) \in G$. If P is empty, then it has no extreme points or extreme rays, and the constraints determining G become simply

$$y \in S \tag{32}$$

and $(z, y^*) \in G$ for any value of z.

$\Rightarrow$: If P is not empty, then it has at least one extreme point, $\hat{u}$, so the constraints of P2 include the inequality

$$z \geq (b - F(y))'\hat{u} + f(y) \tag{33}$$

Then any $(z, y) \in G$ must have a z value satisfying

$$z \geq \min_{y \in S}\{(b - F(y))'\hat{u} + f(y)\} \tag{34}$$

But S is closed and bounded and F and f are continuous on S. Thus by the Weierstrass theorem, the right-hand side of (34) is finite and z has a finite lower bound on G.

Consider now a modified problem P2, MP2, in which only a proper subset of the constraints determining G are included.

$$\left.\begin{aligned} &\text{minimize } z \\ \text{subject to}\quad & z \geq f(y) + (b - F(y))'u_i^p, \quad i \in I_1 \\ & (b - F(y))'u_i^r \leq 0, \quad i \in I_2 \\ & y \in S \end{aligned}\right\} \text{problem MP2} \qquad \begin{aligned} &\\ &(35)\\ &(36)\\ &(37)\end{aligned}$$

where I_1 and I_2 are proper subsets of the integers $1, \ldots, n_p$ and $1, \ldots, n_r$, respectively. Let G' be the set of all (z, y) satisfying the constraints of MP2. Since not all constraints are included,

$$G' \supseteq G$$

so MP2 minimizes the objective of P2 over a constraint set containing that of P2. A solution to MP2 is thus optimal for P2 if and only if it lies in G, i.e., satisfies all constraints of P2. Of course, these constraints are not explicitly available, since the extreme points and rays of P have not been tabulated in advance. This is overcome by using the linear programs (15)–(16) to generate the constraint of P2 which is most unsatisfied by the current solution to MP2, or to show that all constraints are satisfied. This constraint or row generation is dual to the column-generation procedures discussed earlier in Sections 2.6 and 3.3 and Chapter 4.

Under assumptions 1 and 2, if MP2 is feasible it has a finite optimal solution (z^0, y^0). This solution is optimal for P2 if and only if

$$(b - F(y^0))'u_i^p \leq z^0 - f(y^0), \qquad i = 1, \ldots, n_p \tag{38}$$

$$(b - F(y^0))'u_i^r \leq 0, \qquad i = 1, \ldots, n_r \tag{39}$$

Consider now finding either the most unsatisfied constraint in (38) or a constraint of (39) which is unsatisfied. The most unsatisfied constraint of (38) is that with largest left-hand side, i.e., the constraint

$$\max_{1 \leq i \leq n_p} (b - F(y^0))'u_i^p \leq z^0 - f(y^0)$$

But if the linear function $(b - F(y^0))'u$ attains a finite maximum over P, it does so at an extreme point of P, so the above constraint may be generated by solving the linear program

$$\left.\begin{array}{ll} & \text{maximize } (b - F(y^0))'u \\ \text{subject to} & u \in P \end{array}\right\} \begin{array}{l}\text{linear} \\ \text{subproblem}\end{array} \tag{40}$$

This program is the dual, (16), with $y = y^0$. Since z^0 is assumed finite and

$$\min\{z \mid (z, y) \in G\} \geq z^0 \tag{41}$$

then, by Lemma 1, P is not empty. Thus this problem has either a finite optimal solution or an unbounded solution. If unbounded, its objective approaches $+\infty$ along the half line, $u_i^p + \lambda u_i^r$, $\lambda \geq 0$, in which case

$$(b - F(y^0))'u_i^r > 0 \qquad \text{for some } i$$

This implies that one of the constraints (39) is violated. Thus both sets of constraints (38), (39) are satisfied if and only if

$$\max\{(b - F(y^0))'u \mid A'u \leq c, u \geq 0\} \leq z^0 - f(y^0) \tag{42}$$

Note, however, that (38) must hold as an equality for some i; otherwise, z^0 could be reduced without violating any constraints. Thus, if (42) holds, it holds as an equality also, so the optimality test is as follows.

THEOREM 2 (OPTIMALITY TEST). *If* (z^0, y^0) *is optimal for MP2, it is optimal for P2 if and only if*

$$\max\{(b - F(y^0))'u \mid A'u \leq c, u \geq 0\} = z^0 - f(y^0) \tag{43}$$

If the optimality test is not passed, then some constraint of (38)–(39) is not satisfied, so

$$\max\{(b - F(y^0))'u \mid A'u \leq c, u \geq 0\} > z^0 - f(y^0) \tag{44}$$

If the above maximum is attained at an extreme point u^0,

$$(b - F(y^0))'u^0 > z^0 - f(y^0) \tag{45}$$

Referring back to the constraints of the problem P2, (38)–(39), the above states that the new constraint created from the extreme point u^0 is not satisfied by the solution to the current MP2. A new MP2 is then formed by adding this constraint.

If the linear subproblem (40) is unbounded, then the simplex method leads to an extreme point of P, u^0, and an extreme ray of P, v^0, such that $(b - F(y^0))'u$ tends to $+\infty$ along the half-line

$$u = u^0 + \lambda v^0, \qquad \lambda \geq 0 \tag{46}$$

For this to happen, v^0 must satisfy

$$(b - F(y^0))'v^0 > 0 \tag{47}$$

so the new constraint of the form (39) created from the extreme ray v^0 is not satisfied by the current solution (z^0, y^0). This constraint is then added to the current MP2, yielding a new problem.

In the unbounded case, it may happen that (45) holds for the extreme point u^0 in (46). In this case both the constraints (45) and (47) are added to MP2.

Summary of the Algorithm. An iterative procedure based on the preceding results is given below.

1. Initiate the procedure with problem MP2, where only a few (or no) constraints of the form (35)–(36) are included.
2. Solve problem MP2. If MP2 is infeasible, so are P2 and P1. Otherwise, we obtain either a finite optimal solution (z^0, y^0) or the information that the solution is unbounded. If unbounded set $z^0 = -\infty$, let y^0 be an arbitrary element of S and go to step 3.

3. Solve the dual linear program (16) [or the primal, (15), if it is feasible]. If the dual is infeasible then, by Lemma 1, the original problem P1 has an unbounded solution. If the dual is unbounded, go to step 6.
4. If the optimal objective value in step 3 is equal to $z^0 - f(y^0)$, the solution (z^0, y^0) solves P2. If x^0 solves (15), then (x^0, y^0) solves P1.
5. If the optimality test in step 3 is not passed and the dual, (16), has a finite optimal solution, u^0, then

$$z^0 < f(y^0) + (b - F(y^0))'u^0 \tag{48}$$

so the current solution to MP2 does not satisfy the constraint

$$z \geq (b - F(y))'u^0 + f(y) \tag{49}$$

add this constraint to MP2 and return to step 2.
6. If the dual, (16), has an unbounded solution, the simplex method locates an extreme ray v^0 and extreme point u^0 such that the dual objective approaches $+\infty$ along the half-line

$$u = u^0 + \lambda v^0, \qquad \lambda \geq 0 \tag{50}$$

The inequality

$$(b - F(y^0))'v^0 > 0 \tag{51}$$

is satisfied, so y^0 does not satisfy the constraint

$$(b - F(y))'v^0 \leq 0 \tag{52}$$

Add this constraint to MP2. If, in addition,

$$z^0 < f(y^0) + (b - F(y^0))'u^0 \tag{53}$$

for the extreme point u^0 in (50), add a constraint of the form (49) to MP2. Return to step 2.

Finite convergence of the algorithm follows directly from the finite number of constraints of P2.

THEOREM 3 (FINITE CONVERGENCE). *The above iterative procedure will terminate in a finite number of iterations, either with the information that P1 is infeasible or unbounded, or with an optimal solution to P1.*

PROOF. Problem P2 has only a finite number of constraints. If the optimality test is not passed, then one or more new constraints are added to

MP2. Thus, in a finite number of iterations either the optimality test is passed or a full set of constraints will be obtained, whereupon the test will be met at the next step (by Theorem 1). P1 is infeasible if and only if MP2 is infeasible at some stage. P1 is unbounded if and only if the dual linear program (16) is infeasible, which will be detected in the first step.

Upper and Lower Bounds. An attractive feature of this algorithm is the availability of upper and lower bounds on the optimal objective value, which both converge to this value as optimality is achieved. The upper bound is generated by a sequence of feasible solutions to P1, so the best of these may be taken as a solution if the procedure is terminated short of optimality.

To obtain the upper bound, let (z^i, y^i) solve MP2 at step i and assume the primal linear program (15) is feasible when $y = y^i$. If x^i solves (15), then (x^i, y^i) is feasible for P1 and

$$c'x^i + f(y^i) \geq \min\{z \mid (z, y) \in G\} \tag{54}$$

The lower bound arises from the fact that constraints are being added to MP2, so if z^i is the optimal objective value for MP2 at step i,

$$z^i \leq z^{i+1} \leq \min\{z \mid (z, y) \in G\} \tag{55}$$

Equations (54) and (55) yield the best set of bounds at step i as

$$z^i \leq \min\{z \mid (z, y) \in G\} \leq \min_{1 \leq j \leq i} (c'x^j + f(y^j)) \tag{56}$$

As the iterative procedure converges, both right- and left-hand sides of (56) converge to the middle quantity. In fact, the optimality test (43) is simply the condition that these bounds be equal.

Solution of the Problem MP2. Although the procedure applies formally to a very general class of problems, it is practical only when MP2 can be solved efficiently. Instances when this occurs are the following.

1. S is the set of vectors in E^p with nonnegative integral-valued components, $F(y) = By$ and $f(y) = d'y$. Problem P1 is then a linear program with some integer and some continuous variables. Problem MP2 is a pure integer linear program, for which a large number of techniques are available [10]. Since successive problems MP2 differ by the addition of one or two constraints, a cutting-plane method for solving MP2, such as that of Gomory, may be efficient. Benders' algorithm is evidently considered most promising in this application [10].
2. S is determined by a set of linear and nonlinear inequalities

$$S = \{y \mid g_i(y) \leq 0,\ i = 1, \ldots, r,\ a_j'y \leq b_j,\ j = 1, \ldots, s\}$$

with the g_i nonlinear but continuous and differentiable. The functions $f(y)$ and $F(y)$ may be nonlinear but are continuous and differentiable. Then MP2 has a linear objective but nonlinear constraints, and a mixed interior—exterior penalty method (see [11], [12], and Section 1.3.4) seems appropriate. Currently satisfied constraints are maintained satisfied by the interior term, while new constraints are incorporated through an exterior penalty term. In this way, the previous optimal solution to MP2 may be used as a starting point for the new penalty function minimization. If F is concave and f and the g_i are convex, MP2 is a convex program. Then, with mild addition assumptions [12], such a penalty algorithm will converge to an optimal solution, as will a variety of other methods. Note that, in this case, nonbinding constraints may be dropped from MP2, as explained in Section 5.2.

In any application, Benders' procedure has the desirable property that any special structure the matrix A may have is preserved, which is ordinarily not true if the problem is solved in an integrated way (i.e., without some form of partitioning). Thus, if A is of the transportation type, then the primal, (15), will be a transportation problem whose right-hand side changes at each step. In this case it is desirable to solve (15) rather than its dual (16). Since, at each iteration (15) changes only in its right-hand side, the optimal dual solution at iteration i is feasible for the new dual at iteration $i + 1$. A dual linear programming algorithm thus seems most appropriate for its solution. Of course, if (15) is infeasible, then a dual extreme ray must be generated. This is easily done if either the primal or dual simplex methods are used (see Problems 6 and 7).

7.3.2 Relation to the Decomposition Principle and Cutting-Plane Algorithms. To see the relations between Benders' algorithm and the Dantzig–Wolfe decomposition procedure, let the original problem (7.3.1-1)–(7.3.1-3) be a linear program, shown below with its dual:

Primal

$$\text{minimize } c'x + d'y$$

subject to (1)

$$Ax + By \geq b$$
$$x \geq 0, \qquad y \geq 0$$

Dual

$$\text{maximize } b'u$$

subject to (2)

$$B'u \leq d$$
$$A'u \leq c$$
$$u \geq 0$$

For simplicity, assume that the polyhedron

$$P = \{u \mid A'u \leq c,\ u \geq 0\} \tag{3}$$

is bounded. Then Benders' problem P2, given in (7.3.1-24)–(7.3.1-27) becomes

$$\left.\begin{array}{lll} & \text{minimize } z & \\ \text{subject to} & z + (By - b)'u_j \geq d'y, & \text{all } j \\ & y \geq 0 & \end{array}\right\} \begin{array}{l}\text{Benders'}\\ \text{problem}\\ \text{P2}\end{array} \tag{4}$$

where u_j is the jth extreme point of P. The problem MP2 in (7.3.1-35)–(7.3.1-37) is the same as the above with j restricted to some subset of its range, i.e., with only a few constraints. The optimality test in (7.3.1-43) is

$$\max\{(b - By^0)'u \mid A'u \leq c, u \geq 0\} = z^0 - d'y^0 \left.\right\} \begin{array}{l}\text{Benders'}\\ \text{optimality}\\ \text{test}\end{array} \tag{5}$$

where (z^0, y^0) solves MP2. To perform this test, one must obviously solve the linear subproblem

$$\left.\begin{array}{ll} & \text{maximize } (b - By^0)'u \\ \text{subject to} & A'u \leq c \\ & u \geq 0 \end{array}\right\} \begin{array}{l}\text{Benders'}\\ \text{linear}\\ \text{subproblem}\end{array} \tag{6}$$

Now let us apply the decomposition principle to the dual, (2). We write any $u \in P$ as a convex combination of extreme points of P:

$$u = \sum_j \lambda_j u_j, \qquad \sum_j \lambda_j = 1, \qquad \lambda_j \geq 0 \tag{7}$$

and substitute the result into the remaining dual constraints and the dual objective, yielding the following master program:

$$\left.\begin{array}{ll} & \text{maximize } \sum_j (b'u_j)\lambda_j \\ \text{subject to} & \sum_j (B'u_j)\lambda_j \leq d \\ & \sum_j \lambda_j = 1 \\ & \lambda_j = 0 \end{array}\right\} \begin{array}{l}\text{master}\\ \text{program}\end{array} \tag{8}$$

The dual of the master program is

$$\text{minimize } d'y + v$$

subject to

$$v + (By)'u_j \geq b'u_j$$
$$y \geq 0$$

or, letting

$$z = d'y + v \tag{9}$$

$$\left.\begin{array}{ll} & \text{minimize } z \\ \text{subject to} & \\ & z + (By - b)'u_j \geq d'y, \quad \text{all } j \\ & y \geq 0 \end{array}\right\} \begin{array}{l}\text{dual of} \\ \text{master} \\ \text{program}\end{array} = \begin{array}{l}\text{Benders'} \\ \text{program} \\ \text{P2}\end{array} \tag{10}$$

Comparing with (4), Benders' program P2 and the dual of the master program are seen to be the same. The program MP2 is the dual of a restricted master program, containing only a subset of the possible columns.

To bring out additional dual relationships, let (y^0, v^0) be the simplex multipliers corresponding to the optimal solution of some restricted master program. To see if this solution solves the master program, we compute the maximal relative cost factor

$$\bar{c}_j = b'u_j - (y^0)'B'u_j - v^0 \tag{11}$$

Since u_j is an extreme point of P, maximizing $\bar{c}_j$ is equivalent to solving the subproblem

$$\left.\begin{array}{ll} & \text{maximize } (b - By^0)'u \\ \text{subject to} & A'u \leq c \\ & u \geq 0 \end{array}\right\} \begin{array}{l}\text{subproblem of} \\ \text{decomposition} \\ \text{principle}\end{array} = \begin{array}{l}\text{Benders'} \\ \text{subproblem}\end{array} \tag{12}$$

which is Benders' subproblem given in (6). The optimality test for the master program is

$$\max \bar{c}_j \leq 0 \tag{13}$$

or, by (11),

$$\max_j (b - By^0)'u_j \leq v^0 \tag{14}$$

It was shown in Section 3.6 that the above holds with equality. From (9),

$$z^0 = d'y^0 + v^0 \tag{15}$$

and (14) becomes

$$\max\{(b - By^0)'u \mid A'u \leq c, u \geq 0\} = z^0 - d'y^0 \left.\begin{matrix}\text{optimality test}\\ \text{for master}\\ \text{program}\end{matrix}\right\} = \begin{matrix}\text{Benders'}\\ \text{optimality}\\ \text{test}\end{matrix} \tag{16}$$

Comparing with (5), this is seen to be Benders' optimality test.

Summing up, we have shown that Benders' algorithm and the Dantzig–Wolfe decomposition principle are really duals of one another. If one considers the dual of the primal–dual pair (1)–(2) (Benders considers the primal), then solving this dual using the decomposition principle is precisely the same as Benders' algorithm. Of course, when one considers nonlinear programs, there are important differences between the two procedures. The most important is that Benders' algorithm can handle a much wider variety of programs than any current extension of the decomposition principle.

Benders' procedure is also quite similar to the cutting-plane methods of integer programming [10] and of convex programming [13]. Problem P2 is solved by successively cutting off solutions which do not satisfy all constraints, as illustrated in Figure 7-2. Constraints ① and ② correspond to

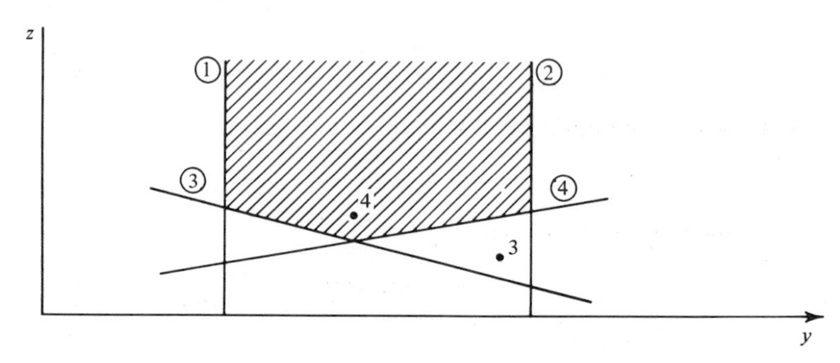

FIGURE 7-2

extreme rays of P. For the case of a single y variable, these simply bound y. Constraint ③ corresponds to an extreme point of P and, under assumptions 1 and 2, ensures that min z is finite. The set S requires that y be a nonnegative integer, so point 3 is the optimal solution for constraints ①–③. Constraint ④ cuts this solution off, yielding the shaded region as the constraint set of MP2. Point 4 is the new optimal solution, with larger z value than point 3.

7.3.3 Application to a Warehouse Location Problem. Balinski and Wolfe [10, 14] have applied Benders' algorithm to a fixed-charge warehouse-location problem. The general problem is: Given a plant producing a single product, a finite set of locations at which warehouses may be built, and a set of customers with known demands, determine which locations to use so that the total cost of shipping to and storage at the warehouses, and of shipping from there to the customers is minimized. The problem considered by Balinski and Wolfe is a specialized version of the above with the following assumptions:

1. The plant and all warehouses have infinite capacity.
2. The cost of shipping to and storage at a warehouse is a fixed positive quantity if a positive amount is shipped and zero otherwise, as shown in Figure 7-3 for warehouse i. The fixed cost represents the capital investment required to construct the warehouse. This would be the only variable cost if costs of shipping to and storing at all warehouses were linear and equal.

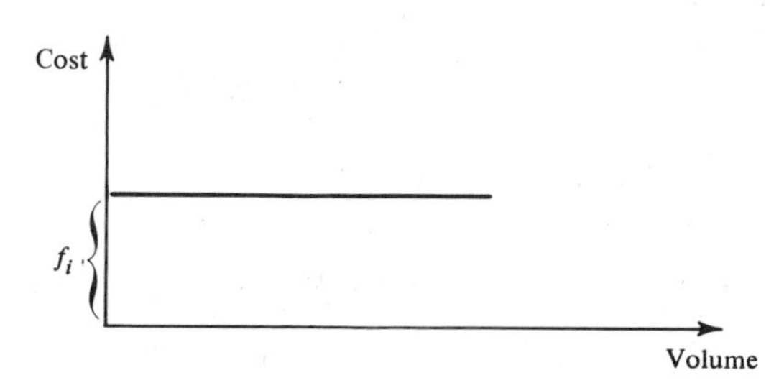

FIGURE 7-3

3. The cost of delivery to customer j from warehouse i is linear, with unit cost $\hat{c}_{ij} \geq 0$.

The problem may be formulated as a mixed integer–continuous variable linear program, with 0–1 variables representing the alternatives of building or not building a warehouse. If $\hat{x}_{ij}$ is the quantity shipped from warehouse i to customer j, then

$$\sum_i \hat{x}_{ij} \geq d_j, \qquad j = 1, \ldots, n, \quad \hat{x}_{ij} \geq 0 \tag{1}$$

Defining

$$x_{ij} = \frac{\hat{x}_{ij}}{d_j}, \qquad c_{ij} = d_j \hat{c}_{ij} \tag{2}$$

the mixed-variable program is as follows.

$$\left.\begin{aligned} &\text{minimize} \sum_i f_i y_i + \sum_{i,j} c_{ij} x_{ij} && (3)\\ &\text{subject to} \quad \sum_i x_{ij} \geq 1, \quad j = 1, \ldots, n && (4)\\ &0 \leq x_{ij} \leq y_i, \quad \text{all } i, j && (5)\\ &y_i = 0 \text{ or } 1 && (6) \end{aligned}\right\} \text{primal problem}$$

If $y_i = 1$, warehouse i is used, since (5) places no additional restriction on x_{ij}, and a cost f_i is incurred. If $y_i = 0$, $x_{ij} = 0$ for all j and the cost contribution is zero.

Let $y = (y_1, \ldots, y_m)$ be fixed, so (3)–(5) is a linear program. Benders' algorithm adds constraints to the pure integer program MP2 by solving the dual of (3)–(5), which is

$$\left.\begin{aligned} &\text{maximize} \sum_j v_j - \sum_{i,j} y_i u_{ij} && (7)\\ &\text{subject to} \quad v_j - u_{ij} \leq c_{ij} && (8)\\ &v_j \geq 0, \quad u_{ij} \geq 0 && (9) \end{aligned}\right\} \text{dual problem}$$

The polyhedron P in (7.3.1-19) is, in this application, the set of all $\{v_j, u_{ij}\}$ satisfying (8)–(9), and the problem P2 is

$$\text{minimize } z$$

$$\text{subject to} \quad z \geq \sum_i v_i^p - \sum_{i,j} u_{ij}^p y_i + \sum_i f_i y_i \tag{10}$$

$$\sum_i v_i^r - \sum_{i,j} u_{ij}^r y_i \leq 0 \tag{11}$$

$$y_i = 0 \text{ or } 1 \tag{12}$$

with (10) holding for all extreme points $\{v_j^p, u_{ij}^p\}$ of P and (11) for all extreme rays.

As mentioned earlier, the inequalities (11) are necessary and sufficient for y to be feasible, i.e., to admit feasible values of x_{ij} in (3)–(5). However, examination of (3)–(5) shows that any $y \neq 0$ is feasible, so (11) can be dropped as long as the condition $y \neq 0$ is enforced.

The algorithm can be initiated with a few (or no) constraints of the form (10) in problem P2. Let the solution to this problem, MP2, be y^0. The primal–dual pair of problems (3)–(5) and (7)–(9) must now be solved, with y

fixed at y^0. In this case, optimal solutions may be obtained by inspection. Let the current set of open sources be

$$I = \{i \mid y_i^0 = 1\} \tag{13}$$

We assume that $y \neq 0$ is enforced, so I is not empty. The primal (3)–(5) then becomes

$$\text{minimize} \sum_{i \in I} \sum_j c_{ij} x_{ij} \tag{14}$$

$$\text{subject to} \quad \sum_{i \in I} x_{ij} \geq 1, \quad j = 1, \ldots, n, \quad x_{ij} \geq 0 \tag{15}$$

Since $c_{ij} \geq 0$, (15) holds with equality in an optimal solution. Since the constraints (15) are independent, the solution is obtained by setting $x_{ij} = 1$ for that c_{ij} which is minimal and zero otherwise; i.e.,

$$x_{k_j j}^0 = 1, \quad \text{where } c_{k_j j} = \min_{i \in I} c_{ij}, \quad j = 1, \ldots, n \tag{16}$$

$$x_{ij}^0 = 0, \quad \text{otherwise} \tag{17}$$

This solution supplies customer j from the cheapest open source, k_j. To obtain the corresponding dual solution, complementary slackness implies

$$v_j = u_{k_j j} + c_{k_j j}, \quad j = 1, \ldots, n \tag{18}$$

The remaining dual constraints are

$$u_{ij} \geq 0 \tag{19}$$

$$u_{ij} \geq v_j - c_{ij}, \quad i \neq k_j \tag{20}$$

Substituting (18) into the dual objective (7), we see that this objective is independent of the $u_{k_j j}$, since $\sum_j u_{k_j j}$ appears twice with opposite signs. Thus we may as well set $u_{k_j j} = 0$. To maximize the remaining terms, u_{ij} should be made as small as possible subject to (19)–(20), i.e., should be set equal to the maximum of the right-hand-side values. This yields the optimal dual solution

$$v_j^0 = c_{k_j j}, \quad j = 1, \ldots, n \tag{21}$$

$$u_{ij}^0 = \max\{0, v_j^0 - c_{ij}\}, \quad i \neq k_j \tag{22}$$

Let d^0 be the optimal dual objective value and z^0 the optimal value obtained in solving MP2. If

$$d^0 = z^0 - \sum_i f_i y_i^0 \tag{23}$$

then, by Theorem 7.3.1-2, the current solution is optimal. If not, form a new constraint from $\{v_j^0, u_{ij}^0\}$,

$$z \geq \sum_i v_i^0 - \sum_{i,j} u_{ij}^0 y_i + \sum_i f_i y_i \tag{24}$$

and add this to MP2. An interpretation of this constraint and a suggested alternative is found in [10].

7.3.4 Numerical Example. The following example, taken from [10, p. 293], is a location problem of the form (7.3.3-3)–(7.3.3-6), with four locations and customers, $f_i = 7$, all i, and

$$(c_{ij}) = \begin{bmatrix} 0 & 12 & 20 & 18 \\ 12 & 0 & 8 & 6 \\ 20 & 8 & 0 & 6 \\ 18 & 6 & 6 & 0 \end{bmatrix}$$

CYCLE 1. Initially MP2 has no constraints. Choose $y^0 = (0 \quad 1 \quad 0 \quad 0)$ and set $z^0 = -\infty$. Solving the primal–dual pair of linear programs (3)–(5) and (7)–(9) yields

$$\sum_{i,j} c_{ij} x_{ij}^0 + \sum_i f_i y_i^0 = 33 \tag{1}$$

$$v_1^0 = 12, \quad v_2^0 = 0, \quad v_3^0 = 8, \quad v_4^0 = 6$$

$$u_{11}^0 = 12, \quad u_{33}^0 = 8, \quad u_{44}^0 = 6, \quad \text{all other } u_{ij}^0 = 0$$

The initial problem MP2 is

$$\text{minimize } z$$

subject to

$$z \geq 26 - 5y_1 + 7y_2 - y_3 - y_4 \tag{2}$$

$$y_i = 0 \text{ or } 1 \tag{3}$$

The optimal solution minimizes the right-hand side of the above constraint, setting $y_i = 1$ for negative coefficients and zero otherwise, and is

$$y^0 = (1 \quad 0 \quad 1 \quad 1) \tag{4}$$

with optimal objective value

$$z^0 = 19 \tag{5}$$

The value in (1) is an upper bound on the optimum and that in (5) a lower bound. Since these are not equal, the current solution is not optimal.

CYCLE 2. Solving (3)–(5) and (7)–(9) with y set to the above value yields

$$\sum_i f_i y_i^0 + \sum_{i,j} c_{ij} x_{ij}^0 = 27$$
$$v_1^0 = v_3^0 = v_4^0 = 0, \qquad v_2^0 = 6$$
$$u_{22}^0 = 6, \qquad \text{all other } u_{ij}^0 = 0$$

The new constraint to be added is

$$z \geq 6 + 7y_1 + y_2 + 7y_3 + 7y_4 \tag{6}$$

and the new problem MP2 is to minimize z subject to (2), (3), and (6). This problem is equivalent to

$$\min_{y_i = 0 \text{ or } 1} \max\{\text{rhs (2), rhs (6)}\} \tag{7}$$

the solution can be found by enumeration to be

$$y^0 = (1 \quad 0 \quad 0 \quad 1) \text{ or } (1 \quad 0 \quad 1 \quad 0), \qquad z^0 = 20$$

again upper and lower bounds are not equal, so another cycle must be initiated.

Table 1 summarizes the behavior of the algorithm on this problem. In a row of this table, y^0 is the optimal solution to the problem MP2 whose constraints are given in all the above lines, with optimal objective value y^0 given in the line immediately above [e.g., $y^0 = (1 \quad 0 \quad 1 \quad 0)$ is the optimal solution to the problem with three inequalities, with minimal value $z^0 = 20$]. The column headed $\sum_{i,j} c_{ij} x_{ij}^0 + \sum_i f_i y_i^0$ is the minimal cost for (3)–(6) given that $y = y^0$ and is an upper bound on the minimal cost. The value z^0 is a lower bound.

7.3.5 Computational Experience. Buzby et al. [15] have applied Benders' algorithm to a number of distribution problems with piecewise linear, possibly discontinuous, costs. The problem is to produce and ship at least cost a single product from m plants to n customers. Customer demands are known and plant capacities are given. Transportation costs are linear, but production costs are piecewise linear in some plants. These problems represent a slight simplification of those in Section 2.8, in that no warehouses are present, and a slight generalization of those in Section 7.3.3, in that general piecewise linear costs are allowed.

Such problems may be solved by solving a finite number of standard transportation problems. One simply takes one piecewise linear cost segment from each plant with such costs and uses the slope of that segment as the overall linear cost of the plant. Taking all possible distinct combinations of

TABLE 1

$y^0 = (y_1^0, y_2^0, y_3^0, y_4^0)$	$\sum_{ij} c_{ij} x_{ij}^0 + \sum f_i y_i^0$	$z \geq (\ )$	$+ (\ ) y_1$	$+ (\ ) y_2$	$+ (\ ) y_3$	$+ (\ ) y_4$	z^0
(0 1 0 0)	33	26	−5	7	−1	−1	19
(1 0 1 1)	27	6	7	1	7	7	20
(1 0 0 1)	26	12	7	1	1	7	20
(1 0 1 0)	28	14	7	−1	7	−1	21
(1 0 0 0)	57	50	7	−29	−29	−31	24
(0 0 1 1)	38	24	−11	−5	7	7	26
(1 0 0 1)	26						

such segments and solving a transportation problem for each combination, the optimal solution of the original problem may be obtained. Let there be N_T combinations. If, for example, one plant had three piecewise linear segments and a second had four, there would be $N_T = 12$ separate transportation problems to solve.

The problem may be formulated as a mixed integer–continuous variable linear program (see Section 2.8) to which Benders' algorithm is applicable. This algorithm also solves a finite number of transportation problems, one each time a new integer program is solved. The ratio of the number of such problems solved by the algorithm, N_a, to the number needed to solve the problem by enumeration, N_T, is a measure of the efficiency of the algorithm.

In the computational results reported by Buzby et al. in [15], 135 problems of the type described above were solved. Let

m = number of plants
n = number of customers
m_p = number of plants with piecewise linear costs
r = total number of piecewise linear segments over all m_p plants

These problems had values of the above parameters in the ranges

$$7 \le m \le 19$$
$$7 \le n \le 30$$
$$2 \le m_p \le 6$$
$$5 \le r \le 24$$

Other data, e.g., transportation costs, plant capacities, etc., were generated randomly between fixed ranges. Results are shown in Table 2, where the

TABLE 2

Range of possible alternatives	No. of samples	Range of alternatives examined	Average number of alternatives examined	Range of % efficiency	Average % efficiency
0–20	14	2–12	5.4	13–100	43
21–40	18	2–20	7.8	6.3–56	32
41–60	16	2–22	7.6	4.4–46	16
61–100	15	2–23	12.7	2.1–32	16
101–150	15	2–30	12.0	1.4–21	9.1
151–300	19	2–45	14.6	0.93–23	6.7
300–600	20	2–78	17.7	0.37–20	3.9
600–1000	10	2–48	15.5	0.2–5.0	1.8
1000–2000	5	5–52	37.4	0.3–3.6	2.3
2000–4000	3	3–24	11.7	0.1–0.8	0.44

word alternative signifies a feasible set of values for the integer variables, each of which singles out one of the N_T transportation problems. The efficiency is simply $100N_a/N_T$.

It is easily seen that the percentage efficiency improves substantially as N_T increases; i.e., the algorithm becomes more efficient as the problem becomes harder. Buzby et al. also report that in problems where many of the proposed facilities are uneconomical, i.e., have a large capitalization cost, these are quickly eliminated from competition by the constraints added to the integer program. Thus, in such problems, only a few alternatives are examined. In the 135 problems solved the number of feasible alternatives was reduced, on the average, by a factor of approximately 1.7 at each step.

A summary of data for a particular sample of these problems has been provided by Balinski [10], and is shown in Table 3. These data, of course,

TABLE 3

	P_1	P_2	P_3	P_4	P_5	P_6	P_7	P_8
m_0	3	4	4	5	5	6	6	6
r	7	19	20	17	22	16	22	23
m	10	13	13	16	16	19	19	19
n	14	18	19	24	26	22	28	30
N_a	12	14	21	3	43	38	52	8
N_T	12	450	405	432	1200	288	1944	2592
$100N_a/N_T$	100	3.1	5.2	0.7	3.6	13	2.7	0.3

yield the same conclusions as Table 2. However, as noted in [10], the true usefulness of Benders' algorithm in mixed-variable linear programs depends heavily on the efficiency of the integer programming algorithm used to solve problem MP2. A type of partial enumeration scheme was used to solve the above problems. Another such algorithm has been proposed, in the context of Benders' algorithm, by Zoutendijk [16]. How well these, or other schemes, can solve the integer program MP2 is largely unknown at present.

Problems

1. Show that, if $b_i(y)$ is convex, then the set S in (7.2.1-3) is convex, and $\psi(y)$ is convex over S.
2. Prove the necessity portion of Theorem 7.2.1-2.
3. Show that, if $h_i(y)$ in (7.2.1-44) is convex for all i, then $y^1 \in S$ and $\psi(y^1) < \psi(y_0)$.
4. Show that an alternative formulation of the problem P2 in Section 7.3.1 replaces the objective (7.3.1-24) by $z + f(y)$ and the constraints (7.3.1-25) by

$$z \geq (b - F(y))'u_i^p, \qquad i = 1, \ldots, n_p$$

Develop an iterative algorithm based on this formulation and compare it with the algorithm in Section 7.3.1. Writing P2 in this way places $f(y)$ in its "natural" position in the objective, rather than in the constraints.

5. Consider the possibility of dropping constraints from MP2 when they are not binding at the current optimum. Give conditions on F, f, and S and a strategy for dropping constraints which guarantees finite termination.

6. Show how to construct a dual extreme ray from the final dual simplex tableau if that algorithm shows the primal (7.3.1-15) to be infeasible. Also suggest a way in which more than one such ray might be generated in each application of the dual simplex method. How would this be used in Benders' algorithm and what advantages would it have?

7. Assume that the primal simplex method is used to solve the primal (7.3.1-15), and that (7.3.1-15) is infeasible with the current vector y. Show how a dual extreme ray may be found from the final phase 1 tableau.

8. Transform problem P1 into the equivalent form

$$\text{minimize } z$$

$$\text{subject to} \quad \begin{aligned} &Ax + F(y) \geq b \\ &z \geq c'x + f(y) \\ &x \geq 0, \qquad y \in S \end{aligned}$$

Apply Farkas' lemma to this system, and obtain problem P2 and the statements of Theorem 7.3.1-1. (See [8] for a derivation of this type.)

9. As an extension of Benders' algorithm consider the problem

$$\text{minimize } g(x) + f(y)$$

$$\text{subject to} \quad \begin{aligned} &G(x) + F(y) \geq b \\ &x \in S_1, \qquad y \in S_2 \end{aligned}$$

where g is a convex function, G a vector of such functions, S_1 a convex set, f and F not necessarily convex, and S_2 an arbitrary subset of E^p. This is a nonlinear program in which the convex and nonconvex parts are separable. Attempt to extend Benders' algorithm to this class of problems by applying the duality theory of nonlinear programming.

10. Consider the application of various integer programming algorithms to the problem MP2, in the case where P1 is a mixed-variable linear program. Which appear most suitable? Is it possible to design a special-purpose algorithm which would be even more efficient?

11. Consider the application of Benders' algorithm to the case where A is block diagonal, i.e., has the structure

$$A = \begin{bmatrix} A_1 & & & \\ & A_2 & & \\ & & \ddots & \\ & & & A_m \end{bmatrix}$$

Show that, in this case, the primal–dual pair of linear programs (7.3.1-15)–(7.3.1-16) breaks up into m smaller subproblems. Also show that the possibility exists here of adding more than one constraint to MP2 at each step. How does the optimality test specialize in this case?

This idea of adding more than one constraint corresponds to the restricted master program of the Dantzig–Wolfe decomposition principle, where more than one of the subproblem proposals may enter the basis. Would you expect increased efficiency in Benders' algorithm in this case? Why?

12. Take the dual of the problem used as an example in Section 3.5. Set up Benders' problem P2 and the primal and dual subproblems, (7.3.1-15)–(7.3.1-16). Note the dual relationship between P2 and the master program of Section 3.5.

13. Compare and contrast Benders' procedure with Rosen's partitioning procedure of Chapter 5 when both are applied to linear programs with dual angular structure.

14. Consider a multi-item scheduling problem with "continuous" setups, of the type discussed in Section 4.2, but with three classes of items, each with its own type of machines. Machines of type j not used by class j items can be used by items of class i if $i < j$. Formulate this problem as a mathematical program having the structure of three separate multi-item scheduling problems linked by coupling variables. Apply Benders' algorithm to the problem. Exhibit P2 and its dual, and show that the (primal) subproblems are multi-item scheduling problems with variable right-hand sides. Can these subproblems be solved efficiently by modifying the dual simplex method, or must the procedure of Section 4.2 be used, and why?

15. Consider a problem as in Problem 14 above, but where class i items can use class j machines only if $i = j - 1$. Repeat the analysis of Problem 14 for this case.

References

1. J. B. Rosen, "Convex Partition Programming," in *Recent Advances in Mathematical Programming*, R. L. Graves and P. Wolfe, eds., McGraw-Hill, Inc., New York, 1963, pp. 159–176.
2. H. W. Kuhn and A. W. Tucker, "Nonlinear Programming," in *Proceedings of the Second Berkeley Symposium on Mathematical Statistics and Probability*, Jerzy Neyman, ed., University of California Press, Berkeley, 1950, pp. 481–492.
3. S. Karlin, *Mathematical Methods and Theory in Games, Programming, and Economics*, Vol. 1, Addison-Wesley Publishing Company, Inc., Reading, Mass., 1959, chap. 7.
4. J. B. Rosen, "The Gradient Projection Method for Nonlinear Programming," Parts I and II, *J. Soc. Ind. Appl. Math.*, **8**, 1960, pp. 181–217; **9**, 1961, pp. 514–532.
5. D. Goldfarb, "Extension of Davidons' Variable Metric Method to Maximization Under Linear Inequality and Equality Constraints," *SIAM J. Appl. Math.*, **17**, No. 4, 1969, pp. 739–764.

6. J. B. Rosen and J. C. Ornea, "Solution of Nonlinear Programming Problems by Partitioning," *Management Sci.*, **10**, No. 1, 1963, pp. 160–173.
7. J. C. Ornea and G. G. Eldredge, "Non-linear Partitioned Models for Plant Scheduling and Economic Evaluation," *Proc. A.I.C.H.E.–Ind Chem. Eng. Joint Meeting, London, June 13–17, 1965*, Symposium on the Application of Mathematical Models in Chemical Engineering, Research, Design, and Production, pp. 101–112.
8. J. F. Benders, "Partitioning Procedures for Solving Mixed Variables Programming Problems," *Numerische Mathematik*, **4**, 1962, pp. 238–252.
9. A. W. Tucker, "Dual Systems of Homogeneous Linear Relations," in *Linear Inequalities and Related Systems* (*Annals of Mathematics Studies* 38), Princeton University Press, Princeton, N.J., 1956, pp. 3–18.
10. M. Balinski, "Integer Programming: Methods, Uses, Computation," *Management Sci.*, **12**, No. 3, 1965, pp. 253–313.
11. A. V. Fiacco and G. P. McCormick, "The Sequential Unconstrained Minimization Technique for Nonlinear Programming, A Primal-Dual Method," *Management Sci.*, **10**, Jan. 1964, pp. 360–366.
12. A. V. Fiacco and G. P. McCormick, *Sequential Unconstrained Minimization Techniques for Nonlinear Programming*, John Wiley and Sons, Inc., New York, 1969.
13. J. E. Kelley, "The Cutting Plane Method for Solving Convex Programs," *J. Soc. Ind. Appl. Math.*, **8**, 1960, pp. 703–712.
14. M. L. Balinski and P. Wolfe, "On Benders Decomposition and a Plant Location Problem," *Mathematica Working Paper ARO–27*, 1963.
15. B. R. Buzby, B. J. Stone, and R. L. Taylor, "Computational Experience with a Nonlinear Distribution Problem," private communication, Jan. 1965.
16. G. Zoutendijk, "Enumeration Algorithms for the Pure and Mixed Integer Programming Problem," *International Symposium on Mathematical Programming, Princeton University, Aug. 14–18, 1967.*

8

Duality and Decomposition in Mathematical Programming

8.1 Introduction

A number of the algorithms discussed in previous chapters have used prices as their coordinating mechanism. An outstanding example is the Dantzig–Wolfe decomposition principle of Chapter 3. There the problem involved a number of independent linear subproblems coupled by limitations on shared resources. This problem is decomposed by setting prices on the common resources, and adding the cost of these resources to the objective function of each subproblem. The prices are the dual variables or simplex multipliers of the master program. By appropriately varying these prices, the subproblems are induced to submit proposals to the master program which, when combined with previous proposals, reduce the total cost. It was noted in Section 3.4 that, in this scheme, the master program (i.e., the "central agency") has the ultimate decision-making responsibility. It determines the optimal solution by taking convex combinations of proposals, and also determines new prices to be sent to the subproblems. The subproblems only submit proposals, and a given proposal need not be included in the final solution. The decision-making process is thus only partially decentralized.

The decomposition principle was extended to nonlinear problems in Section 4.4.2. There the subproblems involved minimization of nonlinear convex functions over general convex sets. The common resource constraints remained linear. By using grid linearization, an analogue of the decomposition principle was developed. The subproblems are nonlinear convex programs, and the algorithm no longer converges in a finite number of iterations. As in the completely linear case, the subproblems have limited authority.

The basic idea of decentralization by means of a pricing system is studied further in this section. A general class of nonlinear programs is considered. It is shown that, if the Lagrangian function of such a program is additively separable, decomposition of the problem may be attained, under certain conditions, by pricing. As in previous work, the prices are Lagrange multipliers, and are the decision variables of a dual program. However, there is an important difference between the schemes studied here and those considered previously. Here we seek complete decentralization of the decision-making problem. By this we mean that optimal values of the decision variables are to be obtained directly by the subproblems. The second-level coordinator has responsibility only for the values of the prices. As might be anticipated, such a scheme will not work in all cases. The critical issue is the existence of a saddle point for the Lagrangian function of the problem. Existence can be guaranteed, loosely speaking, only for convex programs. However, useful results may sometimes be obtained for problems not having the required convexity.

8.2 Decomposition Using a Pricing Mechanism

To introduce the basic ideas, consider the following simple example. Let there be n activities. Operating each activity incurs a cost and uses a certain amount of a single resource. Let

$c_i(x_i)$ = cost of operating activity i at level x_i

a_i = amount of resource used per unit of x_i, assumed constant

b = availability of resource

The problem of operating all activities at minimal cost while using no more of the resource than is available is

$$\text{minimize } z = \sum_{i=1}^{n} c_i(x_i) \tag{1}$$

subject to

$$\sum_{i=1}^{n} a_i x_i \le b, \qquad x_i \ge 0 \tag{2}$$

Let us consider a decentralized solution procedure for this problem, with a pricing system used as the coordinating mechanism. Let $u \ge 0$ be the price of the resource, with dimensions (units of z)/(unit of b). Given a value for u, a natural method of decomposition is to associate with each activity a "manager" who has full responsibility for choosing its level, x_i. The manager can purchase as much of the resource as he likes in order to run the activity but must pay for what he uses. A rational manager would choose that activity level which minimizes the total cost of operation, i.e., direct cost plus resource cost. This leads to the subproblem

$$\left.\begin{aligned} &\text{minimize } \tilde{f}_i(x_i, u) = c_i(x_i) + u(a_i x_i) \\ &\text{subject to} \quad x_i \ge 0 \end{aligned}\right\} \text{subproblem } i \tag{3}$$

For fixed u, these subproblems are independent of one another. Let us (temporarily) assume that they have unique finite optimal solutions for any value of $u \geq 0$, denoted by $x_i(u)$. One feels that by increasing u, the managers can be induced to use less of the resource, and vice versa, so u provides some means of coordinating their actions. Natural questions are:

1. Does there exist a value of u, u^0 (called optimal), for which the subproblem solutions $\{x_1(u^0), \ldots, x_n(u^0)\}$ solve the original problem, (1)–(2)?
2. If the answer to question 1 is yes, how can optimality of $\{x_1(u^0), \ldots, x_n(u^0)\}$ be recognized?
3. How can an optimal value of u be computed, if it exists?

In the remainder of this chapter, partial answers to these questions will be given. The answers to 1 and 2 involve saddle points of Lagrangian functions, while the answer to 3 comes under the heading of duality. We note that the subproblems are being asked to provide the optimal solution directly. Thus the decision making may be said to be completely decentralized.

A clue to existence of optimal prices is provided by the Kuhn–Tucker theorem (see reference [1] and Section 1.3.2). Let the functions c_i be convex and differentiable. Form a Lagrangian function

$$L(x, u) = \sum_i c_i(x_i) + u\left(\sum_i a_i x_i - b\right) \tag{4}$$

$$L(x, u) = \sum_i \{c_i(x_i) + u a_i x_i\} - ub \tag{5}$$

Comparing (5) and (3), we see that the Lagrangian is additively separable and can be written

$$L(x, u) = \sum_i \tilde{f}_i(x_i, u) - ub \tag{6}$$

where $\tilde{f}_i$ is the subobjective function in (3). By the Kuhn–Tucker theorem, a vector $x^0 = (x_1^0, \ldots, x_n^0)$ solves (1)–(2) if and only if there is a multiplier $u^0 \geq 0$ such that x^0 satisfies

$$\frac{\partial L}{\partial x_i} = \frac{\partial \tilde{f}_i(x_i, u^0)}{\partial x_i} \geq 0, \qquad i = 1, \ldots, n \tag{7}$$

$$u^0\left(\sum_i a_i x_i - b\right) = 0, \qquad \sum_i a_i x_i \leq b \tag{8}$$

$$x_i \frac{\partial \tilde{f}_i(x_i, u^0)}{\partial x_i} = 0, \qquad x_i \geq 0, \quad i = 1, \ldots, n \tag{9}$$

Since $\tilde{f}_i$ is convex in x_i for any $u \geq 0$, (7) and (9) are necessary and sufficient that x_i^0 minimize $\tilde{f}_i(x_i, u^0)$ subject to $x_i \geq 0$, i.e., that x_i^0 solve the ith subproblem, (3), when $u = u^0$. Thus the Lagrange multiplier u^0 is the desired price. Conditions (8) say that u^0 is large enough to limit total resource consumption to the amount available and that, if optimal consumption is less than b units, the price u^0 is zero. Thus a resource in excess supply costs nothing.

8.3 Saddle Points of Lagrangian Functions

8.3.1 Basic Theorems. The previous discussion utilized the Kuhn–Tucker conditions to assert the existence of an optimal price. These are stationarity conditions, while the subproblems (3) are meaningfully stated only as minimizations. Stationarity and minimality are equivalent only for convex differentiable functions defined over convex sets. However, in many situations, these requirements of convexity and differentiability are not met, e.g., problems involving optimization over finite sets. Lagrange multipliers can still be useful in such problems. To see how, some theorems regarding saddle points of Lagrangian functions are needed. These results have already been stated and proved in Section 1.3.3. For completeness, the needed definitions and theorems are restated (but not reproved) here.

Consider a mathematical program of the form

$$\left.\begin{aligned} &\text{minimize } f(x) && (1)\\ \text{subject to}\quad &g_i(x) \leq 0, \quad i = 1, \ldots, m && (2)\\ &x \in S && (3)\end{aligned}\right\}\text{primal problem}$$

This is called the primal problem. The quantity x is an n vector, S is an arbitrary subset of E^n, and f and the g_i are arbitrary real-valued functions defined on S. The Lagrangian function associated with this problem is

$$L(x, u) = f(x) + \sum_{i=1}^{m} u_i g_i(x), \qquad u_i \geq 0 \tag{4}$$

DEFINITION. A point (x^0, u^0) with $u^0 \geq 0$ and $x^0 \in S$ is said to be a *saddle point* for L if it satisfies

$$L(x^0, u^0) \leq L(x, u^0) \qquad \text{for all } x \in S \tag{5}$$

$$L(x^0, u^0) \geq L(x^0, u) \qquad \text{for all } u \geq 0 \tag{6}$$

The following theorem gives necessary and sufficient conditions for a saddle point of L.

THEOREM 1. *Let $u^0 \geq 0$ and $x^0 \in S$. Then (x^0, u^0) is a saddle point for L if and only if*

(a) *x^0 minimizes $L(x, u^0)$ over S* (7)

(b) $g_i(x^0) \leq 0, \qquad i = 1, \ldots, m$ (8)

(c) $u_i^0 g_i(x^0) = 0, \qquad i = 1, \ldots, m$ (9)

The usefulness of a saddle point is brought out in the following theorem.

THEOREM 2 (SUFFICIENCY OF SADDLE POINT). *If (x^0, u^0) is a saddle point for L, then x^0 solves the primal problem* (1)–(3).

To relate Theorems 1 and 2 to our previous discussion regarding decomposition, let us apply these theorems to the allocation problem considered earlier. To place this in the general form of the primal problem (1)–(3), we let

$$S = \{x \mid x \geq 0\} \tag{10}$$

Then the problem is rewritten

$$\text{minimize} \sum_{i=1}^{n} c_i(x_i) \tag{11}$$

subject to

$$\sum_{i=1}^{n} a_i x_i - b \leq 0 \tag{12}$$

$$x \in S \tag{13}$$

The Lagrangian function for this problem is

$$L(x, u) = \sum_{i=1}^{n} (c_i(x_i) + u a_i x_i) - ub \tag{14}$$

Condition (a) of Theorem 1 requires that $L(x, u^0)$ be minimized over S to obtain x^0. Since L is additively separable in the x_i for fixed u,

$$\min_{x_1 \geq 0, \ldots, x_n \geq 0} \sum_{i=1}^{n} (c_i(x_i) + u a_i x_i) = \sum_{i=1}^{n} \left\{ \min_{x_i \geq 0} (c_i(x_i) + u a_i x_i) \right\} \tag{15}$$

The quantity

$$\min_{x_i \geq 0} c_i(x_i) + u a_i x_i \tag{16}$$

is the optimal objective value for the ith subproblem (8.2-3). Thus these subproblems arise mathematically from trying to satisfy the first of the necessary and sufficient conditions for a saddle point. The second of these conditions [(b) of Theorem 1] says that the value of the optimal price, u^0, must be such that the subproblem solutions jointly satisfy the resource constraint. Condition (c) repeats our earlier remarks that a resource in excess supply has a price of zero.

The process of minimizing the Lagrangian $L(x, u)$ over $x \in S$ for fixed u leads to a set of independent subproblems whenever L is additively separable in x for fixed u and when S can be written as a Cartesian product. In the primal problem (1)–(3), let x be partitioned as

$$x = (x_1 \mid x_2 \mid \cdots \mid x_p), \qquad p \leq n \tag{17}$$

and assume that f, S, and the vector of constraint functions g may be written

$$f(x) = \sum_{i=1}^{p} f_i(x_i) \tag{18}$$

$$g(x) = \sum_{i=1}^{p} g^i(x_i) \tag{19}$$

$$S = S_1 \times S_2 \times \cdots \times S_p \tag{20}$$

where $\times$ denotes Cartesian product and each g^i is an m vector of functions. This is the important special case where the objective f and constraint functions g_i are additively separable, and when the constraints determining S consist of subsets of constraints involving the subvectors x_i separately. Then

$$L(x, u) = \sum_i f_i(x_i) + u \sum_i g^i(x_i) \tag{21}$$

$$L(x, u) = \sum_i (f_i(x_i) + ug^i(x_i)) \tag{22}$$

so the Lagrangian is additively separable in the x_i and

$$\min_{x_1 \in S_1, \ldots, x_p \in S_p} L(x, u) = \sum_i \min_{x_i \in S_i} (f_i(x_i) + ug^i(x_i)) \tag{23}$$

Thus the Lagrangian can be minimized by solving the p independent subproblems

$$\left.\begin{aligned} &\text{minimize } f_i(x_i) + ug^i(x_i) \\ \text{subject to } \quad &x_i \in S_i \end{aligned}\right\} \text{subproblem } i \tag{24}$$

These subproblems have the same pricing interpretation as in our earlier example, with u_k the price of the "resource" whose limited availability is represented by the kth constraint.

In choosing which constraints of the primal will be used to determine S and which written as $g_i \leq 0$, efforts should be made to achieve separability of L and S. That is, the choice should be made so that, if possible, conditions (18)–(20) hold for some partition of x, $(x_1 \mid x_2 \mid \cdots \mid x_p)$.

All the previous results are easily extended to include equality constraints. If the kth constraint is $g_k(x) = 0$, then the multiplier u_k for this constraint may have any sign and condition (c) of Theorem 1 is superfluous, since $g_k(x^0) = 0$.

8.3.2 Everett's Theorem. Everett [3] has shown that if any set of multipliers $u \geq 0$ is chosen for which $\min_{x \in S} L(x, u)$ exists, any vector $x(u)$ which minimizes $L(x, u)$ solves a problem closely related to the primal problem. The result is as follows:

THEOREM 3 (EVERETT). *If $x(u)$ solves the Lagrangian problem*

$$\left.\begin{array}{ll} & \textit{minimize } L(x, u) \\ \textit{subject to} & x \in S \end{array}\right\} \textit{Lagrangian problem} \tag{25}$$

with $u \geq 0$, then $x(u)$ solves the modified primal problem

$$\left.\begin{array}{llr} & \textit{minimize } f(x) & (26) \\ \textit{subject to} & g_i(x) \leq y_i, \quad i = 1, \ldots, m & (27) \\ & x \in S & \\ \textit{where} & y_i = g_i(x(u)) \quad \text{if } u_i > 0 & \\ & y_i \geq g_i(x(u)) \quad \text{if } u_i = 0 & (28) \end{array}\right\} \textit{modified primal}$$

The modified primal has right-hand-side constraint constants either $=$ or $\geq$ to $g_i(x(u))$. Let the constraints $g_i \leq 0$ be regarded as expressing resource limitations and the u_i be viewed as prices for these resources. Then the theorem says that any vector which minimizes $L(x, u)$ solves a primal problem which uses no more of the valuable resources than the vector itself, while availability of the free resources is at least as much as is used by the vector itself. The multipliers u thus convert a constrained problem to an unconstrained (except for the restriction $x \in S$) Lagrangian problem. As mentioned earlier, this is especially attractive if f, the g_i, and S have the separable forms in (18)–(20). Then the Lagrangian problem decomposes into the smaller independent subproblems (24). Note that no restrictions are posed on the functions f, g_i, or on the set S.

The proof of Theorem 3 follows immediately from the observation that, by Theorem 1, $(x(u), u)$ is a saddle point for the Lagrangian

$$L = f(x) + \sum_{i=1}^{m} u_i(g_i(x) - y_i) \tag{29}$$

One uses this theorem by choosing multipliers $u \geq 0$, solving the Lagrangian problem, and observing for what constraint constants y_i the primal problem has been solved. This procedure may be valuable if a family of primal problems is of interest, with different right-hand sides. Of course, not all right-hand-side vectors can be generated by this procedure. Sets of vectors y whose associated problems cannot be solved are termed "gaps" in [3]. A geometric characterization of these gaps is given in Section 8.4.

Although the right-hand-side values resulting from a given choice of u cannot be predicted in advance, there are a few results which make the decision easier. Since u_k has an interpretation as the price of resource k, one feels that if u_k is increased while all other prices are held fixed, the amount of the kth resource used will decrease. This is true, as is shown by the following theorem.

THEOREM 4 (EVERETT). *Let u^1, u^2 be nonnegative m vectors with*

$$u_k^2 > u_k^1, \qquad u_j^2 = u_j^1, \qquad j \neq k \tag{30}$$

If $x(u^i)$ solves the Lagrangian problem with $u = u^i$, then

$$g_k(x(u^2)) \leq g_k(x(u^1)) \tag{31}$$

Another result, useful in computations, deals with near-optimal solutions to the Lagrangian problem. In cases where this problem must be solved numerically, exact solutions are rarely obtained. The following shows that a "good" solution to the Lagrangian problem is a "good" solution to the modified primal (26)–(28).

THEOREM 5 (EVERETT). *Let x^* come within $\epsilon > 0$ of solving the Lagrangian problem; i.e.,*

$$L(x^*, u) \leq L(x, u) + \epsilon, \qquad \text{for all } x \in S$$

Then x^* comes within ϵ of solving the modified primal problem.

The proofs of Theorems 4 and 5 are left for the problems at the end of this section.

8.3.3 Application to Linear Integer Programs. Consider the linear 0–1 integer program.

$$\left.\begin{aligned} &\text{minimize} \sum_{j=1}^{n} c_j x_j && && (32)\\ \text{subject to} \quad &\sum_{j=1}^{n} a_{ij} x_j \leq b_i, && i = 1, \ldots, m && (33)\\ &x_j = 0 \text{ or } 1, && j = 1, \ldots, n && (34) \end{aligned}\right\} \text{problem IP}(b)$$

Since this is separable, it is natural to try to solve it by Lagrangian methods. The Lagrangian function is

$$\begin{aligned} L(x, u) &= \sum_{j=1}^{n} c_j x_j + \sum_{i=1}^{m} u_i \left(\sum_{j=1}^{n} a_{ij} x_j - b_i \right) \\ &= \sum_{j=1}^{n} \left(c_j + \sum_{i=1}^{m} u_i a_{ij} \right) x_j - \sum_{i=1}^{m} u_i b_i \end{aligned} \tag{35}$$

Defining

$$\beta_j = c_j + \sum_{i=1}^{m} u_i a_{ij}$$

the Lagrangian problem is to minimize

$$\sum_{j=1}^{n} \beta_j x_j$$

subject to $x_j = 0$ or 1. The solution is clearly

$$\begin{aligned} \beta_j > 0 &\Rightarrow x_j = 0 \\ \beta_j < 0 &\Rightarrow x_j = 1 \\ \beta_j = 0 &\Rightarrow x_j = 0 \text{ or } 1 \end{aligned}$$

If such a solution is denoted by $x(u)$, then, by Theorem 3, $x(u)$ solves a modified program IP($y(u)$), whose right-hand-side coefficients $y(u)$ are formed as in (28). As shown in [32], $x(u)$ is also an optimal solution to LP($y(u)$), the program (32)–(34) with right-hand side $y(u)$ and the integer constraints (34) replaced by $0 \leq x_j \leq 1$. This is because $(x(u), u)$ forms a saddle point for $L(x, u)$ in (35) with $b_i = y_i(u)$, where x_j is restricted to be

0 or 1. If these restrictions are changed to $0 \leq x_j \leq 1$, the min of $L(x, u)$ is again obtained by taking $x = x(u)$. Thus $(x(u), u)$ is still a saddle point, and this can be true if and only if $x(u)$ solves LP($y(u)$) and u solves its dual. This means that any linear integer program which can be solved via the Lagrangian problem can also be solved by the simplex method. However, as mentioned earlier, even if the Lagrangian procedure cannot solve the problem at hand, it may solve a problem with right-hand side $y(u)$ close enough to the original to be useful. This is often the case in the capital-budgeting problems considered in [31]. There the objective is to be maximized, the b_i are the amounts of capital available in period i, a_{ij} is the capital used by project j in period i, c_j is the present value of all future returns from project j, and $x_j = 1$ implies that project j is accepted. Since the b_i are rarely known exactly, solutions for $y(u)$ close to b may be as good as optimal.

In obtaining such solutions Theorem 5 is useful, making it possible to generate optimal solutions not obtainable directly. Assume that all c_j are integers and suppose that there is an index set J such that

$$\sum_{j \in J} |\beta_j| < 1$$

Set

$$\begin{aligned} x_j^* &= 1 - x_j(u), \qquad & j &\in J \\ x_j^* &= x_j(u), & j &\notin J \end{aligned}$$

Then

$$L(x^*, u) - L(x(u), u) = \sum_{j \in J} |\beta_j| = \epsilon$$

so x^* comes within ϵ of minimizing $L(x, u)$. By Theorem 5, x^* comes within ϵ of solving IP(y^*), where

$$\begin{aligned} y_i^* &= \sum_{j=1}^{n} a_{ij} x_j^* \qquad & \text{if } u_i &> 0 \\ y_i^* &\geq \sum_{j=1}^{n} a_{ij} x_j^* & \text{if } u_i &= 0 \end{aligned}$$

But $\epsilon < 1$ and, since the c_j are integers, the objective of IP(y^*) can assume only integral values. Thus x^* solves IP(y^*). Vectors y^* generated in this way may lie in "gaps," so this option truly extends the Lagrangian method.

Existence of Saddle Points. Our questions regarding existence of optimal multipliers and of recognizing such multipliers if they exist are partially answered by the preceding results. Any vector u^0 which is the u component of a saddle point of L will serve as the desired multipliers, and conditions (a)–(c) of Theorem 1 show how such a point may be recognized. The following theorem deals with existence of a saddle point for problems with convex objective and constraint functions.

THEOREM 6 (KARLIN [4]). *Let S be a convex subset of E^n, f a convex function defined on S, and $g(x) = \{g_1(x), \ldots, g_m(x)\}$ a vector of convex functions defined on S. Assume that there exists a point $x \in S$ such that $g(x) < 0$. If x^0 is a point at which $f(x)$ assumes its minimum subject to $g(x) \leq 0$, $x \in S$, then there is a vector of multipliers $u^0 \geq 0$ such that (x^0, u^0) is a constrained saddle point of*

$$L(x, u) = f(x) + ug(x)$$

Conversely, if (x^0, u^0) is a constrained saddle point of $L(x, u)$, then x^0 minimizes $f(x)$ subject to $g(x) \leq 0$, $x \in S$.

The proof is found in Section 1.3.3.

8.4 Minimax Dual Problem

Although a number of results regarding existence of saddle points have been proved, nothing has yet been said as to how to find such a point. To answer this question, the behavior of the minimum value of $L(x, u)$ must be studied as a function of u. That is, instead of focusing on a single set of Lagrange multipliers, u^0, we consider all $u \geq 0$.

Let

$$h(u) = \min_{x \in S} L(x, u) \tag{1}$$

$$X(u) = \{x \mid x \text{ minimizes } L(x, u) \text{ over } S\} \tag{2}$$

The function h is called the dual function. Its domain of definition is

$$D = \left\{u \mid u \geq 0, \min_{x \in S} L(x, u) \text{ exists}\right\} \tag{3}$$

That is, D is the set of nonnegative vectors u for which $L(x, u)$ has a finite infimum over S which is attained at some point in S. If, for all $u \geq 0$, $L(x, u)$ is continuous in x for all $x \in S$ and if S is closed and bounded then, by the Weierstrass theorem [5],

$$D = (E^m)^+ = \{u \mid u \geq 0\} \tag{4}$$

However, D need not be convex and may be empty. In example 2 of Section 1.3.3 D consists of the single point $u = 2$.

A primal–dual pair of problems is now defined:

Primal

$$\text{minimize } f(x)$$

subject to (5)

$$g_i(x) \leq 0, \qquad i = 1, \ldots, m$$
$$x \in S$$

Dual

$$\text{maximize } h(u)$$

subject to (6)

$$u \in D$$

We call this dual problem the minimax dual, since it may be phrased

$$\underset{u \in D}{\text{maximize}} \min_{x \in S} L(x, u) \tag{7}$$

Such dual problems have been studied by Falk [2], Mangasarian [6], Stoer [7], Rockafellar [8], and others.

It is easily shown that if the primal is a linear program, then (6) is its dual. Let the primal be written

$$\text{minimize } c'x \tag{8}$$

subject to

$$Ax \geq b, \qquad x \geq 0 \tag{9}$$

Then

$$L(x, u) = c'x + u'(b - Ax), \qquad u \geq 0 \tag{10}$$

and

$$h(u) = \min_{x \geq 0}(c'x + u'(b - Ax)) \tag{11}$$

$$h(u) = \min_{x \geq 0}(x'(c - A'u) + b'u) \tag{12}$$

This minimum exists if and only if

$$c - A'u \geq 0$$

or

$$A'u \leq c \tag{13}$$

since otherwise $h(u)$ can be made arbitrarily large and negative. Thus

$$D = \{u \mid A'u \leq c, u \geq 0\} \tag{14}$$

For any $u \in D$, the minimum is attained by choosing

$$x'(c - A'u) = 0 \tag{15}$$

which is the complementary slackness condition for linear programs. Then

$$h(u) = b'u \tag{16}$$

The problem of maximizing $b'u$ over D is the standard linear programming dual for the primal (8)–(9).

If $S = E^n$ and f and the g_i are convex and differentiable, the dual (6) is equivalent to the dual program of Wolfe [9]. Wolfe's dual is

$$\text{maximize } L(x, u) = f(x) + ug(x) \tag{17}$$

$$\text{subject to} \qquad \nabla_x L(x, u) = \nabla f(x) + \sum_i u_i \nabla g_i(x) = 0 \tag{18}$$

$$u \geq 0 \tag{19}$$

This dual problem has both x and u as decision variables. Its feasible region is

$$D_w = \{(x, u) \mid (x, u) \text{ satisfies (18)–(19)}\} \tag{20}$$

The Wolfe dual is equivalent to the minimax dual in the following sense:

$$D_w = \{(x, u) \mid u \in D, x \in X(u)\} \tag{21}$$

$$L(x, u) = h(u) \qquad \text{for each } (x, u) \in D_w \tag{22}$$

These relations follow from the fact that if f and the g_i are convex and differentiable, $L(x, u)$ is convex and differentiable in x for all $u \geq 0$, and attains an unconstrained minimum with respect to x if and only if $u \in D$ and $\nabla_x L(x, u) = 0$. Thus if $u \in D$ and x minimizes $L(x, u)$, then $(x, u) \in D_w$ and conversely.

Our object in the following pages is to uncover the relationships between the primal and dual problems, (5)–(6). We will show that a saddle point for L exists if and only if the optimal values of primal and dual objectives are equal. In this case, any u component of a saddle point solves the dual. An investigation of the differentiability of h then leads to a two-level decomposition algorithm for solving nonlinear programs.

Example 1. To illustrate, consider the primal problem

$$\text{minimize } (x_1 - 1)^2 + (x_2 - 2)^2$$

subject to

$$x_1 + x_2 \leq 2$$
$$x_1 \geq 0, \qquad x_2 \geq 0$$

whose optimal solution is $x_1^0 = 0.5$, $x_2^0 = 1.5$, $\min f = 0.5$. Letting the non-negativities determine S, the dual function is

$$\begin{aligned} h(u) &= \min\{(x_1 - 1)^2 + (x_2 - 2)^2 + u(x_1 + x_2 - 2) \mid x_1 \geq 0, x_2 \geq 0\} \\ &= \min\{(x_1 - 1)^2 + ux_1 \mid x_1 \geq 0\} + \min\{(x_2 - 2)^2 + ux_2 \mid x_2 \geq 0\} - 2u \end{aligned}$$

Since these minima exist for all u,

$$D = \{u \mid u \geq 0\}$$

Solving each of the minimization problems above yields the minimizing values of x_1, x_2:

$$x_1(u) = \begin{cases} 1 - \dfrac{u}{2}, & 0 \leq u \leq 2 \\ 0, & u > 2 \end{cases}$$

$$x_2(u) = \begin{cases} 2 - \dfrac{u}{2}, & 0 \leq u \leq 4 \\ 0, & u > 4 \end{cases}$$

The function h is

$$h(u) = \begin{cases} \dfrac{-u^2}{2} + u, & 0 \leq u \leq 2 \\ \dfrac{-u^2}{4} + 1, & 2 < u \leq 4 \\ 5 - 2u, & 4 < u \end{cases}$$

The graph of h is shown in Figure 8-1. Max h is attained at $u^0 = 1$. Note that

$$\max h = \min f = \tfrac{1}{2}$$

Also

$$x_1(1) = 0.5 = x_1^0$$
$$x_2(1) = 1.5 = x_2^0$$

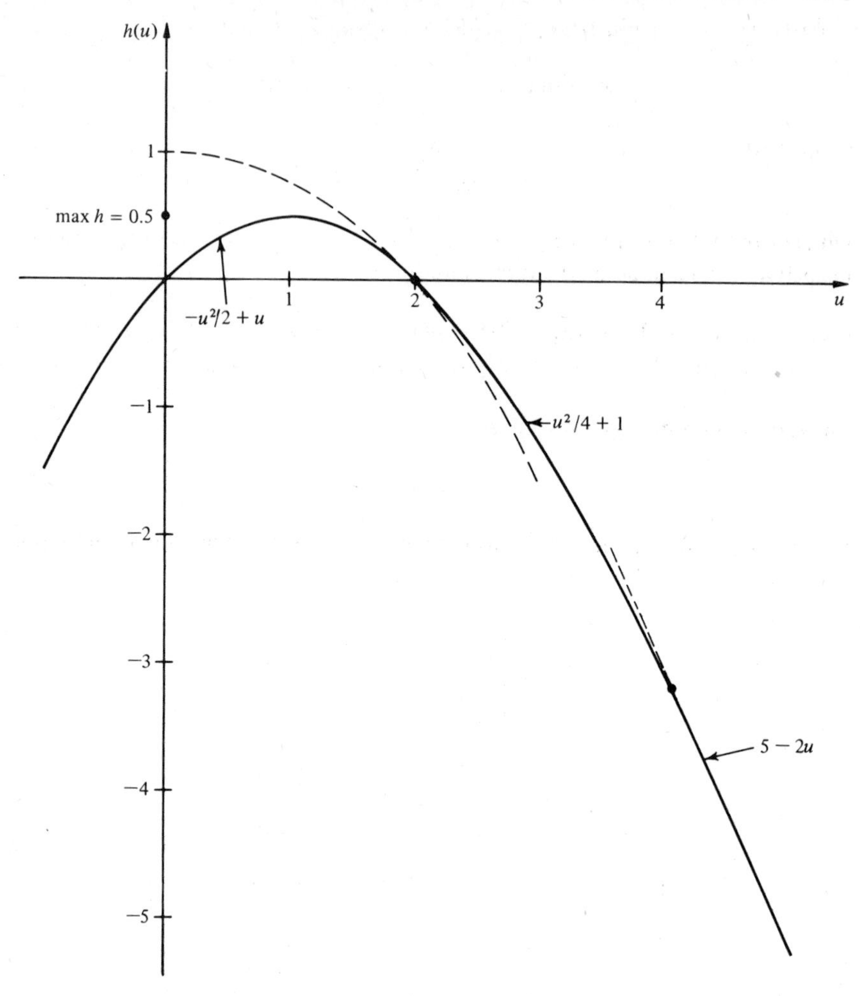

FIGURE 8-1 Graph of h for Example 1.

The point (0.5, 1.5; 1) is a saddle point for L. Further, h is a lower bound on $\min f$ for any $u \geq 0$:

$$h(u) \leq f(x), \qquad \text{all } u \geq 0, \text{ all } x \in S \tag{23}$$

Finally, h is concave for nonnegative u and is differentiable at all points $u \geq 0$, including $u = 2$ and $u = 4$.

Condition (23) and the concavity of h over convex subsets of D will be proved true in the general case. The equality $\max h = \min f$ is true only when a saddle point exists. The function h is not, in general, differentiable everywhere, even for convex problems. The various segments comprising h usually join with discontinuous slope. However, as discussed in Sections 8.5 and 8.7, if f is strictly convex, all g_i are convex, and S is closed and convex, then h has continuous first partial derivatives everywhere in int D.

Example 2. Consider Example 1 of Section 1.3.3, restated here:

$$\text{minimize } \{-x^2 \mid 1 - 2x = 0,\ 0 \le x \le 1\}$$

where

$$h(u) = \min_{0 \le x \le 1} (-x^2 + u(1 - 2x))$$

Performing the minimization

$$h(u) = \begin{cases} u, & u \le -\frac{1}{2} \\ -1 - u, & u > -\frac{1}{2} \end{cases}$$

The graph of h is shown in Figure 8-2. Note that $\max h = -\frac{1}{2} < \min f = -\frac{1}{4}$, and that $x^0 = \frac{1}{2} \notin X(u^0) = \{0, 1\}$. Moreover, h is not differentiable at $u^0 = -\frac{1}{2}$, although it has finite left- and right-hand derivatives there. However, $\max h$ is a lower bound on $\min f$ and h is concave.

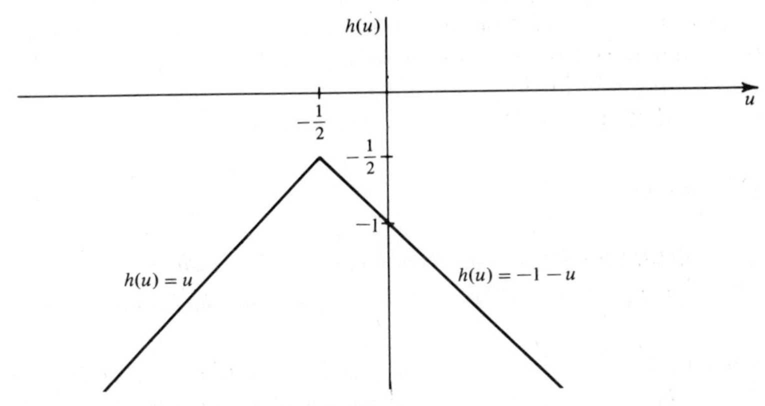

FIGURE 8-2 Graph of h for Example 2.

Proceeding now to show under what assumptions these properties hold, we first prove that $h(u)$ is a lower bound for $\min f$. Thus, given primal feasible and dual feasible points, the associated values of primal and dual objectives bracket the primal optimum. This may be used to terminate computations in an iterative algorithm where such points are available.

THEOREM 1 (LOWER BOUND).

$$h(u) \le f(x) \begin{cases} \textit{for all } x \textit{ satisfying the constraints of } (5) \\ \textit{for all } u \in D \end{cases} \tag{24}$$

PROOF

$$h(u) = \min_{x \in S}(f(x) + ug(x)) \tag{25}$$

$$h(u) \le f(x) + ug(x) \qquad \text{for all } x \in S \tag{26}$$

For all $x \in S$ satisfying $g(x) \le 0$,

$$h(u) \le f(x) \tag{27}$$

and the theorem is proved.

If the feasible regions of both primal and dual are nonempty, then (24) implies that sup h and inf f, both taken over their respective constraint sets, are finite. Of course, neither the sup nor the inf need be attained at any feasible point. If they are, then (24) yields the result that if both primal and dual are feasible, both have optimal solutions. This is always true for linear programs, and would be true here if we were willing to substitute sup for max and inf for min in the statements of primal and dual problems. Other results, immediate from (24), are stated as corollaries.

COROLLARY 1

(a) *If* $\sup\{h(u) \mid u \in D\} = +\infty$, *then the primal is infeasible.*
(b) *If* $\inf\{f(x) \mid x \textit{ is primal feasible}\} = -\infty$, *then the dual is infeasible.*

COROLLARY 2. *If there exist primal feasible* x^0 *and dual feasible* u^0 *such that* $f(x^0) = h(u^0)$, *then* x^0 *solves the primal and* u^0 *solves the dual.*

THEOREM 2. *The dual function* h *is concave over any convex subset of its domain,* D.

PROOF. Let $\hat{D}$ be a convex subset of D and let $u_1 \in \hat{D}$, $u_2 \in \hat{D}$. Then

$$h(\alpha u_1 + (1 - \alpha)u_2) = \min_{x \in S} L(x, \alpha u_1 + (1 - \alpha)u_2) \tag{28}$$

Since L is linear in u,

$$h(\alpha u_1 + (1 - \alpha)u_2) = \min_{x \in S}[\alpha L(x, u_1) + (1 - \alpha)L(x, u_2)] \tag{29}$$

$$\geq \alpha \min_{x \in S} L(x, u_1) + (1 - \alpha) \min_{x \in S} L(x, u_2) \tag{30}$$

$$\geq \alpha h(u_1) + (1 - \alpha)h(u_2) \tag{31}$$

By the above theorem, if D is a convex set, then the dual program requires the maximization of a concave function over a convex set, and thus has no local maxima distinct from the global maximum. In particular, if f and the g_i are continuous on S and if S is closed and bounded, then, as noted earlier, $D = (E^m)^+$, which is convex. If S is closed but not bounded, it may be made so by placing bounds on all components of x, sufficiently large to guarantee that all physically meaningful vectors x satisfy them. Thus the dual problem may, in general, be made well behaved. In that case, the only question which remains is whether or not $\max h = \min f$, i.e., whether equality may be attained in (24).

We now show that equality can be achieved in (24) if and only if a saddle point for L exists. This is closely related to established results. If f and g are continuous and S compact, then it is easily shown (Problem 6) that

$$\min f = \min_{x \in S} \sup_{u \geq 0} L(x, u) \tag{32}$$

Thus primal and dual problems may be stated.

Primal		*Dual*	
$\underset{x \in S}{\text{minimize}} \sup_{u \geq 0} L(x, u)$	(33)	$\underset{u \geq 0}{\text{maximize}} \min_{x \in S} L(x, u)$	(34)

Were it not for the asymmetry introduced by the inner sup in the primal, equality of primal and dual objective values would require that the min max and max min values of L be equal. It is well known (Karlin [4]) that these values are equal if and only if a saddle point exists. A proof that this is true for our primal and dual problems is given below.

THEOREM 3. *(x^0, u^0) is a saddle point for L if and only if x^0 is primal feasible, u^0 is dual feasible, and $f(x^0) = h(u^0)$.*

PROOF. $\Rightarrow$: Since (x^0, u^0) is a saddle point, conditions (a)–(c) of Theorem 8.3-1 hold. Thus x^0 is primal feasible, u^0 is dual feasible, and

$$u^0 g(x^0) = 0 \tag{35}$$

so

$$h(u^0) = f(x^0) + u^0 g(x^0) = f(x^0) \tag{36}$$

$\Leftarrow$: Since x^0 is primal feasible

$$g(x^0) \leq 0 \tag{37}$$

If $x^0 \notin X(u^0)$, then there is an $\tilde{x} \in X(u^0)$ such that

$$h(u^0) = f(\tilde{x}) + u^0 g(\tilde{x}) < f(x^0) + u^0 g(x^0) \tag{38}$$

Since $h(u^0) = f(x^0)$, (38) implies

$$f(x^0) < f(x^0) + u^0 g(x^0) \tag{39}$$

or

$$u^0 g(x^0) > 0 \tag{40}$$

But $u^0 \geq 0$ and $g(x^0) \leq 0$, so this is a contradiction. Thus

$$x^0 \in X(u^0) \tag{41}$$

so

$$f(x^0) = h(u^0) = f(x^0) + u^0 g(x^0) \tag{42}$$

which implies

$$u^0 g(x^0) = 0 \tag{43}$$

By Theorem 8.3-1, relations (43), (41), and (37) imply that (x^0, u^0) is a saddle point for L.

Consider again primal problems for which L is separable in x, thus causing the Lagrangian problem to decompose into subproblems. The preceding results imply that one way to coordinate these subproblems so that their combined solutions solve the primal is to choose the multipliers, u, to solve the dual. The coordinator thus has the task of maximizing h subject to $u \in D$. Solution proceeds as outlined below.

1. Choose initial values $u \in D$ and solve the Lagrangian problem, (8.3-25). In the separable case this may be done by solving the subproblems, (8.3-24).
2. Evaluate the dual function $h(u)$ and choose new multipliers so that it is increased.
3. Return to step 1, stopping when h is maximized.

If, upon termination of this procedure, u^0 solves the dual, some $x^0 \in X(u^0)$ is primal feasible, and $h(u^0) = f(x^0)$, then x^0 solves the primal. How step 2 might be accomplished will be discussed shortly.

Geometric Interpretation Added insight is obtained from the following geometric interpretation [10, 11]. Imbed the primal problem in a family of perturbed problems, with perturbations y_i:

$$\text{minimize } f(x) \tag{44}$$

subject to

$$g_i(x) \leq y_i, \qquad i = 1, \ldots, m \tag{45}$$

$$x \in S \tag{46}$$

That is, consider a family of problems with variable right-hand sides, $y = (y_1, \ldots, y_m)$. The primal problem corresponds to $y = 0$. Let

$$w(y) = \min\{f(x) \mid g(x) \leq y,\ x \in S\} \tag{47}$$

This is the function constructed in the proof of Theorem 1.3.3-3 but with min replacing inf. The function w may not exist for all y, owing either to the program (44)–(46) being infeasible or to the fact that the infimum of f subject to $g(x) \leq y$, $x \in S$, is not attained. To avoid this difficulty, assume that S is closed and bounded and that f and the g_i are continuous on S. Then the domain of w is

$$F = \{y \mid \text{there exists } x \in S \text{ such that } g(x) \leq y\} \tag{48}$$

The function w is nonincreasing over F. If f, g, and S are convex, then F can be shown to be a convex set and w is a convex function over F (see Problem 5). Consider the set of points, R, in E^{m+1} on and above the graph of $w(y)$:

$$R = \{(y_0, y) \mid y \in F,\ y_0 \geq w(y)\} \tag{49}$$

If w and F are convex, this set is convex and might appear as shown in Figure 8-3. This set was used in the proof of Theorem 1.3.3-3. There is an intimate

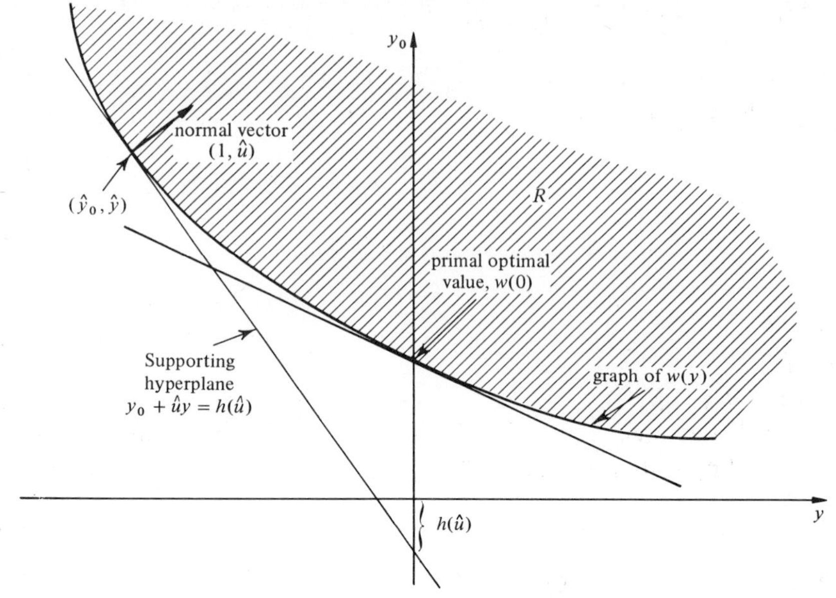

FIGURE 8-3 Supporting hyperplane for the set R.

connection between duality and supporting hyperplanes for this set, as is shown by the following theorem.

THEOREM 4. *Let $\hat{x} \in S$ and $\hat{u} \geq 0$. Then $\hat{x}$ minimizes $L(x, \hat{u})$ over S if and only if*

$$H = \{(y_0, y) \mid y_0 + \hat{u}y = L(\hat{x}, \hat{u})\} \tag{50}$$

is a supporting hyperplane for R at $(f(\hat{x}), g(\hat{x}))$.

PROOF. $\Rightarrow$: We must show

$$y_0 + \hat{u}y \geq L(\hat{x}, \hat{u}) \qquad \text{for all } (y_0, y) \in R$$

If, for some $(y_0^*, y^*) \in R$,

$$y_0^* + \hat{u}y^* < L(\hat{x}, \hat{u})$$

then, since $y_0^* \geq w(y^*) = f(x^*)$, $y^* \geq g(x^*)$,

$$f(x^*) + \hat{u}g(x^*) < L(\hat{x}, \hat{u})$$

Since $x^* \in S$, this contradicts the fact that $\hat{x}$ minimizes $L(x, \hat{u})$ over S.

$\Leftarrow$: Let

$$y_0 + \hat{u}y \geq L(\hat{x}, \hat{u}) \qquad \text{for all } (y_0, y) \in R \tag{51}$$

and consider the set

$$R_1 = \{(f(x), g(x)) \mid x \in S\}$$

Choose $\tilde{x} \in S$ and form the vector $(f(\tilde{x}), g(\tilde{x})) \in R_1$. If we set $y = g(\tilde{x})$, then $y \in F$ and $f(\tilde{x}) \geq w(y)$. Thus $(f(\tilde{x}), g(\tilde{x})) \in R$ and, since $\tilde{x}$ is an arbitrary element of S,

$$R \supseteq R_1$$

Using the above, (51) becomes

$$f(x) + \hat{u}g(x) \geq L(\hat{x}, \hat{u}) \qquad \text{for all } x \in S$$

so $\hat{x}$ minimizes $L(x, \hat{u})$ over S.

According to the above result, the process of minimizing $L(x, \hat{u})$ over S is equivalent to finding a real number

$$h(\hat{u}) = \min_{x \in S} L(x, \hat{u}) = L(\hat{x}, \hat{u})$$

such that the hyperplane H in (50) is a supporting hyperplane to the set R. The value of the dual function, $h(\hat{u})$, is the intercept of this support plane with the y_0 axis, and the slopes of the plane are the numbers $-\hat{u}$. The dual problem is that of finding the support plane with nonpositive slopes having maximal y_0 intercept, while the primal is that of finding a point $(y_0, 0)$ in R with minimal y_0 intercept, the number $w(0)$.

Since w is nonincreasing, all support planes to R have slopes $-u \leq 0$. It is evident from Figure 8-3 that any such support plane has y_0 intercept less than or equal to the primal optimum, $w(0)$. This statement corresponds to the inequality $h(u) \leq \min f$. If R is closed and convex, supporting hyperplanes exist at all boundary points, in particular at $(w(0), 0)$. The intercept of this plane is, of course, equal to $w(0)$, yielding the relation $\min f = \max h$.

Theorem 4 brings out the interpretation of the dual in terms of conjugate functions. Using Definition 9 of Appendix 1 and the comments following it, we see that

$$h(u) = -w^*(-u)$$

where w^* is the conjugate of w. Rockafellar [10] has used the notion of conjugacy to establish an elegant and symmetric duality theory for convex programs.

The geometric viewpoint also shows in what situations the Lagrangian approach can succeed. Consider a primal problem with optimal objective value $w(0) = f(x^0)$. The statement below follows directly from Theorem 4.

COROLLARY. *Let x^0 solve the primal problem. There exist multipliers $u^0 \geq 0$ such that x^0 minimizes $L(x, u^0)$ over S if and only if the set R in* (49) *has a supporting hyperplane at the point $(f(x^0), g(x^0))$ with slopes $-u^0$.*

This corollary applies to primal problems with arbitrary right-hand-side vector y. Thus we have a characterization of the "gaps" discussed in the previous section. Any vector y for which R has a supporting hyperplane at some point $(w(y), v)$ with $v \leq y$ can be solved by solving the Lagrangian problem for some $u \geq 0$. These are the vectors y for which $(w(y), v)$ lies in the intersection of the boundary of the convex hull of R with the boundary of R. Vectors y not having this property constitute the gaps. In Figures 8-4 and 8-5 the sets R are not convex, implying that the primal is not a convex program. Despite this, in Figure 8-4, R has a support plane at $(w(0), 0)$, while there is no support plane at this point in Figure 8-5. There are a number of gaps, one of which is indicated in Figure 8-4.

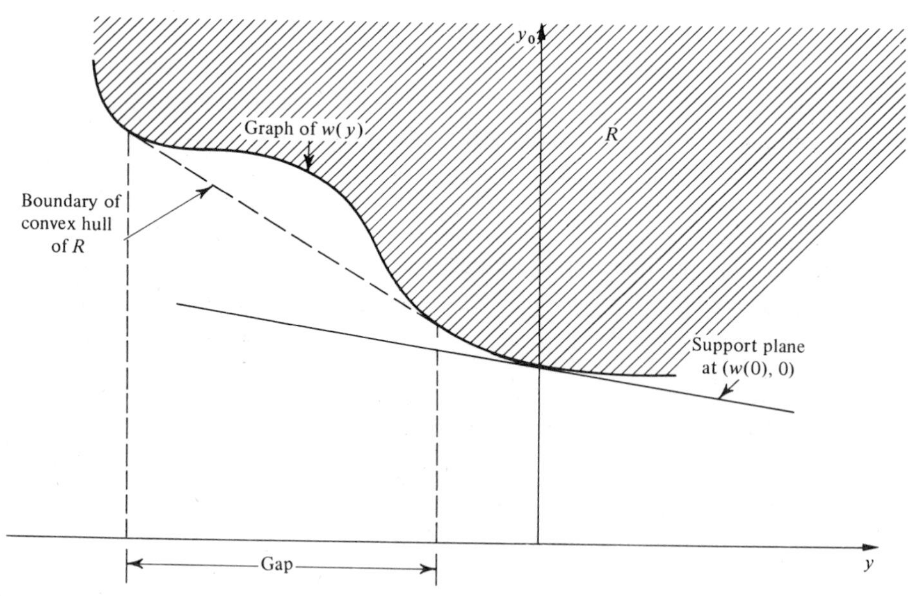

FIGURE 8-4 Support plane at optimal point.

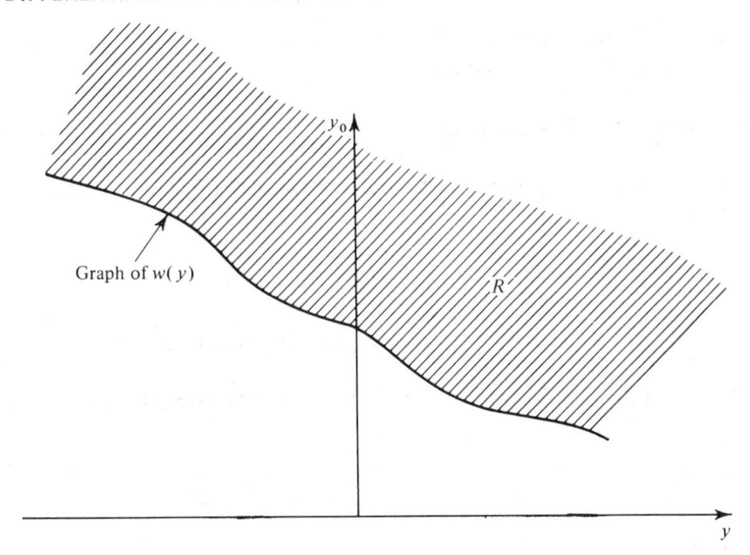

FIGURE 8-5 No support plane at optimal point.

The previous corollary shows that Lagrangian methods can solve some nonconvex problems. In particular, if a class of problems could be delineated for which the associated set R always has a support plane at some point $(f(x^0), g(x^0))$ with $f(x^0) = w(0)$, $g(x^0) \leq 0$, then any problem in this class could be solved by solving the Lagrangian problem for some vector $u^0 \geq 0$.

8.5 Differentiability of the Dual Objective Function[1]

In constructing an algorithm to solve the dual problem, information regarding derivatives of the dual objective function, h, is essential. For example, when the feasible set of the dual, D, is $(E^m)^+$ and h is differentiable a gradient ascent method might well be used. A particularly nice property of the dual problem is that the partial derivatives $\partial h/\partial u_i$ are quite easy to obtain, when they exist. However, as seen in Example 2 of Section 8.4, h is not, in general, differentiable at all points. Since h is concave over convex subsets of D, it is differentiable almost everywhere in the interior of such subsets [13] and possesses (one-sided) directional derivatives in all directions at all points in such subsets.

We will give necessary and sufficient conditions that h be differentiable at a given point and will develop expressions for its partial and directional derivatives. The main tool is a result due to Danskin [12], which holds for any function, ϕ, defined as the minimum of some other function, $F(x, u)$, over $x \in S$. The result is quite general, allowing x to be an element of any compact topological space.

[1] This section uses the results of Appendices 1 and 2.

Let $F(x, u)$ be defined for all u in E^m and for all x in some topological space, S. Assume the following:

Assumption 1. S is compact.

Assumption 2. F and the partial derivatives $\partial F/\partial u_i$ are continuous. Define

$$\phi(u) = \min_{x \in S} F(x, u) \tag{1}$$

$$X(u) = \{x \mid x \text{ minimizes } F(x, u) \text{ over } S\} \tag{2}$$

The (one-sided) directional derivative of ϕ at u in the direction s is defined to be

$$D\phi(u; s) = \lim_{\alpha \to 0+} \frac{\phi(u + \alpha s) - \phi(u)}{\alpha} \tag{3}$$

THEOREM 1 (DANSKIN [12]). *Under assumptions 1 and 2, the (one-sided) directional derivative of ϕ exists in any direction, s, at any point $u \in E^m$, and is given by*

$$D\phi(u; s) = \min_{x \in X(u)} \sum_{i=1}^{m} s_i \frac{\partial F(x, u)}{\partial u_i} \tag{4}$$

PROOF. Let u be any point in E^m and let $\{\alpha^k\}_{k=1}^{\infty}$ be a sequence of positive numbers tending to zero. Define

$$u^k = u + \alpha^k s \tag{5}$$

Let $x^k \in X(u^k)$, $x \in X(u)$ and consider the difference quotient

$$\frac{\phi(u^k) - \phi(u)}{\alpha^k} = \frac{F(x^k, u^k) - F(x, u)}{\alpha^k} \tag{6}$$

$$\frac{\phi(u^k) - \phi(u)}{\alpha^k} = \frac{F(x^k, u^k) - F(x, u^k)}{\alpha^k} + \frac{F(x, u^k) - F(x, u)}{\alpha^k} \tag{7}$$

Since $x^k \in X(u^k)$,

$$F(x^k, u^k) \leq F(x, u^k) \tag{8}$$

so the first term on the right of (7) may be deleted, leaving

$$\frac{\phi(u^k) - \phi(u)}{\alpha^k} \leq \frac{F(x, u^k) - F(x, u)}{\alpha^k}, \qquad \text{for all } x \in X(u) \tag{9}$$

By the mean value theorem,

$$\frac{F(x, u^k) - F(x, u)}{\alpha^k} = \sum_{i=1}^{m} \frac{u_i^k - u_i}{\alpha^k} \frac{\partial F(x, u + \theta^k(u^k - u))}{\partial u_i}, \qquad 0 < \theta^k < 1 \tag{10}$$

$$\frac{F(x, u^k) - F(x, u)}{\alpha^k} = \sum_{i=1}^{m} s_i \frac{\partial F(x, u + \theta^k(u^k - u))}{\partial u_i} \tag{11}$$

Combining (9) and (11),

$$\frac{\phi(u^k) - \phi(u)}{\alpha^k} \le \sum_{i=1}^{m} s_i \frac{\partial F(x, u + \theta^k(u^k - u))}{\partial u_i} \qquad \text{for all } x \in X(u) \tag{12}$$

As $k \to \infty$, $u^k \to u$. By continuity, the partial derivatives on the right of (12) approach $\partial F(x, u)/\partial u_i$. By (12), each element of the sequence of difference quotients is bounded above, and the largest of its limit points is bounded by

$$\limsup_{u^k \to u} \frac{\phi(u^k) - \phi(u)}{\alpha^k} \le \sum_{i=1}^{m} s_i \frac{\partial F(x, u)}{\partial u_i} \qquad \text{for all } x \in X(u) \tag{13}$$

The objective of what follows is to develop an inequality like (13), but with the inequality reversed, for the limit inferior of the sequence of difference quotients. Let $\{u^{k^*}\}$ be a sequence such that

$$\frac{\phi(u^{k^*}) - \phi(u)}{\alpha^{k^*}} \to \liminf_{u^k \to u} \frac{\phi(u^k) - \phi(u)}{\alpha^k} \tag{14}$$

Let $x^{k^*} \in X(u^{k^*})$. Since each $x^{k^*} \in S$ and S is compact, there is a subsequence of $\{x^{k^*}\}$, $\{x^{k'}\}$, which converges to a limit $x' \in S$. The corresponding subsequence of $\{u^{k^*}\}$ is $\{u^{k'}\}$. Since

$$F(x^{k'}, u^{k'}) \le F(x, u^{k'}), \qquad \text{all } x \in S, \text{ all } k' \tag{15}$$

then, letting $k' \to \infty$, continuity implies

$$F(x', u) \le F(x, u), \qquad \text{all } x \in S \tag{16}$$

Thus

$$x' \in X(u) \tag{17}$$

We now examine the behavior of the difference quotients corresponding to the sequences $\{u^{k'}\}$, $\{x^{k'}\}$.

$$\frac{\phi(u^{k'}) - \phi(u)}{\alpha^{k'}} = \frac{F(x^{k'}, u^{k'}) - F(x^{k'}, u)}{\alpha^{k'}} + \frac{F(x^{k'}, u) - F(x, u)}{\alpha^{k'}} \tag{18}$$

Since $x \in X(u)$,

$$F(x, u) \leq F(x^{k'}, u) \tag{19}$$

Thus, removing the second term from the right of (18) yields

$$\frac{\phi(u^{k'}) - \phi(u)}{\alpha^{k'}} \geq \frac{F(x^{k'}, u^{k'}) - F(x^{k'}, u)}{\alpha^{k'}} \tag{20}$$

Again using the mean value theorem,

$$\frac{\phi(u^{k'}) - \phi(u)}{\alpha^{k'}} \geq \sum_{i=1}^{m} \left(\frac{u_i^{k'} - u_i}{\alpha^{k'}}\right) \frac{\partial F(x^{k'}, u + \omega^{k'}(u^{k'} - u))}{\partial u_i}, \qquad 0 < \omega^{k'} < 1 \tag{21}$$

$$\frac{\phi(u^{k'}) - \phi(u)}{\alpha^{k'}} = \sum_{i=1}^{m} s_i \frac{\partial F(x^{k'}, u + \omega^{k'}(u^{k'} - u))}{\partial u_i} \tag{22}$$

Let $k' \to \infty$. Again using continuity, (22) implies

$$\liminf_{u^k \to u} \frac{\phi(u^k) - \phi(u)}{\alpha^k} \geq \sum_{i=1}^{m} s_i \frac{\partial F(x', u)}{\partial u_i} \tag{23}$$

Now consider relations (13) and (23). Since (13) holds for all $x \in X(u)$,

$$\limsup_{u^k \to u} \frac{\phi(u^k) - \phi(u)}{\alpha^k} \leq \min_{x \in X(u)} \sum_{i=1}^{m} s_i \frac{\partial F(x, u)}{\partial u_i} \tag{24}$$

Relations (23)–(24) imply that the sequence $\{(\phi(u^k) - \phi(u))/\alpha^k\}$ has only one limit point, given by

$$D\phi(u; s) = \lim_{u^k \to u} \frac{\phi(u^k) - \phi(u)}{\alpha^k} = \min_{x \in X(u)} \sum_{i=1}^{m} s_i \frac{\partial F(x, u)}{\partial u_i} \tag{25}$$

so the theorem is proved.

To apply Theorem 1 to the function $h(u)$, set $F(x, u) = L(x, u)$. Since

$$\frac{\partial L}{\partial u_i} = g_i(x)$$

assumptions 1 and 2 become as follows:

Assumption 1′. S is a closed and bounded subset of E^n.

Assumption 2′. $f(x)$ and all $g_i(x)$ are continuous on S.

Under these assumptions, $D = (E^m)^+$. However, the dual function h may be defined for all $u \in E^m$ simply by considering the original definition valid for u with negative components. We use this extension of h in what follows.

THEOREM 2 (DIRECTIONAL DERIVATIVES OF h). *Under assumptions* 1′ *and* 2′, *the (one-sided) directional derivative of h exists in any direction, s, at any point $u \in E^m$, and is given by*

$$Dh(u; s) = \min_{x \in X(u)} \sum_{i=1}^{m} s_i g_i(x) \tag{26}$$

Since, under assumptions 1′ and 2′, $-h$ is closed, proper, and convex and $\text{dom } h = E^m$, the existence and finiteness of its directional derivatives can also be deduced from Theorems 9 and 15 of Appendix 2. By analogy with the definition for convex functions, we define a subgradient of a concave function h at u as the vector of slopes of a supporting hyperplane to the graph of h at u. The set of all subgradients of h at u is $\partial h(u)$. Thus we have the following definition.

DEFINITION 1. x^* is a *subgradient* of h at u if

$$h(z) \leq h(u) + x^*(z - u) \qquad \text{for all } z \in E^m$$

The following shows that evaluation of $h(u)$ immediately yields an element of $\partial h(u)$.

THEOREM 3

$$g(x) \in \partial h(u) \qquad \textit{for any } x \in X(u)$$

PROOF. By definition

$$\begin{aligned} h(z) &\leq f(x) + zg(x) \qquad &\text{for all } x \in S \\ h(u) &= f(x^0) + ug(x^0) \qquad &\text{for all } x^0 \in X(u) \end{aligned} \tag{27}$$

Setting $x = x^0$ in (27) and subtracting yields

$$h(z) \leq h(u) + g(x^0)(z - u) \qquad \text{for all } z \in E^m$$

Examples are easily constructed which show that there may be elements of $\partial h(u)$ which are not equal to $g(x)$ for any $x \in X(u)$. Define

$$V(u) = \{g(x) \mid x \in X(u)\}$$

In Example 2 of section 8.4, $\partial h(-\frac{1}{2}) = [-1, 1]$, while $V(-\frac{1}{2}) = \{-1, 1\}$. Thus Theorem 2 is a sharper result than that obtained by direct application of Theorem 16, Appendix 2, to $-h$, which implies

$$Dh(u; s) = \min\{u^*s \mid u^* \in \partial h(u)\} \tag{28}$$

By Theorem 2, the above min is always attained at points in a particular subset of $\partial h(u)$, the set $V(u)$. From this it is easy to show that $V(u)$ must contain all the extreme points of $\partial h(u)$. [In the previous example, $V(u)$ was equal to the set of extreme points.]

THEOREM 4. *Under assumptions* 1′ *and* 2′, *if v is an extreme point of $\partial h(u)$, then $v \in V(u)$.*

PROOF. If v is an extreme point, then, since $h(u)$ is compact, there is a hyperplane which strictly separates it from $\partial h(u)$. That is, there is an $s^0 \in E^n$ such that

$$vs^0 < u^*s^0 \qquad \text{for all } u^* \in \partial h(u), \quad u^* \neq v$$

Thus the min in (28) is attained uniquely at v. Since this min must also be attained in $V(u)$ [by (26)], $v \in V(u)$.

$v \in V(u)$ does not necessarily imply that v is an extreme point of $\partial h(u)$, as is easily shown by counterexample. Consider the problem

$$\text{minimize } \{x \mid 1 - 2x \leq 0, 0 \leq x \leq 1\}$$

where

$$h(u) = \min_{0 \leq x \leq 1} (x + u(1 - 2x))$$

Here $X(\frac{1}{2}) = [0, 1]$, and $V(\frac{1}{2}) = [-1, 1] = \partial h(\frac{1}{2})$.

Further properties of the directional derivative follow from the concavity of $h(u)$ and other results of Danskin.

1. (Appendix 2, Theorem 9). $Dh(u; s)$ is concave and positively homogeneous in s for fixed u.
2. (Danskin [12].) $Dh(u; s)$ is lower semicontinuous in u for fixed s; i.e., if $\{u^k\} \to u$,

$$\liminf_{u^k \to u} Dh(u^k; s) \geq Dh(u; s)$$

(Danskin [12].) If $s = \alpha s_1 + \beta s_2$, α, β any real numbers, then

$$Dh(u; s) \geq \alpha Dh(u; s_1) + \beta Dh(u; s_2) \tag{29}$$

Relation (29), with the equality, is the usual formula which holds when the directional derivative can be written $Dh(u; s) = \nabla h'(u)s$. Since the directional derivative of h cannot always be written in this way, the strict inequality may hold. Thus, in general, derivatives along the coordinate axes are not sufficient to compute the derivative in other directions.

For a given point u, the right-hand partial derivative of h is obtained by setting $s = i$th unit vector $= e_i$ in (26):

$$\left(\frac{\partial h}{\partial u_i}\right)^+ = \min_{x \in X(u)} g_i(x) \tag{30}$$

The left-hand partial derivative is

$$\begin{aligned}\left(\frac{\partial h}{\partial u_i}\right)^- &= \lim_{\alpha \to 0+} \frac{h(u) - h(u - \alpha e_i)}{\alpha} = -Dh(u; -e_i) \\ &= -\min_{x \in X(u)} (-g_i(x))\end{aligned}$$

or

$$\left(\frac{\partial h}{\partial u_i}\right)^- = \max_{x \in X(u)} g_i(x) \tag{31}$$

Formulas (30)–(31) agree with the fact that, for a concave function,

$$\left(\frac{\partial h}{\partial u_i}\right)^+ \leq \left(\frac{\partial h}{\partial u_i}\right)^-$$

A necessary condition for h to be differentiable at u is

$$\left(\frac{\partial h}{\partial u_i}\right)^+ = \left(\frac{\partial h}{\partial u_i}\right)^- = \frac{\partial h}{\partial u_i}, \qquad \text{all } i \tag{32}$$

By (30)–(31), this is true if and only if each g_i is constant over $X(u)$, equal, say, to $c_i(u)$. By (26), the directional derivative is then given by

$$Dh(u; s) = \sum_{i=1}^{m} s_i c_i(u) \tag{33}$$

This is linear in the components of the direction, s. Since h is concave, then, according to the following theorem of Fenchel, (33) implies that h is differentiable at the point u.

LEMMA 1 (FENCHEL [13]). *Let h be a concave function defined over a convex set, D. Let $u^0 \in \text{int } D$, and suppose that $Dh(u^0; s)$ is linear in s. Then h is differentiable at u^0.*

Of course, if each g_i is not constant over $X(u)$, then the right- and left-hand partial derivatives are not equal, and h is not differentiable at u. Thus we have the following theorem.

THEOREM 5. *Assume assumptions 1′ and 2′ hold. The dual objective h is differentiable at a point u^* if and only if each g_i is constant*[2] *over $X(u^*)$. In this case, the partial derivatives of h are given by*

$$\left.\frac{\partial h}{\partial u_i}\right|_{u=u^*} = g_i(x) \qquad \textit{for any } x \in X(u^*) \tag{34}$$

It is interesting to note that these partial derivatives are the same as those obtained when one differentiates L ignoring the dependence of x on u. This derivative is easily evaluated if a minimizing point $x \in X(u)$ is known. Thus, if h can be evaluated and the hypotheses of Theorem 5 hold, the gradient of h may be obtained with little additional effort.

Theorem 5 is borne out by Example 2 of Section 8.4. There $X(u)$ consists of only a single point for all u except $u^0 = -\frac{1}{2}$. The set $X(-\frac{1}{2}) = \{0, 1\}$, the constraint function g is not constant over this set, and h is differentiable everywhere except at u^0. However, right- and left-hand partial derivatives do exist at u^0, given by

$$\left(\frac{\partial h}{\partial u}\right)^+_{u=-1/2} = -1 = \min_{x \in X(u^0)} (1 - 2x) = \min\{1, -1\}$$

$$\left(\frac{\partial h}{\partial u}\right)^-_{u=-1/2} = 1 = \max_{x \in X(u^0)} (1 - 2x) = \max\{1, -1\}$$

An important special case satisfying the hypotheses of Theorem 5 occurs when $L(x, u^*)$ is minimized over S at a unique point, so $X(u^*)$ contains only a single element. This is stated as a corollary.

COROLLARY 1. *Let assumptions 1′ and 2′ hold. If $L(x, u^*)$ is minimized over S at a unique point $x(u^*)$, then h is differentiable at u^*, with partial derivatives*

$$\left.\frac{\partial h}{\partial u_i}\right|_{u=u^*} = g_i(x(u^*)) \tag{35}$$

A number of sufficient conditions may be given to guarantee unique minima for $L(x, u)$. The simplest is strict convexity.

[2] Since $L(x, u^*)$ is constant over $X(u^*)$, this implies that f is also constant over this set.

COROLLARY 2. *Let assumptions* 1′ *and* 2′ *hold, and assume that S is convex, f is strictly convex, and all g_i are convex over S. Then h is differentiable at all points $u \geq 0$.*

A simple argument may be used to obtain relation (35) if enough assumptions are made. Assume f and all g_i are differentiable everywhere and that $S = E^n$. The Lagrangian problem is then unconstrained:

$$\underset{x \in E^n}{\text{minimize}}\ L(x, u) \tag{36}$$

Assume that $L(x, u^*)$ has a unique minimum, $x(u^*)$. Then $x(u^*)$ satisfies

$$\frac{\partial L(x, u^*)}{\partial x_i} = 0, \qquad i = 1, \ldots, m \tag{37}$$

Now assume that there exists a differentiable function $x(u)$, defined in a neighborhood of u^*, which minimizes $L(x, u)$ for u in this neighborhood. Then

$$h(u) = L(x(u), u)$$

$$\left.\frac{\partial h}{\partial u_i}\right|_{u^*} = \left.\left(\frac{\partial L}{\partial u_i} + \sum_j \frac{\partial L}{\partial x_j}\frac{\partial x_j}{\partial u_i}\right)\right|_{u^*} \tag{38}$$

$$\left.\frac{\partial h}{\partial u_i}\right|_{u^*} = \left.\frac{\partial L}{\partial u_i}\right|_{u^*} \qquad \text{(by 37)} \tag{39}$$

$$\left.\frac{\partial h}{\partial u_i}\right|_{u^*} = g_i(x(u^*)) \tag{40}$$

Theorem 5 leads directly to the following theorem.

THEOREM 6. *Let assumptions* 1′ *and* 2′ *hold, let u^0 solve the dual problem, and assume that h is differentiable at u^0. Then any element $x^0 \in X(u^0)$ solves the primal.*

PROOF. By assumptions 1′ and 2′, $D = (E^m)^+$. Since h is differentiable at u^0, the following optimality conditions hold:

$$\text{(a) } u_i^0 > 0 \Rightarrow \left.\frac{\partial h}{\partial u_i}\right|_{u^0} = g_i(x^0) = 0 \tag{41}$$

$$\text{(b) } u_i^0 = 0 \Rightarrow \left.\frac{\partial h}{\partial u_i}\right|_{u^0} = g_i(x^0) \leq 0 \tag{42}$$

The above may be rewritten $g(x^0) \leq 0$ and $u^0 g(x^0) = 0$. These conditions, together with $x^0 \in X(u^0)$ imply, by Theorem 8.3-1, that (x^0, u^0) is a saddle point for L, so, by Theorem 8.3-2, x^0 solves the primal.

Combining Corollary 1 and Theorem 6 shows that, under assumptions 1′ and 2′, if u^0 solves the dual and $L(x, u^0)$ has a unique minimum over S at x^0, then x^0 solves the primal. Thus any class of problems for which this uniqueness can be guaranteed can be solved by minimizing $L(x, u)$ over S for some u. Corollary 2 gives one such class of problems, but others may exist.

8.6 Computational Methods for Solving the Dual

Gradient Algorithms. Assuming, for the moment, that assumptions 1′ and 2′ hold and that h is differentiable,[3] a steepest-ascent algorithm, modified to handle the constraints $u \geq 0$, may be used to maximize h. This leads to a solution procedure which is a more complete version of that specified in Section 8.5:

1. Choose initial values $u^0 \geq 0$. Step i, $i = 0, 1, 2, \ldots$, proceeds as follows.
2. Solve the Lagrangian problem with $u = u^i$, obtaining a solution $x(u^i)$. In the separable case, this may be accomplished by solving the subproblems, (8.3-24).
3. Form the dual function $h(u^i) = L(x(u^i), u^i)$ and its gradient $\nabla h(u^i) = g(x(u^i))$.
4. Define a direction of search, s^i, by

$$s_k^i = \begin{cases} \left.\dfrac{\partial h}{\partial u_k}\right|_{u^i} & \text{if } u_k^i > 0 \quad (1) \\ \max\left\{0, \left.\dfrac{\partial h}{\partial u_k}\right|_{u^i}\right\} & \text{if } u_k^i = 0 \quad (2) \end{cases} \quad k = 1, \ldots, m$$

Choose a new vector u^{i+1} by

$$u^{i+1} = u^i + \alpha_i s^i \tag{3}$$

The step size α_i must be selected so that

$$h(u^{i+1}) > h(u^i) \tag{4}$$

If h is differentiable there exists $\alpha_i > 0$ satisfying the above unless u^i maximizes h. A common procedure is to choose α_i to maximize

$$r(\alpha) = h(u^i + \alpha s^i) \tag{5}$$

[3] As in convex programs with strictly convex objective—see Sections 8.5 and 8.7.

subject to the constraints $\alpha_i \geq 0$ and $u^{i+1} \geq 0$.

5. Return to step (2), stopping when $\alpha_i \approx 0$.

If $r(\alpha)$ is to be maximized, it must be evaluated a number of times, requiring a solution of the Lagrangian problem each time. Suggested procedures for performing this one-dimensional search are described in Section 1.1 (see also [14] and [15]). Step 4 is simply Rosen's gradient projection method for linear constraints [16], specialized to the constraints $u \geq 0$. Convergence of this algorithm to a global solution has been proved in Part II of [16] for concave differentiable objectives.

This gradient procedure may be viewed as a coordination algorithm for a second-level coordinator, whose task is to solve the dual problem given values of h and ∇h. The first-level units solve the subproblems and provide these values. The algorithm has an interesting economic interpretation if the primal problem is viewed as one of minimal-cost resource allocation. Let the primal be

$$\text{minimize} \sum_j f_j(x_j) \tag{6}$$

subject to

$$g_i(x) = \sum_j g_{ij}(x_j) \leq b_i, \qquad i = 1, \ldots, m \tag{7}$$

$$x_j \in S_j \tag{8}$$

where $f_j(x_j)$ is the cost of running activity j at levels x_j, $g_{ij}(x_j)$ is the quantity of resource i used by activity j at levels x_j, b_i is the amount of resource i available, and S_j is determined by technological constraints on each activity. Let $u^* \geq 0$ be viewed as a vector of prices for the resources, and assume that the Lagrangian problem has a unique solution when $u = u^*$, $x(u^*)$. The subproblems are

$$\text{minimize } \tilde{f}_i(x_i, u) = f_i(x_i) + \sum_k u_k^* g_{ki}(x_i) \tag{9}$$

subject to

$$x_i \in S_i \tag{10}$$

The subproblem objective represents the direct cost, f_i, plus the cost of resources used. The dual function is the sum of subproblem minima:

$$h(u^*) = \sum_i \tilde{f}_i(x_i(u^*), u^*) \tag{11}$$

The gradient of h at u^* has components

$$\left.\frac{\partial h}{\partial u_i}\right|_{u^*} = \sum_j g_{ij}(x_j(u^*)) - b_i = e_i(u^*) \tag{12}$$

The summation term may be viewed as the total demand for resource i, while b_i is its supply. Thus $e_i(u^*)$ is the excess demand for this resource. By (1)–(3), the gradient algorithm requires that u_i^* be increased if excess demand is positive and decreased otherwise, unless $u_i^* = 0$. This is a familiar price-adjustment rule of classical economics. Exposing it as a gradient-ascent algorithm in this instance permits its convergence to be analyzed using established results.

Although the steepest-ascent approach is theoretically interesting, other more recent methods are recommended for computational purposes. If all constraints are equalities, implying that the multipliers u are unconstrained in sign, the variable metric method described by Fletcher and Powell [14], or the conjugate gradient method of Fletcher and Reeves [17] are recommended (see Section 1.1). A modification of the Fletcher–Powell method to handle linear constraints is described in [18] and is recommended when any constraints $u_i \geq 0$ occur. All these procedures require only function and gradient values, yet generally converge in far fewer iterations than the method of steepest ascent.

Tangential Approximation. As shown earlier, h need not be differentiable everywhere. If not, then convergence of the various gradient ascent methods cannot be guaranteed. In practice, slow convergence or no convergence at all may result if the corners in h are numerous or occur at critical points (e.g., the maximum). An algorithm which avoids these difficulties has been suggested by Geoffrion [30]. It is based on the fact that if the Lagrangian problem is solved with $u = u^i$, a linear supporting function[4] to the graph of h at u^i is immediately available. This is the function

$$h(u^i) + (u - u^i)g(x^i) = f(x^i) + ug(x^i) \tag{13}$$

where $x^i \in X(u^i)$, so that $g(x^i) \in \partial h(u^i)$. Let $g(x^1), \ldots, g(x^r)$ be subgradients for h at $u^1, \ldots, u^r$. Then, as shown in Figure 8-6, it is natural to use the concave piecewise linear function

$$v^r(u) = \min_{1 \leq j \leq r} \{f(x^j) + ug(x^j)\} \tag{14}$$

as an approximation to h. Assume that S is closed and bounded and f and all g_i are continuous on S, so subgradients exist at all points $u \in E^m$ and $D = (E^m)^+$. The dual problem is then replaced by

$$\text{maximize } v^r(u) \tag{15}$$

$$\text{subject to} \qquad u \geq 0$$

[4] This is a linear function which never lies below the graph of h and contacts it at u^i.

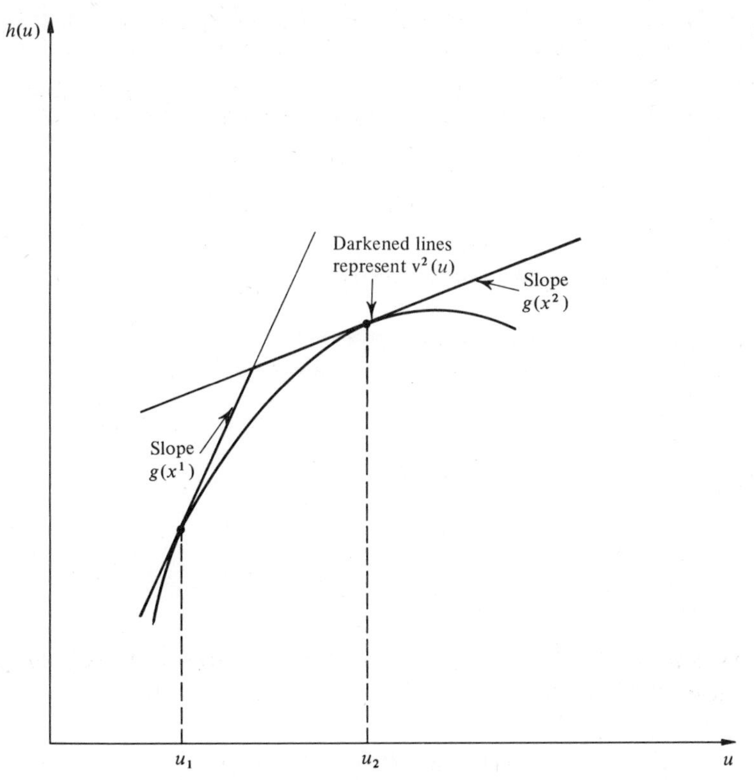

FIGURE 8-6. Tangential approximation to h.

This, in turn, is equivalent to the linear program in the variables (u, σ),

$$\text{maximize } \sigma \tag{16}$$

subject to

$$f(x^j) + ug(x^j) \geq \sigma, \qquad j = 1, \ldots, r \tag{17}$$

$$u \geq 0 \tag{18}$$

The iterative procedure, commencing at cycle r, is

1. Solve the above program, obtaining an optimal solution u^{r+1}.
2. Solve the Lagrangian problem

$$\underset{x \in S}{\text{minimize}} \, (f(x) + u^{r+1}g(x))$$

   obtaining an optimal solution x^{r+1} and a subgradient $g(x^{r+1})$.
3. Add a linear constraint like (17), formed from the point x^{r+1}, to the above program, replace r by $r + 1$, and return to step 1.

Of course, if more than one optimal solution is obtained in step 2, constraints may be formed from each of them and added in step 3.

The above procedure is the dual of the generalized linear programming method of Dantzig and Wolfe, described in Sections 4.3 and 4.4. There the primal problem

$$\text{minimize } \{f(x) \mid g(x) \leq 0,\ x \in S\} \tag{19}$$

was solved by iterative solution of the linear program

$$\text{minimize } \sum_{j=1}^{r} \lambda_j f(x^j) \tag{20}$$

subject to

$$\sum_{j=1}^{r} \lambda_j g(x^j) \leq 0 \tag{21}$$

$$\sum_{j=1}^{r} \lambda_j = 1, \qquad \lambda_j \geq 0 \tag{22}$$

A new column for this program was generated at each cycle by solving the nonlinear subproblem

$$\underset{x \in S}{\text{minimize}}\ (f(x) + u^{r+1} g(x))$$

with u^{r+1} the optimal vector of dual variables for the constraints (21). This is the Lagrangian problem of step 2. Further, if σ is the dual variable for the constraints (22), then the dual of (20)–(22) is precisely the tangential approximation program, (16)–(18). In actual computations, one would generally solve (20)–(22), since it has only $m + 1$ constraints. In this case, tangential approximation becomes identical with the column-generation scheme of generalized linear programming.

Because of these dual relationships, if S and g are convex, then, as shown in Section 4.4, a primal feasible point may be obtained at each cycle. This is the point

$$\hat{x}^{r+1} = \sum_{j=1}^{r} \lambda_j^r x^j$$

where the λ_j^r solve (20)–(22). To see when $\hat{x}^{r+1}$ is optimal note that, since u^{r+1} is feasible for the dual, $h(u^{r+1})$ is a lower bound for its maximal objective value. It is easy to see that σ^{r+1}, the maximal objective value in (16)–(18), is an upper bound, since the objective of this program always overestimates h. Clearly, if

$$h(u^{r+1}) = \sigma^{r+1}$$

then u^{r+1} solves the dual. Then, by Theorem 4.4-1, $\hat{x}^{r+1}$ is optimal. In practice, computations would be terminated when the difference between σ^{r+1} and $h(u^{r+1})$ become less than a specified tolerance. Note that σ^r decreases monotonically, although monotonicity cannot be guaranteed for $h(u^r)$.

As mentioned in Section 4.4, Dantzig has proved convergence of this procedure to an optimal solution as $r \to +\infty$, providing that all columns of (20)–(22) generated in past cycles are retained. The dual interpretation of this is that none of the linear supporting functions comprising the function $v^r(u)$ may be dropped.

As with many of the algorithms of previous chapters, the linear program (16)–(18) may be formulated in a number of different ways if the primal nonlinear program is separable, i.e., has the form

$$\text{minimize} \sum_{i=1}^{p} f_i(x_i)$$

subject to

$$\sum_{i=1}^{p} g^i(x_i) \leq 0$$

$$x_i \in S_i, \qquad i = 1, \ldots, p$$

Then the dual objective h has the form

$$h(u) = \sum_{i=1}^{p} h_i(u)$$

where

$$h_i(u) = \min_{x_i \in S_i}(f_i(x_i) + ug^i(x_i)) \tag{23}$$

Tangential approximation may be applied separately to each function $h_i(u)$. If $X_i(u)$ is the set of x_i which achieve the minimum in (23), then $g^i(x_i)$ is a subgradient of h_i at u for any $x_i \in X_i(u)$. The function

$$f_i(x_i) + ug^i(x_i)$$

is a linear supporting function to the graph of h_i at u, and if

$$x_i^j \in X_i(u_i^j), \qquad j = 1, \ldots, r$$

then the tangential approximation to h_i is the function

$$v_i^r(u) = \min_{1 \le j \le r} \{f_i(x_i^j) + ug^i(x_i^j)\} \tag{24}$$

The dual problem is now replaced with

$$\text{maximize} \sum_{i=1}^{p} v_i^r(u)$$

subject to $u \geq 0$

which is equivalent to the linear program

$$\text{maximize} \sum_{i=1}^{p} \sigma_i$$

subject to $\sigma_i \leq f_i(x_i^j) + ug^i(x_i^j), \quad j = 1, \ldots, r, \; i = 1, \ldots, p$

$$u \geq 0$$

The above program provides the opportunity of adding p new constraints at each cycle, one from each subproblem (23), rather than only a single constraint, as before. The experience with the Dantzig–Wolfe decomposition principle cited in Section 4.2 suggests that this may accelerate convergence. If the dual of the above is solved, then this option corresponds to generating one new column for each subproblem, as discussed in Section 4.4. Of course, subsets of the functions $h_i(u)$ may be grouped together and tangential approximation applied to the sum of the h_i in the subset. This yields approximating linear programs with from 1 to p subsets of constraints.

Before leaving this section, we note an advantage of solution via these dual algorithms not enjoyed by primal methods. Any iterative algorithm which works will produce a sequence of u values, $\{u^i\}$, converging to a u^* which solves the dual. Also available is a sequence of lower bounds $\{h(u^i)\}$ and points $\{x^i\}$, where $x^i \in X(u^i)$. Each such point, by Theorem 8.3-3, solves a modified primal

$$\text{minimize } f(x)$$

subject to $g_j(x) \leq y_j^i, \quad j = 1, \ldots, m$

$$x \in S$$

where

$$y_j^i = g_j(x^i), \qquad (u_j^i) > 0$$

$$y_j^i \geq g_j(x^i), \qquad (u_j^i) = 0$$

Thus we obtain "free" sensitivity information regarding perturbations of the right-hand side. Since the data input to most mathematical models is usually only approximate, such information can be as valuable as the optimal solution itself.

Computational Considerations. In any algorithm for solving the dual, most of the time will be spent solving the Lagrangian problem. If the primal is not convex, then this problem will, in general, be nonconvex also. Since a global minimum must be found, this can be a source of difficulty—see [27] for an example. On the other hand, the subproblems may sometimes be solvable by dynamic programming (see Section 8.8.3), or, if S is finite, by some efficient enumerative procedure. In these instances, global solutions are guaranteed. In any case, it is important that the algorithm which solves the Lagrangian problem makes effective use of the information obtained from minimizing $L(x, u^j)$ in minimizing $L(x, u^{j+1})$. Since u will generally change in small steps, this should reduce solution time considerably. The least that can be done is to use x^j, which is feasible for the Lagrangian problem with $u = u^{j+1}$, as a starting point for the new minimization.

Little is known regarding the relative efficiencies of the tangential approximation and gradient ascent procedures. In preparing computer programs to implement them, the simplicity of a gradient procedure should not be overstated. Even the simpler methods require a one-dimensional search as a subroutine—see Section 1.1. Since this must be done efficiently, the program can be quite complex. Tangential approximation requires a linear programming routine. Even if one is available, communication links with the rest of the program must generally be constructed to permit formulation of the successive linear programs and use of the results.

8.7 Special Results for Convex Problems

In this section a number of duality theorems are presented under various convexity and regularity assumptions. One of the main results has already been proved. Theorem 8.3-6 gives conditions guaranteeing that a saddle point exists, while Theorem 8.4-3 states that (x^0, u^0) is a saddle point if and only if x^0 solves the primal, u^0 solves the dual, and $f(x^0) = h(u^0)$. Combining these two, we obtain the following theorem.

THEOREM 1 (DUALITY THEOREM FOR CONVEX PROGRAMS). *Let S be a convex subset of E^n, f a convex function defined on S, and $g(x)$ an m vector of convex functions defined on S. Assume that there exists a point $x \in S$ such that $g(x) < 0$. Then, if x^0 solves the primal problem, minimize $\{f(x) \mid g(x) \leq 0, x \in S\}$, there is a vector $u^0 \geq 0$ which solves the dual, maximize $\{h(u) \mid u \in D\}$, and $h(u^0) = f(x^0)$.*

Falk [2] has shown that certain additional results hold if the following assumptions hold:

Assumption 1. S is closed and convex, but not necessarily bounded.

Assumption 2. f is strictly convex over S.

Assumption 3. All components of g are convex over S.

THEOREM 2 (FALK [2]). *Under assumptions 1–3, D is open relative to $(E^m)^+$; i.e., D is the intersection of $(E^m)^+$ with an open set.*

PROOF. Some notation to be used in the proof follows:

$N(x, \epsilon)$: open neighborhood of a point x of radius ϵ

$\|x\|$: Euclidean norm of x

$\bar{N}(x, \epsilon)$: closure of N

$\partial N(x, \epsilon)$: boundary of N

$\varnothing$: the empty set

The basic idea of the proof is to choose a point $u^* \in D$ and to find a $\delta > 0$ such that $u \geq 0$ and $u \in N(u^*, \delta) \Rightarrow u \in D$.

Let $u^* \in D$. Since $L(x, u^*)$ is strictly convex over S, it is minimized at a unique point $x(u^*) \equiv x^*$. Let $N(x^*, \epsilon)$ be a neighborhood of x^* of radius $\epsilon > 0$ such that $S \cap \partial N(x^*, \epsilon) \neq \varnothing$. If no such ϵ exists, then S is the single point x^* and the theorem is trivial, since then $D = (E^m)^+$.

Let

$$\mu_1 = h(u^*) = L(x^*, u^*) \tag{1}$$

$$\mu_2 = \min\{L(x, u^*) \mid x \in S \cap \partial N(x^*, \epsilon)\} \tag{2}$$

Note that the feasible set in (2) is nonempty by our previous assumptions. Since $L(x, u^*)$ is strictly convex and $x^* \notin \partial N(x^*, \epsilon)$:

$$\mu_2 > \mu_1 \tag{3}$$

Now let us assume that a $\delta > 0$ can be found such that $u \geq 0$ and $\|u^* - u\| < \delta$ imply the following two inequalities:

$$|L(x^*, u) - \mu_1| < \frac{\mu_2 - \mu_1}{3} \tag{4}$$

$$|L(x, u^*) - L(x, u)| < \frac{\mu_2 - \mu_1}{3} \qquad \text{for all } x \in S \cap \partial N(x^*, \epsilon) \tag{5}$$

Inequality (5) implies

$$L(x, u^*) - L(x, u) < \frac{\mu_2 - \mu_1}{3} \tag{6}$$

or

$$L(x, u) > L(x, u^*) - \frac{\mu_2 - \mu_1}{3} \tag{7}$$

or, since $\mu_2 \leq L(x, u^*)$ for all $x \in S \cap \partial N(x^*, \epsilon)$,

$$L(x, u) > \mu_2 - \frac{\mu_2 - \mu_1}{3} \tag{8}$$

Inequalities (4) and (8) may be sketched as follows:

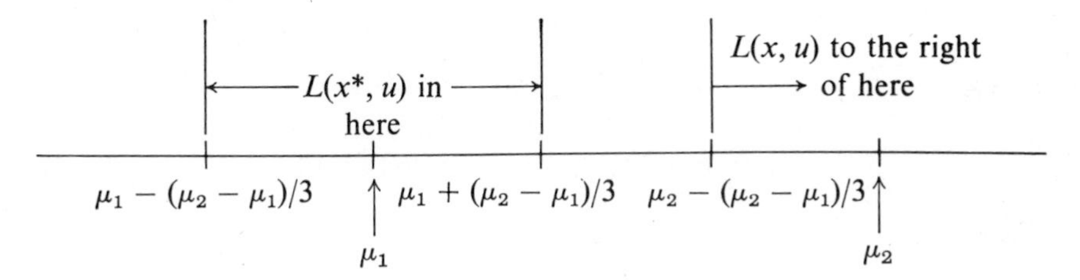

Obviously

$$L(x, u) > L(x^*, u) \qquad \text{for all } x \in S \cap \partial N(x^*, \epsilon) \tag{9}$$

The function $L(x, u)$ must be minimized somewhere in $S \cap \overline{N}(x^*, \epsilon)$, since this set is compact. The minimum cannot occur in $S \cap \partial N(x^*, \epsilon)$ since, by (9), there is a point, x^*, not in $S \cap \partial N(x^*, \epsilon)$, yielding a lower value for $L(x, u)$ than any point in $S \cap \partial N(x^*, \epsilon)$. Thus the minimum is interior to $S \cap \overline{N}(x^*, \epsilon)$ and hence is a local minimum of $L(x, u)$. But $L(x, u)$ is strictly convex, so this minimum is global; i.e., $L(x, u)$ attains a global minimum in S for all $u \geq 0$ such that $\|u^* - u\| < \delta$. This is what we wished to show originally, so the proof will be complete if the δ mentioned previously can be found.

We now construct such a δ. Looking first at (4),

$$|L(x^*, u) - \mu_1| = |f(x^*) + ug(x^*) - f(x^*) - u^*g(x^*)| \tag{10}$$

$$|L(x^*, u) - \mu_1| = |(u - u^*)g(x^*)| \tag{11}$$

By the Schwartz inequality,

$$|L(x^*, u) - \mu_1| \leq \|(u - u^*)\| \, \|g(x^*)\| \tag{12}$$

Let

$$r_1 = \|g(x^*)\| \tag{13}$$

Since we wish $\|u - u^*\| \cdot r_1 < (\mu_2 - \mu_1)/3$, choose

$$\|u - u^*\| < \frac{\mu_2 - \mu_1}{3r_1}, \qquad r_1 > 0 \tag{14}$$

$$\|u - u^*\| < 1, \qquad r_1 = 0 \tag{15}$$

[obviously, any positive number will do in (15)]. Then, by (12), (4) is satisfied.

To satisfy (5),

$$|L(x, u^*) - L(x, u)| = |(u^* - u)g(x)| \tag{16}$$

Again using the Schwartz inequality,

$$|L(x, u^*) - L(x, u)| \leq \|u^* - u\| \; \|g(x)\| \tag{17}$$

Define

$$r_2 = \max\{\|g(x)\| \mid x \in S \cap \partial N(x^*, \epsilon)\} \tag{18}$$

Then

$$|L(x, u^*) - L(x, u)| \leq \|u^* - u\| \cdot r_2, \qquad \text{for all } x \in S \cap \partial N(x^*, \epsilon) \tag{19}$$

To make the above less than $(\mu_2 - \mu_1)/3$. Choose

$$\|u^* - u\| < \frac{\mu_2 - \mu_1}{3r_2}, \qquad r_2 > 0 \tag{20}$$

$$\|u^* - u\| < 1, \qquad r_2 = 0 \tag{21}$$

Again, any positive number will do in (21).

Combining (14)–(15) and (20)–(21) leads to the following specifications for δ:

$$0 < \delta \leq \begin{cases} 1, & r_1 = 0, r_2 = 0 \\ \dfrac{\mu_2 - \mu_1}{3r_1}, & r_1 > 0, r_2 = 0 \\ \dfrac{\mu_2 - \mu_1}{3r_2}, & r_1 = 0, r_2 > 0 \\ \min\left\{\dfrac{\mu_2 - \mu_1}{3r_1}, \dfrac{\mu_2 - \mu_1}{3r_2}\right\}, & r_1 > 0, r_2 > 0 \end{cases} \tag{22}$$

If δ satisfies (22), then $u \geq 0$ and $\|u^* - u\| < \delta$ imply (4)–(5), and the theorem is proved.

The following theorem, together with Theorem 8.4-2, shows that, under assumptions 1–3, the dual problem is a convex program.

THEOREM 3. *Under assumptions* 1–3, *D is convex.*

PROOF. Choose $u_1 \in D$, $u_2 \in D$ and let

$$u_3 = \alpha u_1 + (1 - \alpha)u_2, \qquad 0 < \alpha < 1 \tag{23}$$

Since

$$L(x, u_3) = \alpha L(x, u_1) + (1 - \alpha)L(x, u_2) \tag{24}$$

then

$$\inf_{x \in S} L(x, u_3) \geq \alpha \inf_{x \in S} L(x, u_1) + (1 - \alpha) \inf_{x \in S} L(x, u_2) \tag{25}$$

But both infima on the right of (25) are finite, so $L(x, u_3)$ is bounded below on S. It remains to show that $\inf_{x \in S} L(x, u_3)$ is attained at some point in S.

Let

$$r_i = \inf\{L(x, u_i) \mid x \in S\}, \qquad i = 1, 2, 3 \tag{26}$$

and set

$$r > \max\{r_1, r_2, r_3\} \tag{27}$$

Consider the sets

$$R_i(r) = \{x \in S \mid L(x, u_i) \leq r\} \tag{28}$$

Obviously

$$\inf_{x \in S} L(x, u_3) = \inf_{x \in R_3(r)} L(x, u_3) \tag{29}$$

so if $L(x, u_3)$ attains its infimum over S, it must do so in $R_3(r)$. Since $u_1, u_2 \in D$, the sets $R_1(r_1)$ and $R_2(r_2)$ each consist of a single point and hence are bounded in all directions.[5] Since all nonempty level sets of a convex function are

[5] A subset, R of E^n is said to be bounded in the direction u if there is a hyperplane with outward-pointing normal vector u such that R is contained in a half-space of the hyperplane.

bounded in the same directions [13], $R_1(r)$ and $R_2(r)$ are bounded sets. But, by (24),

$$L(x, u_3) \leq r \Rightarrow \text{either } L(x, u_1) \leq r \text{ or } L(x, u_2) \leq r \tag{30}$$

Thus

$$R_3(r) \subseteq R_1(r) \cup R_2(r) \tag{31}$$

so $R_3(r)$ is a subset of a bounded set and hence is bounded. Since $R_3(r)$ is also closed, it is compact, and $\inf_{x \in S} L(x, u_3)$ must be attained in $R_3(r)$. This completes the proof.

A number of other results follow from assumptions 1–3. These are proved in [2].

1. The minimizing point $x(u)$ is a continuous function of u.
2. The dual objective h has continuous first partial derivatives everywhere in int D.

These results are similar to those obtained in Section 8.5. Examples are easily constructed in which the second partials of $h(u)$ are not continuous everywhere.

According to the theorems in this section, under assumptions 1–3, the dual problem is unconstrained except for the conditions $u \geq 0$, and h is differentiable everywhere in int D. Thus a gradient ascent method is again appropriate for solving the dual. Such techniques have been applied to problems with a strictly convex, separable objective function and linear constraints by Falk [21].

8.8 Applications

8.8.1 Problems Involving Coupled Subsystems.[6] Many systems can be viewed as a set of interacting subsystems. Each subsystem is described by specifying its outputs as functions of its inputs. The subsystems interact because outputs of some are inputs to others. An example is shown in Figure 8-7. Each interconnecting stream is numbered, and the variable x_i associated with stream i represents the quantity or "flow level" of the stream. The arrows labeled y_i are outputs of subsystem i which do not go to other subsystems but instead flow to the "outside world" and presumably have some direct value. The m_i are manipulated or decision variables associated with subsystem i.

[6] See [22]–[27] for earlier work on this topic.

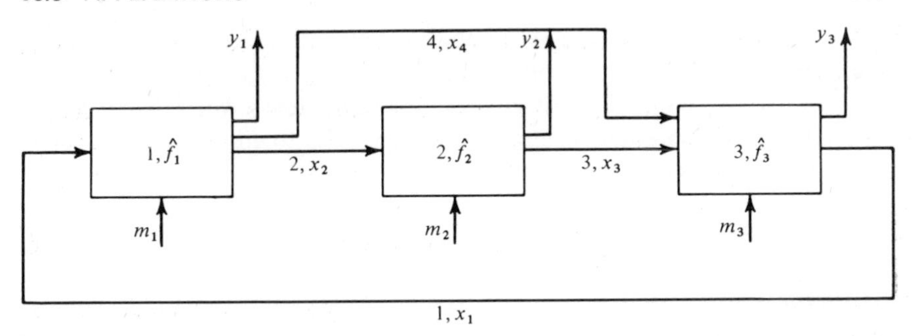

FIGURE 8-7 Coupled system.

Such a structure could model a chemical system, with the subsystems being reactors, separators, etc. In this case, the m_i would be physical variables such as pressures, temperatures, and recycle rates, while the x_i are flows and concentrations of the process streams. If an economic system were being considered, the subsystems could represent the economies of various nations, and the x_i import and export levels.

To model such a system mathematically, we associate two index sets with each subsystem

$$I(j) = \{i \mid \text{stream } i \text{ is an input to subsystem } j\} \tag{1}$$
$$O(j) = \{i \mid \text{stream } i \text{ is an output of subsystem } j\} \tag{2}$$

For example, in Figure 8-7,

$$\begin{aligned} I(1) &= \{1\} & O(1) &= \{2, 4\} \\ I(2) &= \{2\} & O(2) &= \{3\} \\ I(3) &= \{3, 4\} & O(3) &= \{1\} \end{aligned}$$

Let q_j be a vector containing all the inputs to subsystem j:

$$q_j = (x_{k_1}, \ldots, x_{k_{r_j}} \mid m_j), \qquad k_i \in I(j) \tag{3}$$

For example, in Figure 8-7,

$$q_3 = (x_3, x_4 \mid m_3)$$

Each subsystem may now be described by giving its outputs as functions of its inputs:

$$x_i = t_i(q_j), \qquad i \in O(j),\ j = 1, \ldots, r \tag{4}$$
$$y_j = s_j(q_j), \qquad j = 1, \ldots, r \tag{5}$$

where t_i is a function of q_j and s_j is a vector of such functions.

Associated with subsystem j is an objective function, $\hat{f}_j(q_j, y_j)$, representing the return from operating subsystem j. For example, $\hat{f}_j$ might be the value of the vector y_j minus the cost of operating subsystem j when its inputs are at levels q_j. The notation is simplified if we use (5) to eliminate the y_j in $\hat{f}_j$, leaving only a function of q_j:

$$f_j(q_j) = \hat{f}_j(q_j, s_j(q_j)) \tag{6}$$

The total return from operating the system, f, is assumed to be the sum of the subsystem returns:

$$f = \sum_{j=1}^{r} f_j(q_j) \tag{7}$$

The problem considered here is to choose the vectors q_j to maximize (7) subject to the interconnection relations, (4), and to whatever other constraints are present on the vectors q_j. Such constraints may arise, for example, from the presence of bounds on the decision variables m_j and on the interunit flows, x_i. It is assumed that these restrictions apply independently to each subsystem. That is, they may be represented by defining a set S_j and writing

$$q_j \in S_j, \qquad j = 1, \ldots, r \tag{8}$$

Then the problem of coupled system optimization may be written

$$\text{maximize} \sum_{j=1}^{r} f_j(q_j) \tag{9}$$

subject to

$$\left.x_i = t_i(q_j), \qquad i \in O(j),\ j = 1, \ldots, r\right\}\begin{array}{c}m\\\text{constraints}\end{array} \tag{10}$$

$$q_j \in S_j, \qquad j = 1, \ldots, r \tag{11}$$

We assume that this problem has an optimal solution.

The problem (9)–(11) can be attacked in one of two ways. The first is to view only the variables m_j as independent and to use the equalities (10) to eliminate the x_i. Since, in general, each x_i depends on all vectors m_j, the result is a mathematical program in the vector $m = (m_1, \ldots, m_r)$. Although this "integrated" approach reduces the number of independent variables, it has serious drawbacks, especially in cases when there is feedback or recycle in the coupled system. For example, in Figure 8-7, stream 1 is fed back from subsystem 3 to subsystem 1. As a consequence the equalities (10) are implicit; the x_i cannot be evaluated sequentially. Implicit systems of equations must generally be solved numerically, since closed form solutions are possible only in the simplest cases. Thus, to evaluate the objective for a given vector m, an iterative process must be successfully completed.

Despite these difficulties, this approach is widely used and is often successful. An alternative is to view the entire vector $q = (q_1, \ldots, q_r)$ as independent and use penalty functions to handle the constraints (see [28] and [29]). Another alternative, considered here in detail, is to solve the problem by decomposing it into smaller subproblems. To see how this might be done, consider the following plausibility argument, which we justify later. The problem cannot be decomposed unless the subsystems are separated in some way. Imagine doing this by cutting the links between subsystems so that $x_i \neq t_i(q_j)$ in (10). Further, regard each subsystem as being under the jurisdiction of a manager, who sells the subsystem outputs to other subsystems and buys its inputs from other subsystems. Associated with the ith interunit stream is a price, u_i, at which the transactions take place.

The manager of subsystem i views the prices as fixed, and maximizes the net profit of his subsystem. This net profit is

$$\tilde{f}_j(q_j; u) = f_j(q_j) + \sum_{i \in O(j)} u_i t_i(q_j) - \sum_{i \in I(j)} u_i x_i \tag{12}$$

The first term, f_j, is the net profit earned by producing the outputs y_j while using the inputs m_j, the second term is the profit earned by selling the outputs t_i to other subsystems, and the last term is the cost of buying the inputs x_i from other subsystems. The jth subproblem then becomes

$$\left.\begin{array}{ll} & \text{maximize } \tilde{f}_j(q_j; u) \quad (13) \\ \text{subject to} & q_j \in S_j \quad (14) \end{array}\right\} \begin{array}{l}\text{subproblem } j \\ j = 1, \ldots, r\end{array}$$

The task of adjusting the prices is given to a second-level coordinator. Since the inputs $x_i(u)$ and outputs $t_i(u)$ as determined by the managers need not be equal across a cut, it is evident that the coordinator must adjust the prices so that their difference is zero. Considering any one of the interunit stream variables, the value of the variable on the output side of a cut, $t_i(u)$, may be termed a "supply" and that on the input side, $x_i(u)$, a "demand." The difference between these is simply demand minus supply, or excess demand, $e_i(u)$:

$$e_i(u) = x_i(u) - t_i(q_j(u)), \qquad i \in O(j) \tag{15}$$

Then the algorithm

$$\frac{du_i}{dt} = e_i(u), \qquad i = 1, \ldots, m \tag{16}$$

says that when demand exceeds supply the price should increase and vice versa. This price-adjustment rule could serve as a coordination algorithm, although, as we shall see shortly, there are more efficient ones.

Mathematical justification for all this follows from the results of previous sections. The Lagrangian function for the problem (9)–(11) is

$$L = \sum_j f_j(q_j) + \sum_j \sum_{i \in O(j)} u_i(t_i(q_j) - x_i) \tag{17}$$

The term $\sum_j \sum_{i \in O(j)} u_i x_i$ is simply the sum of all possible terms $u_i x_i$ and may be rewritten

$$\sum_j \sum_{i \in O(j)} u_i x_i = \sum_j \sum_{i \in I(j)} u_i x_i \tag{18}$$

Then L may be written

$$L = \sum_j \left[f_j(q_j) + \sum_{i \in O(j)} u_i t_i(q_j) - \sum_{i \in I(j)} u_i x_i \right] \tag{19}$$

This is separable in the vectors q_j for fixed u. The bracketed quantity is, by (12), the objective or "net-profit" function of the jth subproblem, $\tilde{f}_j(q_j; u)$. The dual of the primal problem (9)–(11) is

$$\text{minimize } h(u)$$

subject to

$$u \in D$$

where

$$h(u) = \max_{q_1 \in S_1, \ldots, q_r \in S_r} \left(\sum_{j=1}^{r} \tilde{f}_j(q_j; u) \right) \tag{20}$$

$$D = \{u \mid \text{the above maximum exists}\}$$

This maximization is separable and can be accomplished by solving the subproblems

$$\text{maximize } \tilde{f}_j(q_j; u) \tag{22}$$

subject to

$$q_j \in S_j$$

which is the jth subproblem, (13)–(14), constructed earlier.

The dual function, h, is the sum of these subproblem maxima for given u:

$$h(u) = \sum_j \left(\max_{q_j \in S_j} \tilde{f}_j(q_j; u) \right) \tag{23}$$

By Theorem 8.4-2, h is convex over any convex subset of D (note that the roles of max and min are reversed from previous sections). All other results of previous sections apply here as well. Let

$$Q_j(u) = \{q_j \mid q_j \text{ solves subproblem } j \text{ with multipliers } u\} \tag{24}$$

$$Q(u) = Q_1(u) \times Q_2(u) \times \cdots \times Q_r(u) \tag{25}$$

Then, since the constraints being relaxed here are equalities, Theorem 8.3-2 specializes to the following theorem.

THEOREM 1. *Assume that D is nonempty, and consider a particular vec r of multipliers $u^0 \in D$. A vector $q \in Q(u^0)$ solves the primal problem if and only if q is feasible for the primal, i.e., satisfies the equalities* (10).

Satisfaction of (10) means, of course, that the interconnection error or excess demand in (15), $e_i(u)$, is zero for all i. This verifies the intuitive notion that the perfectly competitive economic system constructed previously yields an overall optimum if and only if supply and demand are equal in all markets.

Theorem 8.4-3, relating saddle points and dual problems, specializes here to the following theorem.

THEOREM 2. *A vector of multipliers $u^0 \in D$ has the property that some $q \in Q(u^0)$ solves the primal problem if and only if u^0 solves the dual and* $\min h = h(u^0) = \max f$.

If each set S_i is closed and bounded and all functions f_j and t_j are continuous, then $D = E^m$ and the dual problem is unconstrained. Solution of the dual can then be attempted by a gradient ascent or tangential approximation algorithm.

All results of Section 8.5 regarding derivatives of h apply here. In particular, assuming $D = E^m$, h is differentiable at a point u^* if $Q(u^*)$ consists of a single point. In that case its partial derivatives are

$$\left.\frac{\partial h}{\partial u_i}\right|_{u^*} = t_i(q_j(u^*)) - x_i(u^*), \qquad i \in O(j) \tag{26}$$

This is the negative of the excess demand function constructed pre iously:

$$\left.\frac{\partial h}{\partial u_i}\right|_{u^*} = -e_i(u^*) \tag{27}$$

Thus the algorithm $du_i/dt = e_i$ is simply a differential version of the method of steepest descent applied to h. In practice, one would use one of the recent, more efficient unconstrained minimizers described in Section 1.1 to minimize h. If h is not differentiable at u^*, the vector

$$-e(u^*) = -(e_1(u^*), \ldots, e_m(u^*))$$

is a subgradient of h at u^* and may be used in the tangential approximation method.

8.8.2 Example—Optimal Control of Discrete-Time Dynamic Systems. A set of interacting subsystems may be coupled in time instead of space, as in the discrete-time dynamic system shown in Figure 8-8. The quantity x_i is a n

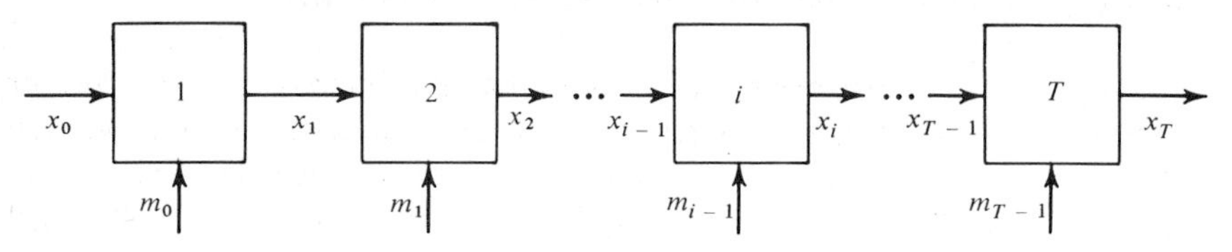

FIGURE 8-8 Discrete-time dynamic system.

vector representing the state of the process, while m_i is an m vector of decision variables, both evaluated at $t = i$. The state evolves in time according to the difference equation

$$x_i = g^i(x_{i-1}, m_{i-1}, i), \qquad x_0 \text{ given}, i = 1, \ldots, T \tag{28}$$

The total return from operating the system over T time periods is assumed to be the sum of the returns at each stage,

$$\text{maximize} \sum_{i=0}^{T-1} f_i(m_i, x_i, i) \tag{29}$$

In many cases, the variables m_i and x_i are required to satisfy certain inequality constraints, which often take the form of upper and lower bounds. These are represented by

$$(m_i, x_i) \in S_i, \qquad i = 0, \ldots, T-1 \tag{30}$$

In addition, the terminal-state vector, x_T, may be required to lie on some terminal manifold:

$$\psi(x_T) = 0 \tag{31}$$

where ψ is a p vector of functions, $p < n$. The problem is to maximize (29) subject to (28), (30), and (31).

As a specific example, consider the following problem:

$$\text{maximize } f = -\sum_{i=1}^{8} x_i'x_i - \sum_{i=0}^{8} m_i'm_i \tag{32}$$

where x_i and m_i are 2-vectors, subject to[7]

$$\begin{aligned} x_i &= g(x_{i-1}, m_{i-1}), \qquad i = 1, \ldots, 9 \\ x_0 &= (-1, 1) \end{aligned} \tag{33}$$

$$|x_i| \leq 1, \qquad i = 1, \ldots, 8 \tag{34}$$

$$|m_i| \leq 1, \qquad i = 0, \ldots, 8 \tag{35}$$

$$x_{1,9} = \tfrac{1}{2} \tag{36}$$

where g is a 2-vector of functions:

$$g_1(x_{i-1}, m_{i-1}) = -x_{1,i-1} + x_{2,i-1} + m_{1,i-1} \tag{37}$$

$$g_2(x_{i-1}, m_{i-1}) = x_{1,i-1} - x_{2,i-1} + m_{2,i-1} \tag{38}$$

This is a linear system with quadratic objective, bounded state and control variables, and one terminal constraint. The objective is constructed to ensure that deviations of the state from zero and expenditures of control energy are both small; i.e., this is a regulator problem with limited control effort.

The Lagrangian for this problem is

$$L = f + \sum_{i=1}^{9} u_i(g(x_{i-1}, m_{i-1}) - x_i) + p_{1,9}(x_{1,9} - \tfrac{1}{2}) \tag{39}$$

and the Lagrangian problem is to maximize L subject to the bounds (34)–(35). Since L is linear in x_9, no maximum exists unless the coefficients of x_9 vanish. However, vanishing of these coefficients is one of the necessary conditions for optimality,

$$\frac{\partial L}{\partial x_{1,9}} = -u_{1,9} + p_{1,9} = 0 \tag{40}$$

$$\frac{\partial L}{\partial x_{2,9}} = -u_{2,9} = 0 \tag{41}$$

These are used to eliminate $u_{2,9}$ and $p_{1,9}$ in L. The separability of L then yields the following subproblems.

[7] Equations (34) and (35) mean that each component of x_i and m_i is bounded.

Subproblem 0

$$\text{maximize } \tilde{f}_0 = -m_0'm_0 + u_1 g(m_0, x_0), \; x_0 = (-1, 1)$$

subject to

$$|m_0| \leq 1 \tag{42}$$

Subproblem i, $i = 1, \ldots, 7$

$$\text{maximize } \tilde{f}_i = -m_i'm_i - x_i'x_i + u_{i+1} g(m_i, x_i) - u_i x_i$$

subject to

$$|x_i| \leq 1, \qquad |m_i| \leq 1 \tag{43}$$

Subproblem 8

$$\text{maximize } \tilde{f}_8 = -m_8'm_8 - x_8'x_8 + u_{1,9} g_1(m_8, x_8) - u_8 x_8$$

subject to

$$|m_8| \leq 1, \qquad |x_8| \leq 1 \tag{44}$$

The dual objective h is

$$h(u) = \sum_{i=1}^{9} (\max \tilde{f}_i) - \tfrac{1}{2} u_{1,9} \tag{45}$$

Since the subproblems have unique optimal solutions for any vector $u = (u_1 \mid \cdots \mid u_9)$, $h(u)$ is defined for all u and is differentiable everywhere. Thus, by Theorem 8.5-6, if u^0 solves the dual, then the subproblem solutions for $u = u^0$ solve the primal. Since each subproblem objective is separable, the subproblems may be solved in closed form. The solutions so obtained are linear functions of the multipliers, u, so the dual function h is quadratic with a positive semidefinite Hessian matrix. As shown in Section 1.1, the gradient descent methods of Fletcher and Powell [14] and Fletcher and Reeves [17] will find the minimum of such a function in as many steps as there are variables (or possibly less). Thus applying either of these procedures to the minimization of h yields a finite algorithm for solving the primal.

In performing these computations, it is possible to obtain primal feasible points, and hence to bound the primal optimum. One simply inserts the m components of the subproblem solutions into the difference equation (33) and computes the corresponding values for x_i. If these x_i satisfy the bounds, (34), then the pair $(\{x_i, m_i\})$ is primal feasible. Figure 8-9 shows the values of primal and dual objectives versus iteration count for the problem (32)–(36), where the primal feasible points are computed as indicated above [20]. The method of Fletcher and Powell [14] was used to minimize h. This method forced ∇h essentially to zero in eight iterations, which is considerably less than the theoretical guarantee of 17 (the number of components of u). Primal feasible points were obtained from iteration 5 on.

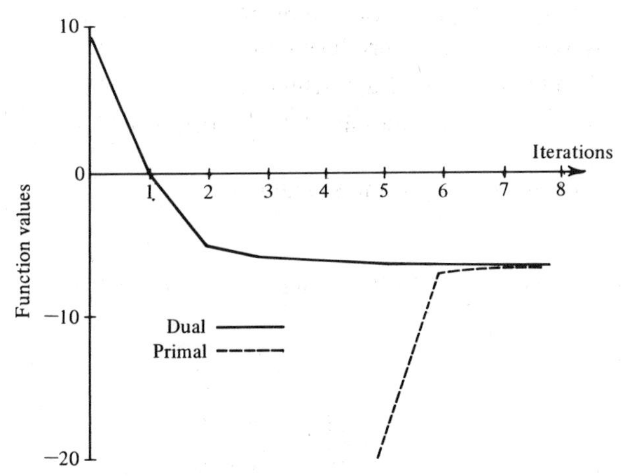

FIGURE 8-9 Primal and dual function values versus number of iterations for a nine-stage sequential system.

8.8.3 Problems in which the Constraint Set is Finite: Multi-item Scheduling Problems. For separable problems involving optimization over a finite (but large) set, decomposition is attractive, since discrete optimization with only a few variables is easily accomplished by dynamic programming or some other combinatorial technique. An example of such a problem is the multi-item scheduling problem of Section 4.2. There, assuming that the number of items is much larger than the number of time periods, an approximate solution was found using linear programming. The algorithm involved solving a set of single-item scheduling problems, using prices provided by a "master" linear program. Here, using the minimax dual of nonlinear programming, the problem is attacked directly. The same subproblems are solved, but the prices are provided by maximizing the dual objective. Any solutions obtained are integral, independent of the number of items or time periods. However, optimal (or even feasible) solutions cannot always be obtained.

The problem is described in Section 4.2. For convenience, the notation and mathematical statement are rewritten here. Let

m_{it} = number of machines used to produce item i in time period t

$m_i = (m_{i1} \cdots m_{iT})'$ = schedule for item i over time

$s_i(m_i)$ = setup cost incurred by schedule m_i (*note:* s_i can be computed given m_i only if the setup cost is independent of the items involved in the changeover)

d_{it} = demand for product i in time period t, assumed known

y_{it} = inventory of item i at the end of time period t

p_s = cost of one setup

n_i = number of pieces of equipment available which cause a given machine to produce item i

k_i = production rate of a machine

b_t = number of machines available in period t

The setup cost incurred by using schedule m_i is

$$s_i(m_i) = p_s \sum_{t=1}^{T} (m_{it} - m_{i,t-1})_+, \qquad m_{i0} \text{ given} \tag{46}$$

where

$$(x)_+ = \begin{cases} x, & x \geq 0 \\ 0, & x < 0 \end{cases} \tag{47}$$

and the inventory cost using m_i is

$$\gamma_i(m_i) = \sum_{t=1}^{T} \gamma_{it}(y_{it}) \tag{48}$$

where the γ_{it} represent holding costs for $y_{it} > 0$ and shortage costs for $y_{it} < 0$ and were chosen to be linear in each of these ranges. The total cost incurred by m_i is

$$c_i(m_i) = s_i(m_i) + \gamma_i(m_i) \tag{49}$$

The scheduling problem is to choose $m_1 \cdots m_I$ to minimize

$$c = \sum_{i=1}^{I} c_i(m_i) \tag{50}$$

subject to

$$y_{it} = y_{i,t-1} + k_i m_{it} - d_{it}, \qquad t = 1, \cdots, T, \qquad h_{i0} \text{ given} \tag{51}$$

$$(y_{it})_{\max} \geq y_{it} \geq (y_{it})_{\min} \tag{52}$$

$$m_{it} \leq n_i, \qquad t = 1, \cdots, T \tag{53}$$

$$m_{it} \text{ a nonnegative integer} \tag{54}$$

$$\sum_{i=1}^{I} m_{it} \leq b_t, \qquad t = 1, \cdots, T \tag{55}$$

The problem (50)–(55) is discrete because of (54). Relation (51) describes the changes in inventory level of item i, (52) bounds the inventory, and (55) requires that the number of machines used in any time period, t, shall not exceed the number available, b_t.

The above problem is quite difficult to solve for large I and T. The problem would decompose into I single-item scheduling problems if (55) were absent. To decompose it in the presence of (55), form the Lagrangian

$$L(m, u) = \sum_{i=1}^{I} c_i + \sum_{t=1}^{T} u_t \left(\sum_{i=1}^{I} m_{it} - b_t \right), \qquad u_t \geq 0 \tag{56}$$

$$L(m, u) = \sum_{i=1}^{I} \left(c_i(m_i) + \sum_{t=1}^{T} u_t m_{it} \right) - \sum_{t=1}^{T} u_t b_t \tag{57}$$

and define

$$S_i = \{m_i \mid m_i \text{ satisfies (51)–(54)}\} \tag{58}$$

The dual problem is

$$\underset{u \geq 0}{\text{maximize}} \quad \min_{m_1 \in S_1 \cdots m_I \in S_I} L(m, u) \tag{59}$$

Again, since the Lagrangian is separable, the inner minimization yields the subproblems

$$\text{minimize } \{c_i(m_i) + u m_i\} \tag{60}$$

subject to

$$m_i \in S_i \tag{61}$$

and the dual problem is

$$\underset{u \geq 0}{\text{maximize}}\ h(u)$$

where

$$h(u) = \sum_{i=1}^{I} h_i(u) - \sum_{t=1}^{T} u_t b_t \tag{62}$$

and where $h_i(u)$ is the minimum value attained in (60). The partial derivatives of the dual function (when they exist) are

$$\frac{\partial h}{\partial u_t} = \sum_{i=1}^{I} m_{it}(u) - b_t \tag{63}$$

Solution proceeds by choosing some initial $u_0 \geq 0$ ($u_0 = 0$ is a good choice) solving the subproblems for this u, evaluating the dual function and its gradient and, for example, changing u_0 by moving in the gradient direction, thus increasing h. Note that the multipliers u_t appear in (60) as costs for the m_{it}. The gradient algorithm would increase the "cost" u_t if (63) were positive, i.e., if the number of machines used by all items in time period t were greater than b_t. Thus the subproblems are each penalized for collectively using more machines than are available.

Note that, since the sets S_i are finite, the minima in (60) exist, and thus the dual function exists (and is concave) for all $u \geq 0$. Additional properties of h follow from the finiteness of the sets S_i. Given any $u_0 \geq 0$, we can imagine that the elements m_i of S_i are ordered along a line according to the value of the subproblem objective in (60) as shown in Figure 8-10. The lowest

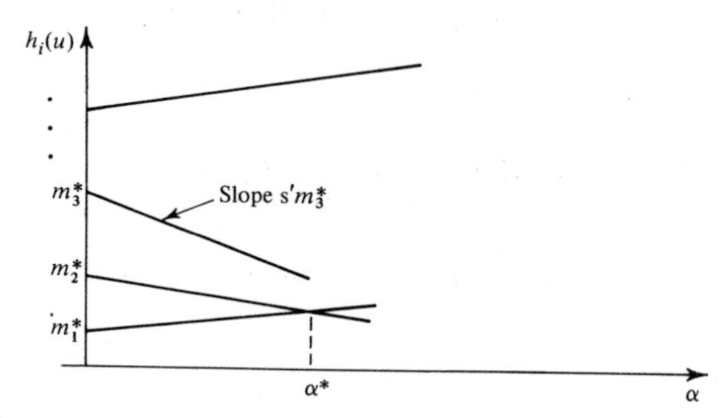

FIGURE 8-10

point represents the minimizing solution for $u = u_0$, and we assume that each optimal solution, $m_i^*(u_0)$ is unique, implying that h is differentiable at u_0. Consider changing u_0 by moving along a direction s; i.e.,

$$u = u_0 + \alpha s, \qquad \alpha \geq 0 \tag{64}$$

As α increases, the function in (60) for each fixed m_i^* is given by

$$h_i^*(\alpha) = (c_i(m_i^*) + u_0' m_i^*) + \alpha(s' m_i^*) \tag{65}$$

which is linear in α for fixed s, m_i^*. Since each of the h_i^* changes as shown in Figure 8-10, the optimal solution does not change until some point $\alpha^* > 0$. Thus the subproblem objective function, $h_i(u(\alpha))$, is linear in α for $0 \leq \alpha \leq \alpha^*$ for any direction s; i.e., the dual function is linear as one travels along any direction in u space until one has traveled some positive distance, whereupon the optimal solution of some subproblem changes. This means the dual function is formed by sections of intersecting hyperplanes. One strategy for maximizing such a function is to follow the gradient along a given hyperplane until another hyperplane is reached, recompute the gradient, etc. (clearly the gradient exists except at the intersection of two hyperplanes, where it is discontinuous). The tangential approximation approach is perhaps even more appealing, since it does not depend on h being differentiable and since here h is piecewise linear. One might suspect that tangential approximation will find the dual optimum in a finite number of steps, although we have no proof of this.

Examples. There are no theoretical guarantees that discrete problems of the type considered here can be solved using duality. Thus solution of a number of sample problems was attempted to help assess the usefulness of the approach. Data for these problems is given below and in Tables 1 and 2

$$T = 6 \text{ weeks}$$

$$I = 15 \text{ items}$$

$$\gamma_{it}(y_{it}) = 0.0015 y_{it}, \; y_{it} \geq 0$$

$$p_s = \$15.20/\text{setup}$$

$$k_i = k = 900 \text{ items/machine/week}$$

TABLE 1

Item	y_{i0}	m_{i0}	$(y_i)_{\min}$	$(y_i)_{\max}$	n_i
1	500	0	50	6,000	4
2	4,000	3	500	6,000	4
3	300	0	100	2,000	1
4	500	0	100	2,000	1
5	5,000	4	2,500	8,000	7
6	2,500	3	1,000	5,000	3
7	1,500	0	100	4,000	1
8	0	0	0	5,000	3
9	5,500	1	500	9,000	1
10	6,000	3	1,000	9,000	5
11	200	1	0	4,000	2
12	100	0	100	6,000	2
13	0	0	0	4,000	1
14	5,000	6	2,500	8,000	7
15	0	2	600	4,000	4

TABLE 2. DEMANDS (in hundreds of units/week)

Item	Week 1	Week 2	Week 3	Week 4	Week 5	Week 6
1	25	28	24	22	20	18
2	15	20	20	18	16	15
3	0.5	1	0.5	0.5	1	1
4	1	1	1	1	1	1
5	40	40	50	60	70	80
6	30	30	30	25	25	25
7	3	3	4	4	4	4
8	25	28	26	25	26	28
9	5	5	5	5	5	5
10	40	42	43	45	40	35
11	10	12	14	16	18	20
12	0	0	0	40	40	0
13	0	0	0	0	3	14
14	45	45	46	48	50	52
15	26	27	24	20	18	12

If each $c_i(m_i)$ is minimized separately subject to $m_i \in S_i$, and if the resultant solutions satisfy (55), these solutions are optimal for (50)–(55). The aggregate machine usage, $\sum_i m_{it}$, for these separate solutions is

Period	1	2	3	4	5	6
Total usage	27	28	31	33	36	28

The number of machines available in time period t, b_t, was chosen equal for all $t (b_t = b$ for all $t)$ and was varied to yield four different problems:

Problem	1	2	3	4
Machines available, b	35	34	33	32

Problem 1 is just barely constrained, as the total number of machines used by the unconstrained solutions is greater than 35 only in period 5. As b is decreased, the problem becomes more tightly constrained until eventually it becomes infeasible.

All problems were solved using a simple steepest ascent on h, modified to account for the constraints $u \geq 0$. Solution data are given in Tables 3–6. Optimal solutions were obtained for problems 1 and 3, as indicated by satisfaction of the machine availability constraints and complementary

TABLE 3 OPTIMAL SOLUTIONS

Problem	Dual optimum	Primal optimum (if attained)
1	2,877.40	2,877.40
2	2,885.14	—
3	2,910.88	2,910.88
4	2,947.60	—

TABLE 4. OPTIMAL MULTIPLIERS

Problem	u_1	u_2	u_3	u_4	u_5	u_6
1	0	0	0	0	6	0
2	0	0	0	1.2495	7.87495	0
3	0	0	0	3	9	0
4	0	0	0.045	6.78	1.3275	0

TABLE 5. MACHINE SLACK AT DUAL OPTIMUM: $\sum_i m_{it} - b$

	Time					
Problem	1	2	3	4	5	6
1	−8	−6	−4	−2	0	−7
2	−8	−6	−2	+2[a]	−2	−5
3	−6	−4	−2	0	0	−3
4	−4	−3	−1	0	+1[a]	−2

[a] Plus sign indicates constraint violation.

TABLE 6. ITERATION-BY-ITERATION SUMMARY FOR $b = 33$

	Multipliers						Machine slack						
Iteration	u_1	u_2	u_3	u_4	u_5	u_6	1	2	3	4	5	6	Dual objectives
1	0	0	0	0	0	0	−6	−5	−2	0	3	−5	2,867.14
2	0	0	0	0	3	0	−6	−5	−2	0	3	−5	2,885.14
3	0	0	0	0	6	0	−6	−4	−2	0	2	−5	2,901.40
4	0	0	0	0	8	0	−6	−4	−2	3	−1	−5	2,901.90
5	0	0	0	3	7	0	−6	−4	−2	0	2	−5	2,905.40
6	0	0	0	3	9	0	−6	−4	−2	0	0	−3	2,910.88

slackness conditions. With problems 2 and 4, the constraints could not be met in all time periods, but infeasibility occurs in only one time period, and its amount is small. Thus the schedules obtained are still useful.

Problems

1. Prove Theorem 8.3-4.
2. Prove Theorem 8.3-5.
3. Prove: If x^i solves the Lagrangian problem with $u = u^i$, $i = 1, 2$, and if

$$g_j(x^2) = g_j(x^1), \qquad j \neq k$$
$$g_k(x^2) > g_k(x^1)$$

then

$$u_k^1 \geq \frac{f(x^1) - f(x^2)}{g_k(x^2) - g_k(x^1)} \geq u_k^2$$

4. Given the problem

$$\text{minimize } f(x)$$

subject to $$g_i(x) \leq b_i, \qquad i = 1, \ldots, m$$

where f and all g_i are differentiable, assume that an optimal solution, $x(b)$, exists for all $b = (b_1, \ldots, b_m)$ in some neighborhood of a point b^*, and that $x(b)$ is differentiable for all b in this neighborhood. Further, assume that a set of multipliers exists such that the Kuhn–Tucker conditions are satisfied when $b = b^*$. Show that

$$\left.\frac{\partial(\min f)}{\partial b_i}\right|_{b^*} = u_i^*, \qquad i = 1, \ldots, m$$

Comment on the pricing interpretation of this result and its use in sensitivity analysis. Also show that the result of Problem 3 tends to the above relation in the limit. See Chapter 9 for a sharpened version of this result, valid for convex programs.

5. Assuming that f, g, and S are convex and that S is compact, prove that $w(y)$ in (8.4-47) has a convex domain of definition and is convex over this domain.
6. Prove that, for the primal problem (8.4-5),

$$\min f = \min_{x \in S} \sup_{u \geq 0} L(x, u)$$

7. Sketch the functions $w(y)$ and sets R for Examples 1 and 2, Section 8.4. Discuss the relations between these sets and the saddle point and duality features of these examples.
8. Consider the quadratic program

$$\text{minimize } \tfrac{1}{2}x'Dx + d'x$$

subject to $$Ax \geq b$$

with D symmetric and positive definite. Show that the dual of this problem is

$$\text{maximize } -\tfrac{1}{2}u'AD^{-1}A'u + u'(AD^{-1}d + b) - \tfrac{1}{2}d'D^{-1}d$$

$$\text{subject to} \qquad u \geq 0$$

This is the dual problem considered in Lemke's method of quadratic programming [19].

9. Prove: If $s = \alpha s_1 + \beta s_2$, then, for any function $\phi(u)$ satisfying the hypotheses of Theorem 8.5-1,

$$D\phi(u; s) \geq \alpha D\phi(u; s_1) + \beta D\phi(u; s_2)$$

Provide an example when the strict inequality holds.

10. Consider a program in which all the hypotheses of Theorem 8.7-1 hold except the existence of $x \in S$ such that $g(x) < 0$. Give an example of such a program for which Theorem 8.7-1 is false.

11. Consider the primal problem

$$\text{minimize } 2x_1^2 - 3x_1 + 4x_2^2 - 3x_2 + 3x_3^2 + 2x_3$$

$$\text{subject to} \qquad x_1 + x_2 + x_3 = 1$$
$$x_1, x_2, x_3 \geq 0$$

Construct analytic expressions for the dual function $h(u)$, the minimizing functions $x_i(u)$, and the first derivative of $h(u)$. Graph these functions, and verify that the assertions of Section 8.7 hold. Is the second derivative of $h(u)$ continuous everywhere?

12. State conditions on the functions f_j, t_i, and on the sets S_i guaranteeing that the decomposition process of Section 8.8.1 be able to solve the primal problem.

13. (Falk [21].) Consider a primal problem of the form

$$\text{minimize } \sum_{i=1}^{n} f_i(x_i)$$

$$\text{subject to} \qquad Ax = b$$
$$x \geq 0$$

where A is an $m \times n$ matrix, $n \geq m$, and x_i is a scalar variable. Assume that each f_i is strongly convex; i.e.,

$$\frac{d^2 f_i}{dx_i^2} > 0, \qquad \text{all } x_i$$

(a) State the dual problem, using multipliers u.

(b) Show that the values of x_i which minimize $L(x, u)$ over $x \geq 0$, $x_i(u)$, are unique for given u. Give explicit formulas for $x_i(u)$.

(c) Show that the domain of the dual, D, is the set $\{u \mid A'u < L\}$, where $L = (L_1, \ldots, L_n)$ and $L_i = \lim_{x_i \to \infty} df_i/dx_i$.

(d) Show that D is divided into polyhedral subregions in the interior of which the dual function, $h(u)$, is continuously differentiable. State inequalities defining these subregions.

(e) Show that $h(u)$ is continuous on the boundaries of these subregions but that second derivatives are generally discontinuous.

(f) Provide a numerical example and illustrate the conclusions of (b)–(e).

References

1. H. W. Kuhn and A. W. Tucker, "Nonlinear Programming," in *Proceedings of the Second Berkeley Symposium on Mathematical Statistics and Probability*, University of California Press, Berkeley, 1950, pp. 481–492.
2. J. E. Falk, "Lagrange Multipliers and Nonlinear Programming," *J. Math. Anal. Appl.*, **19**, No. 1, 1967.
3. H. Everett, "Generalized Lagrange Multiplier Method for Solving Problems of Optimum Allocation of Resources," *Operations Res.*, **11**, 1963, pp. 399–417.
4. S. Karlin, *Mathematical Methods and Theory in Games, Programming, and Economics*, Vol 1, Addison-Wesley Publishing Company, Inc., Reading, Mass., 1959.
5. G. Hadley, *Nonlinear and Dynamic Programming*, Addison-Wesley Publishing Company, Inc., Reading, Mass., 1964.
6. O. L. Mangasarian and J. Ponstein, "Minimax and Duality in Nonlinear Programming," *J. Math. Anal. Appl.*, **11**, 1965, pp. 504–518.
7. J. Stoer, "Duality in Nonlinear Programming and the Minimax Theorem," *Numerische Mathematik*, **5**, 1963, pp. 371–379.
8. R. T. Rockafellar, "Duality and Stability in Extremum Problems Involving Convex Functions," *Pacific J. Math.*, **21**, 1967, pp. 167–187.
9. P. Wolfe, "A Duality Theorem for Nonlinear Programming," *Quart. Appl. Math.*, **19**, 1961, pp. 239–244.
10. R. T. Rockafellar, "Nonlinear Programming," *American Mathematical Society Summer Seminar on the Mathematics of the Decision Sciences, Stanford University, July–Aug. 1967.*
11. D. G. Luenberger, "Convex Programming and Duality in Normed Space," *Proceedings of the IEEE Systems Science and Cybernetics Conference, Boston, Oct. 11–13, 1967.*
12. J. M. Danskin, "The Theory of Max-Min with Applications," *J. SIAM Appl. Math.*, **14**, No. 4, 1966, pp. 641–665.
13. W. Fenchel, "Convex Cones, Sets, and Functions," mimeographed notes, Princeton University, 1963.
14. R. Fletcher and M. J. D. Powell, "A Rapidly Convergent Descent Method for Minimization," *Computer J.*, July 1963.
15. L. S. Lasdon and A. D. Waren, "Mathematical Programming for Optimal Design," *Electro-Technology*, Nov. 1967, pp. 53–71.

16. J. B. Rosen, "The Gradient Projection Method for Nonlinear Programming," Parts I and II, *J. Soc. Ind. Appl. Math.*, **8**, 1960, pp. 181–217; **9**, 1961, pp. 514–532.
17. R. Fletcher and C. M. Reeves, "Function Minimization by Conjugate Gradients," *Computer J.*, July 1964.
18. D. Goldfarb, "Extension of Davidon's Variable Metric Method to Maximization Under Linear Equality and Inequality Constraints", *J. SIAM Appl. Math.*, **17**, No. 4, 1969, pp. 739–764.
19. C. E. Lemke, "A Method of Solution for Quadratic Programming," *Management Sci.*, **8**, No. 4, 1962.
20. L. Munini, "Optimization of Coupled Systems by Decomposition," M.S. Thesis, Systems Research Center, Case Institute of Technology, 1966.
21. J. E. Falk, "An Algorithm for Separable Convex Programming under Linear Equality Constraints," Research Analysis Corp., *Tech. Paper RAC–TP–148, AD 623–092*, 1965.
22. L. S. Lasdon, "A Multi-Level Technique for Optimization," Ph.D. Thesis, Case Institute of Technology, 1964; *Systems Research Center Rept. SRC 50–C–64–19.*
23. L. S. Lasdon and J. D. Schoeffler, "A Multi-Level Technique for Optimization," *Proceedings of the 1965 Joint Automatic Control Conference, Troy, New York.*
24. L. S. Lasdon and J. D. Schoeffler, "Decentralized Plant Control," *ISA Trans.*, **5**, April 1966, pp. 175–183.
25. C. B. Brosilow and L. S. Lasdon, "A Two level Optimization Technique for Recycle Processes," *Proceedings of the A.I.Ch.E.–Ind. Chem. Eng. Joint Meeting, London, June 13–17, 1965*, Symposium on the Application of Mathematical Models in Chemical Engineering Research, Design, and Production.
26. L. S. Lasdon, "Duality and Decomposition in Mathematical Programming," *IEEE Trans. Systems Sci. Cybernetics*, **4**, No. 2, 1968, pp. 86–100.
27. C. B. Brosilow and E. Nunez, "Multi-Level Optimization Applied to a Catalytic Cracking Plant," *Can. J. Chem. Eng.*, **46**, June 1968, pp. 205–212.
28. A. V. Fiacco and G. P. McCormick, *Sequential Unconstrained Minimization Techniques for Nonlinear Programming*, John Wiley & Sons, Inc., New York, 1968.
29. R. Fox and L. Schmit, "Advances in the Integrated Approach to Structural Synthesis," *J. Spacecraft Rockets*, **3**, No. 6, 1966, p. 858.
30. A. M. Geoffrion, "Primal Resource—Directive Approaches for Optimizing Nonlinear Decomposable Systems," *Working Paper 141*, The RAND Corporation, 1968.
31. S. Kaplan, "Solution of the Lorie-Savage and Similar Integer Programming Problems by the Generalized Lagrange Multiplier Method," *Operations Res.*, **14**, 1960, pp. 1130–1136.
32. G. L. Nemhauser and Z. Ullmann, "A Note on the Generalized Lagrange Multiplier Solution to an Integer Programming Problem," *Operations Res.*, **16**, No. 2, 1968, pp. 450–453.

9

Decomposition by Right-Hand-Side Allocation[1]

9.1 Introduction

In Chapter 8 we used a dual program to decompose primal problems of the form

$$\text{minimize } f(x) \tag{1}$$

subject to

$$g(x) \leq 0, \qquad g = (g_1, \ldots, g_m) \tag{2}$$

$$x \in S \subseteq E^n \tag{3}$$

The dual was shown, in Section 8.4, to be intimately related to the primal function

$$w(y) = \min\{f(x) \mid g(x) \leq y, \, x \in S\} \tag{4}$$

In this chapter the primal is solved by working directly with $w(y)$. As in the dual approach, a decomposition results when f and all components of g are additively separable, and S is the direct product of smaller sets. A primal feasible point with monotone-decreasing objective value is obtained at each cycle. This contrasts with solution via the dual, where feasibility occurs only at termination. While the dual method manipulated the prices of scarce resources, allowing the market to determine the allocation, this primal approach allocates the resources directly.

9.2 Problem Formulation

It is convenient to rewrite the constraints (2) with a right-hand-side vector b (which may be zero),

$$g(x) \leq b \tag{5}$$

[1] This chapter is based primarily on the material in [1] and [2].

As in previous chapters, we regard b as a vector of resource availabilities, g as the amount of resources consumed at activity levels x, and f as the cost of operating the activities. The restrictions $x \in S$ are imposed by the technologies of the activities. Further, we again assume that the problem (1)–(3) is separable:

$$\text{minimize} \sum_{i=1}^{p} f_i(x_i) \tag{6}$$

subject to

$$\sum_{i=1}^{p} g_i(x_i) \leq b \tag{7}$$

$$x_i \in S_i, \qquad i = 1, \ldots, p \tag{8}$$

In the above, x_i has n_i components, while each g_i has m components. The functions f_i, g_i and the set S_i are thought of as being associated with the ith subsystem of the overall system.

This primal problem may be viewed as one of optimally allocating the resource vector b to the subsystems. We attempt to solve it by choosing a feasible allocation, testing it for optimality, and improving it if it is not optimal. If the allocation to subsystem i is y_i, then, for feasibility,

$$\sum_{i=1}^{p} y_i \leq b \tag{9}$$

Given the allocation y_i, subsystem i must utilize it as well as possible, so it faces the following subproblem.

Problem $P_i(y_i)$

$$\left.\begin{array}{ll} & \text{minimize } f_i(x_i) \\ \text{subject to} & g_i(x_i) \leq y_i \\ & x_i \in S_i \end{array}\right\} \begin{array}{c}\text{subproblem} \\ i\end{array} \tag{10}$$

Of course, y_i must be chosen such that this subproblem is feasible, so we have the constraint

$$y_i \in V_i = \{y_i \mid \text{there exists } x_i \in S_i \text{ such that } g_i(x_i) \leq y_i\} \tag{11}$$

The minimal subproblem objective value is

$$w_i(y_i) = \min\{f_i(x_i) \mid g_i(x_i) \leq y_i,\ x_i \in S_i\} \tag{12}$$

This is the same kind of function as in (4), a primal function for the ith subproblem. Clearly, to solve the primal, we must choose the y_i to minimize the sum of these functions, subject to (11) and (9). In other words, the primal (6)–(8) is equivalent to the following master program:

$$\left.\begin{aligned} &\text{minimize } w(y) = \sum_{i=1}^{p} w_i(y_i) && (13)\\ &\text{subject to } \sum_{i=1}^{p} y_i \le b && (14)\\ &y_i \in V_i, \quad i = 1, \ldots, p && (15) \end{aligned}\right\} \text{master program}$$

That is, if $y^0 = (y_1^0, \ldots, y_p^0)$ solves the master and x_i^0 solves $P_i(y_i^0)$, then $x^0 = (x_1^0, \ldots, x_p^0)$ solves the primal. Conversely, if x^0 solves the primal, then $(g_1(x_1^0), \ldots, g_p(x_p^0))$ solves the master. Further, the master program is infeasible if and only if the primal is infeasible, and has no optimal solution if and only if the primal has no optimal solution.

Attention is now focused on the above program. It has simple linear constraints (14), nonlinear constraints (15), and a (generally) nonlinear objective. Of course, except perhaps in linear or quadratic problems, explicit expressions for the objective are very difficult, if not impossible, to obtain, unless y_i has only a few components (see [2]). Thus evaluation of w at some point y requires solution of the subproblems $P_i(y_i)$ in (10). These, however, are smaller than the primal and may be solved independently, so some benefits may result, especially when the number of coupling constraints, m, is much smaller than the total number of constraints determining all sets S_i. Further, as we shall show, solution of $P_i(y_i)$ also provides information regarding the directional derivatives of w_i at y_i. This makes the master program easier to solve.

The equivalence between the master and primal problems depends in no way upon convexity of either program. Thus, if means were available for evaluating w_i (and perhaps its derivatives as well), solution could be attempted by any of a number of nonlinear programming methods. However, to make further theoretical progress, we impose the following conditions, which hold throughout the remainder of this chapter.

Convexity and Feasibility Assumptions. For $i = 1, 2, \ldots, p$:

1. S_i is nonempty, compact, and convex.
2. $f_i(x_i)$ is convex and differentiable on S_i.
3. Each component of g_i is convex and differentiable on S_i.
4. The primal problem (6)–(8) has a feasible solution.

The convexity assumptions on f_i, g_i, and S_i ensure that V_i is convex and that w_i is convex over V_i (see Problem 5, Chapter 8), so the master program is a convex program. Feasibility of the primal implies that the master program is feasible. Compactness of S_i, along with continuity of f_i and g_i, imply that subproblem i has an optimal solution whenever it is feasible. Differentiability of all functions will enable us to construct expressions for the directional derivative of w_i using linear programming.

To permit direct application of the theory in Appendixes 1 and 2, we extend each function w_i (and hence w) to all of E^m by defining

$$w_i(y_i) = +\infty, \qquad y_i \notin V_i \tag{16}$$

Thus each w_i becomes a proper convex function with effective domain

$$\operatorname{dom} w_i = V_i$$

We also redefine w by

$$w(y) = +\infty \qquad \text{if } \sum_{i=1}^{p} y_i \nleq b \tag{17}$$

so w is also a proper convex function, whose effective domain is the feasible set of the master program:

$$\operatorname{dom} w = F$$

where

$$F = \left\{ y \,\middle|\, \sum_{i=1}^{p} y_i \leq b,\ y_i \in V_i,\ i = 1, \ldots, p \right\}$$

Despite all these idealizing assumptions, w_i may not be differentiable everywhere on V_i, and the corners may occur at critical points, e.g., the minimum. Even strict convexity of f_i does not imply that w_i is everywhere differentiable, and this is the main source of difficulty in solving the master program. As shown in Section 8.5, assumptions 1–3 plus strict convexity of the objective do imply that the dual function $h(u)$ is differentiable for $u \geq 0$. Thus, in this sense, the primal functions w_i are not as well behaved as their dual counterparts h_i. On the other hand, w_i is strictly convex if f_i is and if assumptions (1)–(3) hold, whereas this is not always true for h_i.

Example 1. To illustrate, consider the problem (Falk [3])

$$w(y) = \min_{x_1, x_2 \geq 0} \{\tfrac{1}{2}x_1^2 + x_1 + \tfrac{1}{2}x_2^2 \mid -x_1 + x_2 = y\}$$

For $y \geq 0$, $x_1 = 0$, so $x_2 = y$, while for $y \leq 0$, $x_2 = 0$, implying $x_1 = -y$. Thus

$$w(y) = \begin{cases} \dfrac{y^2}{2}, & y \geq 0 \\ \dfrac{y^2}{2 - y}, & y \leq 0 \end{cases}$$

This function is strictly convex but has a discontinuity in slope at the minimizing point $y = 0$, as shown in Figure 9-1.

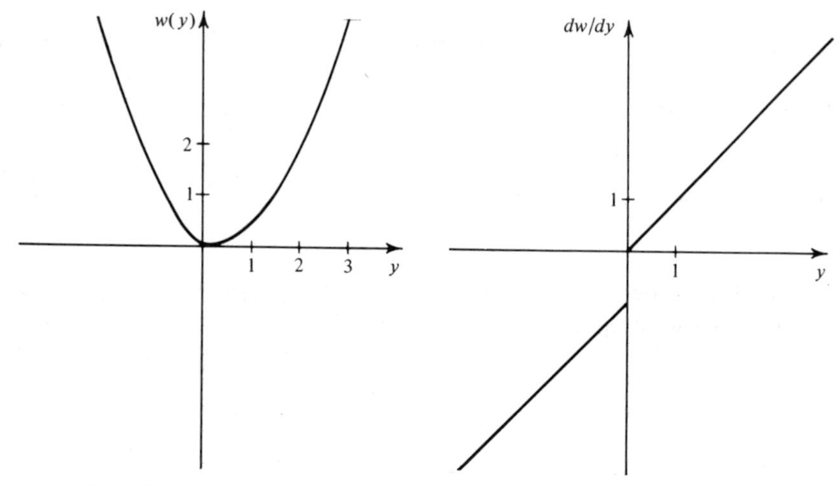

FIGURE 9-1 Primal function and its derivative.

Despite these difficulties, since each w_i is convex, it has finite-valued directional derivatives at all points in int V_i (Theorem 16, Appendix 2). This provides motivation for the algorithm of the next section, which uses these derivatives to successively improve a feasible solution.

9.3 Feasible-Directions Algorithm[2] for the Master Program

The algorithm considered here initiates solution of the master program with a feasible vector y. A direction-finding program is constructed which, by minimizing the directional derivative of the master objective subject to certain feasibility restrictions, either finds a direction, s, in which the allocation y can be improved or shows that y is optimal. If y is not optimal, a step is taken in the direction s, a new allocation is determined, and the process is repeated.

[2] See Section 1.3.4 for a general discussion of feasible-directions algorithms.

Assume, then, that a feasible point $y = (y_1, \ldots, y_p) \in F$ is available. In order that we may take a small step in any direction s_i from y_i and keep $y_i + \alpha s_i$ in V_i, we assume the following.

Assumption 1. Each vector y_i is in the interior of V_i.

Let

$$s = (s_1, \ldots, s_p)$$

be a proposed search direction. To be useful, s should have the following properties [6].

DEFINITION 1. A direction s is said to be *feasible* at a point $y \in F$ if there exists $\epsilon > 0$ such that

$$y + \alpha s \in F \text{ for all } 0 < \alpha \leq \epsilon \tag{1}$$

DEFINITION 2. A feasible direction s is said to be *usable* for w at $y \in F$ if

$$w(y + \alpha s) < w(y) - \alpha\epsilon \qquad \text{for all } 0 < \alpha \leq \epsilon \tag{2}$$

Note that the same ϵ is involved in both definitions. Roughly speaking, a usable direction is one in which a small move can be made which improves the objective without violating any constraint. Relation (2) may be understood geometrically by fixing y and s, defining

$$r(\alpha) = w(y + \alpha s) \tag{3}$$

and rewriting (2) as

$$\frac{r(\alpha) - r(0)}{\alpha} < -\epsilon, \qquad \text{for all } 0 < \alpha \leq \epsilon \tag{4}$$

As seen from Figure 9-2, this requires the slope of the chord through $(\alpha, r(\alpha))$ and $(0, r(0))$ to be less than $-\epsilon$ for all α in the indicated range.

Direction-Finding Problem. For s to be feasible, the following must hold for some range of positive values for α:

$$y_i + \alpha s_i \in V_i, \qquad i = 1, \ldots, p \tag{5}$$

and

$$\sum_i (y_i + \alpha s_i) \leq b \tag{6}$$

or

$$\alpha \sum_i s_i \leq b - \sum_i y_i \tag{7}$$

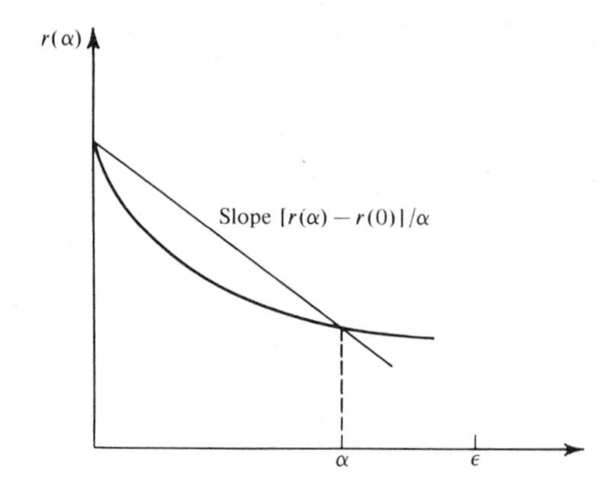

FIGURE 9-2

Since $y_i \in \text{int } V_i$, there is a range of positive α such that (5) is satisfied for any directions s_i. To satisfy (7), if the kth component of $b - \sum_i y_i$ is positive, some range of positive α maintains the kth inequality satisfied for any set of vectors $\{s_i\}$, so such restrictions can be ignored. If this kth component is zero, moves in some directions violate this constraint immediately. Let

$$B = \left\{ j \,\middle|\, b_j - \sum_i y_{ij} = 0 \right\} \tag{8}$$

be the index set of the binding constraints at y. Then for feasibility

$$\sum_i s_{ij} \leq 0, \qquad j \in B \tag{9}$$

To find a "locally best" usable direction, it is intuitively plausible to search for a feasible direction which minimizes the directional derivative of the master objective. Such a direction solves the direction-finding problem

Problem (DF)

$$\text{minimize } Dw(y; s) = \sum_{i=1}^{p} Dw_i(y_i; s_i) \tag{10}$$

subject to the feasibility conditions

$$\sum_{i=1}^{p} s_{ij} \leq 0, \qquad j \in B \tag{11}$$

and the normalization constraint

$$\|s\| \leq 1 \tag{12}$$

where $\|s\|$ is a norm for s.

The constraints (11) are homogeneous and, as shown in Theorem 9, Appendix 2, $Dw_i(y_i; s_i)$ is positively homogeneous in s_i, i.e.,

$$Dw_i(y_i; \lambda s_i) = \lambda Dw_i(y_i; s_i), \qquad \lambda > 0 \tag{13}$$

Thus if the normalization constraint was omitted and there was an s which made the objective (10) negative, then (10) could be forced to approach $-\infty$ by multiplying s by increasingly large positive constants. The inclusion of (12) insures that (DF) always has a finite optimal solution.

It might seem that obtaining an explicit expression for $Dw_i(y_i; s_i)$ would be as difficult as finding such an expression for w_i itself. However, the theory of appendices (1) and (2) will enable us to do this, and to write (DF) as a linear program. First, however, we state a test for optimality of y and show that (DF) always produces a usable feasible direction if one exists.

THEOREM 1. *Let y be feasible for the master program. If $s = 0$ is optimal for the direction finding problem (DF) then y solves the master program.*

PROOF. Since

$$Dw(y; 0) = 0$$

then it must be that

$$Dw(y; s) \geq 0 \tag{14}$$

for all s satisfying the feasibility conditions (11). For all other vectors s, by (9.2-17) $w(y + \alpha s) = +\infty$ for all $\alpha > 0$, so $Dw(y; s) = +\infty$. Thus (14) holds for all $s \in E^m$. Then by Theorem 11 of Appendix 2, y solves the master program.

If the optimality test of the previous theorem is not passed, then a direction s is produced such that $Dw(y; s)$ is negative. This direction is now shown to be usable in the sense of Definition 2.

THEOREM 2. *Let s solve (DF) for some $y \in F$. If $Dw(y; s)$ is negative, s is a usable direction for w at y.*

PROOF. The vector s is feasible by construction. Hence there is an $\bar{\alpha}$ such that

$$y + \alpha s \in F, \qquad \text{for all } 0 < \alpha \leq \bar{\alpha} \tag{15}$$

To prove s usable, we use the fact that

$$Dw(y; s) = \lim_{\lambda \to 0^+} (w(y + \lambda s) - w(y))/\lambda < 0 \tag{16}$$

By Theorem 8 of Appendix 2, the difference quotient

$$d(\lambda) = (w(y + \lambda s) - w(y))/\lambda$$

is nondecreasing. This, coupled with (16), implies the existence of an $\epsilon > 0$ and $\bar{\lambda} > 0$ such that

$$d(\lambda) < -\epsilon, \qquad \text{for all } 0 < \lambda \leq \bar{\lambda} \tag{17}$$

Define

$$\delta = \min(\epsilon, \bar{\lambda})$$

then we can show

$$d(\lambda) < -\delta \qquad \text{for all } 0 < \lambda \leq \delta \tag{18}$$

CASE 1. $\epsilon > \bar{\lambda}$. Here, by (17),

$$d(\lambda) < -\epsilon < -\bar{\lambda} = -\delta, \qquad \text{for all } 0 < \lambda \leq \delta$$

CASE 2. $\epsilon \leq \bar{\lambda}$. Then, by (17),

$$d(\lambda) < -\epsilon = -\delta$$

Since the above holds for all $0 < \lambda \leq \bar{\lambda}$, it certainly holds for all $0 < \lambda \leq \delta$.

Now, let

$$\lambda^* = \min(\bar{\alpha}, \delta)$$

Then, by (15),

$$y + \alpha s \in F, \qquad \text{for all } 0 < \alpha \leq \lambda^* \tag{19}$$

and, by (18),

$$d(\lambda) < -\lambda^*, \qquad \text{for all } 0 < \lambda \leq \lambda^*$$

The above may be written

$$w(y + \lambda s) < w(y) - \lambda\lambda^*, \qquad \text{for all } 0 < \lambda \leq \lambda^* \tag{20}$$

so the theorem is proved, since (19) and (20) are the defining inequalities (1)–(2) for s to be usable at y.

Turning now to the question of obtaining an explicit expression for $Dw(y; s)$, we first characterize the subgradients of the functions w_i. The geometrical ideas of Chapter 8 indicate the result immediately. A subgradient of w_i at y_i is the vector of slopes of a supporting hyperplane to the graph of w_i at y_i. Section 8.4 showed that, if u_i^0 solves the dual of $P_i(y_i)$, $-u_i^0$ is the slope vector of such a hyperplane. Then, if x_i^0 solves $P_i(y_i)$, (x_i^0, u_i^0) is a saddle point for the Lagrangian function of $P_i(y_i)$. Thus the subgradients of w_i at y_i are the negatives of the saddle-point (Kuhn–Tucker) multipliers of $P_i(y_i)$. Further, any optimal solution x_i^0 is sufficient to generate all subgradients, as we now show. In the proof, the subscript i is dropped, since it plays no essential role.

THEOREM 3. *Let*

$$L(x, u) = f(x) + u(g(x) - y)$$

be the Lagrangian function for $P(y)$, and let x^0 solve $P(y)$. Then $-u^0$ is a subgradient of w at y [written $-u^0 \in \partial w(y)$] if and only if (x^0, u^0) is a saddle point for L; i.e.,

$$L(x^0, u) \leq L(x^0, u^0) \leq L(x, u^0) \qquad \text{for all } x \in S,\ u \geq 0$$

PROOF. $\Leftarrow$: By the saddle inequality

$$f(x^0) + u^0(g(x^0) - y) \leq f(x) + u^0(g(x) - y) \qquad \text{for all } x \in S$$

or, since

$$f(x^0) = w(y), \qquad u^0(g(x^0) - y) = 0$$

$$w(y) \leq f(x) + u^0(g(x) - y) \tag{21}$$

Choose $z \in E^m$. If there exists $x \in S$ such that $g(x) \leq z$, then, for such x, we can replace $g(x)$ by z in (21), yielding

$$w(y) \leq f(x) + u^0(z - y)$$

Taking the infimum of the right-hand side of the above over $x \in S$ such that $g(x) \leq z$ yields

$$w(y) \leq w(z) + u^0(z - y) \tag{22}$$

If there is no $x \in S$ such that $g(x) \leq z$, then $w(z) = +\infty$, so the above holds for all $z \in E^m$. This is the defining relation for $-u^0$ to be an element of $\partial w(y)$.

$\Rightarrow$: Assume that (22) holds for all $z \in E^m$. Choose

$$z = g(x^0) \tag{23}$$

which implies

$$w(z) = f(x^0) = w(y) \tag{24}$$

Using (23) and (24) in (22) yields

$$u^0(g(x^0) - y) \geq 0$$

But $u^0 \geq 0$ and $g(x^0) - y \leq 0$, so

$$u^0(g(x^0) - y) = 0 \tag{25}$$

The above, combined with the relation

$$g(x^0) \leq y$$

are two of the three saddle-point criteria of Theorem 8.3-1. To show the third, we use $f(x^0) = w(y)$ and (25) in (22) to obtain

$$f(x^0) + u^0(g(x^0) - y) \leq w(z) + u^0(z - y) \qquad \text{for all } z \in E^m \tag{26}$$

Choose x arbitrarily from S and let

$$z = g(x) \tag{27}$$

Since x is feasible for $P(g(x))$

$$w(z) \leq f(x) \tag{28}$$

Using (27) and (28) in (26),

$$f(x^0) + u^0(g(x^0) - y) \leq f(x) + u^0(g(x) - y) \tag{29}$$

Since x is an arbitrary element of S, (29) is the third of the saddle criteria, and the theorem is proved.

By Theorem 17 of Appendix 2, a convex function is differentiable at a point if and only if it has a single subgradient there. Since, with x^0 optimal for $P(y)$, (x^0, u^0) is a saddle point for L if and only if u^0 solves the dual of $P(y)$, Theorem 3 implies the following theorem.

THEOREM 4. *The function w is differentiable at y if and only if the dual of $P(y)$ has a unique optimal solution.*

In Example 1 of Section 9.2, w was not differentiable at $y = 0$. The dual of $P(0)$ is to maximize

$$h(u) = \min_{x_1, x_2 \geq 0} \left(\frac{x_1^2}{2} + x_1 + \frac{x_2^2}{2} + u(-x_1 + x_2) \right)$$

over all u. It is easily verified that

$$h(u) = \begin{cases} \dfrac{-u^2}{2}, & u \leq 0 \\ 0, & 0 \leq u \leq 1 \\ \dfrac{-(u-1)^2}{2}, & u \geq 1 \end{cases}$$

As shown in Figure 9-3 all values of u in the interval $[0, 1]$ solve this dual. From Figure 9-1 we see that w is not differentiable at $y = 0$, and that the

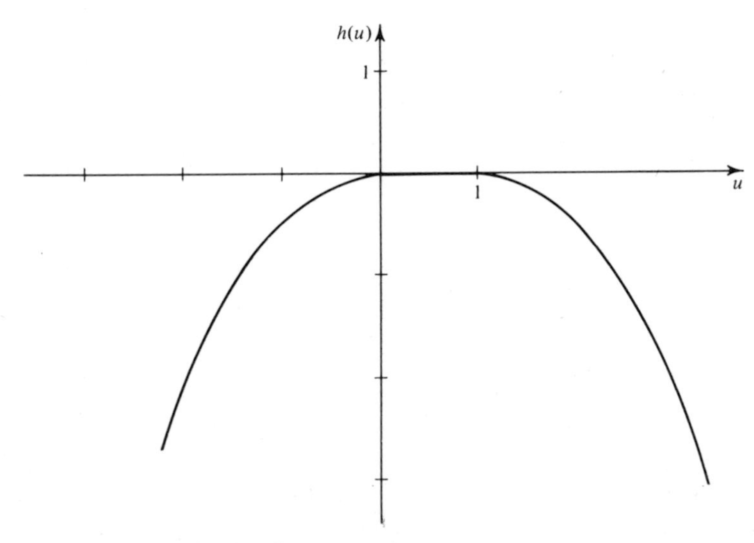

FIGURE 9-3 Dual objective for Example 1.

subgradients at 0 are just the negatives of the optimal multipliers, the set $[-1, 0]$.

As shown by Geoffrion [2], Theorem 3 may be used to obtain an explicit expression for $Dw_i(y_i; s_i)$ at points $y_i \in \text{int dom } w_i$, i.e., $y_i \in \text{int } V_i$. Then, as shown in Appendix 2, $\partial w_i(y_i)$ is nonempty, compact, and convex and, by Theorem 16 of Appendix 2,

$$\begin{aligned} Dw_i(y_i; s_i) &= \max_{x^* \in \partial w_i(y_i)} (x^* s_i) \\ &= \max_{u_i \in U_i} (-u_i s_i) \end{aligned} \tag{30}$$

where U_i is the set of all optimal multiplier vectors for $P_i(y_i)$. Note that, by definition of V_i, the condition $y_i \in \text{int } V_i$ is equivalent to the existence of an $x_i \in S_i$ such that $g_i(x_i) < y_i$. This is the Karlin constraint qualification of Theorem 8.3-6, guaranteeing the existence of an optimal vector of multipliers for $P_i(y_i)$.

Let each set S_i be determined by a vector of constraint functions c_i,

$$S_i = \{x_i \mid c_i(x_i) \le 0\} \tag{31}$$

We assume that each component of c_i is differentiable and convex. Then, if x_i^0 is any optimal solution to $P_i(y_i)$, $u_i \in U_i$ if and only if x_i^0, u_i, and some λ_i satisfy the Kuhn–Tucker conditions for $P_i(y_i)$ (see Section 1.3.2.) Formed from the Lagrangian

$$\tilde{L}_i = f_i + u_i(g_i - y_i) + \lambda_i c_i$$

these are

$$J'_{g_i} u_i + J'_{c_i} \lambda_i = -\nabla f_i \tag{32}$$

$$u_i(g_i - y_i) = 0 \tag{33}$$

$$\lambda_i c_i = 0 \tag{34}$$

$$u_i \ge 0, \qquad \lambda_i \ge 0 \tag{35}$$

In the above, J_{g_i} is the Jacobian matrix of g_i, the matrix whose rows are the gradients of the functions comprising g_i, and similarly for J_{c_i}. All functions of x_i are evaluated at x_i^0. By (30), $Dw_i(y_i; s_i)$ is the optimal objective value of the linear program

$$\text{maximize } \{-u_i s_i \mid (u_i, \lambda_i) \text{ satisfies (32)–(35)}\} \tag{36}$$

Some variables of this program may be dropped by using the complementary slackness constraints (33)–(34). Since $g_i - y_i$ and c_i are both nonpositive vectors, these are equivalent to

$$u_{ij} = 0 \qquad \text{if } g_{ij} - y_{ij} < 0 \tag{37}$$

$$\lambda_{ij} = 0 \qquad \text{if } c_{ij} < 0 \tag{38}$$

Thus (36) may be written

$$\text{maximize } -\sum_j u_{ij} s_{ij} \tag{39}$$

$$\text{subject to} \qquad \sum_j \nabla g_{ij} u_{ij} + \sum_j \nabla c_{ij} \lambda_{ij} = -\nabla f_i \tag{40}$$

$$u_{ij} \geq 0, \qquad \lambda_{ij} \geq 0 \tag{41}$$

where the summations are over j such that $g_{ij} - y_{ij} = 0$ and $c_{ij} = 0$.

Since we wish to find a "best" s_i, the above program is not in ideal form; the objective involves the product of decision variables. Working with the dual avoids this difficulty. The dual (with s_i temporarily fixed) is

$$\text{minimize } -\nabla f_i' z_i \tag{42}$$

$$\text{subject to} \qquad \nabla g_{ij}' z_i \geq -s_{ij} \tag{43}$$

$$\nabla c_{ij}' z_i \geq 0 \tag{44}$$

Since $Dw_i(y_i; s_i)$ is finite, the primal (39)–(41) has a finite optimal solution. By the Corollary to duality theorem 1.2.3-1, the above dual also has a finite optimal solution, and its optimal objective value is equal to $Dw_i(y_i; s_i)$. Thus, for a fixed vector $s = (s_1, \ldots, s_p)$, the directional derivative of $w(y)$ is

$$Dw(y; s) = \sum_{i=1}^{p} \min\{-\nabla f_i' z_i \mid (z_i, s_i) \text{ satisfy (43)–(44)}\} \tag{45}$$

The direction-finding problem asks us to find a vector s which minimizes the above subject to feasibility and normalization constraints. Minimizing the sum of the minima in (45) over a subset of vectors s_i is equivalent to doing the minimization simultaneously over s_i and z_i. That is, we solve the program

$$\text{minimize } -\sum_{i=1}^{p} \nabla f_i' z_i \tag{46}$$

$$\text{subject to} \qquad \sum_{i=1}^{p} s_{ij} \leq 0, \qquad j \text{ such that } \sum_{i=1}^{p} y_{ij} = b_j \tag{47}$$

$$\|s\| \leq 1 \tag{48}$$

and, for $i = 1, \ldots, p$,

$$\nabla g'_{ij} z_i \geq -s_{ij}, \qquad j \text{ such that } g_{ij}(x_i^0) - y_{ij} = 0 \tag{49}$$

$$\nabla c'_{ij} z_i \geq 0, \qquad j \text{ such that } c_{ij}(x_i^0) = 0 \tag{50}$$

The above becomes linear if we use the L_∞ norm

$$\|s\| = \|s\|_\infty = \max_j |(s)_j| \tag{51}$$

Then the normalization constraint (48) becomes

$$-1 \leq s_{ij} \leq 1 \tag{52}$$

We note two facts about this program. First, only the s_{ij} appearing in (49) are essential. This is because they determine the allocation of resources y_i to currently binding subproblem constraints. Other s_{ij} are associated with inactive constraints, and changes in the right-hand side of these are (locally) arbitrary. Thus s_{ij} appearing in (47) but not in (49) are set to their lower bounds [to make (47) as loose as possible] while s_{ij} not in (47) or (49) are arbitrary. Second, the problem has angular structure, with (47) the coupling constraints and (49)–(50) the angular blocks. Thus it may be solved efficiently by applying the methods of Chapters 3, 5, or 6. The constraints (52) may be converted to upper bounds and nonnegativities by the substitution

$$\tilde{s}_{ij} = s_{ij} + 1 \tag{53}$$

The bounds may then be handled without increasing basis size by the procedure of Section 6.3.

Further simplifications occur at points where one or more of the functions w_i are differentiable. This occurs at a point y_i if and only if $\partial w_i(y_i)$ contains only a single element, i.e., if and only if the u_i component of feasible solutions to the linear system (40)–(41) is unique. If this component is u_i^0, then the directional derivatives is given by $-u_i^0 s_i$, and the constraints (49)–(50) may be dropped. When all functions w_i are differentiable, the direction-finding problem reduces to

$$\text{minimize } -\sum_{i=1}^{p} u_i^0 s_i \tag{54}$$

subject to

$$\sum_{i=1}^{p} s_{ij} \leq 0, \qquad j \in B \tag{55}$$

$$-1 \leq s_{ij} \leq 1 \tag{56}$$

This has at most m constraints, plus upper bounds. In practice, this situation can be expected to occur often, since each w_i is differentiable almost everywhere in int V_i. In any case, it is not likely that all functions w_i will fail to be differentiable at a given point y. Those which are differentiable may be dealt with as indicated above; the others require inclusion of a set of constraints (49)–(50) and a term $-\nabla f_i' z_i$ in the objective.

Linear Minimization. Given a usable direction s, a new point y is generated by choosing a step size $\alpha > 0$ in the expression

$$y^n = y + \alpha s \tag{57}$$

There are many schemes for choosing α [6]. Many of these attempt to solve the convex one-dimensional problem

$$\text{minimize } \{w(y + \alpha s) \mid y + \alpha s \in F\} \tag{58}$$

Generally, small steps are taken along s from y until either the unconstrained minimum of $w(y + \alpha s)$ is exceeded or the point $y + \alpha s$ leaves F. The procedures of Section 1.1, formulated there for unconstrained problems, can be adapted to account for the constraint $y + \alpha s \in F$. In explicit form, this constraint is

$$\sum_{i=1}^{p} (y_i + \alpha s_i) \leq b \tag{59}$$

$$y_i + \alpha s_i \in V_i \tag{60}$$

The inequalities (59) are equivalent to an easily computed upper bound on α. Violations of the others, (60), are indicated by infeasibility of some subproblem.

Of course, our formulation of the direction-finding problem is based on the condition $y_i \in \text{int } V_i$. This can be guaranteed by taking a step slightly shorter than optimal in (58) if some y_i^n is on the boundary of V_i. If this perturbation is allowed to approach zero as the iterations proceed, boundary points can be approached as limiting values.

Once the solution of (58) has been satisfactorily approximated, the direction-finding problem is re-solved with $y = y^n$, and the process is iterated. Stop criteria can be based on Theorem 1; i.e., stop when the optimal objective value of (DF) becomes greater than $-\epsilon$ for some number of cycles, or on the behavior of the sequence of $w(y)$ values (which is monotone decreasing).

9.4 Alternative Approach to the Direction-Finding Problem

In [1], Silverman has proposed an alternative formulation of the problem (DF), which extends the ideas in [4]. Although somewhat more complex than the formulation of the previous section, it leads to a smaller direction-finding problem. Also of interest is the way it deals with the subgradients and directional derivatives of w_i.

The development begins with a characterization of the set of optimal multipliers for $P_i(y_i)$, U_i. In this and in what follows we focus on one particular subproblem and drop the subscript i. Thus $P_i(y_i) \equiv P(y)$, $U_i \equiv U$, etc.

THEOREM 1. *Let x^0 solve $P(y)$. Then U is equal to the set of u components of optimal solutions to the following linear program in the variables* (u, λ):

$$\text{maximize } u(g - y) + \lambda c \tag{1}$$

subject to

$$J_g'u + J_c'\lambda = -\nabla f \tag{2}$$

$$u \geq 0, \qquad \lambda \geq 0 \tag{3}$$

where J_g and J_c are as in (9.3-32), *and all functions of x are evaluated at x^0.*

PROOF. By definition, $u \in U$ if and only if x^0, u, and some λ satisfy the Kuhn–Tucker conditions (9.3-32)–(9.3-35). The constraints of the above program are these conditions with the complementary slackness equalities deleted. However, since $g - y$ and c are nonpositive, these are equivalent to the maximization of the objective (1).[3]

As shown in Section 9.3, all optimal solutions of (1)–(3) will have $u_j = 0$ if $g_j - y_j < 0$ and $\lambda_j = 0$ if $c_j < 0$. Thus we may work in a space of reduced dimension by deleting all zero components of u and λ. If

$$B_i = \{j \mid g_{ij}(x_i^0) - y_{ij} = 0\} \tag{4}$$

and we are dealing with subproblem i, u has one component for each index in B_i.

According to the above, U is the set of all convex combinations of u-components of optimal extreme-point solutions to (1)–(3), a compact convex polyhedron. Let $E(U)$ be the set of these extreme points, with elements u_t, $t = 1, 2, \ldots$. For any vector s, the maximum in

$$Dw(y; s) = \max_{u \in U}(-us)$$

occurs at one or more extreme points. Let

$$C_t = \left\{s \mid \max_{u \in U}(-us) = -u_t s\right\} \tag{5}$$

[3] Whose maximum value is zero.

be the set[4] of vectors s for which the maximum is assumed at u_t. For s to be in C_t, it is necessary and sufficient that

$$-u_t s \geq -u_j s, \qquad \text{all } u_j \in E(U) \tag{6}$$

or

$$(u_t - u_j)s \leq 0, \qquad j \neq t \tag{7}$$

Let W_t be the matrix with jth row $u_t - u_j$,

$$W_t = \begin{bmatrix} u_t - u_1 \\ u_t - u_2 \\ \vdots \\ u_t - u_T \end{bmatrix} \tag{8}$$

Then (7) may be written

$$W_t s \leq 0 \tag{9}$$

and C_t is given by

$$C_t = \{s \mid W_t s \leq 0\} \tag{10}$$

This is the intersection of a finite number of half-spaces containing the origin, a convex polyhedral cone. As such, it has a finite set of spanning vectors. Let v_{rt} be the rth spanning vector for C_t. Then any $s \in C_t$ can be written

$$s = \sum_r \beta_{rt} v_{rt}, \qquad \beta_{rt} \geq 0 \tag{11}$$

Further, since for any $s \in C_t$,

$$Dw(y; s) = -u_t s \tag{12}$$

then, by using (11),

$$Dw(y; s) = \sum_r \beta_{rt}(-u_t v_{rt}), \qquad s \in C_t \tag{13}$$

$$Dw(y; s) = \sum_r \beta_{rt} Dw(y; v_{rt}) \tag{14}$$

[4] Note that, if we are considering subproblem i, u_t and s have the same number of components as u, which is the number of elements in B_i.

Thus the directional derivative of w at y is linear over each cone C_t. Since, for any $s \in E^m$, the maximum in (5) occurs at some extreme point u_t, the union of all cones C_t is E^m. Thus, for any $s \in E^m$, there exist $\beta_{rt} \geq 0$ such that

$$s = \sum_{r,t} \beta_{rt} v_{rt} \tag{15}$$

$$Dw(y; s) = \sum_{r,t} \beta_{rt} Dw(y; v_{rt}) \tag{16}$$

Vectors v_{rt} having these properties are called basic proposals in [1] and [4].

The situation is, of course, greatly simplified when U consists of only a single element u, i.e., when w is differentiable at y. Then

$$Dw(y; s) = -us \tag{17}$$

and the basic proposals are the columns of the matrix $[I \mid -I]$, with I the identity matrix.

In what follows, the subproblem subscript i must be reintroduced. For simplicity, we let r_{ij} denote the jth basic proposal and C_t^i the tth cone [of the form (10)] associated with subproblem i. The vector[5] r_{ij} is, of course, a spanning vector of one of the cones C_t^i. In this notation, relations (15) and (16) become

$$s_i = \sum_j \beta_{ij} r_{ij} \tag{18}$$

$$Dw_i(y_i; s_i) = \sum_j c_{ij} \beta_{ij} \tag{19}$$

where

$$c_{ij} = Dw_i(y_i; r_{ij}) \tag{20}$$

The procedure for obtaining the vectors r_{ij} is now summarized.

1. Begin with a vector $y = (y_1, \ldots, y_p)$ with $y_i \in \text{int } V_i$ and

$$\sum_i y_i \leq b$$

For $i = 1, \ldots, p$, do the following:

[5] Note that the r_{ij} have as many components as there are indices in B_i.

2. Solve $P_i(y_i)$, obtaining an optimal solution x_i^0.
3. Solve the linear program (1)–(3), evaluating all functions of x_i at x_i^0, and obtain all optimal extreme-point solutions $\{u_j^i, \lambda_j^i\}$. Let $E(U_i)$ be the set of all elements u_j^i.
   (a) If, for some i, $E(U_i)$ contains only a single element u_i^0, the directional derivative is given by

$$Dw_i(y_i; s_i) = -u_i^0 s_i$$

   (b) For i such that $E(U_i)$ has more than one element, form the matrices W_t^i in (8) and find spanning vectors[6] for the cones C_t^i in (10). The basic proposals, $\{r_{ij}\}$, are the union of all these sets of spanning vectors.

Unfortunately, the sum over i of the linear expressions (19) cannot be used as the objective of a direction-finding program without additional restrictions. The problem is that, while there exist $\beta_{ij} \geq 0$ such that (19) is valid for any s_i, the right-hand side of (19) is not necessarily equal to $Dw_i(y_i; s_i)$ [with s_i given by (18)] for arbitrary nonnegative β_{ij}. For (18)–(19) to be valid, all r_{ij} which have positive coefficients must be in the same cone. That is, the β_{ij} must satisfy the following condition.

Admissible Interpolation Constraint. A set of nonnegative numbers $\{\beta_{ij}\}$ satisfies the admissible interpolation constraint (AIC) if $\beta_{it} > 0$ and $\beta_{i\tau} > 0$ implies that r_{it} and $r_{i\tau}$ are elements of the same cone, say C_q^i.

If the AIC is imposed for each i, we can use (18) and (19) in the problem (DF), stated in (9.3-10)–(9.3-12), to obtain the following formulation of (DF)

$$\text{minimize} \sum_{i,j} c_{ij}\beta_{ij} \tag{21}$$

subject to

$$\Big(\sum_{i,j} \beta_{ij} r_{ij}\Big)_k \leq 0, \qquad k \in B \tag{22}$$

$$\beta_{ij} \geq 0, \qquad \text{all } i, j$$

the normalization

$$\|s\| \leq 1 \tag{23}$$

and

$$\text{AIC for each } i \tag{24}$$

[6] Algorithms for this purpose are described in [4].

The admissible interpolation constraint (24) is distinctly nonlinear. Fortunately, as we now show, if the AIC is dropped and the program (21)–(23) is solved, any optimal solution satisfies the AIC.

THEOREM 2. *Any optimal solution to the direction-finding problem* (21)–(23) *satisfies the* AIC, (24).

PROOF. Assume that (DF) has an optimal solution, β^0, which does not satisfy the AIC. That is, there is an index i and numbers β_{is}^0 and β_{it}^0 such that r_{is} and r_{it} are not in the same cone. By (20),

$$c_{ij}\beta_{ij}^0 = Dw_i(y_i;\, \beta_{ij}^0 r_{ij}) = \max_{u \in U_i}(-u\beta_{ij}^0 r_{ij}), \qquad \text{all } j \tag{25}$$

Let

$$s_i^0 = \sum_j \beta_{ij}^0 r_{ij} \in C_q^i \tag{26}$$

Then

$$Dw_i(y_i;\, s_i^0) = \max_{u \in U_i}(-us_i^0) = -u_q s_i^0 = -\sum_j \beta_{ij}^0(u_q r_{ij}) \tag{27}$$

By definition of c_{ij},

$$c_{ij}\beta_{ij}^0 = \beta_{ij}^0 \max_{u \in U_i}(-ur_{ij}) \geq -\beta_{ij}^0 u_q r_{ij}, \qquad \text{all } j \tag{28}$$

By hypothesis, both r_{is} and r_{it} are not in C_q^i. Assume $r_{is} \notin C_q^i$, which implies $\beta_{is}^0 r_{is} \notin C_q^i$. Then when $j = s$, the max in (28) is not attained at u_q, so

$$c_{is}\beta_{is}^0 > -\beta_{is}^0(u_q r_{is}) \tag{29}$$

By (28) and (29),

$$\sum_j c_{ij}\beta_{ij}^0 > -\sum_j \beta_{ij}^0(u_q r_{ij}) = Dw_i(y_i;\, s_i^0) \tag{30}$$

Now write s_i^0 as a linear combination of spanning vectors of C_q^i

$$s_i^0 = \sum_{r_{ij} \in C_q^i} \hat{\beta}_{ij} r_{ij}, \qquad \hat{\beta}_{ij} \geq 0 \tag{31}$$

Then

$$Dw_i(y_i;\, s_i^0) = -u_q s_i^0 = \sum_{r_{ij} \in C_q^i} c_{ij}\hat{\beta}_{ij} \tag{32}$$

Combining (30) and (32),

$$\sum_j c_{ij}\beta^0_{ij} > \sum_{r_{ij}\in C^i_q} c_{ij}\hat{\beta}_{ij} \tag{33}$$

Let $\hat{\beta}_i$ be the vector whose components are $\hat{\beta}_{ij}$ for $r_{ij} \in C^i_q$ and zero otherwise. Let

$$\begin{aligned} \hat{\beta} &= (\beta^0_1, \beta^0_2, \ldots, \hat{\beta}_i, \ldots, \beta^0_p), \\ \beta^0 &= (\beta^0_1, \beta^0_2, \ldots, \beta^0_i, \ldots, \beta^0_p) \end{aligned} \tag{34}$$

Since s^0_i is given by both (26) and (31), $\hat{\beta}$ satisfies the feasibility and normalization constraints (22)–(23). By (33), $\hat{\beta}$ has a lower objective value than β^0, which contradicts the optimality of β^0. Thus the theorem is proved.

Note that, although no convexity assumptions appear explicitly in this proof, the proof does depend on them, since use of (25) and (27)–(28) for the directional derivative requires these assumptions.

As in Section 9.3, (21)–(23) becomes a linear program if the L_∞ norm is used. Then (23) becomes

$$-e_m \le \sum_j \beta_{ij} r_{ij} \le e_m, \qquad i = 1, \ldots, p \tag{35}$$

where e_m is an m vector with all components unity. This is equivalent to

$$\sum_j \beta_{ij} r_{ij} + v_i = e_m, \qquad i = 1, \ldots, p \tag{36}$$

$$0 \le v_i \le 2e_m \tag{37}$$

The resulting program, (21)–(22) and (36)–(37), has as many linear constraints as there are indices in the union of B and all the sets B_i, plus upper bounds. This is smaller than the program (46)–(50) of Section 9.3. It has angular structure, with (22) the coupling constraints and (36)–(37) the diagonal blocks. Thus any of the special techniques for solving such problems (see Chapters 3, 5, and 6) may be used. As in Section 9.3, the program simplifies when one or more of the functions w_i is differentiable at the current point y. Then, if $u^0_i = \nabla w_i(y_i)$, we may drop the variables β_{ij} for all j, set

$$s_i = \sum_i \beta_{ij} r_{ij}$$

in (22), set

$$\sum_j c_{ij}\beta_{ij} = -u^0_i s_i$$

in (21) and replace the ith set of constraints in (36) and (37) by the upper bounds

$$-e_m \leq s_i \leq e_m$$

If all w_i are differentiable at y, the formulations of this and the previous section become identical.

9.5 Tangential Approximation

Geoffrion [4] has proposed an alternative approach for solving the master program. It is based on the idea of tangential approximation, discussed earlier in Section 8.6. Familiarity with that material is assumed here.

As with the dual function $h(u)$, evaluation of $w_i(y_i)$ provides us (perhaps with some additional effort), with a subgradient $u_i \in \partial w_i(y_i)$. As shown in Section 9.3, $-u_i$ is simply an optimal Lagrange multiplier vector for $P_i(y_i)$. Thus, if $\bar{u}_i \in \partial w_i(\bar{y}_i)$,

$$w_i(\bar{y}_i) + \bar{u}_i(y_i - \bar{y}_i)$$

is a linear supporting function to the graph of w_i at y_i. Let $u_i^1, \ldots, u_i^r$ be subgradients for w_i at $y_i^1, \ldots, y_i^r$. Then the tangential approximation to w_i, shown in Figure 9-4, is the convex piecewise linear function

$$v_i^r(y_i) = \max_{1 \leq j \leq r} \{w_i(y_i^j) + u_i^j(y_i - y_i^j)\} \tag{1}$$

Using this, the master program is replaced with the problem

$$\text{minimize} \sum_{i=1}^{p} v_i^r(y_i) \tag{2}$$

subject to

$$\sum_{i=1}^{p} y_i \leq b$$

$$y_i \in V_i, \qquad i = 1, \ldots, p$$

This, in turn, is equivalent to the program in the variables (y, σ)

$$\text{minimize} \sum_{i=1}^{p} \sigma_i \tag{3}$$

subject to

$$\sum_{i=1}^{p} y_i \leq b \tag{4}$$

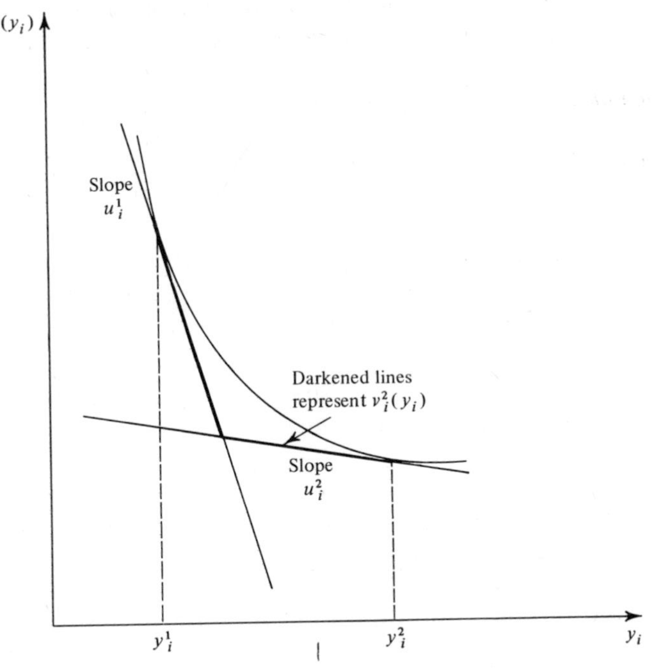

FIGURE 9-4 Tangential approximation to w_i.

and, for $i = 1, \ldots, p$

$$\sigma_i \geq w_i(y_i^j) + u_i^j(y_i - y_i^j), \qquad j = 1, \ldots, r \tag{5}$$

$$y_i \in V_i \tag{6}$$

The above is linear except for the constraints $y_i \in V_i$. The iterative procedure, commencing at cycle r, is

1. Solve the above program, obtaining an optimal solution
$$y^{r+1} = (y_1^{r+1}, \ldots, y_p^{r+1})$$
2. For $i = 1, \ldots, p$, solve $P_i(y_i^{r+1})$, obtaining an optimal objective value $w_i(y_i^{r+1})$ and a subgradient u_i^{r+1}.
3. Add p linear constraints like (5), formed from the points (y_i^{r+1}, u_i^{r+1}), to the above program, replace r by $r + 1$, and return to step 1.

As in Section 8.6, upper and lower bounds on the minimal objective value of the master objective are immediately available. The upper bound is $\sum_{i=1}^{p} w_i(y_i^{r+1})$, while the minimal value of (3) is a lower bound. When these bounds become equal, y^{r+1} solves the master program.

In stating the method we assumed tacitly that subproblem i had an optimal solution and a vector of optimal Lagrange multipliers whenever it was feasible. To ensure the former we assume, for $i = 1, \ldots, p$,

Assumption 1. S_i is compact and convex.

Assumption 2. f_i and all components of g_i are convex and continuous[7] on S_i.

These represent a weakening of the assumptions made in Section 9.2. As shown earlier, existence of the multipliers is assured if $y_i \in \text{int } V_i$, so we need only be concerned with boundary points. Thus we could impose the condition

Assumption 3. For $i = 1, \ldots, p$, subproblem i has an optimal Lagrange multiplier vector for all y_i on the boundary of V_i.

Unfortunately, there appear to be no simple conditions on f_i and g_i which guarantee this. An alternative is to perturb any vector on the boundary slightly so as to move it into the interior. For example, $y_i + \delta$ is in int V_i if $y_i \in V_i$ and δ is a vector with all positive components. If δ is allowed to approach zero, optimal solutions on the boundary may be approached as limiting values.

Since optimal solutions to the master program tend not to occur at points y such that $\partial w_i(y_i) = \varnothing$ for some i, the problem of existence of multipliers is probably not serious. To see this, Theorem 13, Appendix 2, implies that, if $y_i \in V_i$, $\partial w_i(y_i) = \varnothing$ if and only if $Dw_i(y_i; s_i)$ is not proper in s_i. Since $Dw_i(y_i; 0) = 0$, this can happen if and only if $Dw_i(y_i; s_i) = -\infty$ for some s_i, i.e., w_i decreases at an infinite rate as we move from y_i into the interior[8] of V. Thus a minimization algorithm will tend to move away from points y_i such that $\partial w_i(y_i)$ is empty.

A problem not yet dealt with is that of satisfying the constraints $y_i \in V_i$ in (3)–(6). This is the main difficulty in applying tangential approximation to the master program. Similar difficulties did not arise in the dual method of Section 8.6, since there the only constraints were $u \geq 0$. Further, the problem is more severe here than in the feasible-direction procedure considered earlier because now there is no way to approach the boundary of V_i "gradually" by taking small steps along a given direction.

Since explicit constraints defining V_i are difficult to obtain except in one or two dimensions (see [2]), we consider two schemes for constructing polyhedral approximations to V_i. One approximates V_i as the intersection of supporting half-spaces, the other as a convex polyhedron determined by

[7] Since a convex function is continuous in the interior of its domain, the continuity requirement need only be applied to boundary points.

[8] It is shown in [7, p. 79] that if $Dw_i(y_i; s_i) = -\infty$ for some s_i, it is $-\infty$ for all s_i such that $s_i - y_i$ points into int V_i.

points in V_i. The latter maintains feasibility at each step, whereas the former cannot guarantee this.

Outer Linearization of V_i. The generation of supporting half-spaces for V_i (see Figure 9-5) rests on the following theorem.

THEOREM 1. *$y_i \in V_i$ if and only if y_i satisfies the (uncountably infinite) system of linear inequalities*

$$\lambda y_i \geq \min_{x_i \in S_i} \lambda g_i(x_i) \qquad \textit{for all } \lambda \in \Lambda \tag{7}$$

where

$$\Lambda = \left\{ \lambda \,\middle|\, \sum_{j=1}^{m} \lambda_j = 1,\ \lambda_j \geq 0,\ j = 1, \ldots, m \right\} \tag{8}$$

Further, every linear inequality in (7) *defines a supporting half-space for V_i.*

The proof requires the following lemma.

LEMMA 1 ([8], p. 84). *Let S be a compact convex subset of E^n and $f_1(x), \ldots, f_m(x)$ convex continuous functions defined on S. If there is no $x \in S$ such that*

$$f_i(x) \leq 0, \qquad i = 1, \ldots, m$$

then there is an m vector $\lambda \in \Lambda$ such that

$$\min_{x \in S} \sum_{i=1}^{m} \lambda_i f_i(x) > 0$$

PROOF OF THEOREM 1. $\Rightarrow$: If $y_i \in V_i$, there is an $x_i \in S_i$ such that

$$\lambda y_i \geq \lambda g_i(x_i) \qquad \text{for any } \lambda \in E^m, \quad \lambda \geq 0$$

Taking the minimum of the right-hand side over S_i and considering only $\lambda \in \Lambda$ yields (7).

$\Leftarrow$: Assume $y_i \notin V_i$; i.e., there is no $x_i \in S_i$ such that $g_i(x_i) \leq y_i$. By assumptions 1 and 2 of this section, Lemma 1 applies. Thus there is a vector $\lambda \in \Lambda$ such that

$$\min_{x_i \in S_i} \lambda(g_i(x_i) - y_i) > 0$$

That is,

$$\min_{x_i \in S_i} \lambda g_i(x_i) > \lambda y_i$$

so some inequality of (7) is violated.

Consider now the statement on supporting half-spaces. Clearly, for any $\lambda \in \Lambda$, the half-space

$$R_i(\lambda) = \left\{ y_i \mid \lambda y_i \geq \min_{x_i \in S_i} \lambda g_i(x_i) \right\} \tag{9}$$

contains V_i, since $y_i \in V_i$ implies $y_i \in R_i(\lambda)$. This half-space is a support, since its boundary contacts V_i at the point $y_i = g_i(x_i(\lambda)) \in V_i$, where $x_i(\lambda)$ minimizes $\lambda g_i(x)$ over S_i.

Since V_i is defined by an infinite set of linear constraints, it is natural to apply the relaxation strategy of Section 5.2. That is, if in step 1 of the tangential approximation procedure y^{r+1} is such that $y_i^{r+1} \notin V_i$ for some i [signaled by infeasibility of $P_i(y_i^{r+1})$ in step 2] then, by Theorem 1, y_i^{r+1} violates some of the constraints (7). One or more violated constraints must then be generated and added to (3)–(6).[9] An example is shown in Figure 9-5,

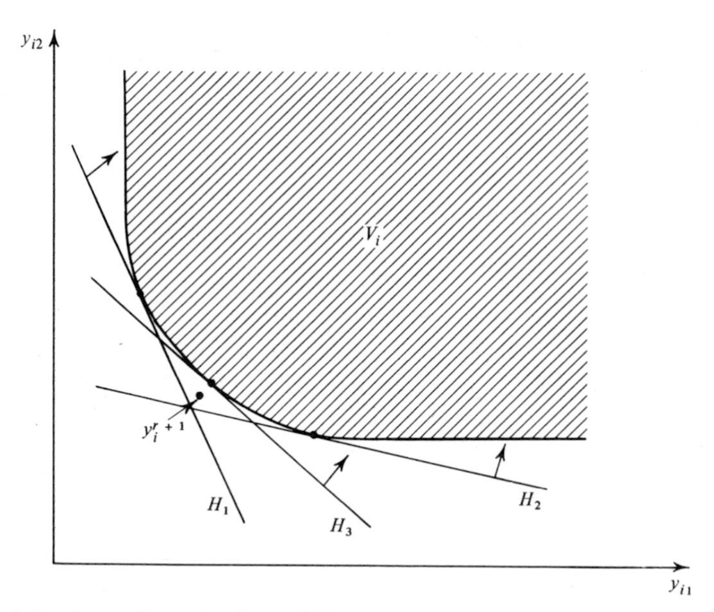

FIGURE 9-5 Outer linearization of V_i.

where V_i is already approximated by two half-spaces, defined by hyperplanes H_1 and H_2. The solution y_i^{r+1} lies outside V_i and the half-space defined by H_3, which excludes y_i^{r+1}, is added. This is the approach used by Kelley in his cutting-plane method [5].

[9] Strictly satisfied constraints may also be dropped.

To implement this strategy, the algorithm which solves subproblem i must be capable of starting from an infeasible point and, if subproblem i is infeasible, must locate a violated constraint of (7). As noted in [2], most dual methods do this automatically, as do phase 1 procedures used to find an initial feasible point for primal algorithms. Note that the addition of these linear constraints to (3)–(6) gives rise to a block-diagonal linear program with coupling constraints, amenable to solution by a number of algorithms discussed in Chapters 3, 5, and 6.

Linear Case. The relaxation procedure is simplified when g_i is linear and S_i polyhedral. Specifically, let

$$g_i(x_i) = A_i x_i, \qquad S_i = \{x_i \mid B_i x_i = b_i,\ x_i \geq 0\} \tag{10}$$

so

$$V_i = \{y_i \mid \text{there exists } x_i \geq 0 \text{ such that } B_i x_i = b_i,\ A_i x_i \leq y_i\} \tag{11}$$

Since V_i is polyhedral, a finite set of constraints defining it may be derived by applying Farkas' lemma, just as in Benders' algorithm, Section 7.3. The result is that a vector y_i is in V_i if and only if

$$b_i' u_{1j}^r + y_i' u_{2j}^r \geq 0, \qquad \text{all } j \tag{12}$$

where (u_{1j}^r, u_{2j}^r) is the jth generator of the convex polyhedral cone

$$C = \{(u_1, u_2) \mid B_i' u_1 + A_i' u_2 \geq 0,\ u_2 \geq 0\} \tag{13}$$

The number of constraints in (12) is generally very large. To see if all are satisfied by a particular vector $\bar{y}_i$ we consider the linear program

$$\text{minimize } b_i' u_1 + \bar{y}_i u_2 \tag{14}$$

subject to $\qquad (u_1, u_2) \in C$

All constraints are satisfied if and only if the optimal objective value of this program is zero. Otherwise, the minimal value is unbounded, and an extreme ray must be located from which a violated constraint is formed. Because of the homogeneity of (14), such an extreme ray is perhaps most easily found by performing a phase one operation on the dual of (14), which has objective coefficients zero and the constraints of (11). A suitable dual extreme ray may easily be obtained from the final simplex tableau if this program is infeasible. The constraint formed from this is added to (3)–(6) (along with similar constraints for other indices i and constraints refining the tangential approximation) and (3)–(6) is re-solved.

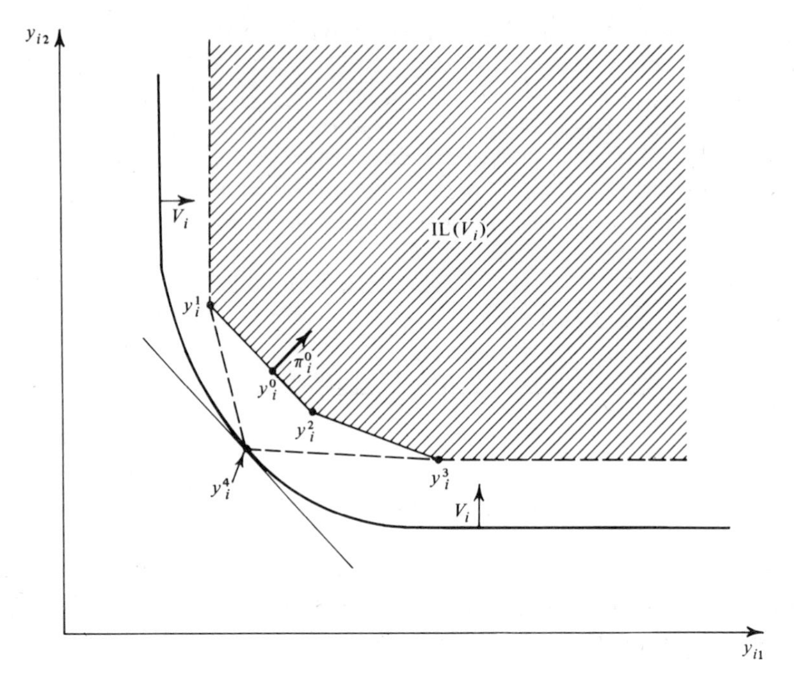

FIGURE 9-6 Inner linearization of V_i.

Inner Linearization of V_i. An alternative approach which maintains master program feasibility approximates V_i as shown in Figure 9-6. There the points $y_i^j, j = 1, 2, 3$ are in V_i and the approximating set, the inner linearization IL(V_i), is the set of all points in, above, and to the right of the convex hull of the y_i^j:

$$\text{IL}(V_i) = \left\{ y_i \mid y_i \geq \sum_{j=1}^{r_i} \alpha_{ij} y_i^j, \text{ for some } \alpha_i \geq 0, \sum_{j=1}^{r_i} \alpha_{ij} = 1 \right\} \tag{15}$$

This is the same tactic as was used in the development of generalized linear programming, Sections 4.3 and 4.4. Since V_i is convex and is unbounded in any direction with nonnegative coordinates, IL(V_i) is contained in V_i. When each V_i is treated in this way the program (2) becomes

$$\text{minimize} \sum_{i=1}^{p} v_i^r(y_i) \tag{16}$$

subject to

$$\sum_{i=1}^{p} y_i \leq b$$

and, for $i = 1, \ldots, p$,

$$y_i \geq \sum_{j=1}^{r_i} \alpha_{ij} y_i^j$$

$$\sum_{j=1}^{r_i} \alpha_{ij} = 1$$

$$\alpha_{ij} \geq 0, \qquad \text{all } j$$

If the v_i^r are handled as in (3)–(6), this is again a block-diagonal linear program with coupling constraints.

An appealing way to improve the approximation of V_i arises from the following result.

THEOREM 2. *Let* $\{(y_1^0, \ldots, y_p^0), (\alpha_1^0, \ldots, \alpha_p^0), (\sigma_1^0, \ldots, \sigma_p^0)\}$ *solve* (16) *and let* π_i^0, $i = 1, \ldots, p$, *be vectors of optimal dual variables for the constraints* $y_i \geq \sum_j \alpha_{ij} y_i^j$. *Then, for* $i = 1, \ldots, p$, y_i^0 *is on the boundary of the current inner linearization* $\mathrm{IL}(V_i)$, π_i^0 *is the normal vector of a supporting hyperplane to* $\mathrm{IL}(V_i)$, *and this hyperplane contains* y_i^0.

PROOF. The vector π_i^0 satisfies the following dual constraints, formed from primal columns associated with α_{ij}:

$$\pi_i^0 y_i^j \geq \gamma_i^0, \qquad j = 1, 2, \ldots, r_i \tag{17}$$

where γ_i^0 is the optimal dual variable for the constraint $\sum_j \alpha_{ij} = 1$. The complementary slackness and nonnegativity conditions are also satisfied:

$$\pi_i^0 \geq 0 \tag{18}$$

$$\pi_i^0 \left(y_i^0 - \sum_{j=1}^{r_i} \alpha_{ij}^0 y_i^j \right) = 0 \tag{19}$$

Let $\alpha_i = (\alpha_{i1}, \ldots, \alpha_{ir_i})$ be a nonnegative vector summing to one. If constraint j of (17) is multiplied by α_{ij} and the result is summed over j we obtain

$$\pi_i^0 \left(\sum_{j=1}^{r_i} \alpha_{ij} y_i^j \right) \geq \gamma_i^0 \tag{20}$$

If

$$y_i \geq \sum_{j=1}^{r_i} \alpha_{ij} y_i^j$$

then, since $\pi_i^0 \geq 0$,

$$\pi_i^0 y_i \geq \gamma_i^0$$

Thus all y_i in $\text{IL}(V_i)$ lie on one side of the hyperplane

$$H_i^0 = \{y_i \mid \pi_i^0 y_i = \gamma_i^0\} \tag{21}$$

By complementary slackness

$$\alpha_{ij}^0(\pi_i^0 y_i^j - \gamma_i^0) = 0, \qquad \text{all } j$$

which, after summing, yields

$$\pi_i^0\left(\sum_{j=1}^{r_i} \alpha_{ij}^0 y_i^j\right) = \gamma_i^0 \tag{22}$$

Using (19),

$$\pi_i^0 y_i^0 = \pi_i^0\left(\sum_{j=1}^{r_i} \alpha_{ij}^0 y_i^j\right) \tag{23}$$

so, combining (22) and (23),

$$\pi_i^0 y_i^0 = \gamma_i^0$$

Thus H_i^0 must be a supporting hyperplane to $\text{IL}(V_i)$. It has normal vector π_i^0 and contains y_i^0. This implies that y_i^0 is on the boundary of $\text{IL}(V_i)$.

To use this result to improve the approximation of V_i we take π_i^0 (suitably normalized) as the normal vector of a supporting hyperplane for V_i, find a point of contact of this plane with V_i, and add this point to the set $\{y_i^j\}$. This is illustrated in Figure 9-6, where y_i^4 is the new point. The new polyhedral approximation is bounded by the dashed lines. This procedure has the advantages that the point added is always a boundary point of V_i and is in some sense near the previous solution y_i^0. Assuming that the true optimal y_i lies near y_i^0, the approximation of V_i is refined where it is most important, in the vicinity of the optimum. To compute the new y_i^j, note that the supporting hyperplane to V_i with normal π_i^0 is, by (7), defined by the linear equation

$$\pi_i^0 y_i = \min_{x_i \in S_i} \pi_i^0 g_i(x) \tag{24}$$

If we solve the convex program

$$\text{minimize } \{\pi_i^0 g_i(x) \mid x_i \in S_i\}$$

obtaining a solution x_i^0, and set

$$y_i^{r_i+1} = g_i(x_i^0)$$

then clearly $y_i^{r_i+1}$ satisfies (24) and is the desired point.

Problems

1. Prove that, if f_i is strictly convex and if assumptions 1 and 3 of Section 9.2 hold, w_i is strictly convex.

2. Verify the computations of $w(y)$ in Example 1 of Section 9.2 and of the corresponding $h(u)$ in Section 9.3.

3. Show that if, for $i = 1, \ldots, p$, $-u_i \in U_i$, i.e., $u_i \in \partial w_i(y_i)$, then $(u_1, \ldots, u_p)$ is a subgradient of

$$w = \sum_{i=1}^{p} w_i$$

at $y = (y_1, \ldots, y_p)$.

4. Derive the inequalities defining V_i in (9.5-12).

References

1. G. J. Silverman, "Primal Decomposition of Mathematical Programs by Resource Allocation," *Tech. Memorandum 116*, Operations Research Department, Case Western Reserve University, 1968.
2. A. M. Geoffrion, "Primal Resource—Directive Approaches for Optimizing Nonlinear Decomposable Systems," *Memorandum RM–5829–PR*, The RAND Corporation, Santa Monica, Calif., 1968.
3. J. M. Falk, "A Constrained Lagrangian Approach to Nonlinear Programming," *Technical Report*, Department of Mathematics, University of Michigan, 1965, pp. 76–83.
4. E. V. W. Zschau, "A Primal Decomposition Algorithm for Linear Programming," Graduate School of Business, Stanford University, *Working Paper 91*, 1967.
5. J. E. Kelley, Jr., "The Cutting Plane Method for Solving Convex Programs," *J. Soc. Ind. Appl. Math.*, No. 8, pp. 703–712, 1960.
6. D. M. Topkis and A. Veinott, "On the Convergence of Some Feasible Direction Algorithms for Nonlinear Programming," *J. SIAM Control*, **5**, No. 2, 1967.
7. W. Fenchel, *Convex Cones, Sets, and Functions*, lecture notes, Princeton University, 1951.
8. C. Berge and A. Ghouila-Houri, *Programming, Games, and Transportation Networks*, John Wiley & Sons, New York, 1965.

APPENDIX 1

Convex Functions and Their Conjugates[1]

We are concerned here with functions which may take on the values $\pm\infty$. This requires the following extended definition of convexity.

DEFINITION 1

Let f be a function from a convex set C into $[-\infty, +\infty]$. Then f is *convex* if

$$f(\alpha x_1 + (1 - \alpha)x_2) \leq \alpha\mu_1 + (1 - \alpha)\mu_2 \tag{1}$$

for all $x_1, x_2 \in C$, all $f(x_1) \leq \mu_1 \in E^1, f(x_2) \leq \mu_2 \in E^1, 0 < \alpha < 1$.

If f does not take on both $+\infty$ and $-\infty$ as values, then the ambiguous combination $\infty - \infty$ cannot arise in (1), and (1) simplifies to the usual definition

$$f(\alpha x_1 + (1 - \alpha)x_2) \leq \alpha f(x_1) + (1 - \alpha)f(x_2) \tag{2}$$

Given a function convex on C, we can extend it to E^n by defining $f(x) = +\infty$ for $x \notin C$. The infimum of this function over E^n is equal to its infimum over C, so this device replaces constrained minimizations with formally unconstrained ones. In the remainder of this appendix, all convex functions will be of this extended type, from E^n into $[-\infty, +\infty]$.

DEFINITION 2

The *epigraph* of a convex function, f, denoted by epi f, is the set of all (finite) points on and above the graph of f:

$$\text{epi } f = \{(x, \mu) \mid x \in E^n, \mu \in E^1, \mu \geq f(x)\} \tag{3}$$

It is easily shown that f is convex if and only if epi f is a convex set.

[1] Appendices 1 and 2 are based mainly on [1] and [3].

DEFINITION 3

The *effective domain* of a convex function f is the set

$$\operatorname{dom} f = \{x \mid f(x) < +\infty\} \tag{4}$$

The set $\operatorname{dom} f$ is the projection of $\operatorname{epi} f$ on E^n; i.e.,

$$\operatorname{dom} f = \{x \mid \text{there exists } \mu \in E^1 \text{ such that } (x, \mu) \in \operatorname{epi} f\} \tag{5}$$

Either from this or from the fact that $\operatorname{dom} f$ is a level set of f it follows that $\operatorname{dom} f$ is convex if f is.

DEFINITION 4

A convex function f is said to be *proper* if $\operatorname{dom} f$ is nonempty and

$$f(x) > -\infty \qquad \text{for all } x \in \operatorname{dom} f$$

A proper convex function is shown in Figure A1-1. The shaded area represents $\operatorname{epi} f$. Note that f has a discontinuity at $x = x_u$, and hence $\operatorname{epi} f$ is not a closed set.

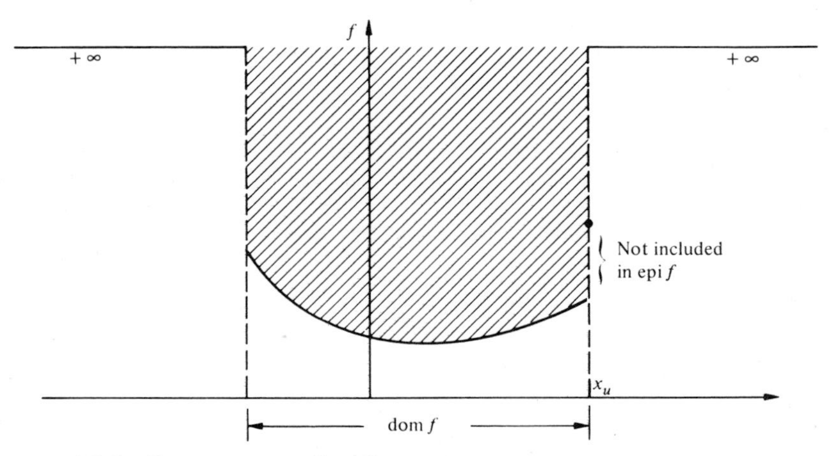

FIGURE A1-1 Proper convex function.

DEFINITION 5

The *relative interior* of a convex set, C, denoted by $r \operatorname{int} C$, is its interior relative to the smallest affine manifold containing it.

An affine manifold is simply the translation from the origin of a linear subspace, i.e., a set of the form $\{x_0\} + V$, with x_0 a fixed point and V a linear subspace. A hyperplane which does not contain the origin is an affine manifold. The smallest such manifold containing C is the intersection of all affine manifolds containing

DEFINITION 10

The *second conjugate* of f, f^{**}, is simply the conjugate of f^*, given by

$$f^{**}(x) = \sup_{x^*}(x^*x - f^*(x^*)) \tag{17}$$

THEOREM 8

Let

$$D^* = \{(x, \mu) \mid x \in E^n, \mu \in E^1, x^*x - \mu \le f^*(x^*) \text{ for all } x^* \in E^n\} \tag{18}$$

Then

(a) $\text{cl}\, f = f^{**}$.

(b) $D^* = \text{epi}\, f^{**} = \text{epi}(\text{cl}\, f)$.

(c) $f^{**}(x) = \sup_{x^*} \inf_{y} (x^*(x - y) + f(y))$.

PROOF

(a)
$$\begin{aligned}(\text{cl}\, f)(x) &= \sup_{(x^*,\mu^*)\in D} (x^*x - \mu^*) \\ &= \sup_{\mu^* \geqslant f^*(x^*)} (x^*x - \mu^*) \\ &= \sup_{x^*}(x^*x - f^*(x^*)) = f^{**}(x)\end{aligned}$$

(b) The proof is just as in Theorem 7.

(c)
$$\begin{aligned}f^{**}(x) &= \sup_{x^*}\left(x^*x - \sup_{y}(x^*y - f(y))\right) \\ &= \sup_{x^*}\left(x^*x + \inf_{y}(f(y) - x^*y)\right) \\ &= \sup_{x^*} \inf_{y} (x^*(x - y) + f(y))\end{aligned}$$

THEOREM 9

Let f be convex. Then

(a) *f^* is closed.*

(b) *f^* is proper if and only if f is proper.*

PROOF

For (a), by definition,

$$\begin{aligned}(\text{cl}\, f^*)(x^*) &= \sup_{(x,\mu)\in D^*} (x^*x - \mu) \\ &= \sup_{\mu \geqslant f^{**}(x)} (x^*x - \mu) \\ &= \sup_{\mu \geqslant (\text{cl}\, f)(x)} (x^*x - \mu) \\ &= \sup_{x}(x^*x - (\text{cl}\, f)(x)) \\ &= (\text{cl}\, f)^*(x^*)\end{aligned}$$

Since, by Theorem 5, $(\text{cl } f)(x) \leq f(x)$

$$x^*x - (\text{cl } f)(x) \geq x^*x - f(x)$$

so

$$\text{cl } f^* = (\text{cl } f)^* \geq f^*$$

But, by Theorem 5,

$$\text{cl } f^* \leq f^*$$

Thus

$$\text{cl } f^* = f^*$$

and f^* is closed.

To prove (b) note that, by the same reasoning used in proving Theorem 6, f^* is proper if and only if $\varnothing \neq \text{epi } f^* \neq E^{n+1}$. (*Note:* f^* is closed, and the closure of a function which is $-\infty$ anywhere is equal to $-\infty$ everywhere.) By Theorem 7, $\text{epi } f^* = D$ and, by Theorem 6, $\varnothing \neq D \neq E^{n+1}$ if and only if f is proper.

THEOREM 10

If f is convex, closed, and proper, then

$$f(x) + f^*(x^*) \geq x^*x \qquad \text{for all } x \in E^n, \quad \text{all } x^* \in E^n \tag{19}$$

Further, if $x \in r$ int dom f, there is an x^* such that the above holds with equality. Dually, if $x^* \in r$ int dom f^*, there is an x such that the above holds with equality.

PROOF. By definition

$$f^*(x^*) \geq x^*x - f(x) \qquad \text{for all } x \in E^n, \text{ all } x^* \in E^n$$

so (19) holds. To prove the remaining assertions, since f is proper, dom f is nonempty and, by Theorem 1, has a nonempty relative interior. Choose $x_1 \in r$ int dom f. Then, by Theorem 2, there is an x^* such that

$$f(x) \geq f(x_1) + x^*(x - x_1) \qquad \text{for all } x \in E^n$$

or

$$x^*x_1 - f(x_1) \geq x^*x - f(x)$$

Since equality holds for $x = x_1$,

$$x^*x_1 - f(x_1) = \sup_x \{x^*x - f(x)\} = f^*(x^*)$$

To prove the dual result for f^*, $x^* \in r \operatorname{int} \operatorname{dom} f^*$ implies that there is an x such that

$$f^*(z^*) \geq f^*(x^*) + x(z^* - x^*) \qquad \text{for all } z^*$$

Repeating the above manipulations leads to

$$(\operatorname{cl} f)(x) = f^{**}(x) = x^*x - f^*(x^*)$$

Since f is closed, $(\operatorname{cl} f)(x) = f(x)$ and the theorem is proved.

APPENDIX 2

Subgradients and Directional Derivatives of Convex Functions[1]

Throughout this appendix, f is assumed to be a convex function defined on E^n with values in $[-\infty, +\infty]$.

Subgradients

DEFINITION 1

A (row) vector x^* is said to be a *subgradient* of f at a point x, if

$$f(z) \geq f(x) + x^*(z - x) \qquad \text{for all } z \in E^n \tag{1}$$

DEFINITION 2

The *set of all subgradients* of f at x is denoted by $\partial f(x)$.

As seen from Figure A2-1, x^* is simply the slope of a supporting hyperplane to the graph of f at x. If f is differentiable at x, the only such plane is the tangent plane, and $x^* = \nabla f(x)$. Then the directional derivative of f at x in the direction s is

$$Df(x; s) = \nabla f'(x)s$$

However, convex functions need not be differentiable everywhere, as at x in the figure. Nonetheless, wherever f is finite, directional derivatives do exist in all directions. Our main goal in this appendix is to develop an expression for the directional derivative in terms of subgradients. This requires that a number of properties of $Df(x; s)$ and $\partial f(x)$ be developed.

THEOREM 1

The set $\partial f(x)$ is closed and convex (possibly empty).

[1] This material draws heavily on the contents of Appendix 1.

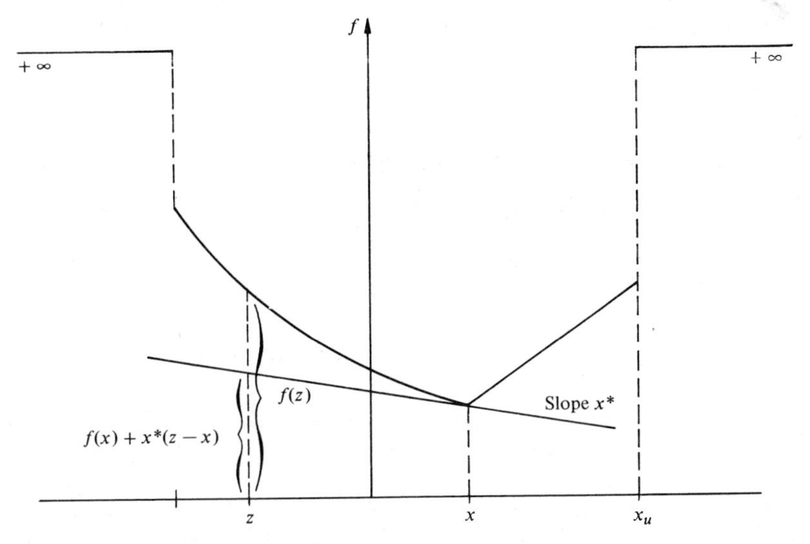

FIGURE A2-1 Subgradient of f at x.

PROOF. To prove convexity, let x_1^* and $x_2^* \in \partial f(x)$. Then

$$f(z) \geq f(x) + x_1^*(z - x) \qquad \text{for all } z \in E^n \tag{2}$$

$$f(z) \geq f(x) + x_2^*(z - x) \qquad \text{for all } z \in E^n \tag{3}$$

Multiplying (2) by α, (3) by $(1 - \alpha)$, $0 < \alpha < 1$, and adding yields

$$f(z) \geq f(x) + (\alpha x_1^* + (1 - \alpha)x_2^*)(z - x) \qquad \text{for all } z \in E^n$$

so

$$(\alpha x_1^* + (1 - \alpha)x_2^*) \in \partial f(x)$$

To show $\partial f(x)$ is closed, let $\{x_i^*\}$ be a convergent sequence of elements from $\partial f(x)$ with limit x^* and assume $x^* \notin \partial f(x)$. Then there is a $\hat{z} \in E^n$ and an $\epsilon > 0$ such that

$$f(\hat{z}) + \epsilon = f(x) + x^*(\hat{z} - x) \tag{4}$$

For any element x_n^* of the sequence $\{x_i^*\}$,

$$f(\hat{z}) \geq f(x) + x_n^*(\hat{z} - x) \tag{5}$$

Subtracting (4) from (5) yields

$$\epsilon \leq (x^* - x_n^*)(\hat{z} - x) \leq \|x^* - x_n^*\| \, \|\hat{z} - x\|$$

the final relation being the Schwartz inequality, with $\|y\|$ the norm of y. Thus

$$\|x^* - x_n\| \geq \frac{\epsilon}{\|\hat{z} - x\|} > 0 \qquad \text{for all } n$$

which contradicts the convergence of $\{x_i^*\}$ to x^*. Thus $x^* \in \partial f(x)$ and $\partial f(x)$ is closed.

THEOREM 2

If f is proper and $x \in r$ int dom f, then $\partial f(x) \neq \varnothing$.

PROOF. This is immediate from Theorem 2 of Appendix 1.

THEOREM 3

If dom f has a nonempty interior and $x \in$ int dom f, then $\partial f(x)$ is compact.

PROOF. Since x is interior to dom f, there is $\epsilon > 0$ such that $N(x, \epsilon) \in$ dom f, where N is an open neighborhood of x of radius ϵ. If any component of x^*, say $(x^*)_i$, could grow without bound (say positively), then

$$f(z) \geq f(x) + (x^*)_i(z - x)_i + \sum_{j \neq i} (x^*)_j(z - x)_j \qquad \text{for all } z \in E^n$$

Choose $z \in N(x, \epsilon)$ such that $(z - x)_i > 0$. Then as $(x^*)_i \rightarrow +\infty$, the above inequality is violated, since $f(z)$ is finite. A similar argument holds if $(x^*)_i \rightarrow -\infty$, so $\partial f(x)$ is bounded.

In Figure A2-1, $x_u \notin$ int dom f and $\partial f(x_u)$ is not bounded, since there are support planes at x_u with slopes approaching $+\infty$. Note that the condition $x \in r$ int dom f is not enough to ensure $\partial f(x)$ bounded, as is seen by considering f in Figure A2-1 as a function defined on a line in E^2, parallel to one axis. Then the tangent line at x becomes a plane and can be pivoted, using the line in Figure A2-1 as an axis, as far as desired in either direction. In so doing, one of the slopes approaches $+\infty$ or $-\infty$.

THEOREM 4

$0 \in \partial f(x^0)$ if and only if x^0 minimizes f.

PROOF. $0 \in \partial f(x^0)$ if and only if

$$f(z) \geq f(x^0) + 0(z - x^0) \qquad \text{for all } z \in E^n$$

which holds if and only if x^0 minimizes f.

THEOREM 5

If $\partial f(x) \neq \varnothing$, then (cl f)$(x) = f(x)$.

PROOF. If $x^* \in \partial f(x)$, then

$$\begin{aligned} f(z) &\geq f(x) + x^*(z - x) \qquad \text{for all } z \in E^n \\ &\geq x^*z - [x^*x - f(x)] = x^*z - \mu^* \end{aligned}$$

When $z = x$, the above holds with equality implying

$$f(x) = x^*x - \mu^* = \sup_{(y,\mu)\in D} (yx - \mu) = (\text{cl } f)(x)$$

where D is the set in (11) Appendix 1.

There is an intimate connection between the subgradients of a convex function and its conjugate, as is shown by the following two results.

THEOREM 6

$x^* \in \partial f(x)$ if and only if

$$f^*(x^*) + f(x) = x^*x \tag{6}$$

PROOF. $x^* \in \partial f(x)$ if and only if

$$f(z) \geq f(x) + x^*(z - x) \qquad \text{for all } z \in E^n$$

i.e., if and only if

$$x^*x - f(x) \geq x^*z - f(z) \qquad \text{for all } z \in E^n$$

with equality holding for $z = x$. This implies

$$x^*x - f(x) = \sup_z\{x^*z - f(z)\} = f^*(x^*)$$

so the theorem is proved.

Note that the above simply requires that the inequality of Theorem 10, Appendix 1, which holds at any points x, x^*, hold with equality when $x^* \in \partial f(x)$.

THEOREM 7

(a) *$x^* \in \partial f(x)$ implies $x \in \partial f^*(x^*)$.*
(b) *If $f(x) = (\text{cl } f)(x)$, then $x \in \partial f^*(x^*)$ implies $x^* \in \partial f(x)$.*

PROOF. By Theorem 6,

$$x^* \in \partial f(x) \qquad \text{if and only if } f^*(x^*) + f(x) = x^*x \tag{7}$$

$$x \in \partial f^*(x^*) \qquad \text{if and only if } f^{**}(x) + f^*(x^*) = x^*x \tag{8}$$

To prove (a), since $\partial f(x)$ is nonempty, Theorem 5 implies $(\mathrm{cl}\, f)(x) = f(x)$. Then, by Theorem 8 of Appendix 1,

$$f^{**}(x) = (\mathrm{cl}\, f)(x) = f(x) \tag{9}$$

so (7) becomes (8) and $x \in \partial f^*(x^*)$. To prove (b), $f(x) = (\mathrm{cl}\, f)(x)$ implies that (9) holds, so (8) becomes (7) and $x^* \in \partial f(x)$.

The following corollary, which follows immediately from Theorem 7, is often useful.

COROLLARY

If $(\mathrm{cl}\, f)(x) = f(x)$, *then* $x^* \in \partial f(x)$ *if and only if* $x \in \partial f^*(x^*)$.

Directional Derivatives[2]

DEFINITION 3

The (one-sided) *directional derivative* of f at x in the direction s is

$$\mathrm{Df}(x; s) = \lim_{\lambda \to 0+} \frac{f(x + \lambda s) - f(x)}{\lambda} \tag{10}$$

DEFINITION 4

A function h defined on E^n is said to be *positively homogeneous* if

$$h(\lambda x) = \lambda h(x) \qquad \text{for all } \lambda > 0 \text{ and all } x \in E^n$$

THEOREM 8

Assume that $f(x)$ *is finite. Then the difference quotient*

$$d(\lambda) = \frac{f(x + \lambda s) - f(x)}{\lambda} \tag{11}$$

is nondecreasing for all $\lambda > 0$.

PROOF. The fact that $f(x)$ is finite implies that the ambiguous combination $\infty - \infty$ cannot appear, so $d(\lambda)$ is well defined. Since f is convex, the one-dimensional function

$$g(\lambda) = f(x + \lambda s) \tag{12}$$

is convex in λ for all x and s. Let $\lambda_2 > \lambda_1 > 0$. Then, as seen from Figure A2-2,

$$g(0) + \lambda_1\left[\frac{g(\lambda_2) - g(0)}{\lambda_2}\right] \geq g(\lambda_1) \tag{13}$$

[2] See [7, pp. 79–88], for related material.

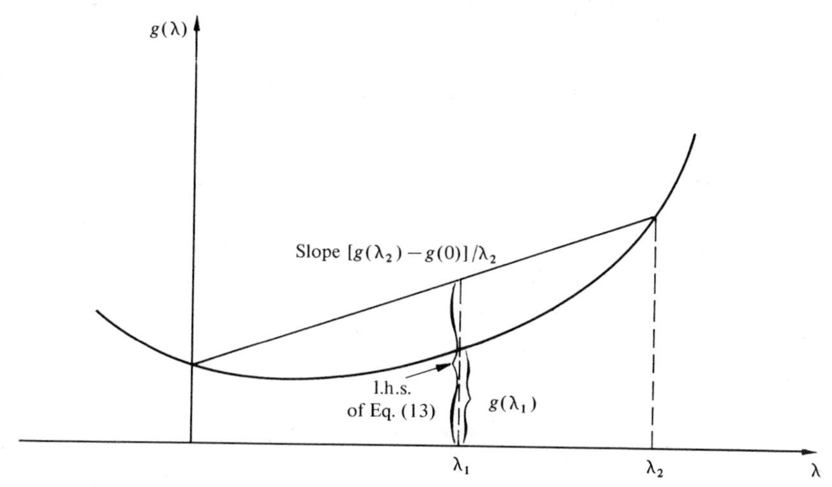

FIGURE A2-2 Illustration of equation (13), proof of Theorem 8.

or, using (12),

$$f(x) + \frac{\lambda_1}{\lambda_2}(f(x + \lambda_2 s) - f(x)) \geq f(x + \lambda_1 s)$$

which, since $\lambda_1 > 0$, $\lambda_2 > 0$, becomes

$$\frac{f(x + \lambda_2 s) - f(x)}{\lambda_2} \geq \frac{f(x + \lambda_1 s) - f(x)}{\lambda_1}$$

The following proves the statements regarding existence of directional derivatives made earlier.

THEOREM 9

Let x be a point such that $f(x)$ is finite. Then $\mathrm{Df}(x; s)$ *exists for all $s \in E^n$ (possibly equal to $\pm\infty$) and is a convex, positively homogeneous function of s.*

PROOF. By Theorem 8, $d(\lambda)$ is nondecreasing for $\lambda > 0$. Thus, for any sequence $\{\lambda_i\}$ of positive numbers converging to zero, the sequence $\{d(\lambda_i)\}$ is nonincreasing. Hence it has a limit (which may be $\pm\infty$), which is $\mathrm{Df}(x; s)$. To prove convexity, let $\alpha > 0$, $\beta > 0$, $\alpha + \beta = 1$. Then

$$\begin{aligned}\mathrm{Df}(x; \alpha s_1 + \beta s_2) &= \lim_{\lambda \to 0+} \frac{f(x + \lambda(\alpha s_1 + \beta s_2)) - f(x)}{\lambda} \\ &= \lim_{\lambda \to 0+} \frac{f(\alpha(x + \lambda s_1) + \beta(x + \lambda s_2)) - f(x)}{\lambda}\end{aligned}$$

Using the convexity of f,

$$\leq \lim_{\lambda \to 0^+} \frac{\alpha f(x + \lambda s_1) + \beta f(x + \lambda s_2) - f(x)}{\lambda}$$

Writing $f(x) = \alpha f(x) + \beta f(x)$,

$$= \lim_{\lambda \to 0^+} \frac{\alpha[f(x + \lambda s_1) - f(x)]}{\lambda} + \frac{\beta[f(x + \lambda s_2) - f(x)]}{\lambda}$$

$$= \alpha \mathrm{Df}(x; s_1) + \beta\, \mathrm{Df}(x; s_2)$$

Positive homogeneity follows from the definition

$$\mathrm{Df}(x; \alpha s) = \lim_{\lambda \to 0^+} \frac{f(x + \lambda \alpha s) - f(x)}{\lambda} \tag{14}$$

where $\alpha > 0$. Let

$$\lambda\alpha = \beta$$

Since $\alpha > 0$, $\lambda \to 0^+$ implies $\beta \to 0^+$, so (14) becomes

$$\mathrm{Df}(x; \alpha s) = \alpha \lim_{\beta \to 0^+} \frac{f(x + \beta s) - f(x)}{\beta} = \alpha\, \mathrm{Df}(x; s)$$

THEOREM 10

Let f be proper and choose $x \in \mathrm{dom}\, f$. Then $x^* \in \partial f(x)$ if and only if

$$\mathrm{Df}(x; s) \geq x^* s \qquad \text{for all } s \in E^n \tag{15}$$

PROOF. By definition, $x^* \in \partial f(x)$ if and only if

$$f(z) \geq f(x) + x^*(z - x) \qquad \text{for all } z \in E^n$$

Choosing $z = x + \lambda s$, $\lambda > 0$ yields

$$\frac{f(x + \lambda s) - f(x)}{\lambda} \geq x^* s$$

and letting $\lambda \to 0^+$,

$$\mathrm{Df}(x; s) \geq x^* s \qquad \text{for all } s \in E^n$$

The following is a useful characterization of minimizing points of a convex function.

THEOREM 11

Let f be proper. Then x^0 minimizes f if and only if

$$\mathrm{Df}(x^0; s) \geq 0 \qquad \textit{for all } s \in E^n \tag{26}$$

PROOF. By Theorem 4, x^0 minimizes f if and only if $0 \in \partial f(x^0)$. By Theorem 10, this is true if and only if (26) holds.

DEFINITION 5

The *indicator function* of a convex set C, denoted by $\delta(\cdot\,|C)$, is given by

$$\delta(x|C) = \begin{cases} 0, & x \in C \\ +\infty, & x \notin C \end{cases} \tag{16}$$

Clearly, $\delta(\cdot\,|C)$ is convex with effective domain C. It is proper if and only if C is nonempty and closed if and only if C is closed.

DEFINITION 6

The *support function* of a convex set C, denoted by $\delta^*(\cdot\,|C)$ is defined by

$$\delta^*(x^*|C) = \sup_{x \in C}(x^*x) \tag{17}$$

Since

$$\sup_{x \in C}(x^*x) = \sup_{x \in E^n}(x^*x - \delta(x|C))$$

$\delta^*(\cdot\,|C)$ is the conjugate of $\delta(\cdot\,|C)$. It is thus convex (Theorem 7, Appendix 1), closed, and proper if and only if C is nonempty (Theorem 9, Appendix 1). Further, it is positively homogeneous, since if $\lambda > 0$,

$$\sup_{x \in C}((\lambda x^*)x) = \lambda \sup_{x \in C}(x^*x)$$

The following result plays an important role in establishing an expression for $\mathrm{Df}(x; s)$.

THEOREM 12

Let h be a proper, convex, positively homogeneous function on E^n. Then the set

$$C = \{x^* \mid h(x) \geq x^*x \textit{ for all } x \in E^n\} \tag{18}$$

is nonempty, closed, and convex and

$$\mathrm{cl}\, h = \delta^*(\cdot\,|C) \tag{19}$$

PROOF. By Theorem 8, Appendix 1, cl h is the second conjugate of h; i.e.,

$$\text{cl } h = f^{**}$$

where

$$f^*(x^*) = \sup_x \{x^*x - h(x)\}$$

is, by Theorems 7 and 9, Appendix 1, closed, proper, and convex. Let $x = \lambda z$, $\lambda > 0$. Then

$$f^*(x^*) = \sup_z (\lambda x^* z - h(\lambda z))$$

Using the positive homogeneity of h,

$$= \lambda \sup_z (x^* z - h(z)) = \lambda f^*(x^*)$$

Thus $f^*(x^*)$ must be either 0 or $+\infty$ for any $x^* \in E^n$. Since it is closed, proper and convex, it is the indicator function of a closed nonempty convex set. This set is the set of all x^* such that

$$f^*(x^*) = \sup_x (x^*x - h(x)) = 0$$

i.e., the set of all x^* such that

$$x^*x - h(x) \leq 0 \qquad \text{for all } x \in E^n$$

i.e., the set

$$C = \{x^* \mid h(x) \geq x^*x \text{ for all } x \in E^n\}$$

Thus $f^* = \delta(\cdot|C)$ and cl $h = \delta^*(\cdot|C)$.

Theorems 10 and 12 would yield immediately an expression for $\text{cl}_s \, \text{Df}(x; s)$ if conditions guaranteeing that $\text{Df}(x; s)$ satisfy the hypotheses of Theorem 12 were known. We already know $\text{Df}(x; s)$ is convex and positively homogeneous in s. The following theorem shows when it is proper.

THEOREM 13

Let f be proper. Then $\partial f(x)$ is nonempty if and only if $x \in \text{dom} f$ and $\text{Df}(x; s)$ *is proper in s.*

PROOF. $\Rightarrow$: Choose $x^* \in \partial f(x)$. Then

$$f(z) \geq f(x) + x^*(z - x) \qquad \text{for all } z \in E^n$$

If we choose $z \in \operatorname{dom} f$, then $f(x) < \infty$, so $x \in \operatorname{dom} f$. By Theorem 10, $\mathrm{D}f(x; s) > -\infty$ for all $s \in E^n$. Since $\mathrm{D}f(x; 0) = 0$, $\mathrm{D}f(x; s)$ is proper in s.

$\Leftarrow$: By Theorem 12, the set

$$C = \{x^* \mid x^*s \leq \mathrm{D}f(x; s)\}$$

is nonempty. By Theorem 10, $\partial f(x)$ is also nonempty.

In the following we approach the main goal of this appendix by deriving an expression for $\mathrm{cl}_s \, \mathrm{D}f(x; s)$.

THEOREM 14

If $\partial f(x)$ is nonempty, then

$$\mathrm{cl}_s \, \mathrm{D}f(x; s) = \sup\{x^*s \mid x^* \in \partial f(x)\} = \delta^*(s \mid \partial f(x)) \tag{20}$$

PROOF. Since $\partial f(x) \neq \varnothing$, $\mathrm{D}f(x; s)$ is proper, convex, and positively homogeneous in s (Theorem 13). Then, by Theorem 12,

$$\mathrm{cl}_s \, \mathrm{D}f(x; s) = \sup\{x^*s \mid x \in C\}$$

where

$$C = \{x^* \mid \mathrm{D}f(x; s) \geq x^*s \text{ for all } s \in E^n\}$$

But by Theorem 10, $C = \partial f(x)$.

The above yields an expression for $\mathrm{D}f(x; s)$ for x such that this derivative is closed in s. Conditions guaranteeing this are supplied by the following.

THEOREM 15

Let f be proper. Then, if $x \in r$ int dom f, $\mathrm{D}f(x; s)$ is closed and proper in s.

PROOF. By Theorem 2, $\partial f(x) \neq \varnothing$ and, by Theorem 13, $\mathrm{D}f(x; s)$ is proper in s. To prove it closed, let

$$V = \{s \mid \mathrm{D}f(x; s) < \infty\}$$

be the effective domain (with respect to s) of $\mathrm{D}f(x; s)$. Choose s_1 and s_2 in V, let α and β be any nonzero real numbers, and consider the point

$$\begin{aligned} \alpha s_1 + \beta s_2 &= |\alpha|(\operatorname{sgn} \alpha)s_1 + |\beta|(\operatorname{sgn} \beta)s_2 \\ &= (|\alpha| + |\beta|)\left(\frac{|\alpha|}{|\alpha| + |\beta|}\,\tilde{s}_1 + \frac{|\beta|}{|\alpha| + |\beta|}\,\tilde{s}_2\right) \\ &= \gamma(\tilde{\alpha}\tilde{s}_1 + \tilde{\beta}\tilde{s}_2) \end{aligned}$$

where γ, $\tilde{\alpha}$, and $\tilde{\beta}$ are positive and $\tilde{\alpha} + \tilde{\beta} = 1$. Then, using the convexity and positive homogeneity of the directional derivative in s,

$$\begin{aligned} \mathrm{D}f(x; \alpha s_1 + \beta s_2) &= \mathrm{D}f(x; \gamma(\tilde{\alpha}\tilde{s}_1 + \tilde{\beta}\tilde{s}_2)) \\ &= \gamma\, \mathrm{D}f(x; \tilde{\alpha}\tilde{s}_1 + \tilde{\beta}\tilde{s}_2) \\ &\leq \gamma[\tilde{\alpha}\, \mathrm{D}f(x; \tilde{s}_1) + \tilde{\beta}\, \mathrm{D}f(x; \tilde{s}_2)] \\ &= \gamma\tilde{\alpha}\, \mathrm{D}f(x; \tilde{s}_1) + \gamma\tilde{\beta}\, \mathrm{D}f(x; \tilde{s}_2) \end{aligned} \tag{21}$$

To see what happens if $\tilde{s}_1 = -s_1$, note that if $\mathrm{D}f(x; s)$ is finite, then $x + \lambda s$ must be in $\operatorname{dom} f$ for all $\lambda > 0$ and sufficiently small. Since $x \in r \operatorname{int} \operatorname{dom} f$, $x - \lambda s$ must also be in $\operatorname{dom} f$ for $\lambda > 0$ and sufficiently small, so $\mathrm{D}f(x; -s)$ is finite (it cannot be $-\infty$ since it is proper). Hence the right-hand side of (21) is finite and $\alpha s_1 + \beta s_2 \in V$. Since $0 \in V$, V is a vector subspace of E^n. Hence all its points are relative interior points, its relative boundary is empty, and, by Theorem 5 of Appendix 1,

$$\mathrm{cl}_s\, \mathrm{D}f(x; s) = \mathrm{D}f(x; s)$$

By Theorems 14 and 15, if $x \in r \operatorname{int} \operatorname{dom} f$, then

$$\mathrm{D}f(x; s) = \sup\{x^*s \mid x^* \in \partial f(x)\} \tag{22}$$

By Theorem 3, if $\operatorname{dom} f$ has a nonempty interior and $x \in \operatorname{int} \operatorname{dom} f$, then $\partial f(x)$ is compact. The sup in (22) above is then attained at some point in $\partial f(x)$. This is stated as a theorem, which is used in Chapters 8 and 9.

THEOREM 16

Let $\operatorname{dom} f$ *have a nonempty interior and choose* $x \in \operatorname{int} \operatorname{dom} f$. *Then*

$$\mathrm{D}f(x; s) = \max\{x^*s \mid x^* \in \partial f(x)\} \tag{23}$$

This result can be verified geometrically by inspection of Figure A2-1. Let s emanate from x toward the right. Then $\mathrm{D}f(x; s)$ is the right-hand derivative of f at x. This is equal to the slope of the supporting line through x, which is turned counterclockwise as far as possible; i.e., $\max\{x^* \mid x^* \in \partial f(x)\}$. Similarly, $\mathrm{D}f(x; -s)$ is $-\min\{x^* \mid x^* \in \partial f(x)\}$, as predicted by (23).

Further inspection of Figure A2-1 suggests that $\partial f(x)$ consists of a single point if and only if f is differentiable at x, in which case $\mathrm{D}f(x; s) = \nabla f(x)'s$ is linear in s. That this is true is proved below.

THEOREM 17

Let f be proper. Then the following statements are equivalent:

(a) $x \in \operatorname{int} \operatorname{dom} f$ *and f is differentiable at* x.

(b) $\mathrm{D}f(x; s)$ *is linear in* s.

(c) $\partial f(x)$ *contains only a single point* x^*.

PROOF. (a) $\Rightarrow$ (b): This is immediate, since, by (a),

$$\mathrm{D}f(x; s) = \nabla f'(x)s$$

(b) $\Rightarrow$ (c): Let

$$\mathrm{D}f(x; s) = a's \qquad \text{for all } s \in E^n \tag{24}$$

Since, by (22)

$$\mathrm{D}f(x; s) = \sup\{x^*s \mid x^* \in \partial f(x)\}$$

if $\partial f(x)$ contained more than one point, say x_1^* and x_2^*, there would be vectors s_1, s_2 such that the sup for s_1 was attained at x_1^* and for s_2 at x_2^*. Then (24) would be violated.

(c) $\Rightarrow$ (a): By (c),

$$\mathrm{D}f(x; s) = x^*s \tag{25}$$

If $x \notin \text{int dom} f$, there is an $\tilde{s}$ such that $x + \lambda\tilde{s} \notin \text{dom} f$ for all $\lambda > 0$. This implies $\mathrm{D}f(x; \tilde{s}) = +\infty$, contradicting (25). Thus $x \in \text{int dom} f$ and the proof is reduced to the classical case, proved in [7, p. 85].

References

1. R. T. Rockafellar, "Convex Functions and Dual Extremum Problems," Ph.D. Thesis, Harvard University, 1963.
2. R. T. Rockafellar, "Nonlinear Programming," *American Mathematical Society Summer Seminar on the Mathematics of the Decision Sciences, Stanford University, July–Aug. 1967.*
3. M. Hamala, "Geometric Programming in Terms of Conjugate Functions," *Discussion Paper 6811*, Center for Operations Research and Econometrics, Catholic University of Louvain, Heverlee, Belgium, 1968.
4. H. G. Eggleston, *Convexity*, Cambridge University Press, New York, 1958.
5. S. Karlin, *Mathematical Methods and Theory in Games, Programming and Economics*, Vol. 1, Addison-Wesley Publishing Company, Inc., Reading, Mass., 1959.
6. F. A. Valentine, *Convex Sets*, McGraw-Hill, Inc., New York, 1964.
7. W. Fenchel, *Convex Cones, Sets, and Functions*, lecture notes, Princeton University, 1951.

List of Symbols

s'	transpose of vector s
$\nabla f(x)$	gradient vector of $\mathbf{f}(\mathbf{x})$
$\left[\begin{array}{c:c} A & C \\ \hdashline B & D \end{array}\right]$	partitioned matrix with submatrices A, B, C, D
$\mathbf{x} \geq 0$	with $\mathbf{x}$ a vector, means each component of $\mathbf{x}$ is nonnegative
$\mathbf{x} > 0$	with $\mathbf{x}$ a vector, means each component of $\mathbf{x}$ is positive
$\{x_i\}$	the set of elements $x_1, x_2, \ldots$
$\Rightarrow$	forward implication of a theorem
$\Leftarrow$	reverse implication of a theorem
$\Leftrightarrow$	if and only if
E^n	n-dimensional euclidean vector space
$(E^n)^+$	$\{x \mid x \in E^n, x \geq 0\}$
$E_y f(x, y)$	expected value of f with respect to the random argument y
(c_{ij})	matrix with i, j element c_{ij}
int S	interior of set S
r int S	relative interior of set S
$\lim_{x \to 0^+} f(x)$	limit of f as x approaches zero through positive values only
minimize $\{f(x) \mid g(x) \leq 0\}$	means minimize $f(x)$ subject to the constraints $g(x) \leq 0$
$\min \{f(x) \mid g(x) \leq 0\}$	means the minimum value of $f(x)$ subject to $g(x) \leq 0$
$[-1, 1]$	the closed interval $-1 \leq x \leq 1$
$\{-1, 1\}$	the set whose elements are -1 and 1

Index

R

S